THE ALKALOIDS

Chemistry and Pharmacology

Volume XXIV

THE ALKALOIDS
Chemistry and Pharmacology

Edited by

Arnold Brossi
National Institutes of Health
Bethesda, Maryland

VOLUME XXIV

1985

ACADEMIC PRESS, INC.
(Harcourt Brace Jovanovich, Publishers)

Orlando • San Diego • New York • London
Toronto • Montreal • Sydney • Tokyo

ACADEMIC PRESS, INC.
Orlando, Florida 32887

United Kingdom Edition published by
ACADEMIC PRESS INC. (LONDON) LTD.
24–28 Oval Road, London NW1 7DX

LIBRARY OF CONGRESS CATALOG CARD NUMBER: 50-5522

ISBN 0-12-469524-8

PRINTED IN THE UNITED STATES OF AMERICA

85 86 87 88 9 8 7 6 5 4 3 2 1

CONTENTS

Chapter 4. Aporphine Alkaloids

TETSUJI KAMETANI AND TOSHIO HONDA

Chapter 5. Phthalideisoquinoline Alkaloids and Related Compounds

D. B. MACLEAN

Chapter 6. The Study of Alkaloid Structures by Spectral Methods

R. J. HIGHET AND JAMES W. WHEELER

CONTRIBUTORS

Numbers in parentheses indicate the pages on which the authors' contributions begin.

I. RALPH C. BICK (113), Department of Chemistry, University of Tasmania, Hobart, Tasmania 7005, Australia

CARSTEN CHRISTOPHERSEN (25), Department of General and Organic Chemistry, University of Copenhagen, The H. C. Ørsted Institute, DK-2100 Copenhagen, Denmark

MOHAMMAD A. HAI* (113), Department of Chemistry, University of Tasmania, Hobart, Tasmania 7005, Australia

R. J. HIGHET (287), Laboratory of Chemistry, National Heart, Lung, and Blood Institute, Bethesda, Maryland 20205

TOSHIO HONDA (153), Hoshi University, Faculty of Pharmaceutical Science, Ebara 2-4-41, Tokyo 142, Japan

TETSUJI KAMETANI (153), Hoshi University, Faculty of Pharmaceutical Science, Ebara 2-4-41, Tokyo 142, Japan

D. B. MACLEAN (253), Department of Chemistry, McMaster University, Hamilton, Ontario L8S 4M1, Canada

W. C. TAYLOR (1), Department of Organic Chemistry, The University of Sydney, Sydney, New South Wales 2006, Australia

JAMES W. WHEELER (287), Department of Chemistry, Howard University, Washington, D.C. 20059

* Present address: Department of Chemistry, Jahangirnagar University, Savar, Dhaka, Bangladesh.

PREFACE

Two important groups of isoquinoline alkaloids, "The Aporphine Alkaloids," reviewed in Vols. IV, IX, and XIV, and "The Phthalideisoquinoline Alkaloids," reviewed in Vols. VII and IX and also discussed in a more general way in Vol. XVII, have now been brought up to date and are presented in detail, including pharmacological properties.

"*Aristotelia* Alkaloids" and "*Eupomatia* Alkaloids," predominantly originating from plants of the Australian continent and mentioned only casually in earlier volumes, are reviewed and presented as two separate groups of alkaloids.

"Marine Alkaloids" represent a large and fast-growing group of natural products of the marine habitat, not explored in depth and possibly containing many novel substances of interesting pharmacological properties. The state of the art of this interesting group of compounds is covered up to 1984.

The chapter on "The Study of Alkaloids by Spectral Methods," the second of a series of articles which appeared recently in this treatise, describes modern physical methods important for the structure determination and characterization of alkaloids.

Arnold Brossi

CONTENTS OF PREVIOUS VOLUMES

Contents of Volume I (1950)

edited by R. H. F. Manske and H. L. Holmes

Contents of Volume II (1952)

edited by R. H. F. Manske and H. L. Holmes

Contents of Volume III (1953)

edited by R. H. F. Manske and H. L. Holmes

Contents of Volume IV (1954)

edited by R. H. F. Manske and H. L. Holmes

Contents of Volume V (1955)

edited by R. H. F. Manske

Contents of Volume VI (1960)

edited by R. H. F. Manske

Contents of Volume VII (1960)

edited by R. H. F. Manske and H. L. Holmes

Contents of Volume VIII (1965)

edited by R. H. F. Manske and H. L. Holmes

Contents of Volume IX (1967)

edited by R. H. F. Manske and H. L. Holmes

Contents of Volume X (1967)

edited by R. H. F. Manske and H. L. Holmes

Contents of Volume XI (1968)

edited by R. H. F. Manske and H. L. Holmes

Contents of Volume XII (1970)

edited by R. H. F. Manske and H. L. Holmes

Contents of Volume XIII (1971)

edited by R. H. F. Manske and H. L. Holmes

Contents of Volume XIV (1973)

edited by R. H. F. Manske and H. L. Holmes

Contents of Volume XV (1975)

edited by R. H. F. Manske and H. L. Holmes

Contents of Volume XVI (1977)

edited by R. H. F. Manske and H. L. Holmes

Contents of Volume XVII (1979)

edited by R. H. F. Manske and H. L. Holmes

Contents of Volume XVIII (1981)

edited by R. H. F. Manske and R. G. A. Rodrigo

Contents of Volume XIX (1981)

edited by R. H. F. Manske and R. G. A. Rodrigo

Contents of Volume XX (1981)

edited by R. H. F. Manske and R. G. A. Rodrigo

Contents of Volume XXI (1983)

edited by Arnold Brossi

Contents of Volume XXII (1983)

edited by Arnold Brossi

Contents of Volume XXIII (1984)

edited by Arnold Brossi

—CHAPTER 1—

EUPOMATIA ALKALOIDS

W. C. TAYLOR

Department of Organic Chemistry
The University of Sydney
Sydney, Australia

I. Introduction

Robert Brown, when a naturalist on the voyage of Mathew Flinders to Australia in 1802–1803, found in the colony of Port Jackson a new species, which he named *Eupomatia laurina*. He observed: "This genus forms a very unexpected addition to the Anonaceae, of which it will constitute a distinct section. . . ." In 1858 a second species of the genus, *E. bennettii,* was recognized by von Mueller in material collected from Moreton Bay (Brisbane), but no other species with close affinities has since been found. The classification of the genus has caused some difficulty, but most taxonomists have followed Diels who in 1912 placed it in the monogeneric family Eupomatiaceae as a very old, isolated family with no surviving close relatives (*1*). The wood and the flowers in particular show many ancient characteristics and *Eupomatia* is thus considered to be one of the most primitive living angiosperms (*2*). This morphological and phylogenetic uniqueness is reflected in the rich variety of secondary metabolites present in *E. laurina,* namely, novel lignans and alkaloids (*3*). The alkaloids form the subject of this review.

Eupomatia laurina R.Br. is a small understory tree occurring intermittently along the entire coastal strip of eastern Australia and of eastern New Guinea in favorable situations combining shelter, warmth, moisture, and good soil. The basic fraction of a methanol extract of the bark yielded (*4*) by chromatography on

THE ALKALOIDS, VOL. XXIV

ISBN 0-12-469524-8

alumina the 7-oxoaporphine alkaloid liriodenine (**1**) (yield, 0.009%), the related base norushinsunine (**2**) (0.0038%), and a new base, EL base-1 (0.0005%), later

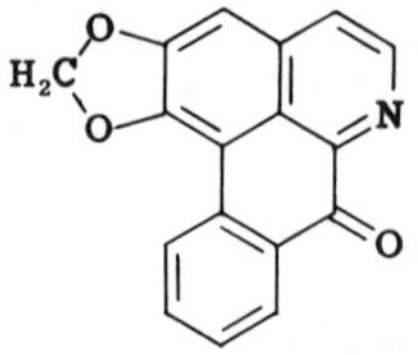

1 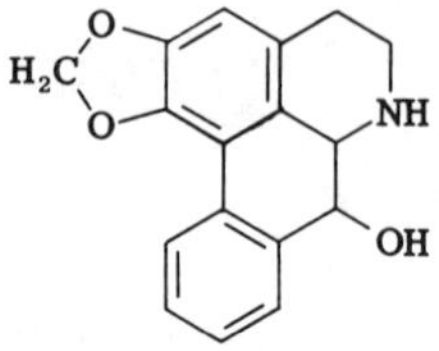 **2**

renamed eupolauridine (*5*). Careful chromatography of the methanol neutral fraction gave, in addition to lignans, small amounts of two yellow substances, which were, in fact, very weak bases: EL base-2 (0.002%), later renamed eupolauramine (*5*); and EL base-3 (0.0004%), later renamed hydroxyeupolauramine (*5*). Subsequently two additional bases, EL base-4 and EL base-5, were isolated (*6*). The leaves and wood have yielded only liriodenine (*7*). Eupolauridine has since been isolated also from *Cananga odorata* Hook. f. et Thomson (family, Annonaceae), along with liriodenine and the related 1,2-methylenedioxyaporphine alkaloids anonaine, roemerine, and ushinsunine (*8*).

The purpose of this chapter is to review the elucidation of the structures of the unique group of alkaloids eupolauridine, eupolauramine, and hydroxyeupolauramine and to summarize related synthetic endeavors. Since the amounts of the alkaloids available were very small, initial structural assignments necessarily relied heavily on spectral arguments. On the assumption that the three bases were structurally related, it was first thought that eupolauridine was 3,4-diazafluoranthene (**3**), but this was disproved by synthesis. However, 1,6-di-

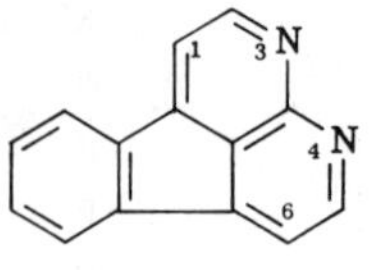

3 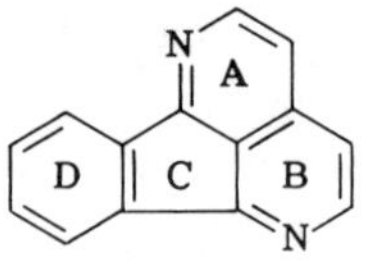 **4** 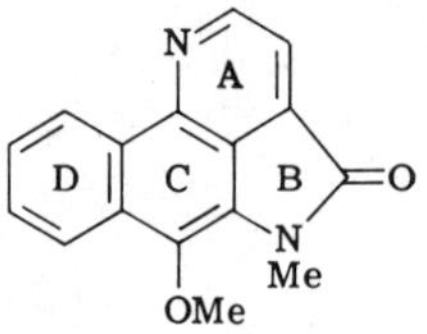 **5**

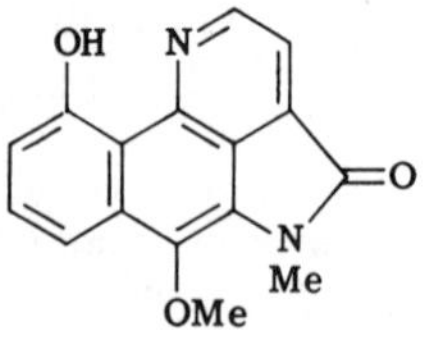

6 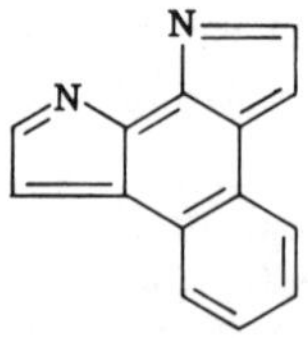 **7** 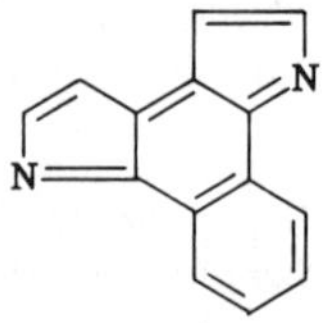 **8**

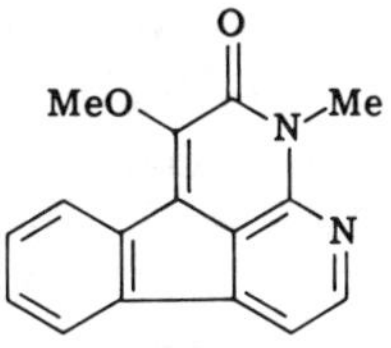

azafluoranthene (**4**),* which was then synthesized, proved to be identical with eupolauridine. With the two synthetic bases available in quantity it was possible to show that eupolauridine was not structurally related to eupolauramine and hydroxyeupolauramine. Thus the evidence bearing on the structures of the latter bases turned out to be not relevant to the structure of eupolauridine, but the assumption that it was certainly guided the choice of the initial synthetic target. Subsequently the structure of eupolauramine was shown to be **5** by a single-crystal X-ray determination, and it followed that hydroxyeupolauramine was **6**.

II. Eupolauridine

A. Structure

Eupolauridine crystallizes from light petroleum in pale-yellow needles, mp 156–157°C; methiodide, red prismatic needles from methanol, mp 267–269°C(d); methosulfate, yellow prisms from methanol–ether, mp 213–215°C(d) (*4*). The hydrochloride forms orange crystals from methanol–ether, mp 248–249°C(d) (*8*). The molecular formula, $C_{14}H_8N_2$, and high-intensity UV absorption (λ_{max} 228, 233, 278, 288, 296sh, 335, 350, and 367 nm; log ϵ 4.33, 4.34, 4.23, 4.20, 3.98, 3.57, 3.81, and 3.80) indicated the presence of an extended aromatic chromophore. The absence of absorption in the region above 3100 cm^{-1} in the IR spectrum showed that both nitrogen atoms must be tertiary. The mass spectrum, which had no high-intensity fragmentation peaks, confirmed the aromatic stability of the molecule but was otherwise of little diagnostic value (*5*).

The NMR spectrum (100 MHz) of eupolauridine—which consisted solely of a pair of AB doublets, each of two protons, and a pair of second-order multiplets, each of two protons also—revealed the symmetry of the alkaloid. By double resonance experiments it was concluded that the two protons resonating at δ8.63 were coupled to the two protons resonating at δ7.29 (J = 6.0 Hz) and that the multiplets at δ8.02–7.80 and 7.52–7.30 arose from an AA′BB′ system,† almost

* The correct systematic names of **3** and **4** are indeno[1,2,3-*de*][1,8]naphthyridine and indeno-[1,2,3-*ij*][2,7]naphthyridine, respectively. "Aza" nomenclature is used for the text because of its brevity and because it draws attention to possible biogenetic relations.

† Iterative analysis of the spectrum measured at 400 MHz has given the following parameters: $\delta_{A,A'}$7.99, $\delta_{B,B'}$7.47; J_{AB} = 7.5, $J_{AB'}$ = 1.1, $J_{AA'}$ = 0.7, and $J_{BB'}$ = 7.7 Hz (*9*).

certainly forming part of a symmetrical, ortho-disubstituted benzene ring. These results suggested the four possible structures **3, 4, 7,** and **8.**

The chemical shifts and the coupling constant of the AB doublets were similar to those observed for α-H and β-H of pyridine. Furthermore, eupolauridine exhibited the same characteristic behavior as pyridine on protonation (addition of CF_3COOH to the $CDCl_3$ solution) in that the shift of the signal from β-H was greater than that of the signal from α-H (β-H, $\Delta = 0.43$ and α-H, $\Delta = 0.12$ for eupolauridine; β-H, $\Delta = 1.06$ and α-H, $\Delta = 0.25$ ppm for pyridine). Comparable data for reasonable analogs of **7** or **8** were not available. Additional support for **3** or **4** was found in the fact that the UV spectrum of eupolauridine was similar to that of fluoranthene.

Since eupolauridine, eupolauramine, and hydroxyeupolauramine occur together, it was reasonable to assume that they were closely related and that an examination of the two last-named bases might provide sufficient evidence to distinguish clearly between the possible structures.

Eupolauramine showed many of the same spectral features as eupolauridine (see Section III,A), suggestive of an extended aromatic structure, but the IR spectrum now showed carbonyl absorption and the NMR spectrum had new peaks for *O*-methyl and *N*-methyl groups. The remainder of the NMR spectrum resembled that of eupolauridine except that the AB doublets arose from only one pair of protons, and the AA′BB′ pattern was replaced by an ABCD pattern, with one proton markedly deshielded, possibly by the methoxy group in an adjacent peri position. A $C_{14}N_2$ skeleton was therefore indicated as in eupolauridine but with the positions α and β to one of the nitrogen atoms substituted, thereby removing the symmetry present in eupolauridine. Six structures, such as **9** and variations based on **3, 4, 7,** and **8** were considered possible for eupolauramine. Simultaneously hydroxyeupolauramine was considered to have a structure in which the phenolic hydroxy group occupied the position of the strongly deshielded proton in eupolauramine. Attractive as some of the structures may have appeared, there were objections of one kind or another to each of them, but the balance was considered in the first instance to favor **3** (3,4-diazafluoranthene) for the structure of eupolauridine. The synthesis of this new substance was therefore undertaken, but the product was found to be different from eupolauridine. Structure **4** (1,6-diazafluoranthene) was then synthesized and this proved to be identical with eupolauridine.

B. Synthesis

The general plan adopted for the synthesis of eupolauridine (1,6-diazafluoranthene) (**4**) consisted of a Hantzsch synthesis of a suitable pyridine derivative, its conversion to a 4-azafluorene derivative, and finally the construction of the second heterocyclic ring (*5*) (Scheme 1).

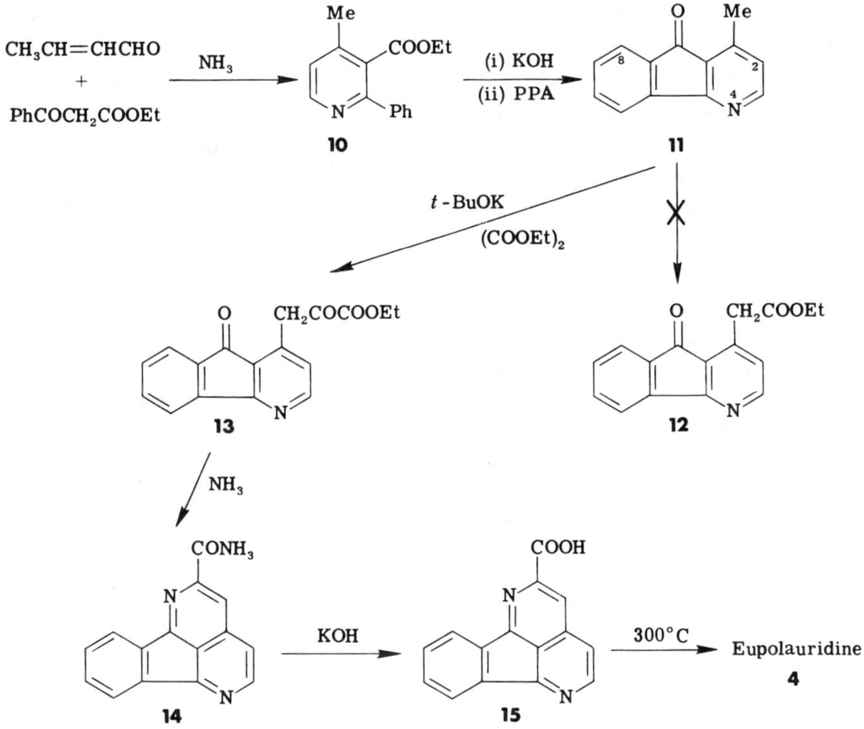

$CH_3CH{=}CHCHO$ + $PhCOCH_2COOEt$ →(NH_3) **10** →((i) KOH, (ii) PPA) **11**

11 →(*t*-BuOK, $(COOEt)_2$) **13**; **11** ✗→ **12**

13 →(NH_3) **14** →(KOH) **15** →(300°C) Eupolauridine **4**

SCHEME 1

In contrast to the report of Tsuda *et al.* (*10*) that crotonaldehyde condenses readily with ethyl 3-aminocrotonate in the presence of piperidine to give a good yield of the dihydropyridine derivative, which on oxidation with dilute nitric acid affords the pyridine derivative (**10** with Me instead of Ph), it was found that condensation involving ethyl 3-aminocinnamate under the same conditions gave a complex mixture, which on oxidation gave ethyl 4-methyl-2-phenylnicotinate (**10**) in only 5% yield. However, the overall yield was raised to 10% and the procedure was simplified when ethyl benzoylacetate itself was condensed with crotonaldehyde in the presence of concentrated ammonia. Hydrolysis of the ester gave the related acid, the structure of which was established by its decarboxylation to the known 4-methyl-2-phenylpyridine and by its cyclization by polyphosphoric acid to 1-methyl-4-azafluoren-9-one (**11**).

Attempts to convert **11** to **12** (which it was planned to condense with ammonia to obtain the desired ring system) by treatment with potassium *tert*-butoxide or potassium hydride, followed by diethyl carbonate or ethyl chloroformate, were fruitless. However, condensation of **11** with diethyl oxalate in the presence of potassium *tert*-butoxide gave the ester (**13**) in good yield. The ester (**13**) reverted to the parent compound (**11**) very readily under alkaline conditions, but ethanolic ammonia afforded the amide (**14**) in reasonable yield. Hydrolysis of **14** furnished the acid (**15**), which was smoothly decarboxylated to 1,6-diazafluoranthene (**4**), identical in all respects with eupolauridine.

Another route from **11** also gave eupolauridine (Scheme 2). Treatment of the substance with formaldehyde and dimethylammonium chloride yielded the Mannich product (**16**), which was isolated as the dihydrochloride. When **16** was heated with ammonia under pressure, eupolauridine was formed as the main product together with 3-methyl-1,6-diazafluoranthene (**17**). Presumably **16** and ammonia by base exchange yield **18,** which cyclizes to **19.** Dehydration of the latter would then yield **20,** a tautomeric form of **17.** Loss of formaldehyde from **18** or **19** by retroaldol reaction, followed by cyclization, in the former case would yield 2,3-dihydroeupolauridine, which could undergo disproportionation or oxidation (by formaldehyde) to yield **4.**

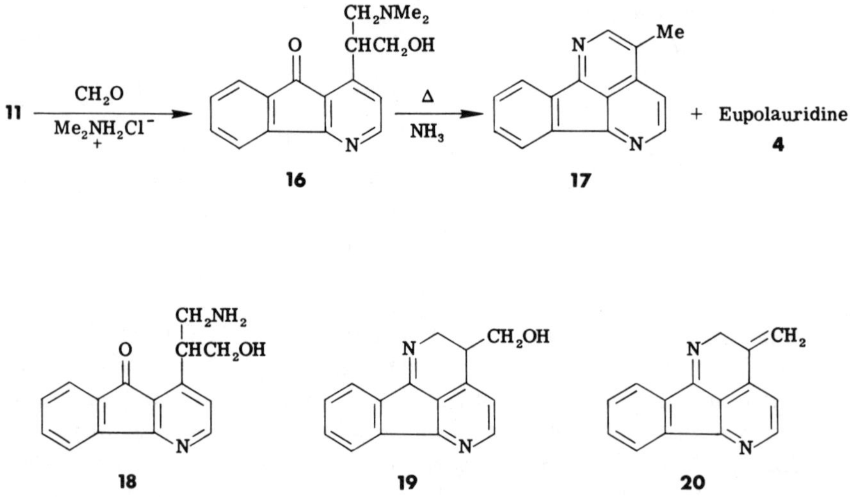

11 $\xrightarrow[Me_2\overset{+}{N}H_2Cl^-]{CH_2O}$ **16** (CH_2NMe_2, $CHCH_2OH$) $\xrightarrow[NH_3]{\Delta}$ **17** (Me) + Eupolauridine **4**

18 (CH_2NH_2, $CHCH_2OH$) **19** (CH_2OH) **20** ($=CH_2$)

SCHEME 2

The synthesis of 3,4-diazafluoranthene was more troublesome, and two initial approaches failed (*5*). The successful approach (Scheme 3) was based on the work of Powers and Ponticello (*11*) who prepared 2-oxo-4-phenyl-1,2-dihydronicotinonitrile (**21**) by the ethoxide-catalyzed condensation of ethyl cyanoacetate with 3-amino-1-phenylprop-2-en-1-one. By modification of their procedure, the yield was raised from 10 to about 20%.

Cyclization of **21** by polyphosphoric acid gave 2*H*-2-azafluorene-1,9-dione (**22**) as previously described. Condensation of **22** with cyanoacetamide under Cope conditions gave **23,** which, it was hoped, could be converted to a 3,4-diazafluoranthene derivative. However, cyclization by polyphosphoric acid

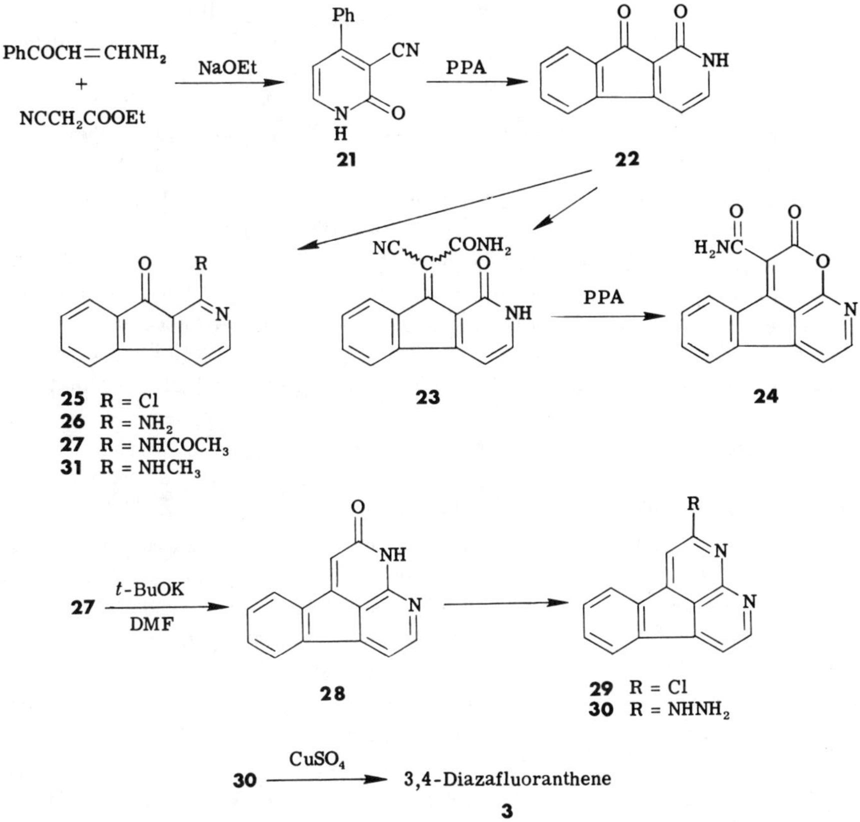

$PhCOCH{=}CHNH_2$ + $NCCH_2COOEt$ —NaOEt→ **21** —PPA→ **22**

22 → **25**, **23**; **23** —PPA→ **24**

25 R = Cl
26 R = NH_2
27 R = $NHCOCH_3$
31 R = $NHCH_3$

27 —*t*-BuOK, DMF→ **28** → **29** R = Cl; **30** R = $NHNH_2$

30 —$CuSO_4$→ 3,4-Diazafluoranthene **3**

SCHEME 3

yielded 2-oxo-2*H*-4-aza-3-oxafluoranthene-1-carboxamide (**24**). Treatment of **22** with phosphorus oxychloride gave 1-chloro-2-azafluoren-9-one (**25**), which was converted by ammonia with copper sulfate catalyst to the amine (**26**). This substance did not undergo a Friedlander cyclization with a range of substrates, but a Camps synthesis on the acetyl derivative (**27**) was successful. Treatment of **27** with potassium *tert*-butoxide in dimethylformamide gave the required 3,4-diazafluoranthen-2(3*H*)-one (**28**). Phosphorus oxychloride smoothly afforded the 2-chloro-3,4-diazafluoranthene (**29**), but removal of the chlorine was unexpectedly troublesome. Catalytic hydrogenation of **29** in the presence of a variety of catalysts gave mainly green polymeric material, or at best 3,4-diazafluoranthene in very low yield. A number of chemical methods also failed, but finally it was found that the hydrazine derivative (**30**) could be converted in 37% yield to the parent system (**3**). It was then, of course, found that this product was not identical with eupolauridine.

III. Eupolauramine

A. Structure

Eupolauramine (*4,5*) crystallizes from ethanol as golden-yellow prisms, mp 190–191°C. The molecular formula, $C_{16}H_{12}N_2O_2$, and high-intensity UV absorption (λ_{max} 230sh, 236, 291, 301, and 400 nm; log ϵ 4.67, 4.69, 4.42, 4.39, and 3.81) again indicated an extended aromatic chromophore; the fact that the spectrum was virtually unaltered by the addition of acid was consistent with the low basicity of the alkaloid. The mass spectrum had no high-intensity peaks apart from the base peak (*m/e* 249) (loss of methyl radical from the molecular ion, *m/e* 264). The IR spectrum showed carbonyl absorption [ν_{max} (Nujol) 1718, 1698, and 1652 cm^{-1}; ν_{max} (CCl_4) 1718 and 1652 cm^{-1}]. In the NMR spectrum (100 MHz), two three-proton singlets at δ3.63 and 4.00 were assigned to *N*-Me and *O*-Me groups, respectively, and thus it was concluded that eupolauramine had a $C_{14}N_2$ skeleton as in eupolauridine. A one-proton doublet (J = 4.7 Hz) at δ9.0 was attributed to an α-pyridine proton and a matching one-proton doublet at δ7.75 to the β-proton. A one-proton multiplet at δ8.01 and a matching one at δ8.86 were assigned to the A and A′ protons of the AA′BB′ system observed in eupolauridine and the remaining multiplet at δ7.62 to the B and B′ protons.

The total evidence was taken to indicate that in eupolauramine the positions α and β to one nitrogen atom were substituted, and that the marked deshielding effect on one proton (δ8.86) was due to a peri effect from the newly introduced methoxyl group. This interpretation led to structures such as **9** being considered, as well as others, based on the possible parent systems. However, objections were raised against each of the structures for one reason or another. Nevertheless, it was structure **9** for eupolauramine that guided the choice first to synthesize 3,4-diazafluoranthene as the possible structure of eupolauridine. With this objective achieved, it was possible to undertake the preparation of **9.**

Reaction of the chloroazafluorenone intermediate (**25**) with methylamine gave the corresponding methylimino derivative (**31**). Treatment of **31** with excess potassium hydride, followed by methoxyacetyl chloride, resulted in successive acylation and cyclization to 1-methoxy-3-methyl-3,4-diazafluoranthene-2(3*H*)-one (**9**). This substance was clearly not identical with eupolauramine. Chemically, the methoxyl group was very susceptible to nucleophilic displacement: treatment of **9** with ethanolic sodium ethoxide at room temperature gave the corresponding 1-ethoxy derivative. Spectroscopically, the IR spectrum of **9** had a carbonyl band at 1650 cm^{-1} as expected. In the NMR spectrum, although the coupling patterns were similar, the chemical shifts were quite different from those of eupolauramine. In particular, the two downfield signals from the ABCD system occurred markedly upfield from the corresponding signals observed for eupolauramine. It was concluded that the peri effect of the methoxy group is small and could not account for the shifts observed for eupolauramine.

When the structure of eupolauridine was determined to be 1,6-diazafluoranthene (**4**) by synthesis, it followed from the assumptions made that eupolauramine would have to have structure **32**. This structure was clearly incompatible with the IR and NMR evidence, and it had to be concluded that eupolauramine and hydroxyeupolauramine were not based on the same ring system as that in eupolauridine. Nevertheless, with ample supplies of synthetic eupolauridine available, it was of interest to attempt the synthesis of **33,** the demethoxy derivative of **32.** Eupolauridine formed without difficulty the quaternary methiodide and methosulfate salts. However, efforts to oxidize either derivative by alkaline potassium ferricyanide under a variety of conditions gave intractable mixtures. The failure of the standard reaction to produce **33** was considered excellent evidence that **32** is an unstable substance and reinforced the conclusion that the structures of eupolauramine and hydroxyeupolauramine were not based on it.

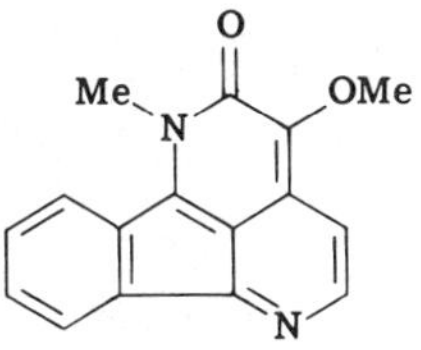

32 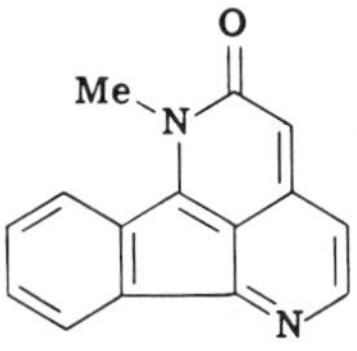 **33**

At this point the eupolauramine problem was solved by Bowden, Freeman, and Jones (*12*) who performed a single-crystal X-ray diffraction analysis. The structure of eupolauramine was shown to be 6-methoxy-5-methylbenzo[*h*]pyrrolo[4,3,2-*de*]quinolin-4(5*H*)-one (**5**). In this structure there has been superficially an interchange of particular 5- and 6-membered rings compared with **4**; the NMR data so far discussed can be interpreted equally well on this basis.

Later, ^{1}H-NMR data at 400 and ^{13}C-NMR data at 100.62 MHz were obtained, which further characterized the eupolauramine structure and assisted in the determination of the structure of hydroxyeupolauramine (*13*). Iterative analysis of the 4-spin system arising from H-7–H-10, which was now almost first order, yielded the following complete assignment: H-7, δ8.12; H-8, δ7.73; H-9, δ7.67; H-10, δ8.98; $J_{7,8}$ = 8.22, $J_{8,9}$ = 7.14, $J_{9,10}$ = 8.04, $J_{7,9}$ = 1.21, $J_{8,10}$ = 1.34, and $J_{7,10}$ = 0.61 Hz. The starting point for the analysis was the assignment of H-7, based on a specific nuclear Overhauser enhancement (NOE) (Fourier transform difference mode) on selective irradiation of 6-OMe (δ4.00); an NOE was also observed to 5-Me, whereas on irradiation of this group (δ3.63) the only NOE observed was to 6-OMe. The connectivity of the other signals was then determined by double-resonance experiments. H-10, being at a peri position, undergoes strong anisotropic deshielding by the C-10b—N-1 bond.

The ^{13}C-NMR spectrum was unequivocally assigned by means of noise-decoupled and coupled spectra at 100.62 MHz. All C—H connectivities (direct and

TABLE I
^{13}C NMR of Eupolauramine Derivatives[a]

Carbon	5	6	67	Carbon	5	6	67
2	150.2	147.6	150.5	9	126.5	114.1*	109.7
3	116.9	116.4	115.8	10	124.0	159.5	159.7
3a	132.7	132.8	132.6	10a	129.8	119.9	119.0
4	166.7	166.7	167.2	10b	142.5	144.2	143.5
5a	124.7	124.6	125.6	10c	122.3	122.6	122.6
6	137.4	138.6	137.5	NMe	28.3	28.6	28.2
6a	132.3	133.5	134.9	6-OMe	63.3	63.7	63.4
7	122.5	114.5*	115.4	10-OMe	—	—	56.7
8	129.1	131.0	129.6				

[a] Asterisk indicates some ambiguity regarding the assignment.

long range) were established by specific low power ^{1}H irradiations in coupled spectra. Spin–lattice relaxation time (T_1) values were also measured.

Several difficulties were originally experienced in detecting signals from quaternary carbons embedded in the polycyclic aromatic system when spectra were measured at 20 MHz. This was because of the inordinately long T_1 values of the nonprotonated carbons. However, at 100.62 MHz, signals were readily observed because of markedly shorter T_1 values (increased contribution from chemical-shift anisotropy relaxation mechanism). The shifts of the carbon atoms of eupolauramine are given in Table I.

B. Synthesis

Three syntheses of eupolauramine have been reported, two of them by Karuso and Taylor (*14*) and the third by Levin and Weinreb (*15*). The approaches of the two groups are quite different. Levin and Weinreb used contemporary chemistry, a key step in their synthesis being an intramolecular Diels–Alder cycloaddition of an olefinic dienophile onto an oxazole to generate eventually an annulated pyridine ring. Karuso and Taylor used a more classical approach based on benzoquinoline chemistry, but one of their syntheses is highly direct and efficient and makes gram quantities of eupolauramine readily accessible for biological assay.

1. Benzoquinoline Approach

The general approach was to construct first the benzo[*h*]quinoline (**38**) having appropriate functionalization for subsequent manipulation to generate the lactam ring (*14*).

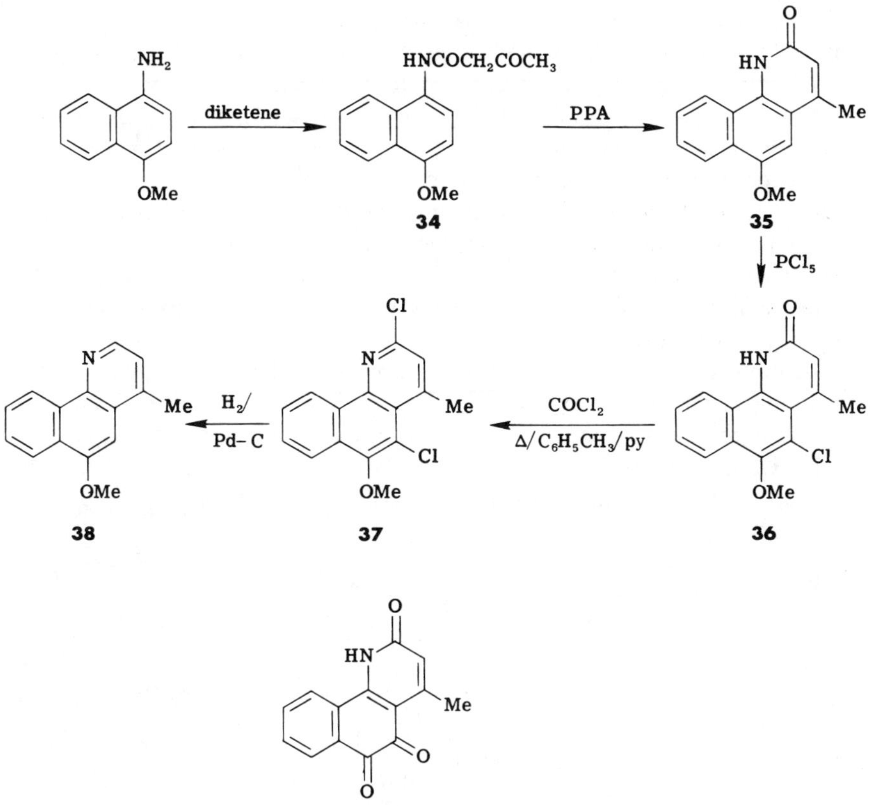

SCHEME 4

The acetoacetamidonaphthalene (**34**) formed from 4-methoxynaphthylamine and diketene (Scheme 4) was quantitatively cyclodehydrated in warm polyphosphoric acid to 6-methoxy-4-methylbenzo[*h*]quinolin-2(1*H*)-one (**35**). The next step, conversion of **35** to the benzoquinoline (**38**), proved extraordinarily difficult.

Treatment of **35** with phosphorus halides under a wide range of conditions did not in the slightest effect the conversion of the pyridone system to a chloropyridine. At best, phosphorus pentachloride gave a good yield of the 5-chloro derivative (**36**), which nevertheless was eventually to prove a useful substrate for the desired reaction. Forcing conditions caused O-demethylation and formation of the quinone (**39**) by oxidation.

Finally, the rarely used reagent phosgene (in toluene at reflux temperature) was found to be the reagent of choice. With the parent benzoquinolinone, reaction was still very slow and low yielding. It was found that the addition of ~0.1–

0.5 mol of pyridine decreased the reaction time, but did not markedly increase the yield above 30%. Surprisingly, the above-mentioned 5-chloroquinolinone (**36**) proved to be a much better substrate, giving a 94% yield of the 2,5-dichlorobenzoquinoline (**37**) within 24 hr (pyridine catalysis). The 5-chloro substituent conceivably interferes with some electronic interaction of 6-OMe with the pyridone system (steric inhibition of resonance?). Its presence caused no problem since hydrogenolysis of **37** readily afforded the desired benzo[*h*]-quinoline (**38**) in high yield. It is interesting to note that pyrazine and quinoxaline derivatives of the quinone (**39**) also reacted readily with phosgene to give the corresponding chloro derivatives, which were then hydrogenolyzed to the parent heterocycle systems.

Attempts to condense various aldehydes with the 4-Me group in **38** as the preliminary step prior to oxidation were fruitless. Direct oxidation with selenium dioxide in refluxing xylene led to moderate yields of the aldehyde (**40**) and the acid (**41**), the former being converted to the latter by alkaline silver oxide. Early attempts to nitrate the methyl ester (**42**) at C-5 were unsuccessful and this route was abandoned.

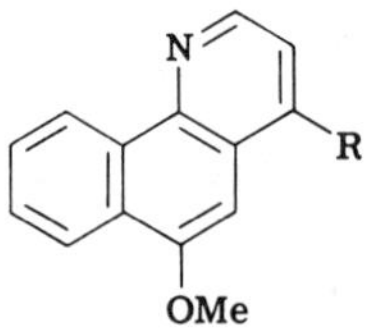

N R OMe

40 R = CHO
41 R = CO_2H
42 R = CO_2Me

In another approach (Scheme 5), dibromination of 4-Me and bromination at C-5 was achieved with *N*-bromosuccinimide, followed by bromine in carbon tetrachloride, to give **43.** Silver ion-facilitated solvolysis of the gem dibromide yielded the bromo aldehyde (**44**), which was readily converted directly to the *N*-methylcarboxamide (**45**) by the action of nickel peroxide in the presence of methylamine. This bromoamide is well constituted to undergo ring closure, but initial attempts to effect the reaction were unsuccessful. Finally, a mediocre yield (15%) of eupolauramine was achieved on treatment of **45** with sodium hydride and cuprous bromide in dimethylformamide. The major product was aldehyde **40.** This can be explained by the *in situ* production of a copper hydride, which would be capable of reducing both the aryl halide and the amide.

After this limited success through initial functionalization of 4-Me, the alternative approach requiring initial introduction of nitrogen at C-5 of **38** was examined (Scheme 6). Precisely defined conditions were required to form the 5-nitro derivative (**46**). Otherwise, a variety of other products such as the quinone (**47**), the 5,7-dinitro derivative, or the *N*-nitronium salt of **38** were formed. Potassium

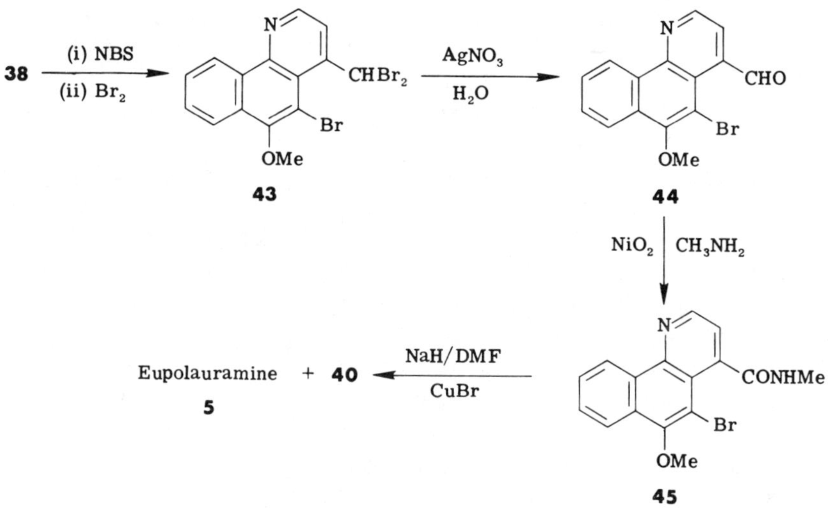

38 —(i) NBS; (ii) Br_2→ **43** (4-$CHBr_2$, 5-Br, 6-OMe) —$AgNO_3$, H_2O→ **44** (4-CHO, 5-Br, 6-OMe) —NiO_2 | CH_3NH_2→ **45** (4-CONHMe, 5-Br, 6-OMe) —NaH/DMF, CuBr→ Eupolauramine **5** + **40**

SCHEME 5

38 —KNO_3/H_2SO_4, $MeNO_2$→ **46** R = Me, **48** R = H (4-Me, 5-NO_2, 6-OR) —H_2/Pd–C→ **49** (4-Me, 5-NH_2, 6-OMe) —Ac_2O→ **50** (4-Me, 5-NHCOMe, 6-OMe) —SeO_2, Py→ **51** (lactam, N–H, C=O, OMe) —NaH/DMF, CH_3I→ Eupolauramine **5**

47 (4-Me, 5,6-dione)

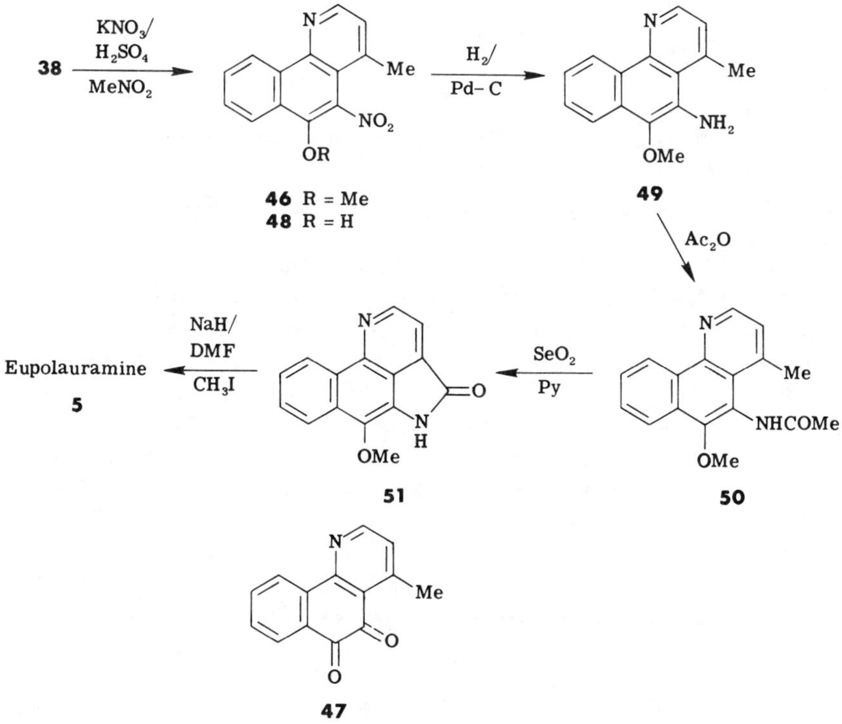

SCHEME 6

nitrate in absolute sulfuric acid at −60 to −30°C with nitromethane as diluent gave a crude product contaminated with a small amount of the nitrophenol (**48**), but this was of no consequence since treatment of the material with diazomethane effected remethylation, thereby ensuring an excellent overall yield of the methoxynitro compound (**46**).

Functionalization of C-4–Me in **46** was unsatisfactory: for example, selenium dioxide oxidation was sluggish, and bromination caused the hard-gained nitro group to be replaced by bromine to give again the tribromide (**43**).

In contrast the acetamide (**50**), prepared by medium-pressure catalytic hydrogenation of **46** to the amine (**49**), followed by immediate acetylation, proved to be a highly suitable substrate for the desired reaction. Treatment of **50** with selenium dioxide in pyridine effected three reactions *in situ* in high overall yield: oxidation of 4-Me, deacetylation, and ring closure to the bright-yellow lactam (**51**, de-*N*-methyleupolauramine). Methylation of **51** readily afforded eupolauramine. The overall yield in the nine steps from 4-methoxynaphthylamine was 38%.

2. Oxazole Approach

The Levin and Weinreb synthesis (*15*) (Scheme 7) commenced with the readily accessible methoxyoxazoline (**52**), which was treated with the Grignard reagent (**53**) to give the adduct (**54**). Nickel peroxide oxidation of **54** gave the desired oxazole derivative (**55**). The required dienophile system was generated by hydrolysis of the acetal function in **55** and condensation of the corresponding aldehyde with methyl (dimethoxyphosphoryl)acetate to yield exclusively the trans-unsaturated ester (**56**).

Unexpectedly, **56** was converted on heating in refluxing *o*-dichlorobenzene to a mixture of pyridinols **58** and **59.** It was suggested that these products probably arise from the Diels–Alder adduct (**57**) via oxidative fragmentation. However, when the thermolysis of **56** was done in the presence of 0.75 equivalents of diazabicyclononene, the reaction took a different course, and the desired pyridine (**60**) was the exclusive product (76% yield). Undoubtedly the presence of the base favors loss of water from the initial adduct (**57**), but its role may be more complex. Dehydrogenation of **60** to **61** was effected by *N*-bromosuccinimide.

Compound **61**, as a benzo[*h*]quinoline, is reminiscent of derivatives prepared by Karuso and Taylor (Section II,B,1), but of course it lacks the required 6-OMe group. Arene oxide chemistry provided the key to the generation of oxygen functionality at C-6 and simultaneously to the formation of the lactam ring of eupolauramine.

Exposure of **61** to aqueous sodium hypochlorite under phase-transfer conditions produced the arene oxide (**62**), which was then treated with dimethylalumi-

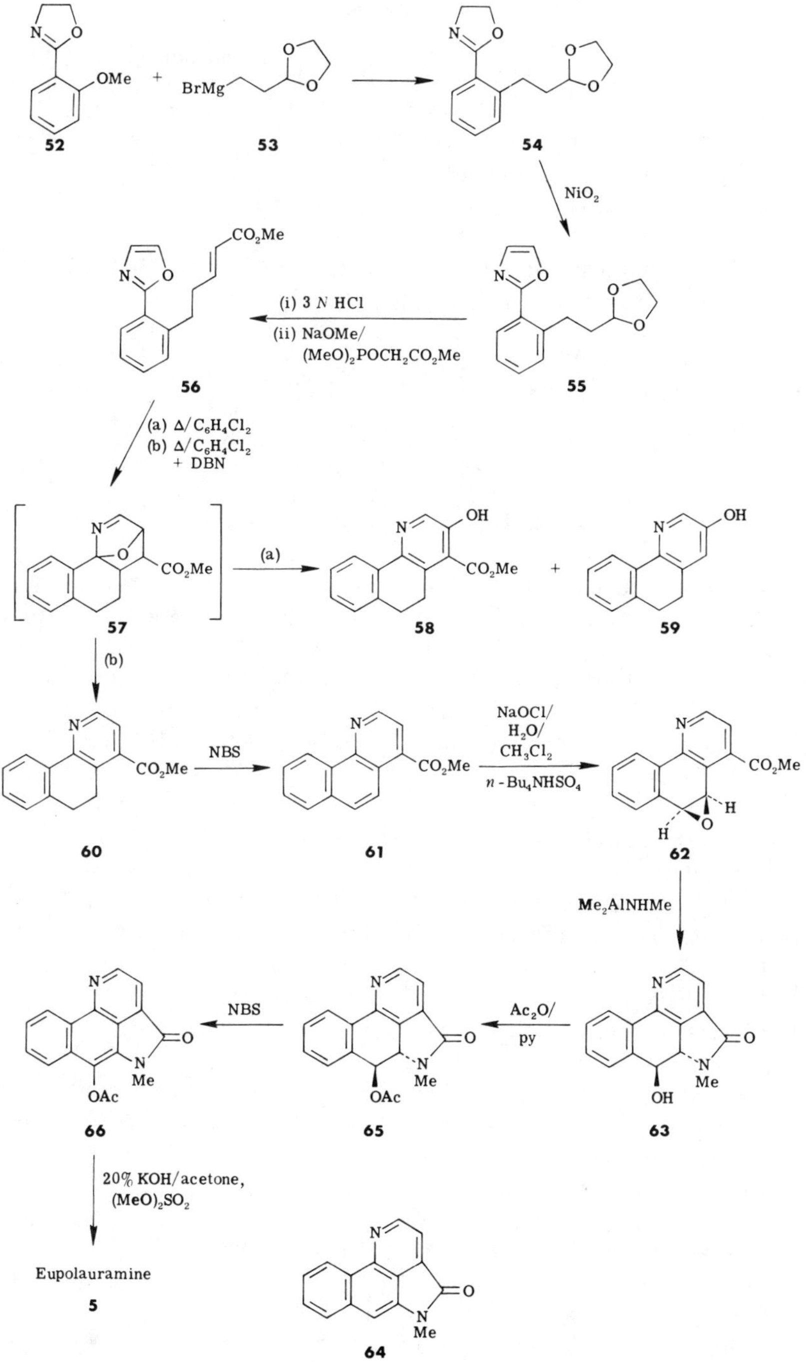

52 + 53 → 54

NiO_2

55

(i) 3 *N* HCl
(ii) NaOMe/ $(MeO)_2POCH_2CO_2Me$

56

(a) $\Delta/C_6H_4Cl_2$
(b) $\Delta/C_6H_4Cl_2$ + DBN

57

(a)

58 + 59

(b)

60

NBS

61

$NaOCl/H_2O/CH_3Cl_2$
n-Bu_4NHSO_4

62

$Me_2AlNHMe$

63

Ac_2O/py

65

NBS

66

20% KOH/acetone, $(MeO)_2SO_2$

Eupolauramine
5

64

SCHEME 7

num *N*-methylamide to yield the tetracyclic hydroxylactam (**63**) (49%). The precise course of events in this reaction is uncertain: the reagent is capable of converting the ester to an *N*-methylamide, which could cyclize to **63,** and also of directly opening the oxirane ring prior to lactam formation. It was reported that no products of regioisomeric opening of the oxirane ring were detected.

Surprisingly, oxidation of the secondary hydroxyl group in **63** to produce *O*-demethyleupolauramine could not be effected under any conditions, the only product obtained being the anhydro derivative (**64**). However, oxidation (*N*-bromosuccinimide) of the acetyl derivative (**65**) gave the dehydro derivative (**66**), which was hydrolyzed and methylated in one step [20% KOH–acetone, $(MeO)_2SO_2$] to yield eupolauramine. The overall yield in the 11 steps from oxazoline (**52**) was 4.5%.

IV. Hydroxyeupolauramine

Hydroxyeupolauramine crystallizes from ethanol as fine, orange needles, mp 282–283°C (*4*). Intense UV absorption similar to that of eupolauramine, and the molecular formula, $C_{16}H_{12}N_2O_3$, suggested that the substance was a hydroxy derivative of eupolauramine (*5*). Infrared absorption for OH was not apparent in a Nujol mull or $CHCl_3$ solution, but in a KBr disk broad absorption was present at 3400 cm^{-1}. Carbonyl absorptions [ν_{max} (Nujol) 1700 and 1652 cm^{-1}; ν_{max} ($CHCl_3$) 1704 and 1653 cm^{-1}; ν_{max} (KBr) 1715, 1702, and 1656 cm^{-1}] were observed at frequencies similar to those for eupolauramine. The NMR spectrum (100 MHz) showed signals for *N*-Me (δ3.76) and *O*-Me (δ4.08) and two one-proton doublets (J = 4.7 Hz) at δ8.99 and 7.91 (pyridine α and β protons, respectively). Three aromatic protons resonated as an apparent doublet of doublets at δ7.18 and a multiplet at δ7.6. The remaining proton signal could not be detected in the spectrum measured in deuteriochloroform solution presumably because of the low concentration and the broadness of the signal. However, in deuteriopyridine solution a signal from an exhangeable proton was observed at δ12.58, indicative of a hydrogen-bonded phenolic proton.

Original structural considerations (*5*) were confused by the attempts to relate eupolauramine and hydroxyeupolauramine with eupolauridine. When the structure of eupolauramine was established as **5,** it followed that the most likely structure for hydroxyeupolauramine was **6,** with the OH located at C-10.

NMR studies at 400 MHz (^1H) and 100.62 MHz (^{13}C) support structure **6** (*13*). In particular at 400 MHz the pattern from H-7 to H-9 was clearly that of an ABX system, and iterative analysis yielded the following assignments: H-7, δ7.68; H-8, δ7.66; H-9, δ7.23; $J_{7,8}$ = 8.08, $J_{8,9}$ = 8.05, and $J_{7,9}$ = 1.00 Hz. Irradiation of 6-OMe (δ4.08) gave NOEs to H-7 and 5-Me (δ3.76); irradiation of 5-Me gave an NOE to 6-OMe only. With H-7 assigned, it followed that the hydroxyl group was located at C-10. The absence in the ^{1}H-NMR spectrum of a strongly deshielded proton (H-10) was in accord with this proposal, as was the strong hydrogen bonding experienced by 10-OH (with N-1).

Hydroxyeupolauramine was not appreciably soluble in aqueous sodium hydroxide. However, treatment with sodium hydride in dimethylformamide gave a red anion, and addition of methyl iodide then afforded the methyl ether (**67**), orange-yellow prisms from ethanol, mp 213–215°C.

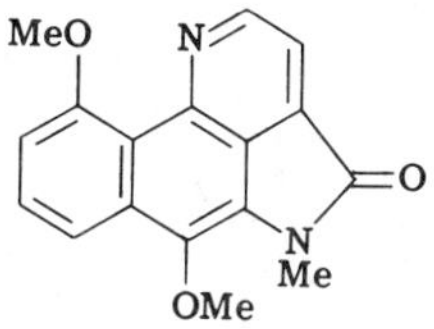

MeO N O N Me OMe

67

The ^{1}H-NMR spectrum (400 MHz) of **67** showed singlets at δ3.70 (5-Me), 4.00 (6-OMe), and a fine doublet at δ4.19 (10-OMe). H-2 resonated at δ9.26 and H-3 at δ7.84 as doublets (J = 4.7 Hz). H-7–H-9 gave an ABX pattern with long range coupling from 10-OMe to H-9. Precise parameters were obtained by iterative analysis: H-7, δ7.79; H-8, δ7.66; H-9, δ7.21; $J_{7,8}$ = 8.19, $J_{7,9}$ = 1.08, $J_{8,9}$ = 8.08, and $J_{9,OMe}$ = 0.35 Hz. Double irradiation experiments confirmed the assignments as did NOE experiments: irradiation of 5-Me gave an NOE only to 6-OMe, as expected; irradiation of 6-OMe gave NOEs to 5-Me and to H-7, and irradiation of 10-OMe gave selectively an NOE to H-9. These results unequivocally defined the substitution pattern in **67,** and hence in hydroxyeupolauramine.

The ^{13}C-NMR spectrum of the methyl ether (**67**) was fully assigned with the aid of coupled spectra and specific-frequency, low-power ^{1}H decouplings. The results provided further cogent evidence for the hydroxyeupolauramine structure. Whereas in the ^{1}H-coupled spectrum of eupolauramine the signals arising from C-6a, C-8, and C-10b are, respectively, a triplet (δ132.3, $^3J_{H\text{-}8}$ = $^3J_{H\text{-}10}$ = 7.2 Hz), a doublet of doublets (δ129.1, $^1J_{H\text{-}8}$ = 160.5 and $^3J_{H\text{-}10}$ = 8.2 Hz), and a doublet of doublets (δ142.5, $^3J_{H\text{-}2}$ = 12.2 and $^3J_{H\text{-}10}$ = 4.3 Hz), the corresponding signals in the spectrum of **67** are all doublets (δ134.9, $^3J_{H\text{-}8}$ = 8.8 Hz; δ129.6, $^1J_{H\text{-}8}$ = 161.6 Hz; δ143.5, $^3J_{H\text{-}2}$ = 12.9 Hz) owing to the absence of H-10. The chemical shifts of the carbon atoms of the methyl ether (**67**) and by analogy of hydroxyeupolauramine are given in Table I (Section II,A).

V. Biogenesis

Structure **4** for eupolauridine is quite remarkable for an alkaloid, in respect of the ring system and the absence of oxygenation. Considered in isolation, its biogenetic origins are so well concealed that biogenetic considerations were not featured in the structural analysis for eupolauridine. However, after the determination of the structures of eupolauramine and hydroxyeupolauramine, the biogenetic picture became clearer because it is reasonable to postulate a relationship between the latter substances and the 7-oxoaporphine class of alkaloids. Indeed, it is now considered highly likely that liriodenine or its putative precursor, 1,2-dihydroxy-7-oxoaporphine (liriodendronine) (**68**) is the progenitor of the three alkaloids (*13*).

Considering first the eupolauramine structure, the lactam ring of **5** is reminis-

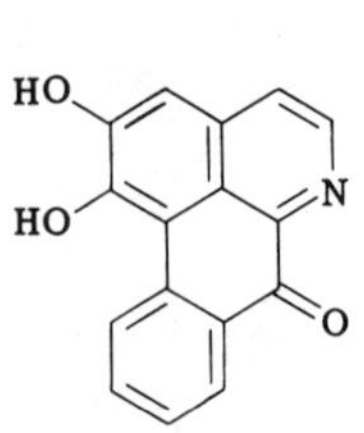

HO, HO, N, O
68

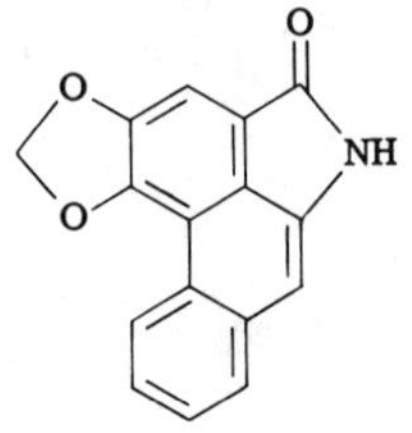

O, O, O, NH
69

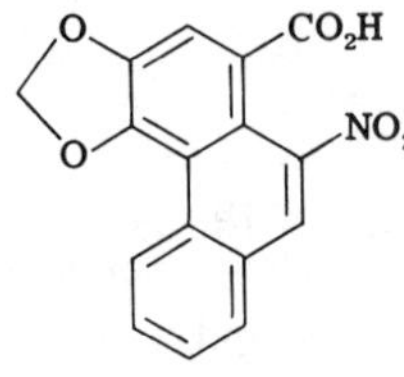

O, O, CO_2H, NO_2
70

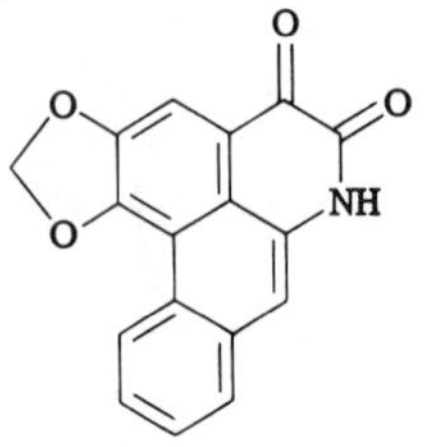

O, O, O, O, NH
71

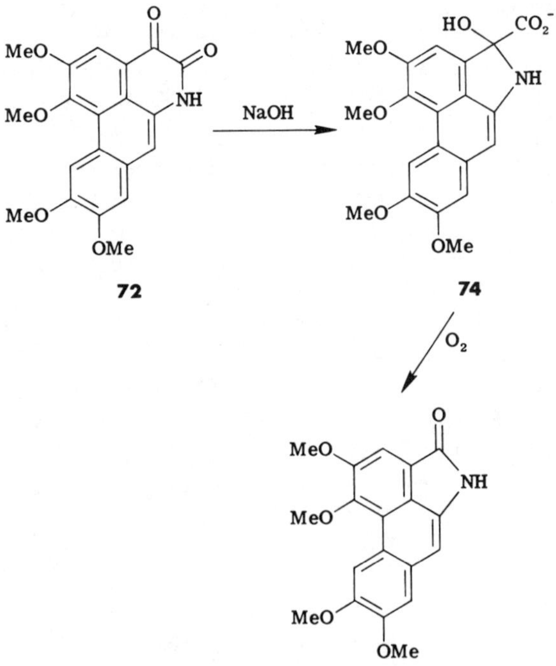

MeO, MeO, MeO, OMe, O, O, NH
72

NaOH

HO, CO_2^-, MeO, MeO, MeO, OMe, NH
74

O_2

MeO, MeO, MeO, OMe, O, NH
73

cent of the aristololactam* structure typified by cepharanone-A (**69**) (*16*). This lactam structure is generally believed to result from oxidation of the heterocyclic ring of an aporphine system. The end result of these processes is the aristolochic acid structure such as aristolochic acid II (**70**). The sequence 4-hydroxyaporphine → 4,5-dioxoaporphine → aristololactam has been proposed (*16, 17*) and there are some biosynthetic results to support this idea (*18*). Also suggestive of this pathway is the cooccurrence in *Stephania cepharantha* (Menispermaceae) of cepharadione A (**71**) and cepharanone A (**69**) (*19*). Furthermore, the 4,5-dioxoaporphine alkaloid pontevedrine (**72**) on treatment with methanolic sodium hydroxide gave the aristololactam (**73**) via a benzilic acid rearrangement, followed by air oxidation of the intermediate α-hydroxy acid anion (**74**) (*17*). It is curious, but presumably of no great significance that no naturally occurring aristololactam is N-methylated. Of more interest is the fact that no aristololactam is oxygenated at C-7; such a system would probably require a 7-oxoaporphine precursor.

Novel as the lactam structure is, the most remarkable feature of the eupolauramine and eupolauridine structures is the aza A ring, which has no parallel in the realm of alkaloid chemistry. It seems inconceivable that this ring stems from a pyridine precursor. Rather, it is believed (*13*) the ring most likely arises from oxidative cleavage of the benzenoid ring A of an aporphine, followed by loss of a carbon atom, introduction of an amine group via transamination, and ring closure to reform ring A (not necessarily in this order).

Scheme 8 shows the full pathway suggested for eupolauramine. The precise order of events for ring A versus ring B–C changes cannot be defined at this stage. The acid (**76**), which could be formed by oxidative cleavage of the dihydroxyaporphine (**68**) at C-1—C-11b, is in fact the homolog of the acid (**77**) prepared originally by W. I. Taylor in his proof of the structure of liriodenine (*20*). Transamination of **76,** followed by ring closure to a pyridone and deoxygenation, would give the 1-azaoxoaporphine (**81**) via **78–80.** An attractive alternative pathway could be transamination of keto acid **75** to **82,** followed by cyclization to **83**; oxidative decarboxylation of this would lead directly to **81.**

There is possibly better precedent for cleavage of the C-1—C-2 bond in **68**—cf. oxidation of catechols to muconic acids (*21*). This would require subsequent loss of the carboxyl group produced from C-1. However, good analogy for the pathway proposed here is provided by the "extradiol" cleavage of dihydroxyphenylalanine in the biosynthesis of betalamic acid and related compounds (*21, 22*). Recently Dagne and Steglich (*23*) reported another example of an alkaloid containing a pyridine ring formed from a benzene ring, namely, erymelanthine, in the *Erythrina* alkaloid series; an extradiol cleavage pathway has been proposed.

* This name, coined by original workers in the field, would appear to be preferred to "aristolactam," which has appeared in certain reviews and papers.

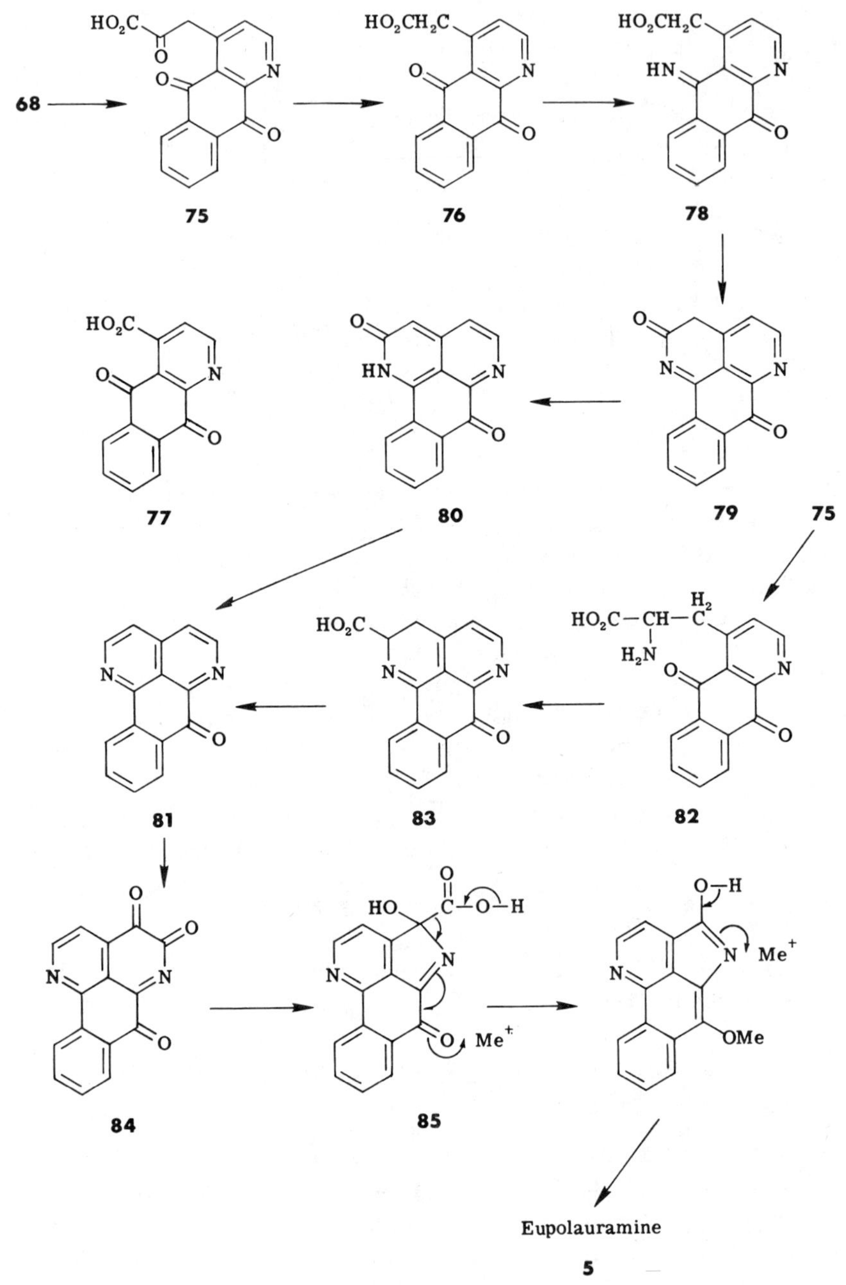

68 → 75 → 76 → 78 → 79 → 80 → 81 → 84 → 85 → Eupolauramine 5

75 → 82 → 83 → 81

77

HO_2C HO_2CH_2C HN N O OH H_2N $HO_2C-CH-CH_2$ Me$^+$ OMe O–H

Eupolauramine

5

Scheme 8

Benzilic acid rearrangement of a 4,5-dioxo derivative (**84**) would produce the hydroxy acid (**85**). An attractive feature of this last structure is that there is no need for an oxidative decarboxylation step, as for pontevedrine (**72**), because of the presence of the 7-oxo function. Vinylogous decarboxylation can occur in **85** to give directly the lactam system (see arrows). O-Methylation could occur simultaneously, followed by N-methylation as shown.

The hydroxyl group at C-10 in hydroxyeupolauramine (**6**) is in a relatively uncommon biogenetic position, in view of the shikimate origin of ring D, but precedents do exist in structures of aporphine alkaloids, e.g., oxopukateine (**86**).

On the basis of the proposals for the biogenesis of eupolauramine, ideas can be put forward for formation of the unique structure present in eupolauridine.

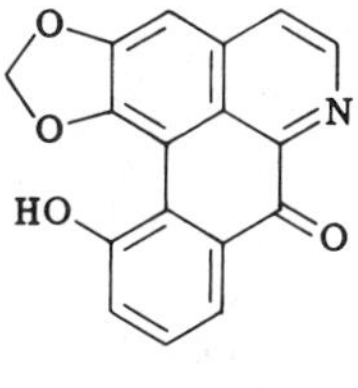

86

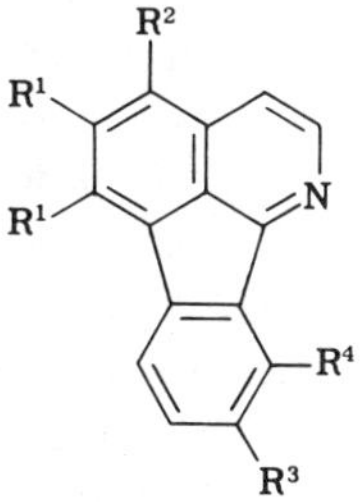

87 R^1 = OMe; R^2 = R^3 = R^4 = H
88 R^1 = R^2 = R^3 = OMe; R^4 = H
89 R^1 = R^2 = OMe; R^3 = OH; R^4 = H
90 R^1 = R^2 = R^3 = R^4 = OMe

Precedent for the 5-membered ring C is provided by azafluoranthene alkaloids discovered recently in members of the Menispermaceae, namely, triclisine (**87**) (*24*), rufescine (**88**) (*25*) and norrufescine (**89**) (*26*), and imeluteine (**90**) (*25*). Superficially, these substances can be considered to arise from the cyclization of 1-phenylisoquinolines in the same manner as aporphines come from benzylisoquinolines. There is an early suggestion to this effect (*25*). However, against this idea is the fact that 1-phenylisoquinoline alkaloids are extremely rare, there being at present only three known examples—the cryptostylines, isolated from a single genus of the Orchidaceae (*27*). If phenylisoquinolines were indeed precursors of the azafluoranthenes, a wider distribution of them could be expected. On the contrary, in the present case and in the cases of **87–90,** *7-oxoaporphines* cooccur. A ring-contraction pathway from an oxoaporphine is therefore favored, with concurrent extrusion of C-7 as carbon monoxide or its equivalent by an as yet undefined process. Eupolauridine could then arise from the azaoxoaporphine (**81**).

An interesting new proposal has been made by Cava *et al.* (*28*) that rufescine (**88**) could arise by extrusion of carbon monoxide from the unusual tropoloisoquinoline alkaloid imerubine (**91**), itself the product of a ring-expansion process

OMe

MeO

MeO

N

O

OMe

91

involving a 7-oxygenated species. However, in general the direct decarbonylation mechanism seems preferable at present. Certainly it better explains the formation of compounds not containing oxygen substituents in the benzene ring, inasmuch as the Cava mechanism requires oxygen-substituent participation. It is hoped that work on other alkaloids produced by *E. laurina* will throw more light on this puzzling area of biogenesis (*9*).

The above-mentioned isolation of eupolauridine along with liriodenine from a member of the Annonaceae (*Cananga*) (*8*) is an interesting chemotaxonomic result, which supports the view that the genus *Eupomatia* does have certain affinities with the family Annonaceae, even if the balance of evidence favors the separate family Eupomatiaceae. Moreover the result signals that further examples of the novel biogenetic pathways involved here could well come to light in the future.

References

1. A. T. Hotchkiss, *J. Arnold Arbor., Harv. Univ.* **36,** 385 (1953).
2. A. J. Eames, "Morphology of the Angiosperms," Chapter 11. McGraw-Hill, New York, 1961.
3. R. W. Read and W. C. Taylor, *Aust. J. Chem.* **34,** 1125 (1981), and previous papers cited therein.
4. B. F. Bowden, E. Ritchie, and W. C. Taylor, *Aust. J. Chem.* **25,** 2659 (1972).
5. B. F. Bowden, K. Picker, E. Ritchie, and W. C. Taylor, *Aust. J. Chem.* **28,** 2681 (1975).
6. R. W. Read and W. C. Taylor, *Aust. J. Chem.* **32,** 2317 (1979).
7. K. Picker, E. Ritchie, and W. C. Taylor, *Aust. J. Chem.* **26,** 1111 (1973).
8. M. Leboeuf and A. Cavé, *Lloydia* **39,** 459 (1976).
9. W. C. Taylor, unpublished results.
10. K. Tsuda, Y. Satch, N. Ikekawa, and H. Mishima, *J. Org. Chem.* **21,** 800 (1956).
11. J. C. Powers and I. Ponticello, *J. Am. Chem. Soc.* **90,** 7102 (1968).
12. B. F. Bowden, H. C. Freeman, and R. D. G. Jones, *J. Chem. Soc., Perkin Trans. 2,* 658 (1976).
13. W. C. Taylor, *Aust. J. Chem.* **37,** 1095 (1984).
14. P. Karuso and W. C. Taylor, *Aust. J. Chem.* **37,** 1271 (1984).
15. J. I. Levin and S. M. Weinreb, *J. Am. Chem. Soc.* **105,** 1397 (1983).
16. M. Shamma and J. L. Moniot, "Isoquinoline Alkaloids Research 1972–1977," Chapter 17. Plenum, New York, 1978.

17. L. Castedo, R. Suau, and A. Mouriño, *Tetrahedron Lett.* 501 (1976).
18. F. Comer, H. P. Tiwari, and I. D. Spenser, *Can. J. Chem.* **47,** 481 (1969).
19. M. Akasu, H. Itokawa, and M. Fugita, *Tetrahedron Lett.* 3609 (1974).
20. W. I. Taylor, *Tetrahedron* **14,** 42 (1961).
21. O. Hayaishi, *in* "Molecular Mechanisms of Oxygen Activation" (O. Hayaishi, ed.), Chapter 1. Academic Press, New York, 1974.
22. C.-K. Wat and G. H. N. Towers, *in* "Biochemistry of Plant Phenolics" (T. Swain, J. B. Harborne, and C. V. Van Sumere, eds.), Chapter 12. Plenum, New York, 1979.
23. E. Dagne and W. Steglich, *Tetrahedron Lett.,* 5067 (1983).
24. R. Huls, J. Gaspers, and R. Warin, *Bull. Soc. R. Sci. Liege* **45,** 40 (1976).
25. M. P. Cava, K. T. Buck, and A. I. da Rocha, *J. Am. Chem. Soc.* **94,** 5931 (1972).
26. M. P. Cava, K. T. Buck, I. Noguchi, M. Srinivasan, M. G. Rao, and A. I. da Rocha, *Tetrahedron* **31,** 1667 (1975).
27. S. Agurell, I. Granelli, K. Leander, B. Lüning, and J. Rosenblum, *Acta Chem. Scand., Ser. B* **B28,** 239 (1974).
28. J. V. Silverton, C. Kabuto, K. T. Buck, and M. P. Cava, *J. Am. Chem. Soc.* **99,** 6708 (1977).

——CHAPTER 2——

MARINE ALKALOIDS

CARSTEN CHRISTOPHERSEN

Department of General and Organic Chemistry
University of Copenhagen, The H. C. Ørsted Institute
Copenhagen, Denmark

I. Introduction

In reviewing an area as nonhomogeneous as marine alkaloids, the reviewer is primarily faced with the problem of definition of alkaloids. This is a difficulty

THE ALKALOIDS, VOL. XXIV

ISBN 0-12-469524-8

normally not encountered in dealing with compounds from limited taxons or established classes of alkaloids. The reader of such texts usually wishes to acquire a detailed survey of the area and is therefore interested in certain structural digressions as long as they support the main theme. This is presumably not the case in a review such as this, where such digressions may very well cloud the issue to such an extent that the main objective is lost.

Many nitrogen-containing metabolites, except for primary metabolites, have been named alkaloids. Literally they should be basic substances; however, several examples are known where this requirement is not fulfilled. The majority of these compounds are known to originate from amino acids; however, these units may be extensively derivatized and rearranged to a degree where resemblance to the original building block is far from evident. At least in principle it may be argued that all the cellular nitrogen may at some time or other pass through the amino-acid pool, thus defining all secondary nitrogen compounds as alkaloids. Even recognition of primary metabolites is often impossible owing to the lack of knowledge of many metabolic processes in marine organisms.

As a consequence of these difficulties, the selection of compounds to be dealt with in this review is somewhat arbitrary and to some extent reflects the choice and interest of the reviewer. Thus, for example, pahutoxin (**1**) of the boxfish (*1*), caulerpicin (**2**) from the green alga *Caulerpa racemosa* (*2*), 1-methylisoguanosine (**3**) from the sponge *Tedania digitata* (*3*) and other unusual nucleosides, 2-isocyanopupukeanane (**4**) from the nudibranch *Phyllidia varicosa* (*4*) and a variety of isocyanides, isothiocyanates, and formamides from sponges, neopterin (**5**) from the eyes of the polychaete *Platynereis dumerilii* (*5*) and other pterins are not dealt with in this review.

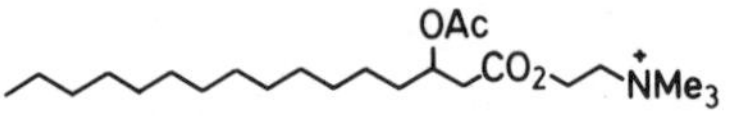

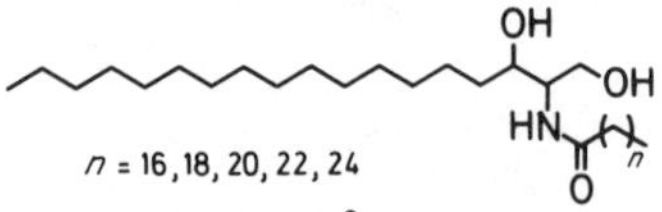

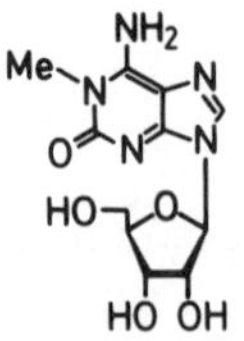

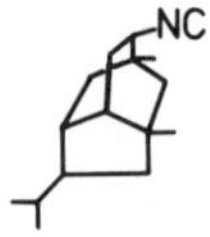

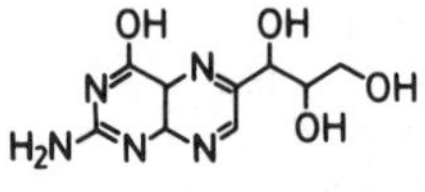

Many sponges contain conspicuous amounts of organically bound bromine (*6*). In several cases the bromine has been located in typical secondary metabolites; however, in some cases it has been demonstrated to occur in the amino acids, forming spongin, the structural proteinaceous matter of the sponges (Porifera) (*7*). In the latter case these unusual amino acids must be primary metabolites. Another related example is the presence of aminophosphonic acids in coelenterates, where these compounds were found also to be incorporated in lipids and proteins (*8*).

Biosynthetic studies in the area of marine alkaloids are extremely rare and, in the few cases where they exist, mainly inconclusive. This sad state of affairs is at least partly connected with difficulties involved in culturing marine organisms. Another inherited problem is the question of the origin of the metabolites. Most marine animals are associated with parasitic, commensal, or symbiotic flora and fauna, the best-known example being the zooxanthellae of the corals. The sponge is a group of marine animals continuously delivering new and unique natural products. They are often heavily infected with algae or bacteria. For example, *Vergonia* (Section V,B) contains intracellular bacteria, which may in turn be infected by bacteriophages (*9*), and an extracellular blue-green alga (Chroococcales) (*10*). The bacteria may constitute nearly 40% of the tissue volume (*11*). The sponge *Agelas oroides* (Section II,E) has huge intercellular bacterial populations, while *Oscarella lobularis* (Section IV,F) has a low intercellular bacterial density (*12*).

In cases where alkaloids from marine organisms have any resemblance to known structure, these are very often of microbial origin. Although evidence is lacking, it is presumably safe to predict that several natural products obtained from sponges are actually synthesized by the associated flora.

In connection with the above-mentioned observations, it is thought provoking to note that many sponges have attracted the attention of the chemist because of screening of their antimicrobial activity (*13*), e.g., the above-mentioned *Verongia* (*14*). Many sponges with a high density of bacteria contain several (4–7) morphologically different, associated bacteria, while species with low bacterial density usually have only one form (*12*). If the symbiont populations are species specific, it seems that they have developed chemical means to keep their habitat free from competitors, or that they alone have overcome the chemical defenses of their host.

What has been said about sponges undoubtedly holds for other marine organisms as well; however, investigations in this area are even more scarce (for a possible example, see Section III,E).

A fair amount of work has been done on the pharmacological properties of marine natural products; however, the bulk of publications deals with few compounds. Therefore, most of the compounds are either pharmacologically unknown or have at best been assayed for a few biological activities.

There is a growing interest in the ecological role of marine natural products. At present few chemical–ecological relationships have been adequately investigated. The near future will undoubtedly add more detailed knowledge to this interesting and important field.

Several texts cover marine natural products. The classic book by Scheuer (*15*) is still worth consulting, while the one by Baslow (*16*) gives a good introduction to marine pharmacology. Halstead (*17*) gives a monumental treatise on all aspects of poisonous and venomous marine animals. A readable account of the chemistry and pharmacology of marine metabolites is given by Hashimoto (*18*). Tabulations of all known marine natural products are found in the books by Baker and Murphy (*19*). Many aspects are covered in a volume edited by Faulkner and Fenical (*20*). A book covering marine pharmacognosy has appeared (*21*). The most detailed and up-to-date reviews are found in the treatise edited by Scheuer (*22–26*).

Numerous more or less detailed reviews covering general or specialized topics in marine natural products research have appeared. General texts include those by Faulkner and Andersen (*27*), Thomson (*28*), Fenical (*29*), Chang (*30*), and the compilations from the First (*31*), Second (*32*), Third (*33*), and Fourth (*34*) Symposia on Marine Natural Products. Among the reviews covering more specialized topics, Scheuer's on metabolites of marine molluscs (*35*), marine invertebrate toxins (*36*), and marine toxins (*37*) can be mentioned. Natural products from Porifera are treated in detail by Minale *et al.* (*38*), from microalgae by Shimizu (*39*), and as antibiotics by Faulkner (*40*). Toxins from blue-green algae are dealt with by Moore (*41*). Marine drugs are reviewed by Youngken and Shimizu (*42*), marine pharmaceuticals by Der Marderosian (*43*), drugs from the sea by Ruggieri (*44*), and substances of potential biomedical importance by Nigrelli *et al.* (*45*). Marine toxins from a medical point of view are treated by Southcott (*46*) and Halstead (*47*). Biochromes are treated by Fox (*48*) and Kennedy (*49*). Highlights of marine natural products research for the year 1979 are discussed by Christophersen and Jacobsen (*50*), and marine alkaloids are currently included in *Specialist Periodical Report, the Alkaloids* (*51*). Of historical interest, but still thought provoking, is the book by von Fürth (*52*).

II. Guanidine Alkaloids

A. Introduction

Tetrodotoxin and saxitoxin of this group of alkaloids are among the best-known marine natural products. The phenomenology associated with these toxins has been known for centuries. For example, the passages in the Old Testament (Deuteronomy 14: 9, 10) "Of all that are in the waters you may eat these: Whatever has fins and scales you may eat. And whatever does not have fins and scales you shall not eat; it is unclean for you" are believed to refer to the tetrodotoxic fish, while (Exodus 7: 20, 21) ". . . All the water that was in the Nile turned to blood. And the fish in the Nile died; and the Nile became foul so that the Egyptians could not drink water from the Nile . . ." is believed to refer to "red tides."

The zoanthoxanthins form a well-defined group of marine pigments derived from two related tetrazacyclopentazulene skeletons. With so far a single exception, this class of brilliantly colored compounds originates from coelenterates of order Zoanthidea. The zoanthoxanthins comprise the only new group of pigments so far identified from marine sources.

The remaining guanidine natural products comprise a heterogeneous group with only occasional structural resemblence, the only uniting feature being the guanidino group. A number of linear guanidine derivatives have been left out of this section; the interested reader is referred to the excellent review on marine guanidine derivatives by Chevolot (*53*) who also treats many compounds mentioned in Section II.

The bioluminescence of marine organisms has been treated thoroughly by Goto (*54*). Accounts of the zoanthoxanthins by Prota (*55*) and the dinoflagellate toxins by Shimizu (*56*) have appeared.

Most of these compounds result probably from variations in arginine metabolism; however, no definite information has been published.

B. Tetrodotoxin

Studies of the toxic component of puffer fish, tetrodotoxin (**6**), have attracted numerous researchers for more than 100 years. These investigations culminated with the presentation of the structure by four different groups at the Third International Symposium on Natural Products in Kyoto, 1964.

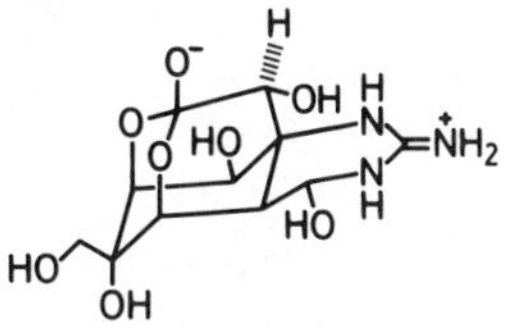

6

This unique toxin is by far the best-studied marine toxin. An extensive and voluminous literature covers its mode of action and general physiology, e.g., *Chemical Abstracts* Ninth Collective Index (1972–1976) has around 330 entries on tetrodotoxin dealing mainly with physiology. Clearly a discussion of these aspects is beyond the scope of the present context. Suffice it to mention that tetrodotoxin blocks the sodium-ion channel, thus preventing the formation of action potentials in excitable cells in much the same way as do the saxitoxins (Section I,C) (*57*).

In spite of the accumulation of heterocylic systems and the unusual hemilactal function, total synthesis of DL-tetrodotoxin has been achieved (e.g., ref. *58*). The physiological activity of structural modifications of the parent molecule has been investigated (e.g., ref. *59*).

Tetrodotoxin is present in a variety of puffer species of the family Tetraodontidae, many of which are used for food in Japan. It is also found in the goby *Gobius criniger* (Gobiidae), totally unrelated to Tetraodontidae (*60*) and in the octopus *Hapalochlaena maculosa* (*61*). One of the four groups who originally solved the structure of tetrodotoxin worked on material isolated from the Californian newt *Taricha torosa* (*62*). Tetrodotoxin seems to be limited to members of the family Salamandridae. The Costa Rica frog *Atelopus chiriquiensis* has also yielded **6** (*63*). Tetrodotoxin occurs in the Japanese ivory shell *Babylonia japonica* (*64, 65*) and in a trumpet shell, *Charonia sauliae* (*66*), as well.

The origin of tetrodotoxin remains a mystery. Feeding experiments with *T. torosa* and *T. granulosa* resulted in recovery of unlabeled toxin, although many metabolites carried a label (*57*). To add to the puzzle, it has been found that cultured puffer fish are devoid of toxicity (*57, 67*).

C. Saxitoxins and Gonyautoxins

Paralytic shellfish poisoning is contracted by ingestion of shellfish having accumulated toxins from deleterious dinoflagellates. The phenomenon is connected with the dinoflagellate blooms known as "red tides," although a visible coloration of the water is not necessarily evident. The threshold concentration of organisms capable of making shellfish toxic is less than can be recognized visually. Accumulation of the toxic principles in shellfish made them the first target for chemical investigations, e.g., the Alaska butter clam *Saxidomus giganteus* from which saxitoxin derives its name. In some cases it has not been demonstrated beyond doubt that the toxins originate from the dinoflagellates; however, modern methods have allowed the culturing of these organisms, thus demonstrating the identity of the shellfish toxins and the dinoflagellate toxins. Most work deals with members of the genus *Gonyaulax*.

The structure elucidation of saxitoxin culminated in 1975, when Schantz (*68*), having been involved in this research for more than 20 years, published a structure based on X-ray diffraction studies of the di-*p*-bromobenzenesulfonate of saxitoxin. Independently, another X-ray structural analysis based on a stable hemiketal appeared (*69*). Both reports assign the structure **7** to saxitoxin. Variations on the saxitoxin structure have appeared later. Neosaxitoxin (**8**) has an *N*-hydroxy substituent at position 1 (*70*). Gonyautoxin-II (**9**) and gonyautoxin-III (**10**) were formulated as the epimeric 11-hydroxysaxitoxins (*71*), but were later found to be the corresponding 11-sulfates (*72*). Gonyautoxin-I (**11**) and gonyautoxin-IV (**12**) are the epimeric neosaxitoxin 11-sulfates (*73*). Gonyautoxin-V (**13**) and gonyautoxin-VI (**14**) are the carbamoyl-*N*-sulfo derivatives of **7** and **8** and consequently generate these toxins on mild hydrolysis (*57, 74*). Gonyautoxin-VIII (**15**) and epigonyautoxin-VIII (**16**) are disulfonated (*75*).

Saxitoxin has pK_a values of 8.3 and 11.3. The former abnormally low value

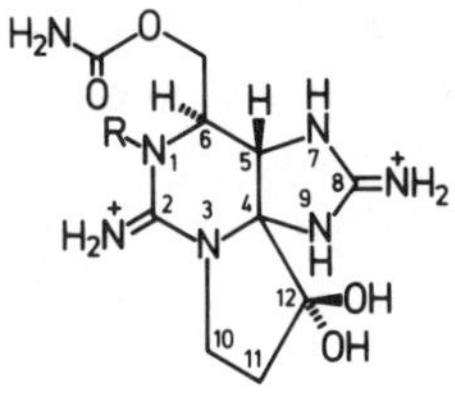

7 R=H
8 R=OH

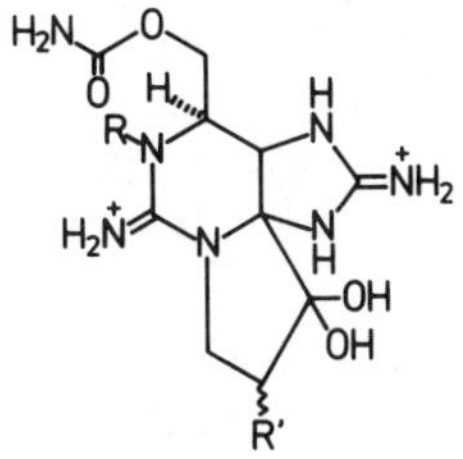

9 R=H, R'=OSO_3^- (11α)
10 R=H, R'=OSO_3^- (11β)
11 R=OH, R'=OSO_3^- (11α)
12 R=OH, R'=OSO_3^- (11β)

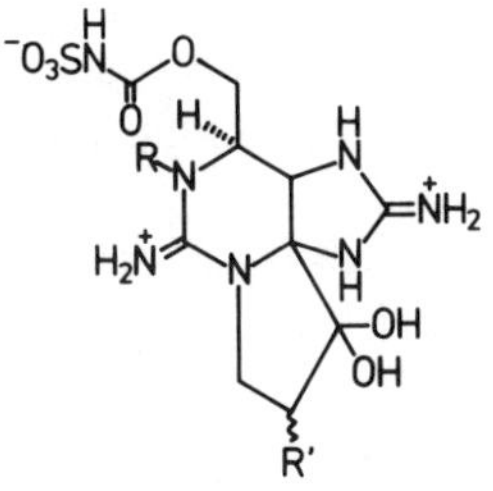

13 R=H, R'=H
14 R=OH, R=H
15 R=H, R'=OSO_3^- (11β)
16 R=H, R'=OSO_3^- (11α)

for a guanidinium ion has given rise to much speculation. It has, e.g., been attributed to one of the hydroxyl groups of the hydrate ketone structure. Based on potentiometric and ^{13}C-NMR studies of saxitoxin and derivatives, Rogers and Rapoport (*76*) have convincingly demonstrated that pK_a 8.3 is associated with the guanidinium group involving C-8, while pK_a 11.3 arises from the guanidinium group involving C-2. The classification of these details is of importance for the deeper understanding of the mechanism by which saxitoxin selectively blocks sodium-ion influx in excitable cells, thus preventing the buildup of action potentials (*57*).

Saxitoxin and saxitoxin-like substances have been isolated from the freshwater blue-green alga *Aphanizomenon flos-aqua* (*41, 77*).

A stereospecific total synthesis of racemic saxitoxin is shown in Scheme 1 (*78*). Lactam **17** was prepared in 74% yield from methyl 2-oxo-4-phthalimidobutyrate by ketalization and hydrazinolysis. The corresponding thiolactam, obtained by phosphorus pentasulfide treatment of **17,** on reaction with methyl 2-bromo-3-oxobutanoate, followed by base treatment, gave **18** in 50% yield from **17.** The thiourea (**19**) was prepared from the vinylogous carbamate (**18**) by condensation with benzyloxyacetaldehyde and silicon tetraisothiocyanate in 75% yield. Hydrazinolysis followed by NOCl treatment, heating, and reaction with ammonia produced **20** in 75% yield. Transformation of ketal **20** to thioketal **21** was effected in 63% yield by boron trifluoride-catalyzed exchange with 1,3-propanedithiol. Acid-catalyzed ring closure gave **22** in 50% yield. Triethyloxonium tetrafluoroborate alkylation, followed by heating with ammonium propionate, gave the diguanidine (**23**) in 33% yield. The decarbamoylsaxitoxin thioketal (**24**) was formed in 75% yield by boron trichloride treatment. The hexaacetate of **24** on NBS treatment, followed by methanol, gave a 30% yield of decarbamoylsaxitoxin, which on reaction with chlorosulfonyl isocyanate, followed by workup with hot water, gave saxitoxin (50% yield after purification).

Biosynthetic studies in this area have been unrewarding (*57*). Feeding cultures of *Gonyaulax tamarensis* labeled acetate or amino acids resulted in mainly un-

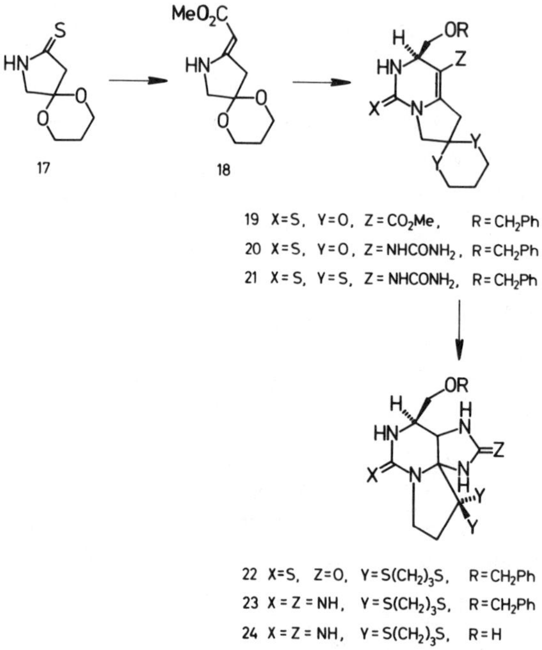

17 18

19 X=S, Y=O, Z=CO_2Me, R=CH_2Ph
20 X=S, Y=O, Z=$NHCONH_2$, R=CH_2Ph
21 X=S, Y=S, Z=$NHCONH_2$, R=CH_2Ph

22 X=S, Z=O, Y=$S(CH_2)_3S$, R=CH_2Ph
23 X=Z=NH, Y=$S(CH_2)_3S$, R=CH_2Ph
24 X=Z=NH, Y=$S(CH_2)_3S$, R=H

SCHEME 1

specifically labeled toxin. The currently accepted view is that these toxins are all synthesized *de novo* by dinoflagellates, although the matter has not been satisfactorily settled.

As paralytic shellfish poison continues to be a public health hazard, analytical methods have been developed to trace the toxins in, e.g., marine snails (*79*), crabs (*80, 81*), and oysters (*82, 83*). The lethality of toxins extracted from *Gonyaulax excavata (tamarensis)* to marine fish has been demonstrated (*84*).

D. ZOANTHOXANTHINS

Colonial anthozoans of the order Zoanthidea have yielded a variety of yellow highly fluorescent pigments—the zoanthoxanthins. The pigments are based either on the 1,3,5,7-tetrazacyclopent[*f*]azulene or the 1,3,7,9-tetrazacyclopent-[*e*]azulene skeleton, the latter occurring in two types, depending on the alkylation pattern of the nuclear nitrogen atoms (Table I).

The first representative of this new heterocyclic system to be isolated was zoanthoxanthin [**25,** mp 275–276°C(d)], the structure of which was verified by X-ray analysis of the 2-chloro derivative (**25,** R_2 = Cl) (*85*). The source of **25** was the Mediterranean zoanthid *Parazoanthus cfr. axinellae* tentatively identified as *P. a. adriaticus* (O. Schmidt, 1862), which later yielded parazoanthox-

TABLE I
ZOANTHOXANTHINS

Structure	No.	R^1	R^2	R^3	R^6
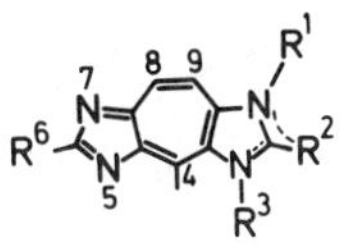	**25**	—	NH_2	Me	NMe_2
	26	—	NH_2	H	NH_2
	27	—	NH_2	H	NMe_2
	28	—	NH_2	Me	NH_2
	29	—	NHMe	Me	NMe_2
	30	Me	= NH	Me	NMe_2
	31	H	NHMe	—	NMe_2
	32	—	NH_2	Me	NHMe
	33	Me	NMe_2	—	NHMe
	36	Me	NH_2	—	NMe_2
	37	H	NMe_2	—	NMe_2
	38	—	NMe_2	Me	NMe_2

Structure	No.	R^1	R^2	R^3	R^8
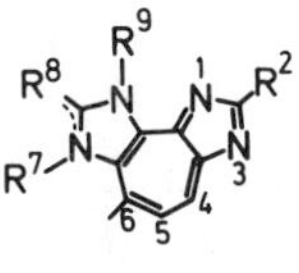	**34**	Me	= NH	Me	NH_2
	35	Me	NH_2	—	NH_2
	41	H	NMe_2	—	NHMe

Structure	No.	R^2	R^7	R^8	R^9
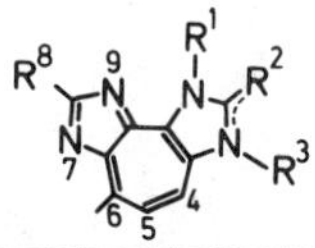	**39**	NMe_2	—	NH_2	Me
	40	NHMe	—	NH_2	Me

anthin A (**26,** mp >300°C, 3 × 10^{-4}% of wet weight) and paraanthoxanthin D [**27,** mp 303–304°C(d), 0.012% of wet weight] and traces of paraanthoxanthin B (**28**) and C. The structures of **26** and **27** were assigned on the basis of chemical of spectroscopical studies (*86*). A methylation study of **25** and **27** afforded parazoanthoxanthin E (**29**) and F (**30**), which was also isolated from *P. a. adriaticus* (**29**) in 4 × 10^{-4}% and **30** in 3 × 10^{-4}% of wet weight (*87*). In the same paper the presence of **31** in *Parazoanthus axinellae* was reported. Parazoanthoxanthin G (**32**) was isolated from *P. a. adriaticus,* while *P. axinellae* (originally considered *Epizoanthus arenaccus*) gave epizoanthoxanthin A (**31**, mp 191–192°C, 1.5 × 10^{-3}% of wet weight) and B (**32,** 6 × 10^{-4}% of wet weight) (*88*). In the same report, the presence of pseudozoanthoxanthin (**34,** mp >310°C, 0.03% of wet weight) and 3-norpseudozoanthoxanthin (**35,** mp >230°C, 10^{-3}% of wet weight) was established (*88*). Subsequently palyzoanthoxanthin A (**36**), B (**37**), and C (**38**) were obtained from *Palythoa mammilosa* or *P. tuberculosa* (*89*).

The gold coral (*Gerardia sp.*), one of the few bathyal marine organisms to be investigated, collected at −350 m, has yielded **39** [mp ~200°C(d), 0.28% of wet weight]. The structure was determined by using X-ray diffraction analysis (*90,*

91). A related metabolite from the same animal was found to be **40** [mp ~180°C(d), 0.019% of wet weight].

Paragracine [**41,** mp 258–262°C(d), 0.059% of wet weight] isolated from parasitic *Parazoanthus gracilis* (Lwowsky) on *Dentitheca habereri* (Stechow), was structure elucidated by X-ray crystallographic analysis of the dihydrobromide (*92*). An account of the chemistry of **41** has appeared (*93*).

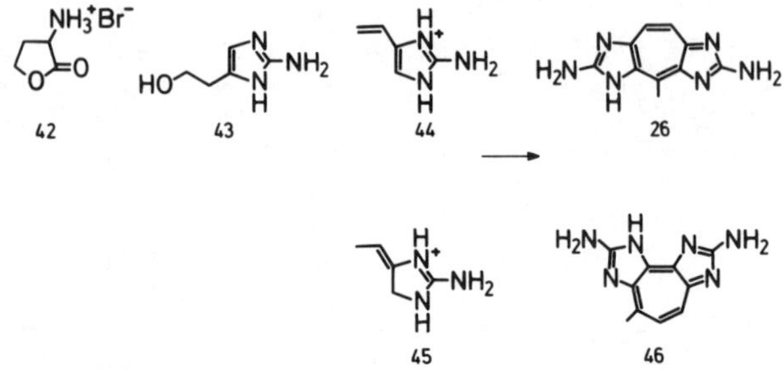

42 43 44 26

45 46

A synthetic approach to the zoanthoxanthins has been published (*94*). Reduction of commercially available **42,** followed by addition of cyanamide and acidification, gave **43** in 64% overall yield. Heating a solution of the hydrochloride of **43** in concentrated sulfuric acid gave a mixture of parazoanthoxanthin A (**26**), 15% yield, and an 8% yield of **46,** named pseudozoanthoxanthin A. Undoubtedly the reaction proceeds via **44** and **45,** Since methods for methylation of both ring and side-chain nitrogen atoms have been devised (*87*), the prototypes **26** and **46** allow the synthesis of homologous zoanthozanthins.

The alkaloids exhibit biological activity, e.g., paragracine (**41**) has papaverine-like activity (*92*) and antihistamine activity (*93*).

A consistent nomenclature system has been suggested (*91*).

E. Other

The major antimicrobial agent of the sponge *Agelas sceptrum* (Lamarck) has been identified as sceptrin [**47,** mp 215–225°C(d), $[\alpha]_D$ −7.4° (MeOH)]. The structure of this constituent, being present in 2.7% of dry weight, was solved using X-ray crystallographic techniques (*95*).

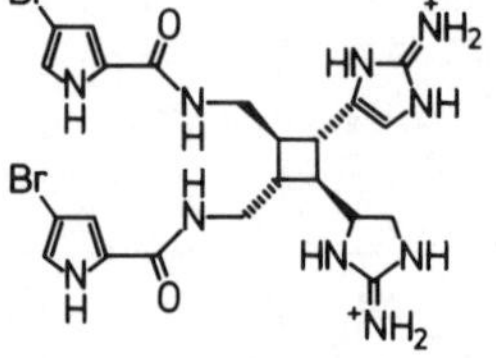

47 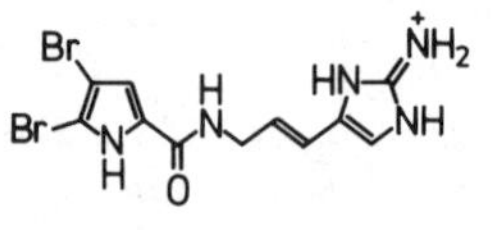 48

Oroidin (**48**) was present in the same material in 0.5% of dry weight. The structure of **48** was verified by X-ray diffraction analysis. Formally, sceptrin is related to debromooroidin by a head-to-head cycloaddition reaction of the latter. This reaction would be allowed photochemically but is disregarded for the biosynthesis of sceptrin, since debromooroidin must be achiral while sceptrin is optically active (*95*).

Oroidin was first isolated from the sponge *Agelas oroides* (*96*) and was later assigned structure **48** (*97*). The Mediterranean sponge *Axinella verrucosa* and the Red Sea sponge *Acanthella aurantiaca* both contain considerable amounts of oroidin together with a new yellow bromo compound (**49**) in 0.5% of dry weight

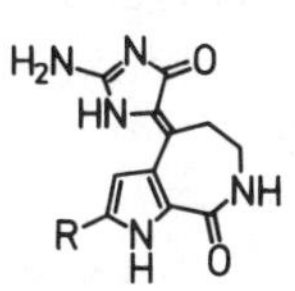

49 R = Br
50 R = H

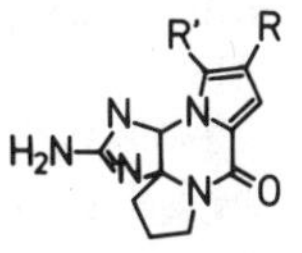

51 R = R' = Br
52 R = Br, R' = H

from *A. verrucosa* and 0.4% of dry weight from *A. aurantiaca*. The structural assignment is the result of X-ray crystallography (*98*).

A yellow compound isolated from the Great Barrier Reef sponge *Phakellia flabellata* (*99*) as a hydrochloride [mp 230–235°C(d)] gave the free base as the dihydrate [mp 220–225°C(d)]. The structure, as determined from chemical and spectroscopic investigations, is identical to debromo-**49** (**50**).

Other interesting guanidine alkaloids, namely, dibromophakellin (**51**) and monobromophakellin (**52**) have been isolated from this sponge. The structure of **51** [mp 237–245°C(d), $[\alpha]_D^{25}$ −203°] has been confirmed by X-ray diffraction analysis of the monoacetyl derivative and reported in a preliminary note (*100*). The full paper discusses the chemical properties of these compounds and the details of the X-ray analysis (*101*). Monobromophakellin hydrochloride [mp 215–220°C, $[\alpha]_D^{25}$ −123° (MeOH)], could be brominated to give dibromophakellin hydrochloride [mp 220–221°C, $[\alpha]_D^{25}$ −205° (MeOH)]. The yields of **51** and **52**, each as the hydrochloride, were 0.0085 and 0.035% of wet weight, respectively.

Like saxitoxin (Section II,C), the phakellins **51** and **52** exhibit abnormally low pK_a values (<8) for the guanidinium ions. A possible explanation has been offered (*101*).

Racemic dibromokephallin has been synthesized from dihydrooroidin (*102*). Commercially available L-(+)-citrulline (**53**) was converted to the ethyl ester, which on sodium-amalgam reduction gave a crude aldehyde, forming **54** on reaction with cyanamide, followed by acid-catalyzed cyclization. The overall yield was 73%. Basic hydrolysis released the amine (**55**) (>70% yield), which

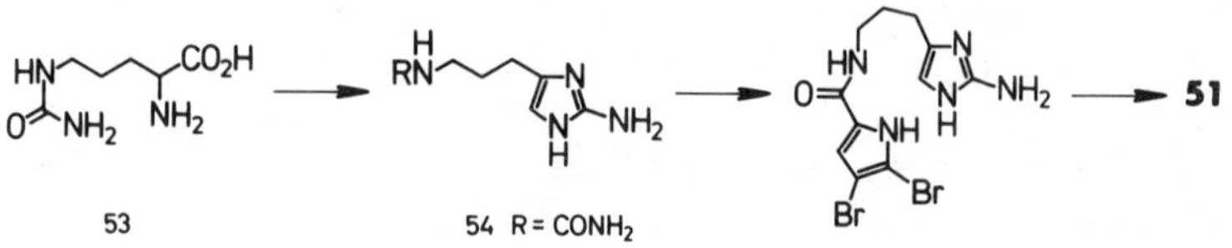

53

54 R = $CONH_2$
55 R = H

56

condensed, base-catalyzed, with 2-trichloroacetyl-4,5-dibromopyrrole in dimethylformamide to give ~50% yield of dehydrooroidin (**56**). An unstable intermediate was formed when the hydrochloride of **56** was exposed to bromine in acetic acid. This intermediate generated racemic dibromophakellin (**51**) quantitatively on treatment with potassium *tert*-butoxide in 2-butanol.

Some of the building blocks forming these alkaloids are found in other organisms, e.g., 2-aminoimidazole occurs in *Reniera cratera* (*103*). Compound **57** has been isolated from the sponge *Agelas* sp. (**53**) and midpacamide (**58**) from the sponge, *Agelas cf. mauritiana* (*104*).

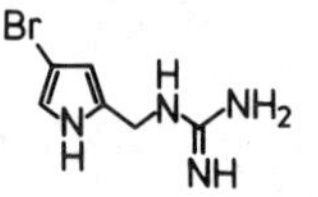

57

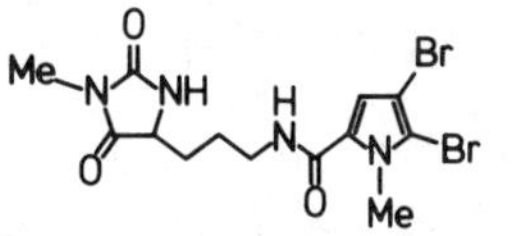

58

The structural formulas of the polyandrocarpidines from an encrusting tunicate, *Polyandrocarpa* (*Eusynstela*) sp., (*105*) have been revised (*106*). The polyandrocarpidines constitute a 9:1 mixture of homologs, each homolog being a mixture of isomers, so that polyandrocarpidine A (**59**) and B (**61**) correspond to

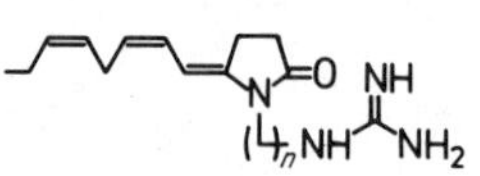

59 $n = 5$
60 $n = 4$

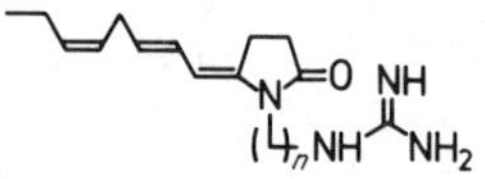

61 $n = 5$
62 $n = 4$

the former polyandrocarpine I, and the minor polyandrocarpine C (**60**) and D (**62**) to the former polyandrocarpine II. The stereochemistry of the side chains was tentatively assigned based on NMR data.

The guanidine structural unit of **49** and **50** is also present in the antineoplastic aplysinopsin (**63**) isolated from five *Thorecta* species (*107*) and from *Verongia spengelii* (*108*). The structural assignment was confirmed by comparison with an authentic sample prepared by base-catalyzed condensation of 3-formylindole with 2-imino-1,3-dimethyltetrahydroimidazole-3-one. The natural as well as the synthetic sample consisted of a 9:1 mixture of geometrical isomers (*107*). On the other hand, NOE experiments on the diacetate of aplysinopsin strongly indicated

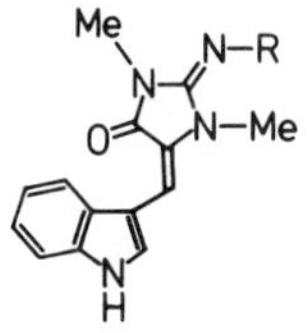

63 R=H
64 R=Me

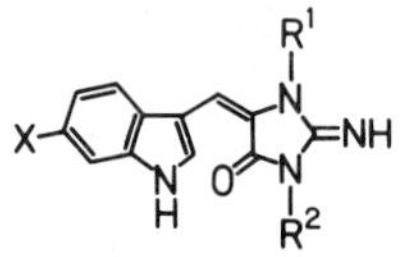

65 R^1=X=H, R^2=Me
66 R^1=H, R^2=Me, X=Br
67 R^1=R^2=Me, X=Br

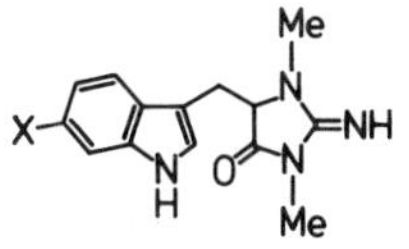

68 X=H
69 X=Br

the configuration to correspond to **63** and therefore indicated this to be the major isomer (*108*). Also, methylaplysinopsin (**64**) has been isolated (*53*).

Aplysinopsin, 2.7% of dry weight, was the major metabolite of a sponge, *Dercitus* sp. (*109*). This animal also contained 2′-de-*N*-methylaplysinopsin [**65,** mp 235°C(d)] in 1.0% of dry weight and 6-bromo-2′-de-*N*-methylaplysinopsin [**66,** mp 186–188°C(d)] in 1.0% of dry weight. The structure of **65** has been confirmed by synthesis. Thus reaction between 3-formylindole and 2-methyl-2-iminoimidazolidin-4-one gave 78% of **65,** identical in all aspects with the natural compound.

In an investigation of the nudibranch *Phestilla melanobranchia* and the coral on which it feeds, *Tubastrea coccinea,* compounds **65** and **66** were found to be common to both organisms, while **67, 68,** and **69** were found only in the coral (*110*).

The structure of the luminescent principle of ostracod crustaceans has been a problem of long standing in marine natural products research. In 1966 *Cypridina* luciferin (**70**) eventually yielded to chemical structure elucidation (*111*) confirmed by total synthesis (*112*). Other luciferins without guanidine groups have

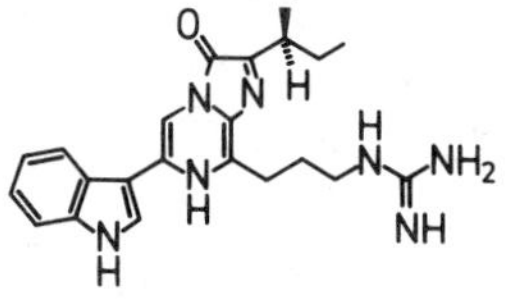

70

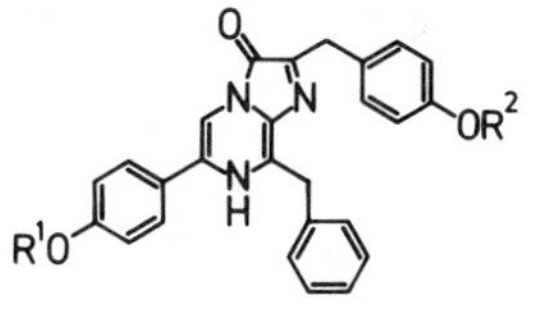

71 R^1=R^2=H
72 R^1=R^2=SO_3H

later appeared. The luciferins of *Renilla reniformis* (cnidarian) (*113*), the decapod *Olophorus* (*114*), and the fish *Neoscopelus microchir* (*115*) are identical with structure **71,** while the derivative **72** occurs in the squid *Wataenia scintillans* (*116*).

Spectroscopic studies revealed the structure of the antimicrobial and cytotoxic constituents of the sponge *Ptilocaulis* aff. *P. spiculifer* (Lamarck, 1814) to be

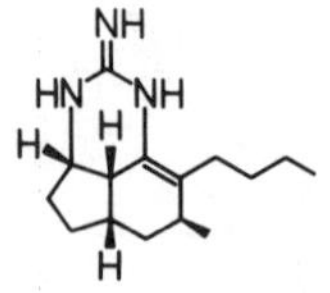

73 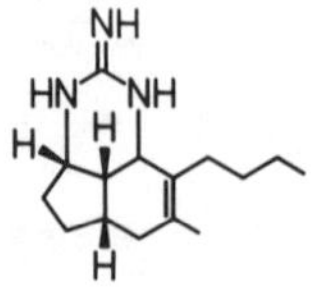 74

ptilocaulin (**73**) and isoptilocaulin (**74**). The structure of ptilocaulin nitrate, mp 183–185°C, has been verified by X-ray crystallographic analysis (*117*).

Two other sponges, as yet unidentified, have very similar antimicrobial spectra and are believed to contain the same antimicrobial compounds (*117*). Ptilocaulin is the more bioactive of the two compounds.

From the sponge *Acarnus erithacus* (de Laubenfels) three antiviral guanidines have been isolated, the acarnidines **75, 76,** and **77.**

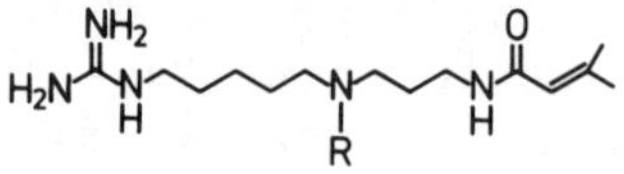

75 R = $CO(CH_2)_{10}CH_3$

76 R = $CO(CH_2)_3CH{=}CH(CH_2)_5CH_3$-(*Z*)

77 R = $COC_{13}H_{21}$

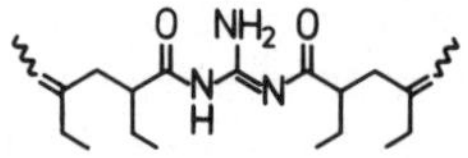

78

The structures were determined by chemical degradation, spectroscopic studies, and synthesis of a model. The $C_{14:3}$ acarnidine (**77**) is tentatively assigned a (5Z,8Z,11Z)-5,8,11-tetradecatrienoyl group (*118*).

The diacylguanidine trophamine (**78**) originates from a dorid nudibranch, *Thiopha catalinae* (Cooper) (*119*).

The marine tunicate *Dendrodoa grossularia* has yielded a cytotoxic alkaloid, dendroine (**79**). X-Ray structural investigation of the acetyl derivative confirmed

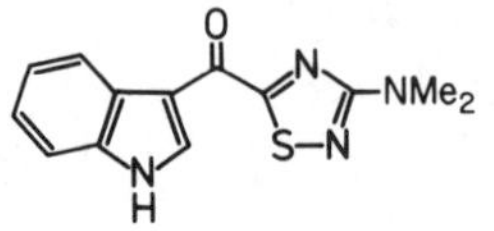

79

the structure of this unique compound, where the guanidine moiety is part of a 1,2,4-thiadiazole ring (*120*).*

* Dendrodoine has been prepared by the 1,3-dipolar addition of *N,N*-dimethylaminonitrile sulfide and indolyl-3-oxalyl nitrile [I. T. Hogan and M. Sainsbury, *Tetrahedron* **40,** 681 (1984)]. It is unfortunate that the tunicate from which dendrodoine was isolated in this study is named *Dendroda grossular*.

III. Indole Alkaloids

A. Introduction

The majority of marine indole alkaloids are rather simple compounds. Many carry unique structural features marking them effectively as of marine origin, e.g., halogenated nuclei. Bacteria and algae have given rise to halogenated simple indoles, while more complicated structures have been isolated from animals (snails, sponges, bryozoans, etc.). However, this trend is presumably only apparent and may very well reflect only coincidences. At present, speculations concerning any possible taxonomic significance of structural diversity seem fruitless, since so little is known about the biogenetic origin of the alkaloids. As touched upon earlier, many animals harbor a wealth of associated organisms and furthermore may accumulate compounds of dietary origin. Even most marine plants carry a covering of epibionts (*121*). Undoubtedly many of these alkaloids play a role in an extremely complicated ecological network, where the participating organisms are so intimately interacting with each other that the full extent of the relationships is still to be elucidated. Consequently, when a compound is described as isolated from a certain organism, this statement is only referring to the current literature and reflects no opinion concerning the actual biogenetic origin of the compound.

Some of the indole alkaloids, namely, those also containing a guanidine moiety, have been treated in Section II.

Most of the compounds in Section III undoubtedly originate via tryptophan metabolism; however, definitive information is lacking.

Marine indoles have been treated by Christophersen (*122*)

B. Simple Indoles

A very complex mixture of polyhalogenated indoles from the red alga *Rhodophyllis membranacea* Harvey has been studied (*123*). Partial separation was effected by column chromatography, yielding six crystalline fractions. Further purification of three of these fractions by GLC gave samples suited for NMR analysis.

Structures **80–83** were determined by comparison with synthetic samples (*124*). NMR spectroscopic studies combined with MS served to identify **84–89** (*123*). The indoles identified all show substitution in position 2 and 3 and 4 or 7 or both. The presence of pentasubstituted derivatives, namely, Br_5 (2 isomers), Br_4Cl (2 isomers), Br_3Cl_2 (2 isomers), Br_2Cl_3 (2 isomers), and hexasubstituted compounds Br_6, Br_5Cl, Br_4Cl_2, and Br_3Cl_3 was demonstrated by MS. Apart from the structures assigned (**80–89**), MS revealed the presence of unassigned

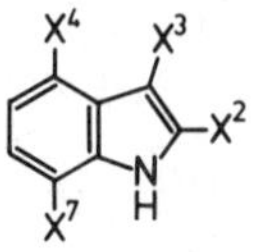

80 $X^2 = X^3 = X^7 = Cl$, $X^4 = H$
81 $X^2 = X^3 = Cl$, $X^7 = Br$, $X^4 = H$
82 $X^2 = X^7 = Br$, $X^3 = Cl$, $X^4 = H$
83 $X^2 = X^3 = X^7 = Br$, $X^4 = H$
84 $X^2 = X^3 = X^4 = Cl$, $X^7 = H$
85 $X^2 = X^3 = X^4 = X^7 = Cl$
86 $X^2 = X^3 = Cl$, $X^4 = Br(Cl)$, $X^7 = Cl(Br)$
87 $X^2 = X^3 = Cl$, $X^4 = X^7 = Br$
88 $X^4 = X^7 = Br$, $X^2 = Br(Cl)$, $X^3 = Cl(Br)$
89 $X^2 = X^3 = X^4 = X^7 = Br$

derivatives with Br_4, Br_3Cl, Br_2Cl_2, Br_2Cl, $BrCl_3$, $BrCl_2$ and Cl_3. A nearly total scrambling of the possible chloro and bromo substitution patterns is indicated in these strongly antifungal agents.

Strong antibacterial and antiyeast activity associated with the red alga *Laurencia brongniartii* J. Ahardh led to the identification of the active principle **92** together with inactive **90, 91,** and **93** (*125*).

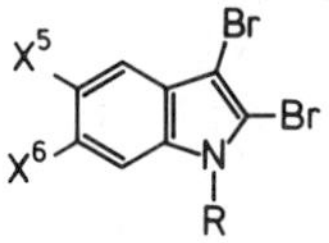

90 R=Me, $X^5 = H$, $X^6 = Br$
91 R=Me, $X^5 = Br$, $X^6 = H$
92 R=H, $X^5 = X^6 = Br$
93 R=Me, $X^5 = X^6 = Br$

The structures were inferred from ^{1}H- and ^{13}C-NMR experiments, taking advantage of the observation that only protons in positions 2 and 7 exhibit substantial solvent-dependent, downfield shifts (*126*). For example, in the case of **90** a resonance at 7.44 ppm was assigned to H-7 since a 0.31 ppm downfield shift was observed by changing the solvent from deuteriochloroform to hexadeuterioacetone. The same diagnostics were used in the case of the *Rhodophyllis* indoles (*123*).

A yellow hemichordate, *Ptychodera flava laysanica* Spengel (Enteropneusta), gave the odorous derivatives **94,** a trace of **95,** and **96** (*127*). The same organism also contains 3,5,7-tribromoindole (**97**) and 5,7-dibromo-6-methoxyindole (**98**) (*128*). Animals collected in another location were green and had only a faint odor. These animals had a mixture of **98** and indigotin derivatives (*129, 130*).

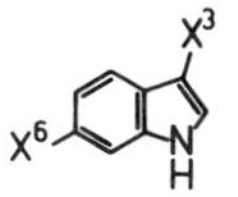

94 X^3=Cl, X^6=H
95 X^3=Br, X^6=H
96 X^3=Cl, X^6=Br

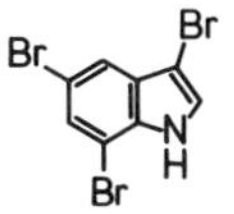

97

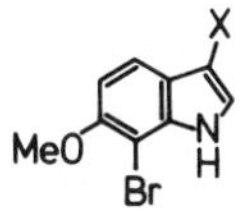

98 X=H
99 X=Br

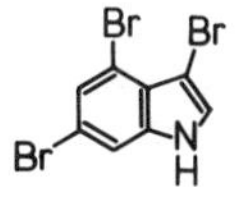

100

Four species of acorn worms have been subjected to a comparative study (*131*). This study further demonstrated the presence of **99** in *P. flava* and **100** in *Balanoglossus carnosus,* while concluding that genera *Phytodera* and *Glossobalanus* owe their characteristic odors to haloindoles and *Balanoglossus,* to brominated phenols. A tentative biogenetic scheme was presented and the presence of 3,6-dibromoindole and N-methylated halogenoindoles was indicated. Compounds **96** and **97** are synthetically available (*128*).

A dihydroxyindole, thought to be either 4,6- or 6,7-dihydroxyindole on the basis of color tests, has been isolated from several species of sponges from the genus *Agelas* (*132*). The structure of this antibiotic indole needs reinvestigation.

During the investigation of a yellow marine pseudomonad a 9:1 mixture of indole-3-carbaldehyde and 6-bromoindole-3-carbaldehyde was identified by comparison with synthetic compounds (*133*). The synthetic 6-bromoindole-3-carbaldehyde was prepared by bromination of the corresponding aldehyde. This work seems to guarantee the origin of these extracellular metabolites.

The green alga *Undaria pinnatifida* has yielded indole-3-carboxylic acid and 3-indolylacetic acid (*134*).

The lactam **101** (mp 194–195°C, 8.3×10^{-5}% of wet weight) has been identified from an algae-infested sponge, *Halichondria melanodocia* (*135*). The

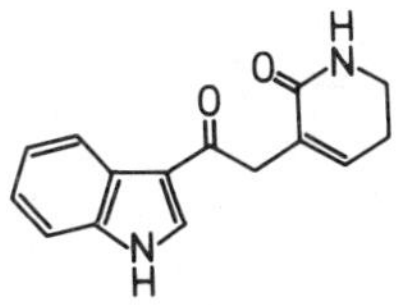

101

structure was deduced from spectroscopic data on comparison with those of 3-acetylindole. A related compound having the ketonic part of the molecule derived from 4-hydroxyacetophenone was also isolated (mp 235–235.5°C, 2.8×10^{-4}% of wet weight). Whether these compounds are metabolites of the associated flora, synthesized by the sponge, or of dietary origin, is unknown.

The procaryotic blue-green algae are interesting organisms from a biochemical point of view. In a search for the agents responsible for the antiinflammatory effect associated with the crude extract from the blue-green alga *Rivularia firma* Womersley, six unique biindoles were isolated (*136*).

The major components, 0.04% of wet weight, were **102** (mp 239–240°C) and

102 103 106

104 X=H, Y=Br
105 X=Br, Y=H

107

103 [mp 178–179°C, $[\alpha]_D^{20}$ +71° ($CHCl_3$)] in a mixture with **104** [mp 220–223°C, $[\alpha]_D^{20}$ +8.5° ($CHCl_3$)], **105** [foam, $[\alpha]_D^{20}$ +11.3° ($CHCl_3$)], **106** [mp 196–200°C, $[\alpha]_D^{20}$ −6.0° (CH_3CN)], **107** [mp 263–264°C, $[\alpha]_D^{20}$ +18.7° (CH_3CN)], and three unidentified derivatives in trace amounts. The major component **102** is responsible for the antiinflammatory effect.

The structure analyses were carried out from a combination of data obtained from ^{13}C-NMR spin–lattice relaxation experiments and ^{13}C–^{1}H coupling constant data. Optical activity is exhibited by compounds **103–107** owing to restricted rotation around the bond connecting the indole nuclei.

Single-crystal X-ray analysis established the absolute configuration of **103** as *R* and of **107** as *S* (*137*).

C. Simple Indoles Related to Tryptophan

Tryptamine (**108**) and tryptamine derivatives are of wide occurrence in nature. The gorgonian *Paramuricea chamaeleon* gave **108, 109,** and **110,** identified by comparison with authentic samples, and **111**, and **112**, and **113**, identified spectroscopically (*138*). Marine worms and sea anemones have been assayed for

108 $R=R^1=R^2=H$
109 $R=R^1=H$, $R^2=Me$
110 $R=OH$, $R^1=R^2=H$
111 $R=H$, $R^1=R^2=Me$
112 $R=OH$, $R=H$, $R^2=Me$
113 $R=OH$, $R^1=R^2=Me$

serotonin (**110**) (*139*). An unidentified 5-hydroxyindole was detected in *Pseudoactinia varia* and *Pseudoactinia flagellifera* (*139*). Serotonin has also been identified from molluscs, cephalopods, tunicates, arthropods, amphibians, and plants (*140*). Bufotenin (**113**) is also known from *Bufo* species and *Amanita* species (*141*).

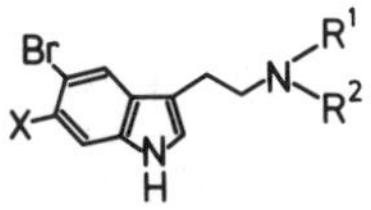

114 X=Br, $R^1=R^2=H$
115 X=Br, $R^1=H$, $R^2=Me$
116 X=H, $R^1=R^2=Me$
117 X=Br, $R^1=R^2=Me$

Sponges have yielded brominated tryptamines, e.g., 5,6-dibromotryptamine (**114**) and N_b-methyl-5,6-dibromotryptamine (**115**) from *Polyfibrospongia maynardii* (*142*). Catalytic hydrogenation of **115** produced N_b-methyltryptamine. Tryptamines **116** and **117** were obtained from *Smenospongia aurea, S. echina* (*143*), and an unidentified Carribean sponge (*144*).

Compounds **114** and **115** showed *in vitro* but not *in vivo* activity against Gram-negative as well as Gram-positive bacteria (*142*). The pharmacological activity of these tryptamine derivatives is unknown. Compound **114** is a dibromo derivative of the well-known hallucinogen N_b,N_b-dimethyltryptamine (*145*).

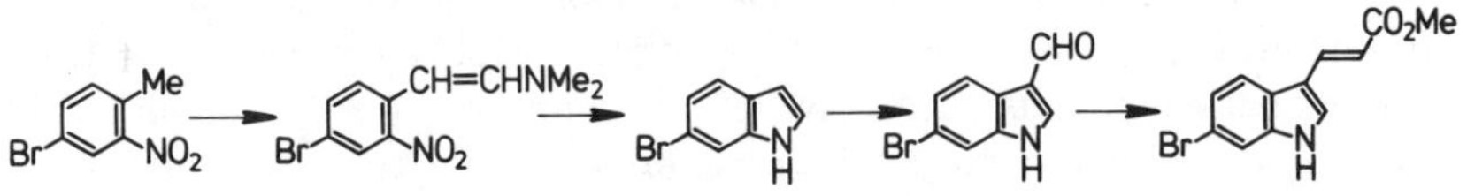

118 119

Synthesis verified the structure of (*E*)-3-(6-bromo-3-indolyl)-2-propenoate (**119**, mp 186°C, 0.19% of dry weight) isolated from a sponge, *Iotrochota* sp. (*146*). Reaction between *N,N*-dimethylformamide dimethyl acetal and 2-nitro-4-bromotoluene gave **118,** which was transformed to 6-bromoindole on catalytic reduction. Vilsmeier–Haack formylation gave 6-bromoindole-3-carbaldehyde, which in a Doebner reaction with monomethyl malonate produced the desired product (**119**).

The structure of L-6-bromohypaphorine [**120,** mp 275–280°C(d), $[\alpha]_D^{15}$ +58° (MeOH + CF_3CO_2H)] from the sponge *Pachymatisma johnstoni* was solved by X-ray techniques since the mass spectrum indicated complex thermal reactions in the inlet system (*147*).

An orange-red pigment, caulerpin, has been isolated from several green algae of the genus *Caulerpa* (*148, 149*). The proposed structure has been revised and proved to be **121** by synthesis (*150, 151*).

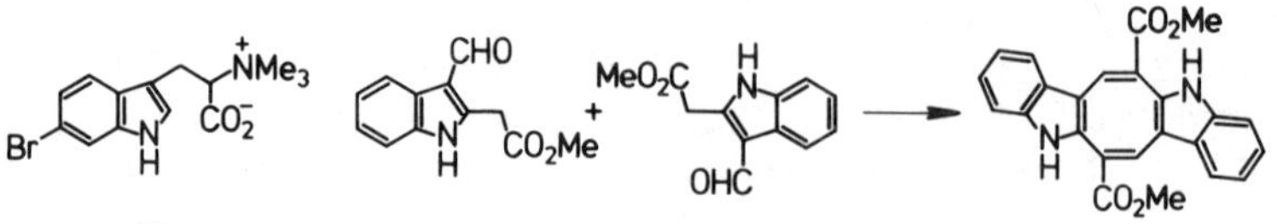

120 121

D. *Cliona* ALKALOIDS

The cosmopolitan sponge *Cliona celata* (Grant) belongs to the family Clionidea. The sponge may be found either free living or burrowing into calcium carbonate shells of molluscs, e.g., the giant barnacle *Balanus nubilus* or the rock scallop *Hinnites multirugosus*. Chemical investigation has been carried out on individuals collected from British Columbia, Canada, as well as from La Jolla, California. Clionamide (**122**) and celenamide A (**127**) and B (**128**) have been treated in a review (*152*).

Clionamide (**122**), isolated as an unstable yellow powder [$[\alpha]_D$ +32.1° (MeOH)] in about 0.025% of wet weight (*153*), was assigned structure **122** as a

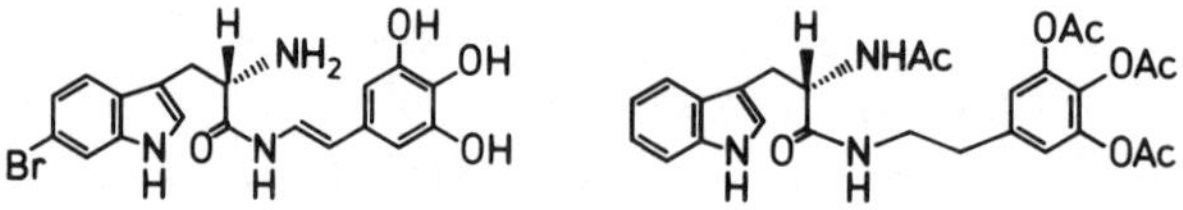

122 123

result of spectroscopic and chemical investigations of the tetraacetate (*154*). The structure was determined by the demonstrated identity of the hydrogenation product of the tetraacetyl derivative of **122** with an authentic sample of **123** prepared from 5-hydroxydopamine and (*S*)-*N*-acetyltryptophan, followed by acetylation. The *E* configuration of the double bond was inferred from ^{1}H-NMR evidence, while the *S* configuration was assigned based on comparisons between data from a degradation products and literature values.

A synthesis of tetraacetylclionamide has been published (*155*).

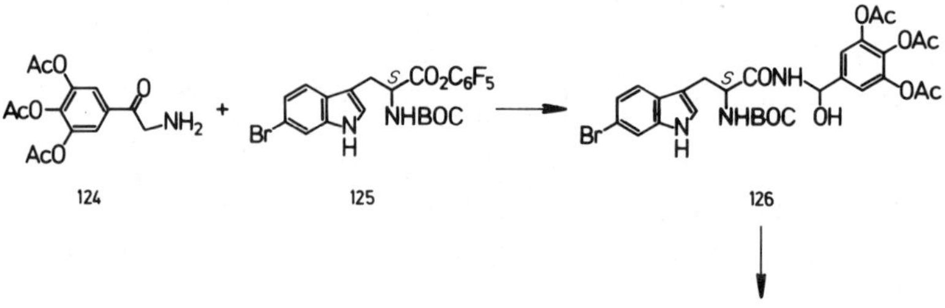

124 125 126

122-Ac$_4$

The amino ketone **124** was obtained from triacetylgallic acid chloride by reaction with, successively, diazomethane, hydrogen chloride, and sodium azide, followed by hydrogenation of the azido ketone. Acylation with (*S*)-benzyloxycarbonyl-6-bromotryptophan pentafluorophenyl ester (**125**), followed by sodium cyanoborohydride reduction, gave the alcohol (**126**). Oxidative elimination of the selenide formed by redox condensation with *p*-nitrophenylselenocyanate and tributylphosphine gave triacetylbenzyloxycarbonylclionamide. Tetraacetylclionamide was formed in an overall yield of 45% from tetraacetylgallic acid chloride by removal of the protecting group with trifluoroacetic acid, followed by acetylation.

Because of the instability and separation problems encountered, the alkaloids mentioned below were isolated as acetyl derivatives. Studies of trideuterioacetylated samples demonstrated that the alkaloids occur in the sponge in their free phenolic forms.

Celenamide A (**127**) and B (**128**) were isolated as hexaacetyl derivatives (0.03% and 0.02% of wet weight, respectively). The structure assignments were the results of spectroscopic studies combined with degradation experiments (*156*).

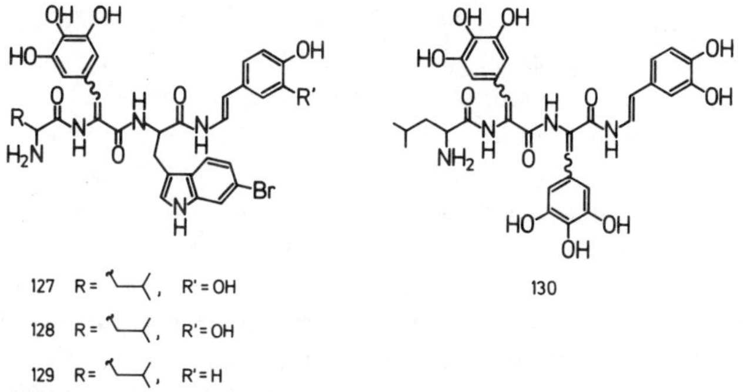

Celenamide C (**129**), 0.003% of wet weight, and celenamide D (**130**), 0.002% of wet weight, were obtained as pentaacetyl and nonaacetyl derivatives, respectively. The chemical degradation experiments and the spectroscopic properties were thoroughly discussed (*157*).

The functions of these linear peptide alkaloids are unknown. It has been proposed that they may play a role as calcium-chelating agents in the burrowing activity of the sponge. Clionamide is mildly antibiotic.

Celenamide D (**130**), although not an indole alkaloid, is clearly a member of the same class of peptide alkaloids as the other *Cliona* compounds. Related alkaloids are encountered in higher plants (for examples, see ref. *158*).

E. Bryozoan Alkaloids

Phylum Bryozoa (syn. Polyzoa, Ectoprocta, or moss animals) comprises about 4000 living species of aquatic colonial animals. The economic importance of these organisms stems mainly from members belonging to fouling communities. An introduction to the biology of bryozoans is available (*159*).

With the notable exception of *Alcyonidium gelatinosum* (L.) and the species mentioned in this section, the phylum is chemically unexplored. The former species contain a dermatitis-producing hapten (*160*).

The bryozoan alkaloids identified so far all originate from the marine bryozoan *Flustra foliacea* (L.) and are, with one exception, formally derived from 6-bromotryptamine, which is synthetically available (*161*). Free-growing colonies of *F. foliacea,* characterized by a lemonlike odor owing to an allelochemical mixture of monoterpenes (*162*), were the starting material for the chemical investigation.

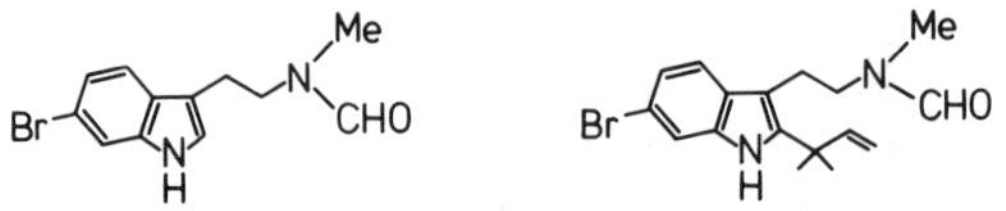

131 132

Spectroscopic structure elucidation revealed the configuration of the, as yet, structurally least complicated alkaloid as 6-bromo-N_b-methyl-N_b-formyltryptamine (**131**) (*163*). Hindered rotation around the formamide C—N bond results in a roughly equimolar mixture of the *Z* and *E* forms. The same phenomenon was encountered in the case of flustrabromine (**132**) where a further complication arises because of an equilibrium for each rotamer where the nitrogen atom associates intramolecularly with the benzene moiety, as evidenced by a doubling of all ^{13}C resonances. Assignments of all signals were possible by a combination of temperature, solvent polarity, and NOE difference experiments (*164*).

A series of tricyclic derivatives have been isolated. Based on extensive spectroscopic measurements, including in particular magnetic circular dichroism and ^{1}H-[^{1}H]NOE difference techniques, the structures of flustramine A (**133,** 0.035% of dry weight) and flustramine B (**134,** 0.035% of dry weight) were

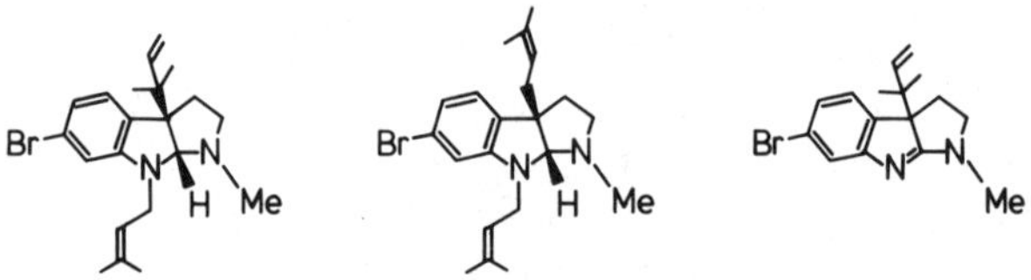

133 134 135

determined (*165, 166*). The absolute configuration is still unknown, while the ring junction was shown to be cis.

Spectroscopic parameters are available from a synthetic study of debromoflustramine B according to Scheme 2 (*167*).

Spectroscopic investigations have served to define the structures of an additional four alkaloids, flustramine C (**135**), flustraminol A (**136**), flustraminol B

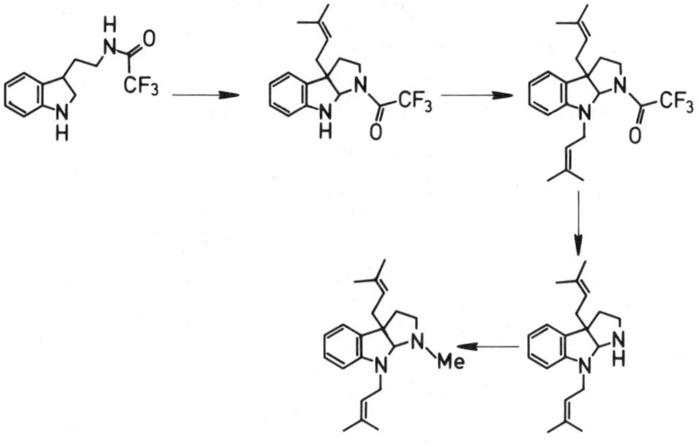

SCHEME 2

(**137**), and flustramide A (**138**). The yields based on dry weight were: **135,** $3 \times 10^{-4}\%$; **136,** $6 \times 10^{-4}\%$; **137,** $8 \times 10^{-5}\%$ (*168*); and **138,** 0.005% (*163*).

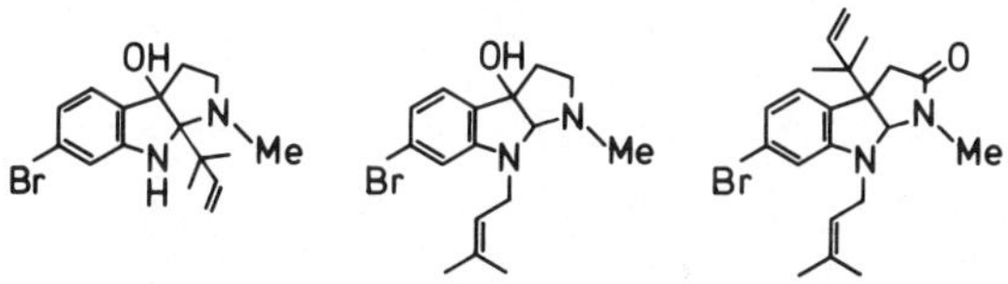

136 137 138

Flustramine C (**135**) was transformed into the debromodihydro analog by lithium aluminum hydride reduction.

Bryozoans from the family Flustrallidae seem to be promising candidates for the isolation of new alkaloids since *Securiflustra securifrons* (Pallas) as well as *Chartella papyracea* (Ellis and Solander) have yielded new bromoindole alkaloids of as yet unknown structures (P. Wulff, J. S. Carlé, and C. Christophersen, unpublished results).

Although the bryozoan alkaloids seem to have no closely related counterparts among the known alkaloids, several mold metabolites possess the structural features met in the flustra compound. Roquefortine (**139**) from *Penicillium*

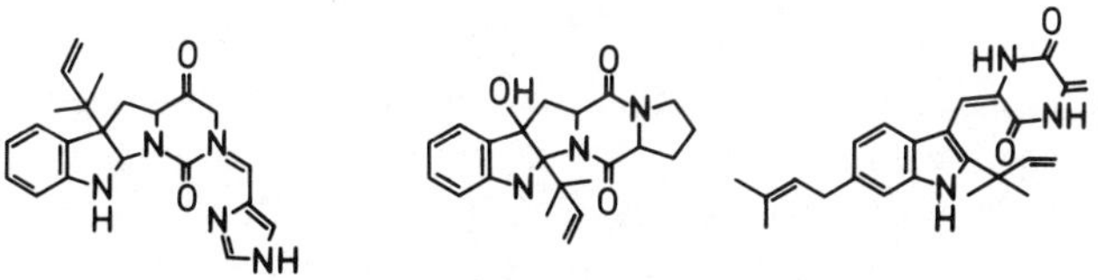

139 140 141

roqueforti (*169*), brevianamide E (**140**) from *Penicillium lanosum* (*170*), and neoechinolin C (**141**) from *Aspergillus amstelodami* (*171*) have the 3a-inverted isoprene unit of **133,** the 3a-hydroxyl, the 8a-inverted isoprene, the unsubstituted N-8 of **136,** and a substituent in position 6 and the inverted isoprene unit in position 2, as in **132,** respectively.

In the light of the admittedly superficial resemblance between the mold metabolites and the *Flustra* alkaloids, the question of the origin of the latter substances gains actuality. It is known that *F. foliacea* and *S. securifrons* incidentally harbor a green alga *Epicladia flustra* (R. Nielsen, personal communication). However, except for the case of caulerpin, eukaryotic algae are not known for their ability to synthesize indole alkaloids.

Certain bryozoans have evolved structures containing dense populations of microorganisms (*172, 173*), among these, *F. foliacea*. The microorganisms look very much like bacteria (*174*). These microorganisms may well be species specific. Actually an association between bacteria and bryozoan larvae has been demonstrated in three species (*175*). This phenomenon may represent a way to transmit the bacterial population to the adult bryozoan. These relationships deserve a closer study from the ecological as well as from the chemical point of view.

F. Lyngbyatoxin and Surugatoxins

The blue-green alga *Lyngbya majuscula* Gomont has given rise to interesting compounds. Lyngbyatoxin A (**142**), the agent responsible for a severe dermatitis named "Swimmer's itch," was isolated from a shallow-water variety of *L. majuscula* (*176, 177*). The structure of the toxin {$[\alpha]_D$ $-171°$ ($CHCl_3$), 0.02% of dry weight} was solved by spectroscopic examination and comparison with teleocidin B (**143**), a metabolite of several *Streptomyces* strains (*178*). Teleocidin B is reported to cause severe irritations and eruptive vesications on human skin (*179*) and shares with lyngbyatoxin A a strong tumor-promoting effect (*180–182*). Both compounds exhibit high toxicity toward fish.

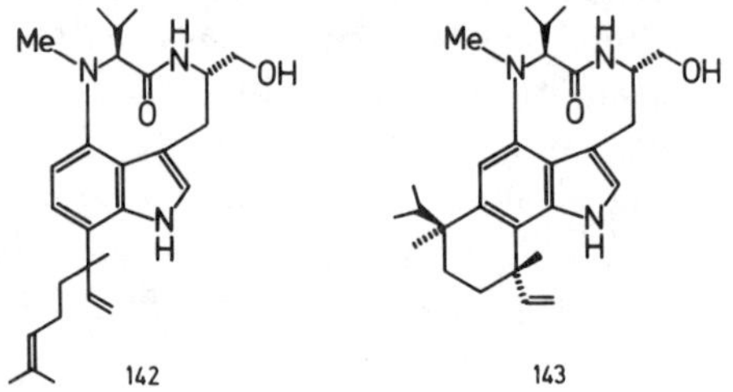

Tetrahydrolyngbyatoxin A on comparison with **143,** where the absolute configuration is known from an X-ray study (*183*), showed that the two toxins have the same absolute configuration of the nine-membered ring. In the NMR spectrum of **142,** several resonances were found to be doubled. This phenomenon was attributed to the presence of either a C-5 linalyl isomer of conformational isomers (*176*). The observation is similar to the one described for the less rigid system flustrabromine (**132**), where participation of rotational isomers has been identified (*164*).

Whether **142** is a true metabolite of the blue-green alga is unknown; however, in the case of surugatoxin (**144**) and neosurugatoxin (**145**), isolated from the midgut gland of the Japanese Ivory shell *Babylonia japonica* from Suruga Bay, the toxins almost certainly do not originate from the snail.

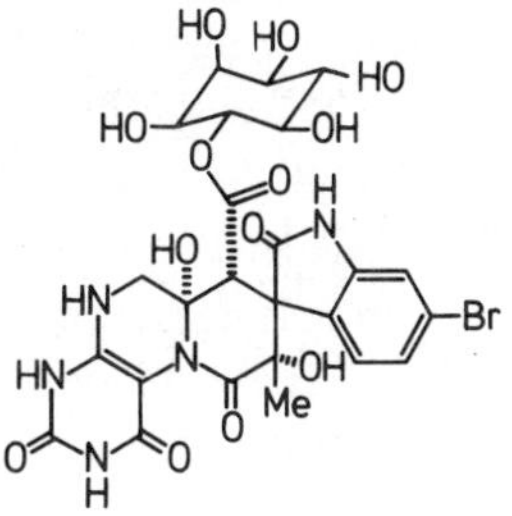

144 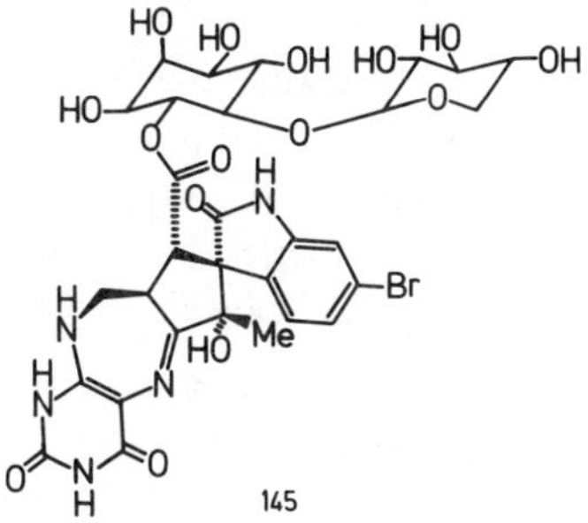 145

Toxicity was present only in gastropods from a limited area of Suruga Bay, and furthermore, the toxicity disappeared and reappeared on displacing the animals to other areas and vice versa. There are thus strong indications that the carnivorous shells ingest the toxins or precursors.

The structure of surugatoxin (**144,** mp >300°C, yield $\sim 10^{-3}\%$ of wet midgut gland) was determined by X-ray crystallographic analysis of the heptahydrate (*184*). The skeleton is unique, containing both indole and pteridine subunits. Neosurugatoxin (**145**) was obtained in a yield of about 4 mg from 20 kg of shellfish. Again, X-ray analysis served to determine the structure of neosurugatoxin hydrate (*185*).

Intoxications following ingestion of the gastropods have been reported. Both compounds are specific inhibitors of nicotinic receptors in autonomic ganglia, **145** exhibiting a 100-fold greater antinicotinic activity than does **144.**

An excellent review covering surugatoxin has appeared (*18*).

G. Tyrian Purple and Hyellazoles

The dye Tyrian purple was the first marine indole to be identified (*186*). Numerous reviews treating the history, biology and chemistry have appeared (*15, 27, 28, 35, 49, 52, 55*). The most extensive review covering all aspects is the one by Baker (*187*). The etymology of the word purple has been discussed (*188*).

The actual dye (**146**) is the product of a complicated series of reactions of precursors from the hypobranchial gland of molluscs from the families Muricidae and Thaisidae. The structure and stereochemistry were verified by X-ray crystallographic analysis (*189, 190*).

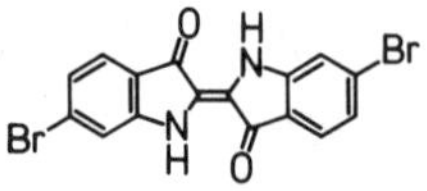

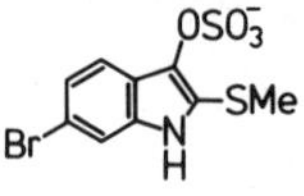

The precursor present in the hypobranchial gland is **147,** which on enzymatic hydrolysis gives rise to **148.** Oxidation of **148** gives **149,** isolated from the gland. Tyriverdin, an intermediate in the formation of 6,6′-dibromoindigotin (**146**), is presumably formed by addition of **148** to **149.** The structure of tyriverdin (**150**) was inferred from model studies of the didebromo analog (*191*) and proved by comparison of a synthetic sample prepared according to Scheme 3 (*188*) with an authentic sample from *Nucella lapilus* (C. Christophersen and F. Wätjen, un-

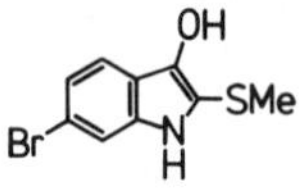

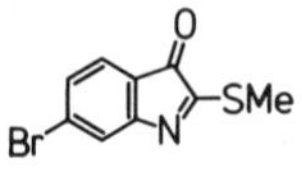

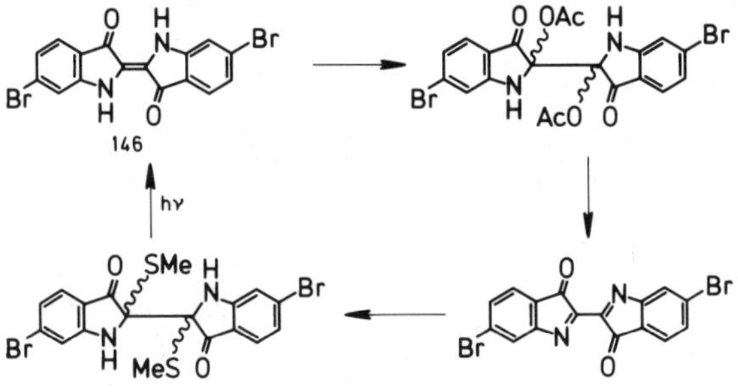

SCHEME 3

published results). The photochemical transformation of **150** to **146** proceeds with a quantum yield of >5, indicating a chain reaction (*188*). The latter observation paired with the fact that tyriverdin crystals obtained from the molluscs are of poor quality precluded a rigorous X-ray study; however, preliminary data suggest tyriverdin to be the meso form (*122*). The gross structure of **150** has been confirmed (*192*).

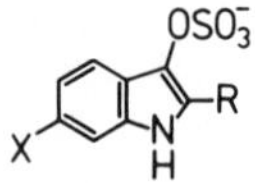

151 X=H, R=H
152 X=H R=SMe
153 X=Br, R=H
154 X=Br, R=SO_2Me

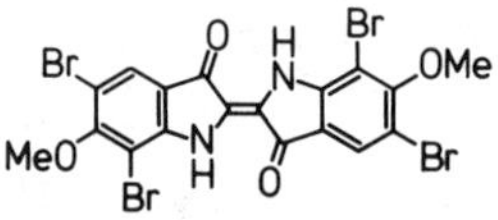

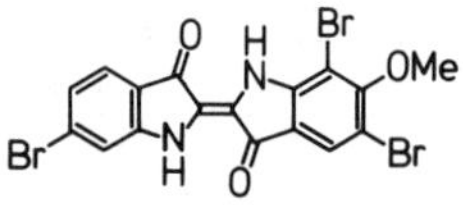

As a result of a preliminary investigation, other precursors (**151–154**) have been reported from Mediterranean species of snails (*193*).

Two new indigotin derivatives, **155** and **156,** have been identified from *Ptychodeva flava laysanica* Spengel together with **146** (*127, 129*).

Two unusual nonbasic carbazole alkaloids, hyellazole (**157** mp 133–134°C, 0.012% of dry weight) and 6-chlorohyellazole (**158** mp 163–164°C, 0.009% of dry weight), have been identified from a supralittoral variety of the blue-green alga *Hyella caspitosa.* The structure of **158** was determined by X-ray analysis (*194*).

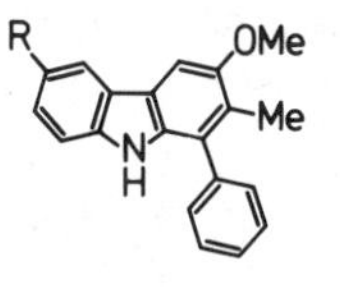

157 R = H
158 R = Cl

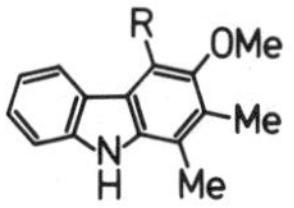

159 R = OMe
160 R = OH

Carbamycin A (**159**) and B (**160**) originate from a *Streptomyces* species (*195*). The structure of **160** was deduced from an X-ray analysis (*196*). Acylation of **160** with acetic anhydride gave the 6-acetyl derivative. The 4-substituent of **160** could be removed to produce a structure identical to **157** except that the phenyl group of **157** is replaced by a methyl group.

Two independent synthetic preparations of **157** have appeared (*197, 198*). The synthesis of **158** (*199*) is shown in Scheme 4.

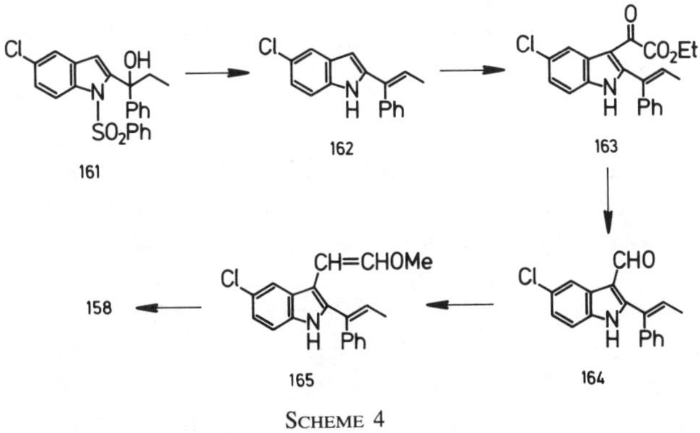

SCHEME 4

Hydrolysis of the alcohol **161,** prepared from *N*-benzenesulfonyl-5-chloroindole and propiophenone gave **162** in an overall yield of 65.4%. Oxalyl chloride, followed by ethanol, transformed **162** to the keto ester **163** in 69.1% yield. Hydrolysis and decarboxylation gave aldehyde **164** (38.9%), which by reaction with (methoxymethylene)triphenylphosphorane gave the divinylindole **165.** Heating the reaction mixture in decalin with Pd–C afforded **158** in 47.4% yield.

IV. Pyrrole Alkaloids

A. Introduction

The pyrrole alkaloids are a heterogenous group ranging from very simple brominated indoles, simple although unusual amino acids, and peptides to the lipophilic malyngamides and the porphyrins and other tetrapyrrole pigments. The latter compounds are universally distributed among aerobic organisms and will only be superficially treated in this text.

Although many structures in this section lend themselves to more or less sophisticated speculation regarding the biochemical routes responsible for their formation, the necessary experimental basis for the construction of a valid hypothesis has not been provided.

What was pointed out in the first sections concerning the important question about the true biological origin of the compounds also holds true in the majority of cases treated in the present section.

B. Simple Pyrroles

Substituted bromopyrroles have been isolated and identified from the sponge *Agelas oroides* (*200*). Methanol extraction of fresh material gave the methyl ester of 4,5-dibromopyrrole-2-carboxylic acid in 0.012% of dry weight, identical with

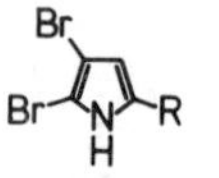

166 R=CO_2H
167 R=$CONH_2$
168 R=CN

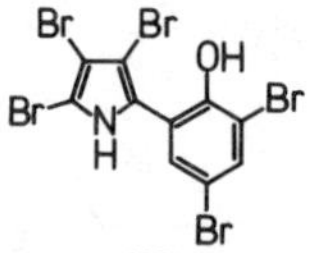

169

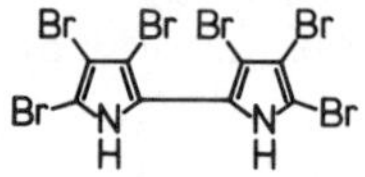

170

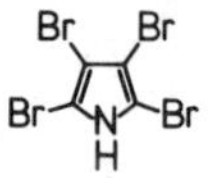

171

a synthetic sample (*201*). In addition, the free acid (**166**) was isolated (0.21%). Acetone extraction demonstrated the ester to be an artefact since the free acid was isolated by this procedure. From the methanol extract the amide (**167** trace, mp 164–166°C) and the nitrile (**168** 0.02%, mp 172–173°C) were isolated. The nitrile could be hydrolyzed to the amide and the acid. A new bromopyrrole, oroidin, was obtained (Section II,E). The major secondary metabolite of the sponge *Agelas cf. mauritiana* was identified as 4,5-dibromo-1-methyl-2-pyrrolecarboxylic acid (*N*-Me-**166**). This animal also contains midpacamide (Section IV,F) (*202*).

The marine bacterium *Pseudomonas bromoutilis* gave the remarkable pentabromo compound **169** (*203*). X-Ray structural analyses have confirmed the structure (*204*). The synthesis of **169** is shown in Scheme 5 (*205*).

The bromopyrrole **169** seems to have widespread ecological significance as an antimicrobial agent since it was later isolated as the major metabolite of yellow

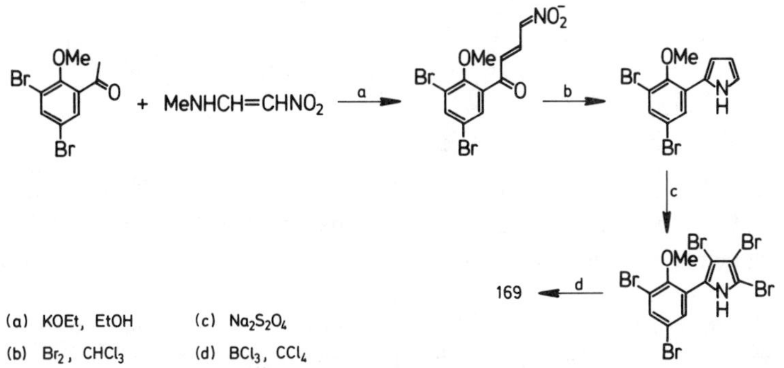

(a) KOEt, EtOH (c) $Na_2S_2O_4$
(b) Br_2, $CHCl_3$ (d) BCl_3, CCl_4

SCHEME 5

and off-white strains of *Chromobacter* (*40*). This metabolite (**169**) may be responsible for the autotoxicity of *Chromobacterium marinum* (*206*) and other marine bacteria as well (*40, 207*). Another *Chromatobacter* sp. produced **169** present in cells and medium (*208*), hexabromo-2,2′-bipyrrole (**170**) intracellular, and 4-hydroxybenzaldehyde extracellular. Tetrabromopyrrole (**171**) was isolated from the same organism in one experiment, while this organism grown on agar seems to lack **171.** The products **170** and **171** were identical to synthetic samples.

The bromopyrrole **172** has been reported from an *Agelas* sp. (*209*).

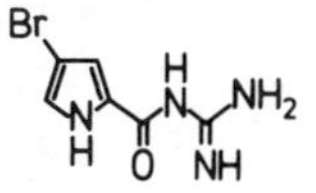

172 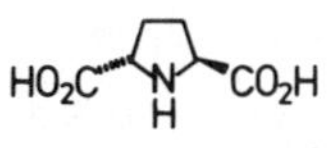 173 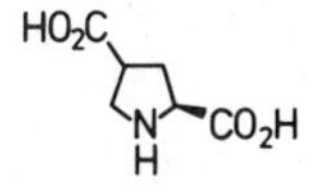 174

Pyrrolidine-2,5-dicarboxylic acid {**173,** 0.028% of fresh weight, mp 340–345°C(d), $[\alpha]_D$ −112° (H_2O), −88.1° (5 *N* HCl)} was isolated from the red alga *Schizymenia dubyi* (*210*). The configuration was assumed to be L,L. Investigation of 50 other species of red algae failed to detect **173,** with the exception of an encrusting species, *Haematocelis rubens* (*211*), which is possibly the tetrasporophyte of *S. dubyi* (*212*).

Pyrrolidine-2,4-dicarboxylic acid {**174,** 0.012% of fresh weight, mp 223–225°C, $[\alpha]_D$ −46.0° (H_2O), −29.7° (5 *N* HCl)} is a consistuent of *Chondria coerulescens* (Rhodomelaccae), *Chondria dasiphylla* (Rhodomelaccae), and *Ceramium rubrum* (Ceramiaccae) (*213*). The flowering plant *Afzelia bella* (Leguminosae) contains **174** in the seeds (*214*).

C. KAINIC ACIDS

The red alga *Digenea simplex* Agardh (Corsican weed) has been known as an anthelmintic for more than 1000 years. The gross structure of the main active

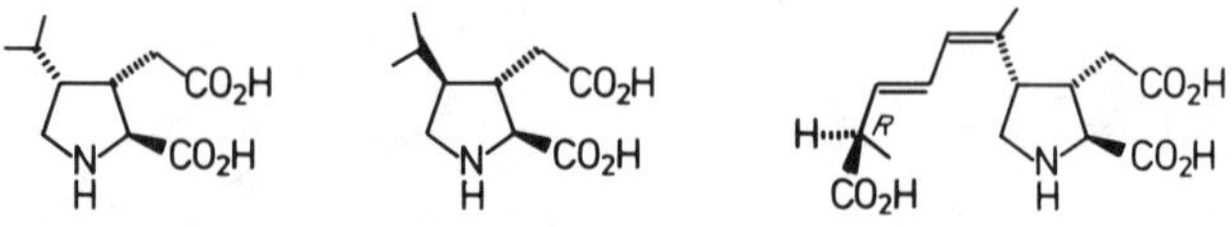

175 176 177

principle α-kainic acid {**175,** mp 251°C(d), $[\alpha]_D^{29}$ −14.8° (H_2O)} was elucidated by a combination of classical degradation experiments and synthesis of degradative fragments. The literature regarding this subject is voluminous; however, an excellent review exists (*215*).

The isolation of **175** from a red alga, *Centroceras clavulatum* from the same order as *D. simplex* (Ceramiales), has been reported (*210*).

A minor constituent of the *D. simplex* extract was identified as α-allokainic acid {**176,** mp 237°C, $[\alpha]_D^{10}$ +8° (H_2O)}.

Another red alga, *Chondria armata* Okamura belonging to the same family (Rhodomelaccae) as *D. simplex* has yielded domoic acid {**177,** mp 217°C, $[\alpha]_D^{12}$ −109.6° (H_2O)}. Domoic acid (**177**) is also a constituent of *Alsidium coralinum* (Rhodomelaceae) (*216*).

Pronounced anthelmintic effects was associated with **175** and **177,** while **176** is only moderately active. Strong insecticidal activity is exhibited by **177.**

The structures of **175** (*217*) and **176** (*217–219*) have been verified by X-ray structure determinations. Careful ^{1}H-NMR analysis of **177** and its methyl ester coupled with chemical degradation and comparison with **175** and derivatives resulted in a proposed structure of domoic acid, which has later been revised and based on an X-ray crystallographic analysis (*220*). The revised structure **177** shows the structure of the side chain to be 1′*Z*,3′*E*,5′*R*.

Enantioselective synthesis has been achieved for (−)-α-kainic acid (*221*), (+)-α-allokainic acid (*222*), and (−)-domoic acid (*223*).

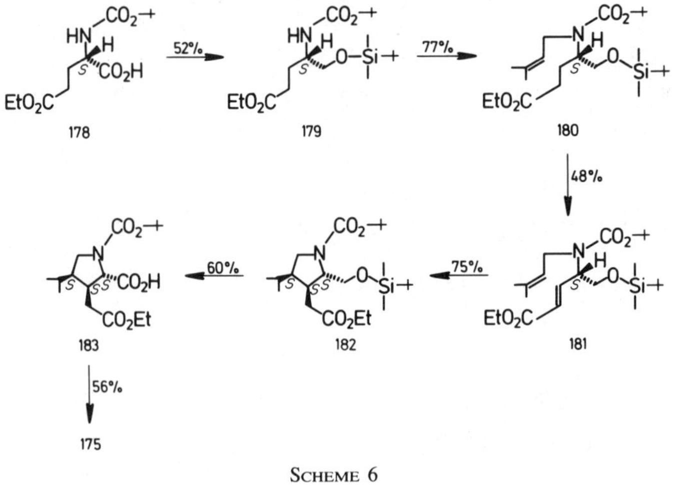

178 → (52%) 179 → (77%) 180 → (48%) 181 → (75%) 182 → (60%) 183 → (56%) 175

SCHEME 6

The synthesis of **175** starts with **178** prepared from *S*-(+)-5-ethyl glutamate. Selective reduction with diborane (57%) and silylation of the resulting primary alcohol gave **179** (92%). Alkylation with 1-bromo-3-methyl-2-butene gave **180.** The conjugated enoate **181** was prepared by deprotonation of **180** with lithium 2,2,6,6-tetramethylpiperidide, selenation of the enolate, oxidation, and selenoxide elimination. Closure of the five-membered ring was effected by heating in toluene to generate **182.** The carboxylic acid **183** was obtained by cleavage of the silyl ether with tetrabutylammonium fluoride and subsequent oxidation with Jones's reagent. Enantiomerically pure **175** {mp 237–243°C(d), $[\alpha]_D^{20}$ −15.0° (H_2O)} was now obtained by saponification with lithium hydroxide, removal of the *tert*-butoxycarbonyl group with trifluoroacetic acid, treatment with an ion-exchange resin, and recrystallization from water (Scheme 6). The overall yield from (*S*)-(+)-5-ethyl glutamate was 5%. Other synthetic approaches to racemic α-kainic acid have been published (*224, 225*).

The enantioselective synthesis of (+)-α-allokainic acid (**176**) is depicted in Scheme 7.

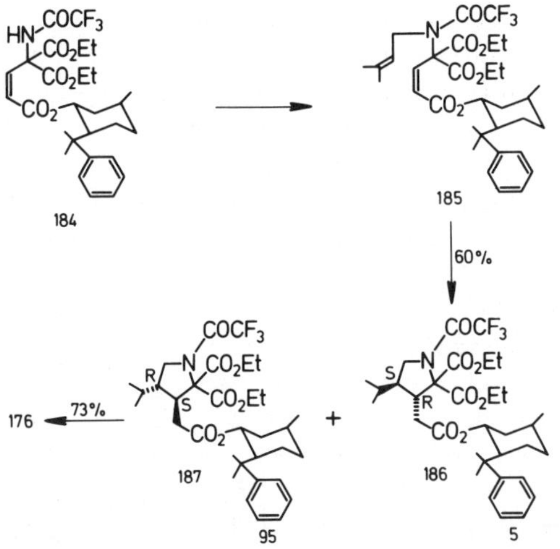

SCHEME 7

Treatment of trifluoroacetylaminomalonic ester with *cis*-β-chloroacrylic (−)-8-phenylmenthol ester in the presence of one equivalent of *t*-BuOK gave the (*Z*)-ester **184.** N-Alkenylation with 1-bromo-3-methyl-2-butene produced **185** in 40% overall yield. Lewis acid (Et_2AlCl or Me_2AlCl)-catalyzed cyclization furnished a 95:5 ratio of **187** and **186.** Saponification of **187,** decarboxylation, precipitation of the copper salt, and decomposition of the latter with hydrogen sulfide gave the enantiomerically pure **176** in 73% yield. The overall yield was more than 15%. Other synthetic pathways have been explored (*226*).

The synthesis of (−)-domoic acid, as depicted in Scheme 8, takes advantage of the [4 + 2] cycloaddition of **188** and 2-trimethylsilyloxy-1,3-pentadiene

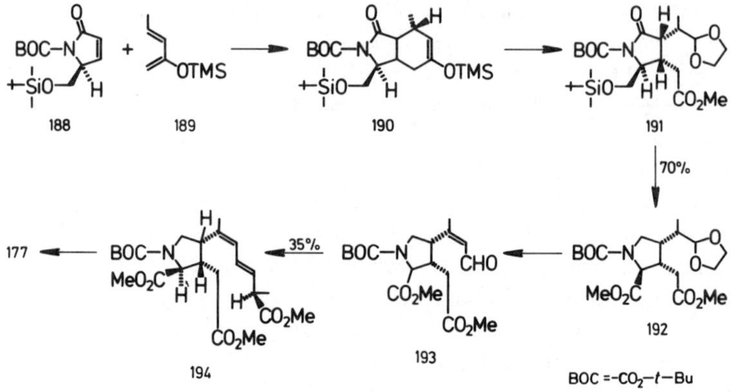

SCHEME 8

(**189**). Compound **188** was prepared from *N-tert*-butoxycarbonyl-L-pyrroglutamic acid via $NaBH_4$, reduction of the intermediately prepared anhydride with carbonic acid ethyl ester, and subsequent silylation, followed by selenenylation–diselenenylation dehydrogenation in an overall yield of 70% without racemization. The cycloaddition proceeded stereospecifically to the single adduct **190** without racemization. Ozonolysis of **190** followed by methylation with diazomethane and subsequent reaction with 2-methyl-2-ethyl-1,3-dioxolane produced **191** in 40% yield. Reduction of the amide and methyl ester functions succeeded with the borane–dimethyl sulfide complex in 70% yield. Selective deprotection of the silyl ether and oxidation with pyridinium dichromate, followed by diazomethane methylation, gave **192.** Removal of the ethylene acetal group gave the corresponding aldehyde (64%) with concomitant epimerization of the C-1′ methyl group. Introduction of the methyoxymethylene group in a Wittig reaction followed by hydroxyselenation with PhSeCl gave the α-seleno aldehyde in 90% yield. Removal of the selenide by bromination gave **193** as the major isomer (67% yield). Reaction with a Wittig reagent prepared from (*R*)-3-*tert*-butoxy-2-methyl-1-bromopropane gave **194.** Removal of the N- and O-protecting groups gave **177** {mp 213°C(d), $[\alpha]_D^{25}$ −111° (H_2O)}.

The wide interest in α-kainic and domoic acids is connected with the potent neuronal excitatory activity associated with these compounds (*227, 228*).

D. MALYNGAMIDES

Investigations of varieties of the blue-green alga *Lyngbya majuscula* have yielded interesting examples of pyrrolin-derived metabolites.

Malyngamide A (**195**), isolated from several shallow-water varieties of *L. majuscula,* was identified by a combination of chemical and spectroscopic techniques (*229*). Malyngamide A could be hydrolyzed to yield **196,** which, together

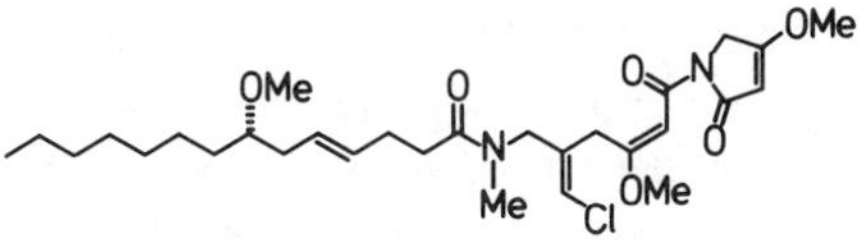

195

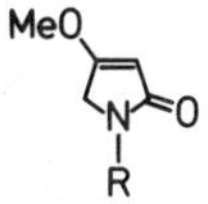

196 R = H
197 R = Ac

with **197** was a constituent of the cyanophyte. NOE experiments established the geometry of the alkenyl chloride function.

On comparison with data for **195,** malyngamide B was assigned structure **198** (*230*). Malyngamide B also originates from a Hawaiian shallow-water variety of

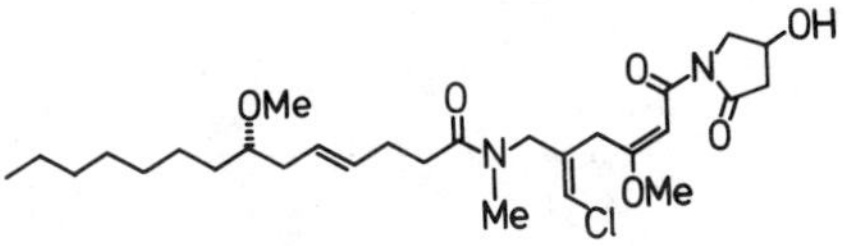

198

L. majuscula. Mild acid hydrolysis led to a ketoamide believed to be the β-ketoamide corresponding to **198.**

Malyngamide C (**199**) and malyngamide C acetate (**200**) from a shallow-water variety of *L. majuscula* (*231*) are reminiscent of stylocheilamide (**200**) and

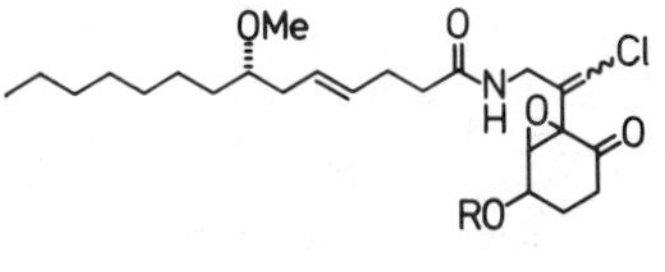

199 R=H 200 R=Ac

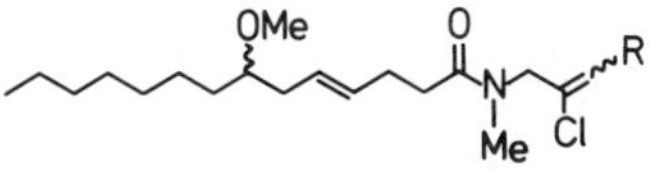

201 R= 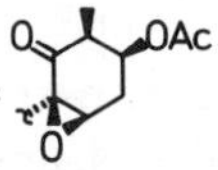 202 R =

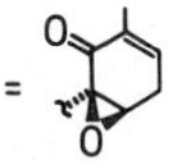

deacetoxystylocheilamide (**199**) from the sea hare *Stylocheilus longicauda* (*232*) where they undoubtedly have a dietary origin and presumably are constituents of *L. majuscula*.

A deep-water *L. majuscula* has yielded malyngamide D (**203**) and E (**204**) (*233*).

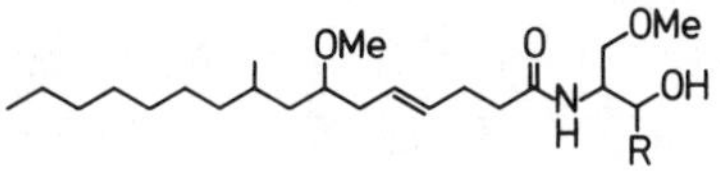

203 R =

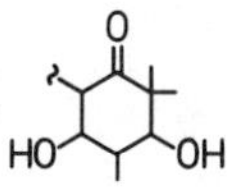

204 R =

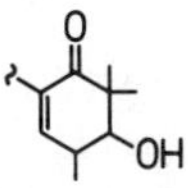

E. PUKELEIMIDES

The 4-methoxy-Δ^3-pyrrolin-2-one subunit of malyngamide A (**195**) is present in the seven pukeleimides (from the Hawaiian word *pukele*: to gather thickly in the water). All compounds were isolated from a shallow-water variety of *L. majuscula* collected at Kahala Beach, Oahu. Pukeleimide C was assigned structure **205** as the result of an X-ray structural analysis (*234*). Pukeleimides A, B, D, E, F, and G (**206–211**) yielded to spectroscopic structure determination.

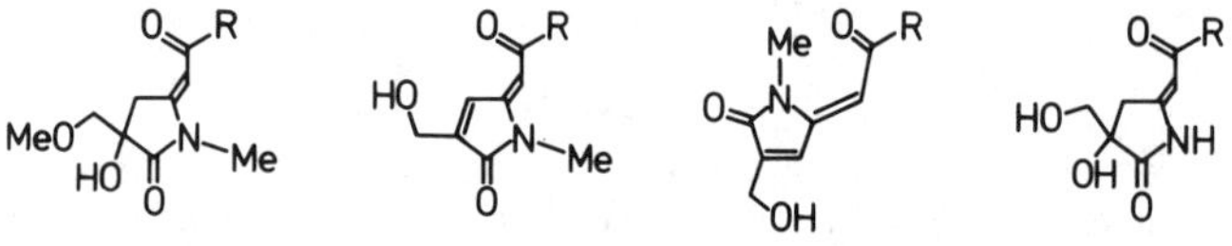

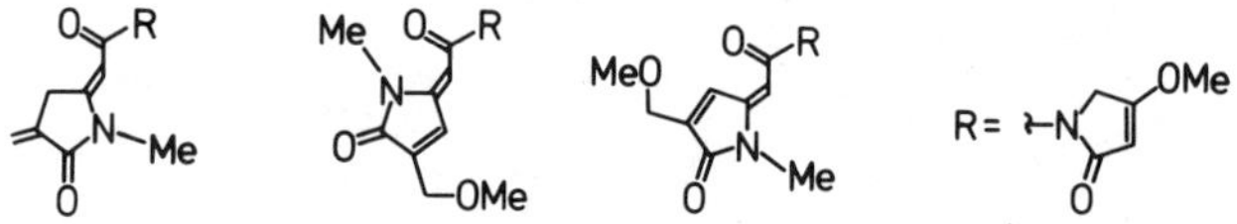

205 206 207 208

209 210 211

F. OTHER

A series of 2,3-disubstituted pyrroles (**212–215**) have been isolated from the sponge *Oscarella lobularis* (*235*). The same material also contained two 3-alkenylpyrrole-2-carboxylic acids, with chain lengths of C_{21} and C_{23}. In addition, seven 3-substituted pyrrole-2-carboxylic acid methyl esters were isolated, four with saturated side chains corresponding to **212** with n = 18, 19, 20, and 22, two with a single double bond (C_{21} and C_{23} side chain), and one with two isolated double bonds (C_{23} side chain). Interestingly enough, another series of formylpyrroles (*236*) originates from a sponge, *Laxosuberites* sp. (**216–218**).

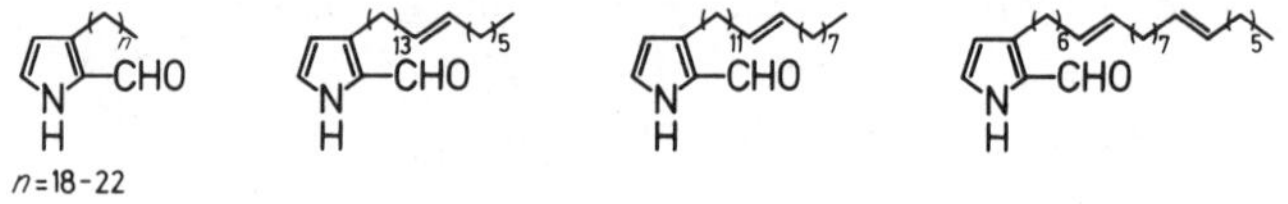

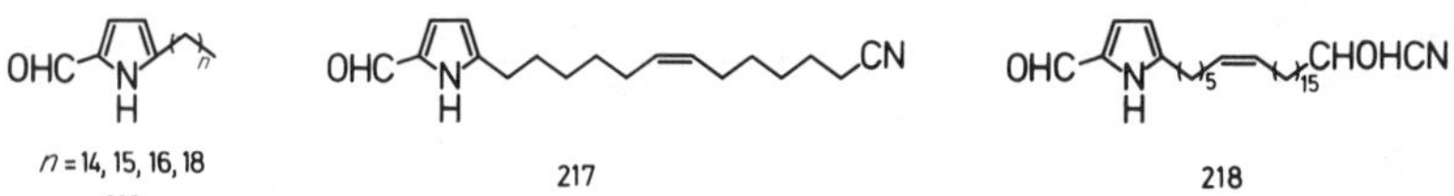

n = 18–22
212 213 214 215

n = 14, 15, 16, 18
216 217 218

The structure assignments were based on spectral analysis and degradative studies. The constituents of the mixture **216** (0.16% of dry weight) were not isolated, but GC–MS showed the mixture to consist of approximately 46% $n = 14$, 12% $n = 15$, 23% $n = 16$, and 19% $n = 18$. The nitrile **217** was present in 0.09% of dry weight, while the unusually stable cyanohydrin **218** (mp 38–40°C, $[\alpha]_D$ 0°) was isolated in 0.07% yield.

Structures **212** were assigned mainly on the basis of ^{1}H-NMR data. The substitution pattern of **212** was inferred from the coupling constant $J_{4,5} = 2.4$ Hz, measured in a D_2O-equilibrated sample (*235*); however, compounds **216** exhibited, after D_2O exchange, $J_{3,4} = 3,7$ Hz. The coupling constant $J_{4,5} = 1.9$ Hz was observed for 2-formyl-3-methylpyrrole and $J_{3,4} = 3.4$ Hz for 2-formyl-5-methylpyrrole after D_2O exchange. The value for **212** is thus between those found for 2,3- and 2,5-substitution, and a reinvestigation of the *O. lobularis* constituents is indicated (*236*).

X-Ray structural analysis confirmed the structure of dysidin {**219,** mp 127–129°C, $[\alpha]_D^{25}$ +141° ($CHCl_3$), 1–1.2% of dry weight} from the sponge *Dysidea*

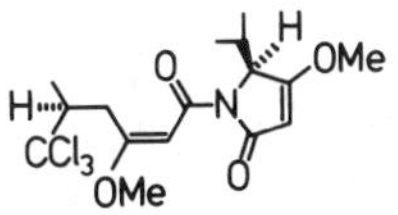

219

herbacea (*237*). Dysidin has several structural features in common with malyngamide A (**195**). The pyrrolinone ring has the same absolute configuration as that of L-valine. *D. herbacea* has yielded related compounds (Section V,E).

Midpacamide (**58**), from the sponge *Agelas cf. mauritiana* (*202*), contains a methylhydantoin residue joined through a side chain to 1-methyl-2,5-dibromopyrrol-2-carboxylic acid (Section IV,B) in an amide linkage. A certain structural similarity to oroidin (Section II,E) is evident. The pyrrole unit of **58** could be derived from proline and the remainder of the molecule from ornithine, amidated on the α-amino group (*202*).

The first example of a quinoid isoindole (**220,** mp 153–154°C, 0.001% of dry weight) was isolated from a sponge, *Reniera* sp. (*238*). The structure was con-

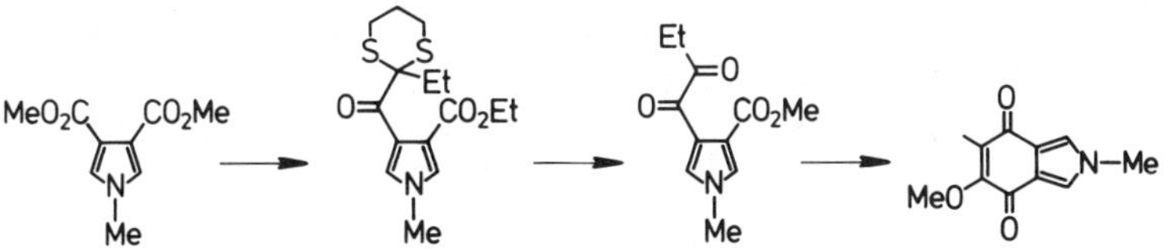

221 222 223 220

firmed by synthesis. Condensation of 2-lithio-2-ethyl-1,3-dithiane with **221** gave a 27% yield of keto ester **222.** Removal of the dithiane group was effected with four equivalents of *N*-chlorosuccinimide and five equivalents of silver nitrate in aqueous acetonitrile to give **223** in 94% yield. Sodium hydride in dimethylformamide caused cyclization to the hydroxyquinone in 30% yield. Since the pyrrole ring is sensitive to excess diazomethane, the methylation was carried out with 0.5 equivalent of diazomethane to give a 30% yield of **220.**

Brown algae have yielded two tripeptides. One, eisenine {**224** mp 225–226°C(d), $[\alpha]_D^{14}$ −54.3°}, was obtained from *Eisenia bicyclis* (*239*). The structure determination (*239, 240*) was verified by synthesis (*241*). Since the brown alga *Ecklonia cava,* which had earlier yielded **224** (*242*) under mild extraction conditions, gave L-glutaminyl-L-glutaminyl-L-alanine and only negligible amounts of **224,** the latter may well be an artifact with the pyrrolidone ring formed from the terminal glutaminyl residue (*243*).

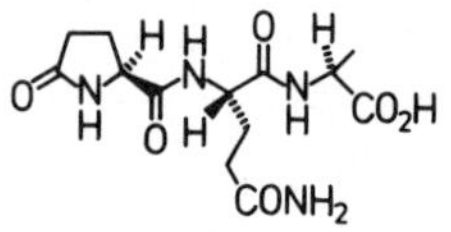

224

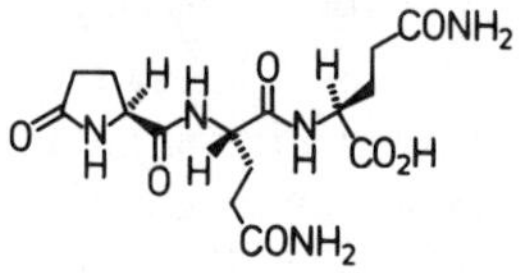

225

Fastigiatine (**225,** mp 190–195°C, $[\alpha]_D$ −43.7 to −46.5°) was isolated from *Pelvetia fastigiata* (*244*). The proposed structure **225** may be an artifact in the same way as **224**; however, it may be erroneous since the physical properties do not correspond with those of a synthetic sample (*241*, mp 203–204°C, $[\alpha]_D^{20}$ −35.3°).

The occurrence of marine tetrapyrrole pigments has been reviewed (*49, 55*).

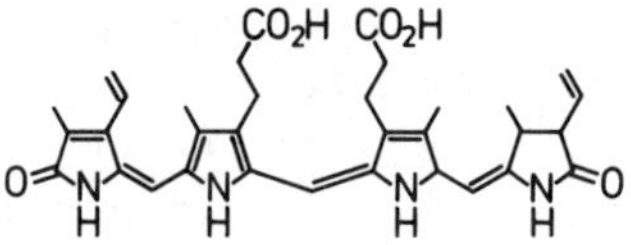

227

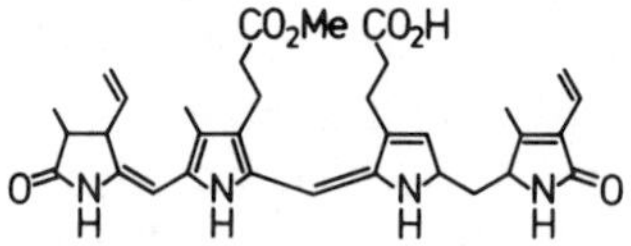

228

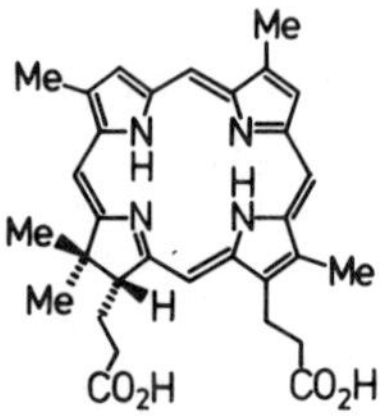

226

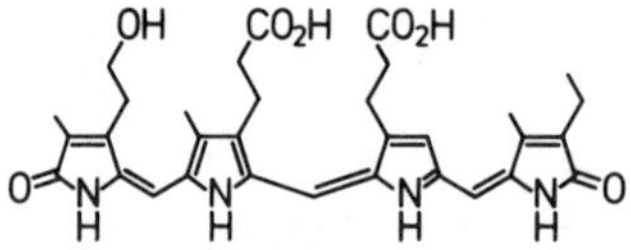

229

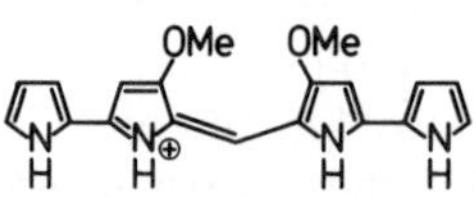

230

The green pigment bonellin (**226**), from the echurian worm *Bonellia viridis,* has especially attracted much attention because of the biological activity associated with this chlorin (*245*). Bonellin seems to determine the sex of the indifferent larvae in such a way that they differentiate into males.

The major constituent of the pigment of the calcareous skeleton of the blue coral *Heliopora coerula* was shown to be the known biliverdin **227** (*246*).

The purple discharge from various sea hares of the genus *Aplysia* has aroused the curiosity of natural products chemists for a long time. This pigment, aplysioviolin (**228**), was proposed to be identical to the monomethyl ester of phycoerythrobilin known from the algal phycoerythrins (*247*).

Turboverdin (**229**) from the ovary of an edible turban shell, *Turbo cornutus*, differs from mesobiliverdin only by substitution of an ethyl group in ring A of the latter by an ethanol side chain (*248*). Turboverdin is the chromophore of a biliprotein.

A deep-blue pigment, **230**, isolated from a compound ascidian was identified as the protonated form of a purple pigment known from a mutant strain of the bacterium *Serratia marcescens* (*249*).

Modern isolation and structure-determination methods have greatly accelerated research in this area of natural products chemistry. Many investigations have important aspects in other disciplines, e.g., the study of photosensitivity in humans after ingestion of photodynamic agents from abalone, for example, *Haliotis discus hannei* (*250*) has yielded the chlorophyll derivative pyropheophorbide as the causative agent (*18*).

V. Miscellaneous Alkaloids

A. Introduction

The bromotyrosine alkaloids have all been isolated from marine sponges of the family Verongiidae. The order Verongida is at present included in the subclass Ceractinomorpha, class Demospongia, but should presumably be removed to a separate subclass (*7*). According to Wiedenmayer (*251*), *Verongia* is synonymous with *Aplysina*, the latter name having preference; however, the taxonomy is complex and difficult within this group. Consequently no effort has been made to correct the names in Section V,B. Members of the genera *Verongia, Psammaplysilla*, and *Ianthella* have been investigated.

One of the rare cases of controlled biosynthetic experiments belongs in this section. Characteristically enough, the feeding experiments were successful only by using an unusual feeding technique. Still, even if some basic biosynthetic pathways are now known for a *Verongia* species, the question of which organism has this capacity is unsettled since *Verongia* species are known to harbor massive populations of microorganisms (Section I).

Organisms from such different taxonomic groups as opisthobranch molluscs, nemerteans, and sponges have yielded pyridine alkaloids. In one case the biosynthesis of the pyridine ring seems to occur via a phenol.

Except for the case of a marine pseudomonad and a marine bryozoan, all of the known quinoline and isoquinoline alkaloids originate from sponges. In one case there is a strong structural resemblance between the sponge alkaloids and metabolites from a microorganism.

In the case of the heterogenous groups of alkaloids treated in Section V,E, the organisms from which they are gained are taxonomically as diverse as the structures encountered.

B. Tyrosine-Derived Alkaloids

Simple tyrosine-derived metabolites are known from marine organisms, e.g., hordenine (*N,N*-dimethyltyramine) from red algae (*252–254*), candicine (*N,N,N*-trimethyltyramine) (*255*), candicine *O*-sulfate (*256*) from a red alga, and cardioactive 3-hydroxy-4-methoxyphenethylamine (*257*) from a soft coral. Halogenated tyrosine has repeatedly appeared in marine material. The coral *Gorgonia carolinni* yielded 3,5-diiodotyrosine (**231**), one of the first marine natural products to be identified (*258*). A number of corals yield **232** (*259*),

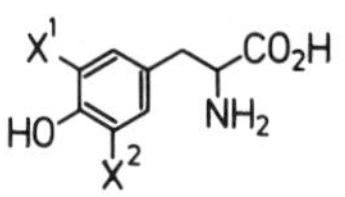

231 $X^1 = X^2 = I$
232 $X^1 = X^2 = Br$
233 $X^1 = H$, $X^2 = I$
234 $X^1 = H$, $X^2 = Br$
235 $X^1 = Cl$, $X^2 = Br$

which could also be obtained from hydrolyzed spongin (the fibrous protein from Porifera) together with **231** (*260*). The bath sponge *Spongia officinalis obliqua* also contains **233** and **234** (*261*). Scleroprotein from the operculum of *Buccinum undatum* contains chlorobromotyrosine, presumaby **235** (*262*).

An array of structures, which may be considered to originate from 3,5-dibromotyrosine (**232**), have subsequently been identified from sponges. Members of the family Verongiidae have been a rich source of these compounds.

Attempts to incorporate radioactivity in the brominated metabolites of *V. aerophoba* by feeding [U-^{14}C]-L-tyrosine or [U-^{14}C]-L-ornithine have failed (*263*). Using precursors encapsulated in multilamellar lipid vesicles prepared from phosphatidylcholine and cholesterol, Tymiak and Rinehart succeeded in demonstrating the conversion of phenylalanine and tyrosine to brominated metabolites (*264*). *Aplysina fistularis* (Pallas) *sensu* Wiedenmayer, 1977 (identical to *Verongia aurea sensu* De Laubenfels, 1948) was shown to biosynthesize **238** and the rearranged dibromogentisamide (**239**) from phenylalanine and tyrosine. The biosynthetic pathway shown in Scheme 9 was proposed by these investigators.

The oxime has never been isolated from natural sources; however, 4-hydroxyphenylpyruvic acid oxime (the debrominated analog) is present in the sponge *Hymeniacidon sanguinea* (*265*).

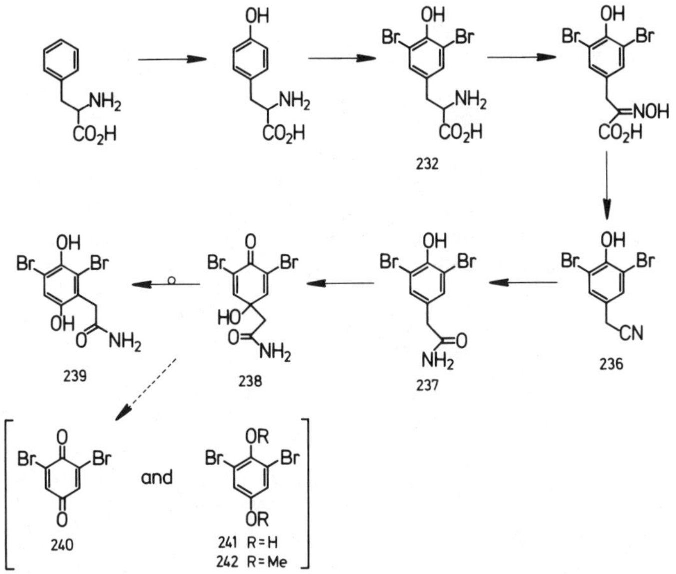

SCHEME 9

Compounds **236** and **237** have been isolated from *A. fistularis* (*264*) together with **243** (*266*). The key dienone **238** has been reported from several species including *V. fistularis* and *V. cauliformis* (*267, 268*). Lithium borohydride re-

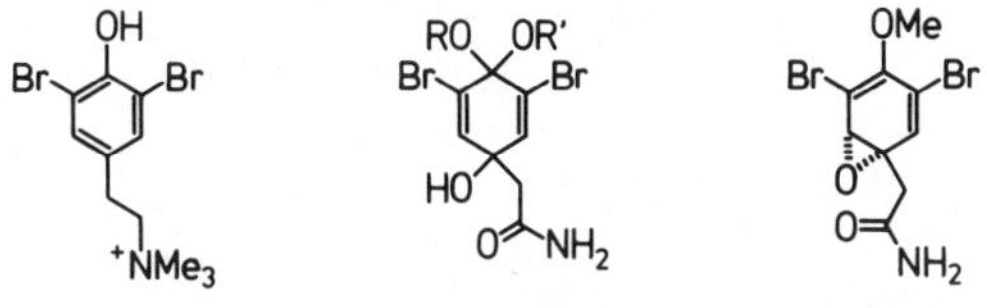

243 244 R=R'=Me 245 R=Me, R'=Et 246

duction of the acetate of **238** generates **237,** which could in turn be oxidized to **238** (*269*).

During the isolation of **238** from a methanol extract, the dimethyl acetal **244** was obtained as well (*268*). Acid-catalyzed deketalization produced **238,** but attempts to induce **238** to react with methanol failed. Subsequent reisolation of **238** from an unidentified *Verongia* sp. by ethanol extraction in addition also gave **245** as a mixture of stereoisomers, as judged from different methoxy signals in ^{1}H NMR (*270*). These results led to the postulation of a common precursor for **238, 244,** and **245,** namely, the arene oxide **246.** 1,4-Addition of water, methanol, or ethanol to **246** would give **238, 244,** or **245,** respectively. The acid-catalyzed addition of methanol to 1,4-dimethylbenzene 1,2-oxide (*271*), afford-

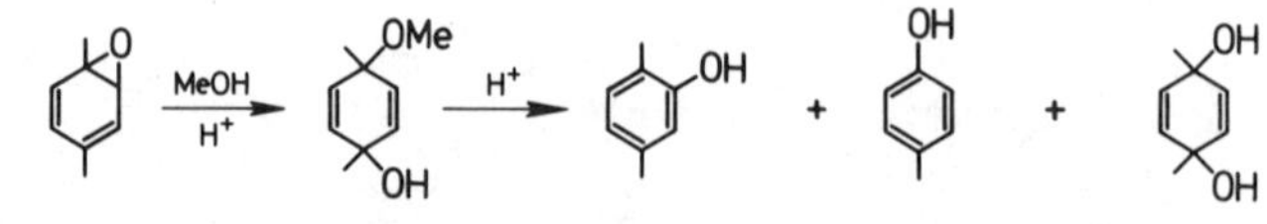

247 248 249 250

ing 4-methoxy-1,4-dimethyl-2,5-cyclohexadienol (**247**), was quoted in support of the arene oxide hypothesis. In acidic solution, **247** gave **248, 249,** and **250** (*270*). Analogously, **246** might be expected to give **244 (245), 238, 251,** and **252,** the latter being the methyl ether of **239.** Compound **251** is known from *A.*

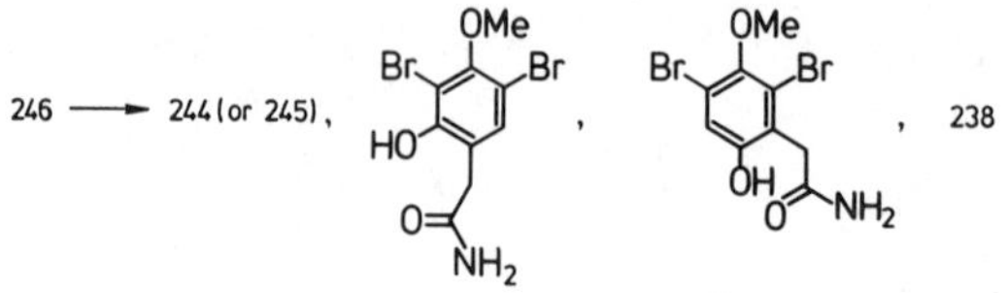

246 → 244 (or 245), 251, 252, 238

aerophoba (*272*) and 2,4-dibromogentisic acid amide (**252**) from *V. aurea* (Hyatt, 1875) (*273*). A synthesis of *252* has been reported (*274*). In short, **246** may still be an intermediate in Scheme 9 between **237** and **238.** Removal of the side chain may produce **240, 241,** and **242,** isolated from *A. fistularis* (*264*).

The tyrosine nucleus is retained in the carbamate **253** {mp 222–225°C, $[\alpha]_D^{25}$ +8.9° (MeOH)}, first isolated from *A. lacunosa* in approximately 0.05% yield

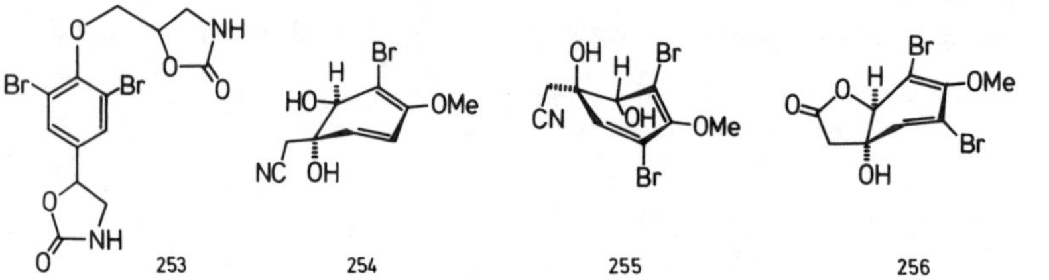

253 254 255 256

based on fresh tissue (*275*). Comparative analysis of several sponges revealed the presence of **253** in nine species of Aplysiniidae (*276*). Interestingly enough, some sponges contained (+)-**253,** others (−)-**253,** and some (±)-**253** (*276, 277*).

Similar observations have been published in the case of Aeroplysinin-1 (**254**), where the dextrorotatory isomer was isolated from *V. aerophoba* (*278*) and the levorotatory enantiomer **255** was detected in *Ianthella ardis* (*279*). The absolute stereochemistry proposed for the two antipodes has been confirmed by X-ray studies of **254** (*280*) and **255** (*281*). Aeroplysinin-2 (**256**) was found in *V. aerophoba* and *Ianthella* sp. (*282*).

Cavernicolin-1 (**257**) and cavernicolin-2 (**258**) originate from *Aplysina* (*Verongia*) *cavernicola* (*283*). The two epimeric compounds present in 0.025% of dry weight were isolated as a 3:1 equilibrium mixture of **257** and **258.** A small

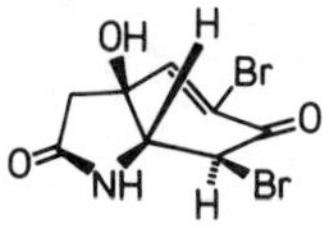

257 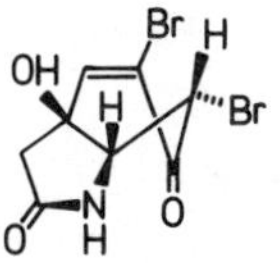 258

optical rotation was attributed to a slight enantiomeric excess. The sponge contains large amounts (0.3% of dry weight) of **238,** from which **257** and **258** could formally be formed by conjugate addition. The intermediacy of an arene oxide is, however, also conceivable.

Aerothionin {**259,** mp 134–137°C(d), $[\alpha]_D$ +252°} is the major component in *V. aerophoba* and *V. thiona* (*284, 285*). Homoaerothionin (**260**) originates from the same sources. An X-ray structural analysis of **259,** isolated from *Aplysina*

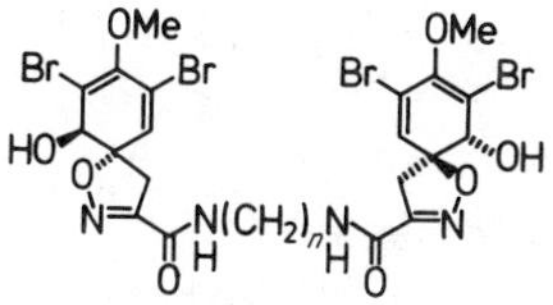

259 $n=4$
260 $n=5$

fistularis (Pallas), combined with CD measurement established the absolute configuration (*286*).

Aplysina fistularis forma *fulva* in addition to **259** and **253** gave aerothionins with the oxygenated (hydroxy and keto) 2-position of the tetramethylenediamine chain (*287*). The same source yielded three new alkaloids. Fistularin-1 {**261,** 0.005% of wet weight, $[\alpha]_D$ +93.5° (MeOH)} yielded to spectroscopic analysis by comparison with the diacetate as did fistularin-2 (**262**). Fistularin-3 {**263,**

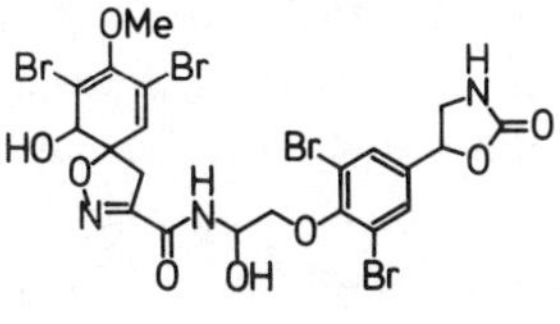

261 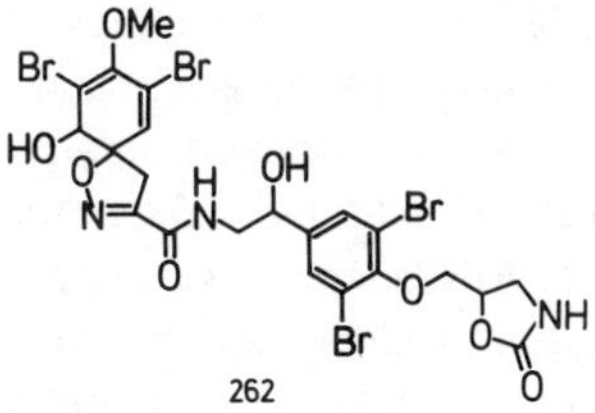 262

0.06% of wet weight, $[\alpha]_D$ +104.2° (MeOH)} was identified also by comparison of spectral data of **263** and the tetraacetate.

Based on spectroscopic evidence paired with chemical degradation experiments, psammaplysin-A and -B from the sponge *Psammaplysilla purpurea* were assigned structures **264** and **265,** respectively (*288*). These antibacterial compounds are unique in that the isoxazole ring normally encountered in these metabolites is exchanged with the 1,3-oxazolin moiety.

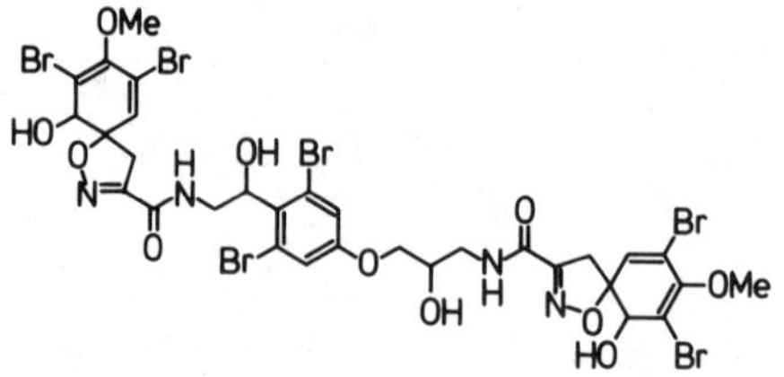

263

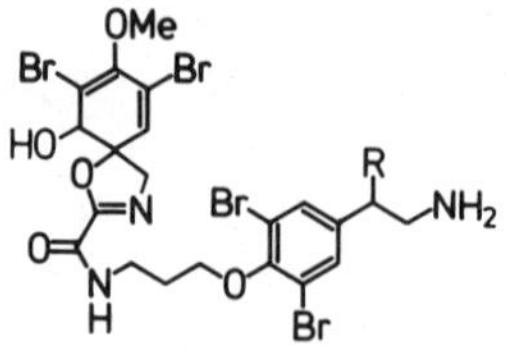

264 R=H
265 R=OH

In a preliminary communication, the isolation and structure elucidation of bastadin-1 (**266**) and bastadin-2 (**267**), from *Ianthella basta,* were reported

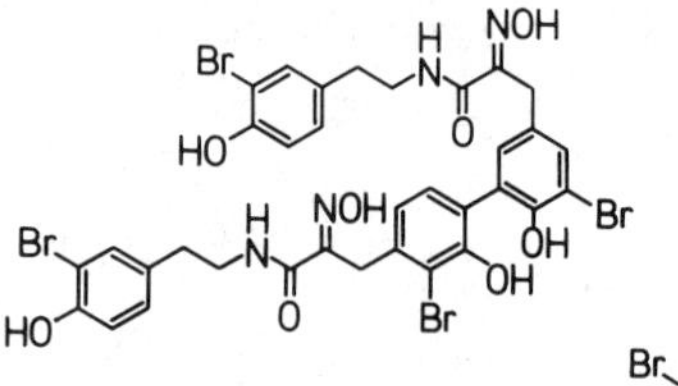

268

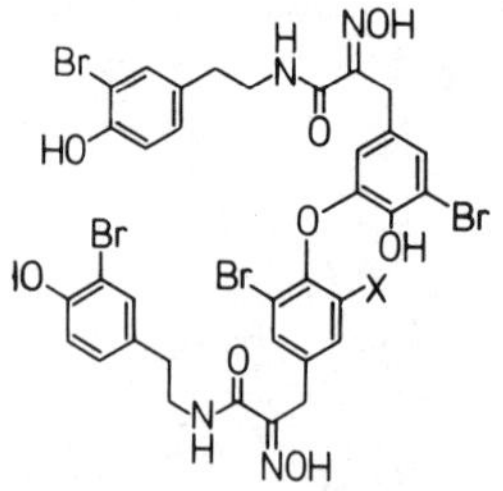

266 X=H
267 X=Br

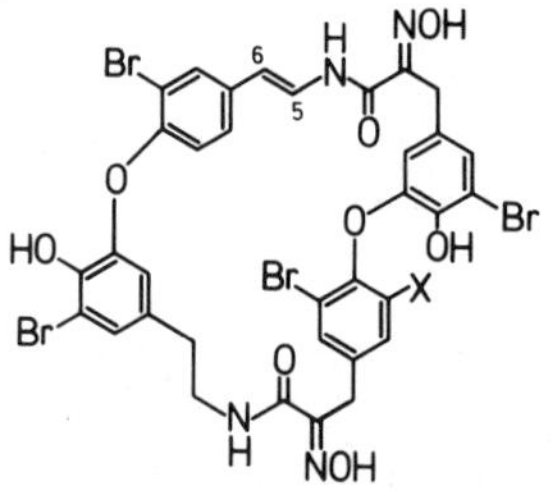

269 X=Br
270 X=H, 5,6-dihydro
271 X=Br, 5,6-dihydro
272 X=H

(*289*). The full paper (*290*), in addition to elaboration regarding the structure elucidation, presented the structures of bastadin-3 (**268**), bastadin-4 (**269**), bastadin-5 (**270**), bastadin-6 (**271**), and bastadin-7 (**272**) from the same source. Bastadin-5 (**270**) was isolated only as the tetramethyl ether. The structures of bastadin-4 (**269**) and bastadin-5 tetramethyl ether were confirmed by X-ray structural analysis. All structures were solved mainly by spectroscopic examination of the pure natural products and derivatives.

All compounds showed potent *in vitro* antimicrobial activity against Gram-positive organisms.

Bastadin-1, -2, and -3 have been synthesized (*291*) according to Scheme 10.

The key compound **279** was prepared from methyl 3,5-dibromo-4-hydroxyphenylpyruvate oxime (**273**) by thallium(III) oxidation to the spiroisoxazol (**275**), yielding **279** almost quantitatively by Zn reduction. Analogously, the spiroisoxazol (**277**), together with other products including **276,** was formed from **274.** Reduction of **276** and **277** gave **280** and **278,** respectively.

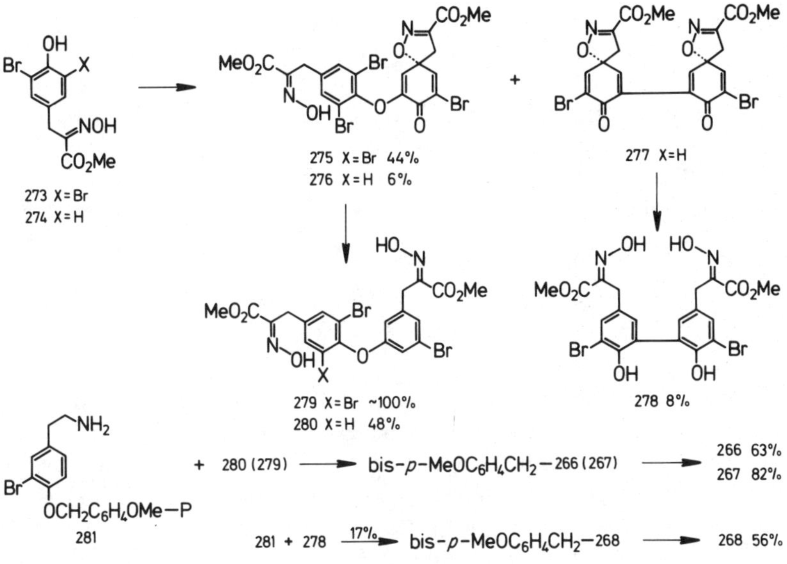

SCHEME 10

Alkylation of 3-bromo-4-hydroxybenzaldehyde with *p*-methoxybenzyl chloride, followed by nitromethane, gave the expected hydroxynitro compound (71%), which was directly dehydrated to the corresponding α,β-unsaturated nitro compound in 79% yield. Reduction with $NaBH_4$ gave the saturated nitro compound, which could be reduced almost quantitatively to 3-bromotryptamine *p*-methoxybenzyl ether (**281**).

Bastadin-1 (**266**) was now synthesized by reaction between **280** and excess **281** to give the diamide (34%), which was deprotected with trifluoroacetic acid to give **266.**

Bastadin-2 (**267**) was prepared in the same way from **281** and **279,** giving 36% diamide.

Bastadin-3 (**268**) was obtained from the corresponding diamide (17%) prepared from **281** and **278.**

The synthetic study indicated that the geometry of the oxime functions in the bastadins is anti to the amide functions.

C. Pyridine-Derived Alkaloids

Several unusual pyridine-derived amino acids have been identified from marine organisms (*15, 215*). Homarine (**282**) was isolated as early as 1933 (*292*) from *Homarus americanus,* and its occurrence in marine invertebrate phyla, as opposed to its apparent absence from freshwater invertebrates, has been described (*293*).

The biosynthesis and physiological role in marine shrimp have been investigated (*294*). The presence of **282** in the fungus *Polyporus sulphureus* has been demonstrated (*295*). Trigonellin (**283**) is known from, e.g., the sponge *Calyx nicaensis* (*296*). Baikiain (**284**), pipecolic acid (**285**), and 5-hydroxypipecolic acid (**286**) have been identified in red algae (*297, 298*), sponges (*299*), red (*298*) and brown algae (*300*), and *Undaria pinnatifida* (*301*), respectively. All compounds are known from terrestrial sources as well (*215*).

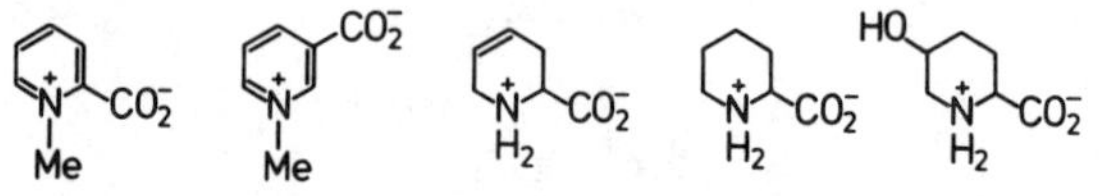

282 283 284 285 286

The marine opisthobranch mollusc *Navanax inermis* Cooper (syn. *Chelidonura inermis*; Cephalaspidea), when heavily molested, secretes a bright-yellow, glandular alkaloid mixture. In the natural habitat these animals locate prey and each other by following the slime trail laid down by numerous opisthobranchs and *Navanax* itself. The trail-following behavior is expeditiously terminated as a *Navanax* encounters the yellow secretion. The major constituent (~90%) of the alarm pheromone is navenone A–C (**287–289**) (*302*) in a 4:2:1 ratio. Navenone A (**287**) (mp 144–145°C), the only major alkaloid, was identified by spectroscopic methods and reduction ($NaBH_4$) to the corresponding alcohol, which could be hydrogenated catalytically to the expected octahydro alcohol. Addition of $Eu(fod)_3$ to **287** allowed the low-field signals to be resolved into one-proton bands and allowed the assignment of the unshifted spectrum. All olefinic protons exhibit coupling constants of 14 or 15 Hz, and thus the configuration must be *E* (*302*).

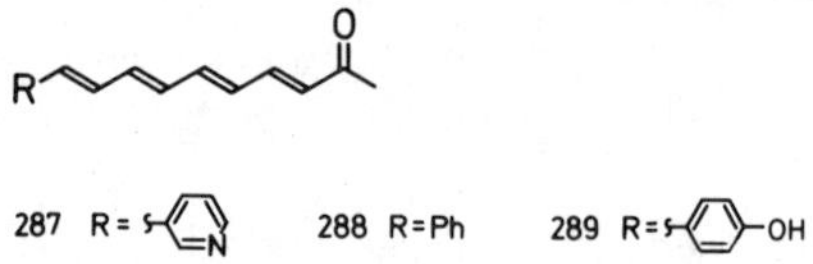

287 R = 3-pyridyl 288 R = Ph 289 R = 4-hydroxyphenyl

Navenone A, B, and C have been synthesized (*303*). In the case of navenone A, the synthesis was accomplished according to Scheme 11. Reaction of 3-

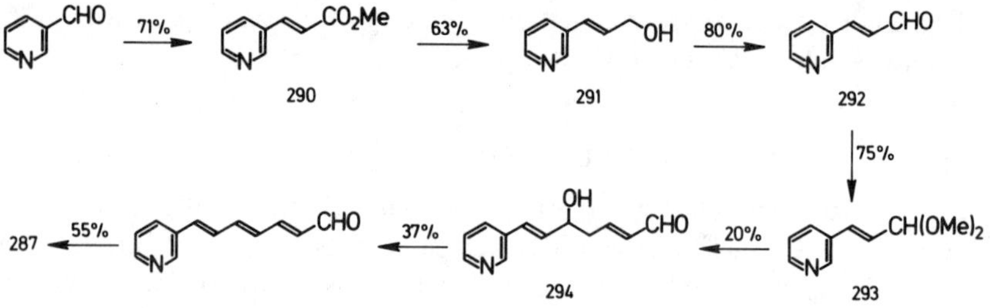

SCHEME 11

formylpyridine with methoxycarbonylmethylenetriphenylphosphorane gave **290,** which on lithium aluminum diethoxydihydride reduction gave **291.** Oxidation with manganese dioxide produced the acrolein (**292**), the acetal of which (**293**) reacted with 1-trimethylsilyloxybutadiene in the presence of titanium tetrachloride to generate **294.** Elimination followed by a Wittig reaction with acetylmethylenetriphenylphosphorane gave a fair yield of navenone A (**287**).

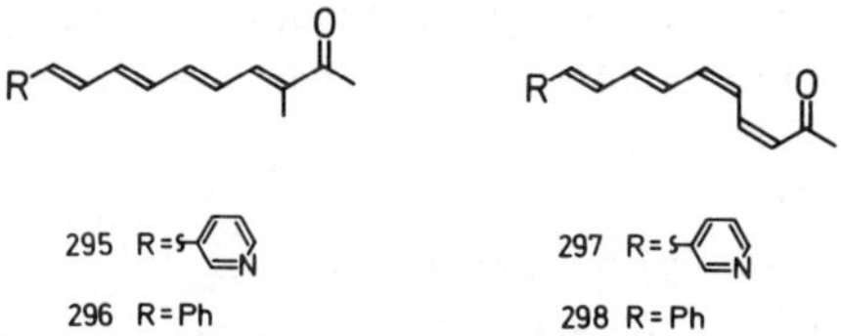

295 R=3-pyridyl

296 R=Ph

297 R=3-pyridyl

298 R=Ph

Investigation of the minor constituents (*304*) revealed the presence of 3-methylnavenone A (**295**) and the cis isomer of navenone A (**297**) together with **296** and **298.** Several unstable compounds, believed to be unknown cis isomers of **287** and **288,** were isolated but were transformed to **287** and **288** before NMR data could be obtained.

Pure **287** and **288** were transformed to **297** and **298,** respectively, on irradiation with visible light. The equilibrium mixture of **287** and **297** seems to favor **287** in a 9:1 ratio.

Feeding experiments indicated that after depletion the regeneration of the navenones proceeds with biosynthesis of **289** with the amount of **287** slowly increasing. This was taken as evidence that the phenol in **289** is the precursor of the benzene ring in **288** and the pyridine **287**, as also supported by ^{14}C-acetate feeding experiments. This hypothesis clearly warrants further investigations.

Nemertines (syn. nemerteans) are members of the invertebrate phylum Rhynchocoela. Nemertines are carnivorous, feeding on annelids, crustaceans, molluscs, and occasionally fish. Some belonging to class Enopla order Hoplonemertini paralyze the prey organism by injecting a toxic fluid from the proboscis while others from class Anopla (e.g., order Heteronemertini) are nonvenomous (*305*). Many nemerteans secrete a toxic integumentary mucus. Early investigations revealed the presence of nicotine-like toxins in nemertines (*306*). Modern isolation and structure determination techniques later allowed one of the toxins from the hoplonemertine *Paranemertes peregrina* to be identified as anabaseine (**299**) (*307*).

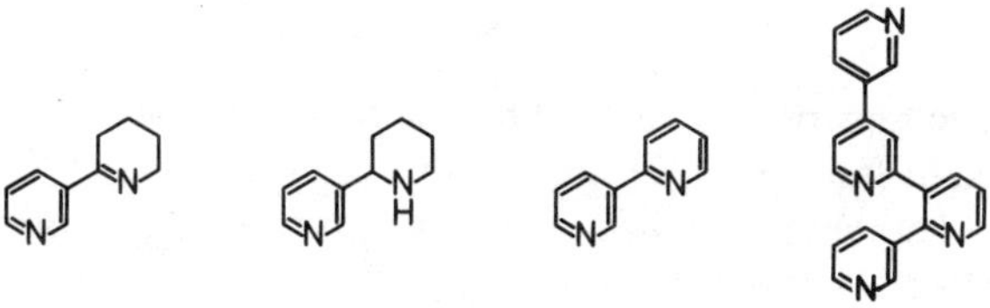

299 300 301 302

Reaction of *N*-benzoylpiperidone and ethyl nicotinate was earlier shown to yield **299** (*308*), isolated as an intermediate in the synthesis of Anabasine (**300**) (*309*), isolated from *Anabis aphylla* L. (Chenopodiacee).

The hoplonemertine *Amphiporus angulatus* yielded 2,3′-bipyridyl (**301**), nemertelline (**302**), anabaseine (**299**), and methylbipyridyl (*310*).

Investigation of the heteronemertine *Cerebratulus lacteus* (Leidy) identified two types of toxins from the integumentary mucus, both polypeptides. Cerebratulus A toxins (11,000 daltons) are lethal to a variety of animal species including mammals, while Cerebratulus B toxins (6,000 daltons) are selectively toxic to crustaceans (*311*).

An interesting example of higher-molecular-weight toxins, designated halitoxin, has been investigated by Schmitz and co-workers (*312, 313*). Halitoxin originates with the sponge *Haliclona rubens* but is also present in *H. viridis* and *H. erina*. Three toxic fractions have been studied, namely, with molecular weight 500–1,000, 1,000–25,000, and >25,000. All three fractions show the same spectral properties and biological activity. Of the numerous methods applied for purification of halitoxin, ultrafiltration was found to be the most effective.

Careful spectroscopic analyses of different fractions of the halitoxin complex resulted in identification of a repeating subunit, the 1,3-dialkylated pyridinium ion. The 1,3-disubstituted pyridinium ions are interlinked via the alkyl groups. In spite of much effort, no other functionalities could be demonstrated. Of the variety of attempts to degrade the toxin, only pyrolysis proved successful. Decomposition at 140–160°C yielded a mixture of 3-alkenylpyridines and 3-(ω-

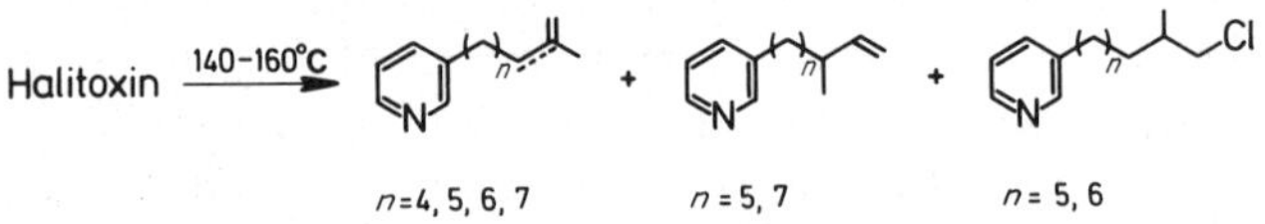

Halitoxin $\xrightarrow{140-160°C}$ 303 (n=4, 5, 6, 7) + 304 (n = 5, 7) + 305 (n = 5, 6)

chloroalkyl)pyridines exemplified by **303, 304,** and **305.** All available data are consistent with the general formulation **306** for halitoxin.

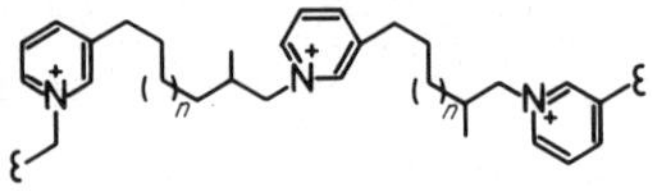

306 n = 2, 3, 4, 5

A tentative tetrameric cyclic structure has been proposed for the 500–1000-molecular-weight fraction (*312*). The methyl-branched 3-alkyl substituent predominating in halitoxin is known from muscopyridine (**307**) isolated from the musk deer (Moschus moschiferus) (*314*).

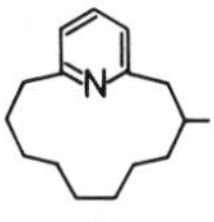

307

Halitoxin is cytotoxic, hemolytic, and toxic to fish and mice. The depolarizing effect of impure samples of halitoxin on endplate and muscle membranes has been studied (*315*).

D. Quinoline and Isoquinoline Alkaloids

By far the most yielding organism with respect to structural diversity is the bright-blue sponge *Reniera* sp. The principal antibacterial agent from this organism, collected near Isla Grande, Mexico, between −10 and −20 m, was renierone (**308**). The structure of this yellow isoquinolinequinone (mp 91.5–92.5°C), present in 0.03% dry weight, was solved by X-ray crystal analysis (*316*). The similarity between renierone (**308**) and mimosin **310** from *Streptomyces lavendulae* No. 314 (*317*) is striking. Renierone has been synthesized (*318*).

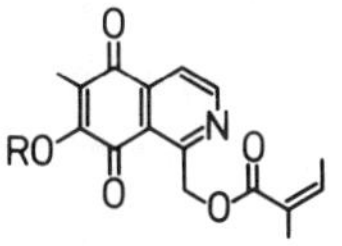

308 R = Me
309 R = H

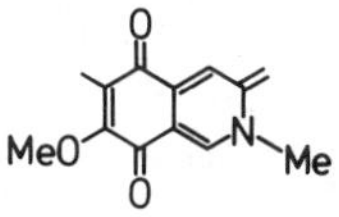

311

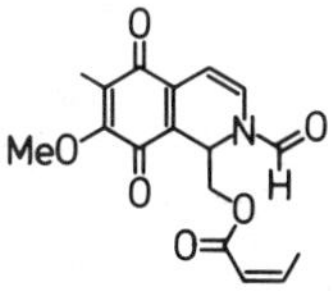

313

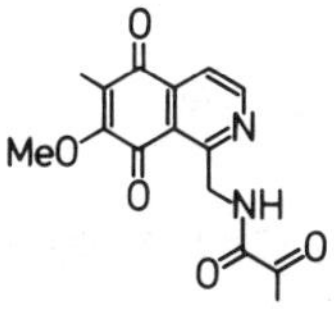

310

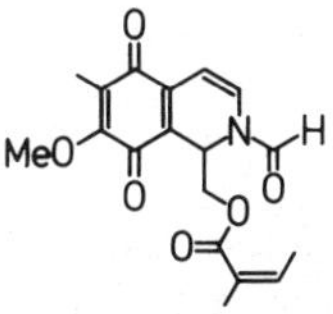

312

A competent and extensive review of isoquinolines from actinomycetes and sponges by Arai and Kubo (*319*) has appeared. The reader is referred to this review for a detailed discussion of all aspects of the chemistry and biological activity of these interesting compounds. In the following, a brief outline of what has so far been learned about the *Reniera* alkaloids is given in order not to omit entirely this intriguing and important area.

Several new alkaloids in addition to mimosamycin (**311**) (*320*), known from *Streptomyces lavendulae* No. 314, have now been identified from the same source (*238*).

A noncrystalline red solid {0.027% dry weight, $[\alpha]_D$ −227° (MeOH)} was found to be a 2:1 mixture of rotational isomers of *N*-formyl-1,2-dihydrorenierone

(**312, 313**). The dominant isomer was shown to be **312.** Hydrolysis of *N*-formyl-1,2-dihydrorenierone gave renierone as the major product.

Renierone (**308**), on treatment with methanolic sodium hydroxide solution, gave *O*-demethylrenierone (**309**), which is also a natural product (0.002% dry weight). Ethereal diazomethane transforms **309** (beige crystals, mp 135–136°C) to **308.**

The simple isoquinolinequinone 1,6-dimethyl-7-methoxy-5,8-dihydroisoquinoline-5,8-dione [**314,** 0.007% dry weight, mp 188–190°C (d)] was identified by comparison of spectral data.

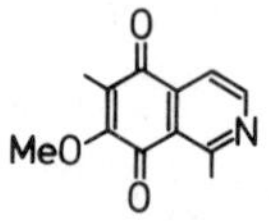

MeO O O N

314

A quinoid isoindole, 2,5-dimethyl-6-methoxy-4,7-dihydroisoindole-4,7-dione (**220**), is the first example of a naturally occurring isoindole. This unique natural product was dealt with in Section IV,F.

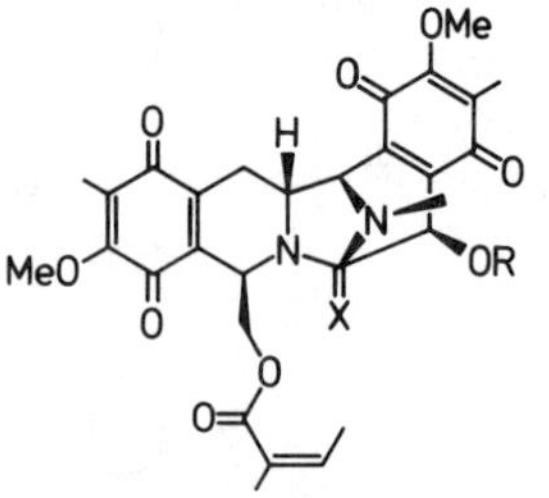

OMe O O H N N MeO O X OR O O

315 R=H, X=H_2
316 R=Et, X=H_2
317 R=H, X=O
318 R=Et, X=O

Four dimeric alkaloids, renieramycin A–D (**315–318**), were found to be major constituents of the sponge. The structures yielded to spectroscopic, mainly ^{1}H NMR, identification. Two of the compounds, namely, renieramycin B (**316**) and D (**318**) are ethyl ethers of reniaramycin A (**315**) and C (**317**), respectively. Renieramycin B was considered an artifact from the storage of the sponge in ethanol. If correct, this is an unusual example of a secondary and a primary alcohol forming an ether in dilute solution. It seems much more likely either that all four compounds are metabolites or, alternatively, that water or ethanol have replaced a more reactive group.

Renieramycin A {**315,** 0.008% dry weight, $[\alpha]_D$ −36.3° (MeOH)} was investigated using extensive nuclear Overhauser enhancement difference spectroscopic (NOEDS) experiments. These data combined with arguments based on chemical shifts convincingly define the relative stereochemistry as depicted in **315.** The positions of the methyl and methoxy groups of rings A and E were assigned on the assumption that **315** was derived from one molecule of *N*-

formyl-1,2-dihydrorenierone (**312, 313**) and one molecule of mimosamycin (**310**).

Renieramycin B {**316,** 0.002% dry weight, $[\alpha]_D$ −32.2° (MeOH)} seems to have the same relative stereochemistry (based on NOEDS) as that of **315.**

Renieramycin C {**317,** 0.002% dry weight, $[\alpha]_D$ −89.2° (MeOH)} and renieramycin D {**318,** 0.001% dry weight, $[\alpha]_D$ −100.7° (MeOH)} correspond closely to **315** and **316** except for the presence of the amide function.

The renieramycins bear a strong resemblance to the saframycins isolated from *Streptomyces lavendulae* No. 314. Saframycin A (**319**) (*321*), B (**320**) (*322*), and D (**321**) (*322*) and the renieramycins all have identical ring systems. The only

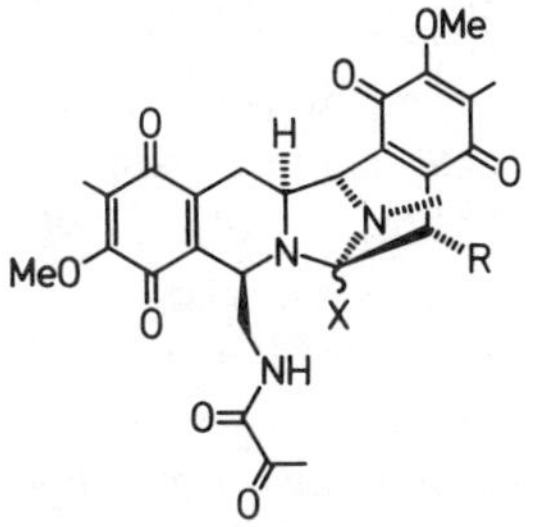

319 X=CN, R=H
320 X=H, R=H
321 X=H, R=OMe

other difference, apart from the cyano group in **319,** is the relative stereochemistry of the point of attachment of the side chain and the replacement of the angelate ester of the renieramycins with the pyruvamide side chain of the saframycins. The renieramycins have not yet been synthesized, but a stereocontrolled total synthesis of the closely related racemic saframycin B has appeared (*323*).

Since the sponge contained relatively small quantities of the alkaloids, the possibility of a symbiotic microorganism origin is left open. Preliminary examination of the same *Reniera* sp. collected in a marine lake in Palau, Western Caroline Islands, suggest that the same metabolites are present (*238*). Clearly, comparative investigations of symbionts from the same species of sponge from different localities are warranted.

All alkaloids showed antimicrobial activity against selected terrestrial and marine microorganisms. Renierone (**308**) and *N*-formyl-1,2-dihydrorenierone (**312, 313**) both inhibit cell division in the fertilized sea-urchin egg assay (*324*). It is still unknown whether the renieramycins exhibit antitumor properties similar to those reported for the saframycins.

A Mediterranean sponge, *Aplysina* syn. *Verongia aerophoba* (Section V,B) gave strongly fluorescent 3,4-dihydroxyquinoline-2-carboxylic acid [**322,** mp 253–254°C(d)] in high yield (2,5% of dry weight) (*325*). The structure determination was done by a combination of chemical degradation, spectroscopic studies, and comparison with a synthetic sample prepared by reaction of 3-bromo-4-hydroxyquinoline-2-carboxylic acid with potassium hydroxide solution. The synthetic product is reported to have mp 261–262°C(d) (*326*). The authors speculate that the quinoline (**322**) is related to tryptophan metabolism, presumably via the kynurenine or kynurenic acid pathways.

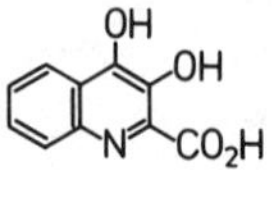

322

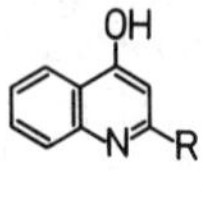

323 R = C_5H_{11}
324 R = C_7H_{15}

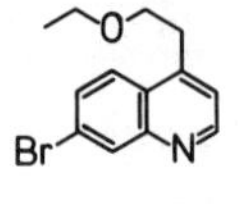

325

Two simple quinolines have been isolated as active parts of the antibacterial extracts of a yellow marine pseudomonad (*133*). The major component, 2-pentyl-4-quinolinol (**323,** mp 141–142°C) comprised 9% of the crude extract and was identified by spectroscopic analysis and by comparison with an authentic sample prepared from ethyl 3-oxooctanoate, aniline, and hydrochloric acid. The minor component, 2-hexyl-4-quinolinol (**324,** mp 146–147°C, 4% of crude extract), was identical to authentic material (*327*). The extracts also yielded 4-hydroxybenzaldehyde and two indoles (Section III,B).

The terrestrial bacterium *Pseudomonas aeruginosa* has yielded **324** (*328*).

A minor constituent of the alkaloid mixture originating with the marine bryozoan *Flustra foliacea* (L.) (Section III,E) was 7-bromo-4-(2-ethoxyethyl)quinoline (**325**) (*329*). The structure was determined by spectroscopic methods including ^{1}H-[^{1}H] NOEDS techniques. Since the natural product had been in contact with ethanol during the extraction procedure, the ethoxy group could not be unambiguously assigned to the natural product. If the ethoxy group has been introduced during the isolation procedure it must, however, have replaced a strikingly reactive group.

E. Other

The sponge *Dysidea herbacea* from The Great Barrier Reef, besides dysidin (**219**) has yielded several related compounds. Dysidenin {**326,** mp 98–99°C, $[\alpha]_D^{21}$ −98° ($CHCl_3$)} was identified by chemical and spectral methods (*330*). The

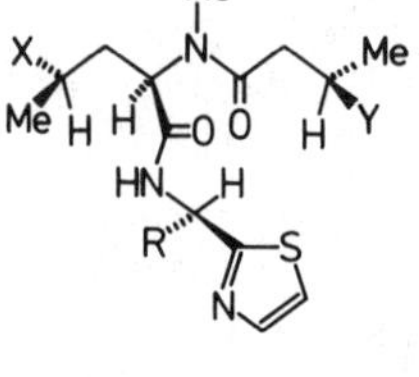

326 X = Y = CCl_3, R = Me
329 X = Y = CCl_3, R = H

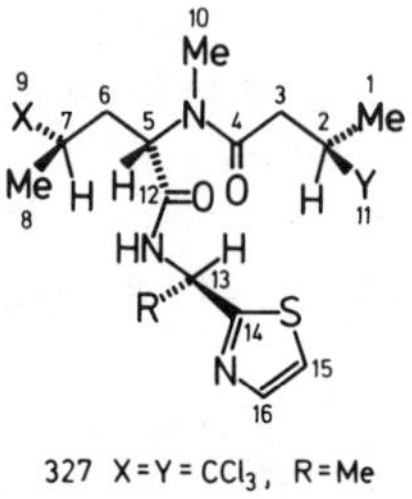

327 X = Y = CCl_3, R = Me
328 X = Y = CCl_3, R = H
330 X = CCl_3, Y = $CHCl_2$, R = H
331 X = $CHCl_2$, Y = CCl_3, R = H

absolute configuration was determined by chemical correlation (*331*). A sample of *D. herbacea* Keller, collected at the North Coast of New Guinea, in addition to small amounts of dysidenin yielded a new toxic stereoisomer, isodysidenin {**327,** >2% of dry weight, $[\alpha]_D^{22}$ +47° ($CHCl_3$)}, which on reduction with the diborane–THF complex, followed by methylation, gave a crystalline derivate suited for X-ray structural analysis (*332*). The result of this analysis defines isodysidenin with absolute configuration as depicted in **327** (2*R*,5*S*,7*R*,13*R*).

Later a Great Barrier Reef collection of the sponge gave four new alkaloids, namely, 13-demethylisodysedenin {**328,** $[\alpha]^{D}_{20}$ +52° ($CHCl_3$)}, 13-demethyldysidenin {**329,** $[\alpha]^{D}_{20}$ −97° ($CHCl_3$)}, 11-monodechloro-13-demethylisodysidenin {**330**, $[\alpha]^{D}_{20}$ +85° ($CHCl_3$)}, and 9-monodechloro-13-demethylisodysidenin {**331,** $[\alpha]^{D}_{20}$ +69° ($CDCl_3$)}, all isolated as colorless gums (*333*). The same organism has yielded the diketopiperazine derivative {**332,** mp 106–107°C, $[\alpha]_D^{20}$ −144° ($CHCl_3$)} in 1% of dry weight (*334*). All the alkaloids isolated may

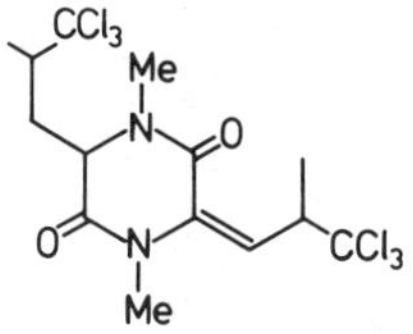

332

originate from chlorinated leucine. The thin incrusting sponge *D. herbacea* harbors blue-green algae, constituting at times half the cellular weight of the sample. Those symbionts may be responsible for the different pattern of metabolites identified in the various samples of the sponge (*334*).

The thiazole nucleus found in dysidin and the dysidenins also occurs in antineoplastic cyclic peptides from the marine tunicate *Lissodinum patella.*

Ulicyclamide {**333,** $[\alpha]_D^{25}$ +35.7° (CH_2Cl_2), 0.05% of dry weight} and ulithiacyclamide {**334,** $[\alpha]_D^{25}$ +62.4° (CH_2Cl_2), 0.04% of dry weight} were identified by a combination of chemical degradation and spectral analysis (*335*) as were patellamide A {**335,** $[\alpha]_D$ +113.9° (CH_2Cl_2)}, patellamide B {**336,** $[\alpha]_D$ +29.4° (CH_2Cl_2)}, and patellamide C {**337,** $[\alpha]_D$ +19° (CH_2Cl_2)} (*336*). The valine-derived part of the molecules are masked by oxazoline formation. Many marine ascidians harbor procaryotic algae (*337, 338*); however, their role in the synthesis of secondary metabolites is not known. Another tropical ascidian, *Didemnum ternatanum,* has yielded *N,N'*-diphenethylurea (*339*).

The herbivorous sea hare *Dolabella auricularia* has yielded a series containing thiazole-derived cyclic peptides, the dolastatins (*340*) with very potent anticancer activity. The structure of dolastatin 3 {**338,** mp 133–137°C, $[\alpha]_D^{26}$ −35.5° (MeOH)} in a yield of approximately 1 mg from 100 kg wet sea hare, has been published (*341*).

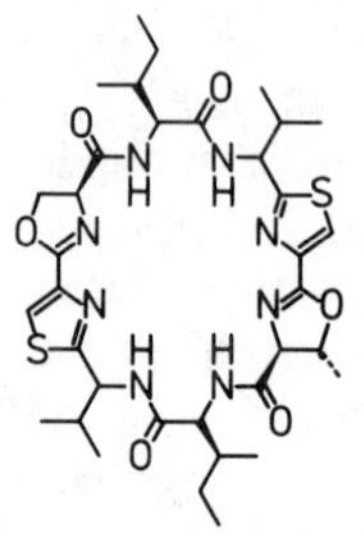

335

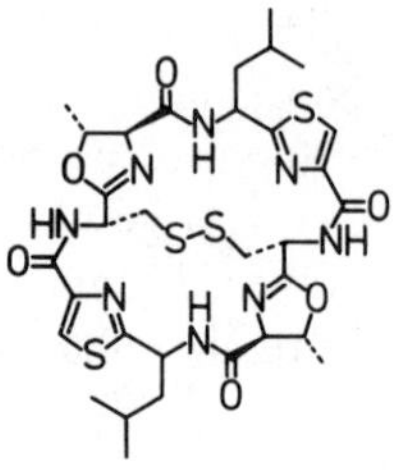

334

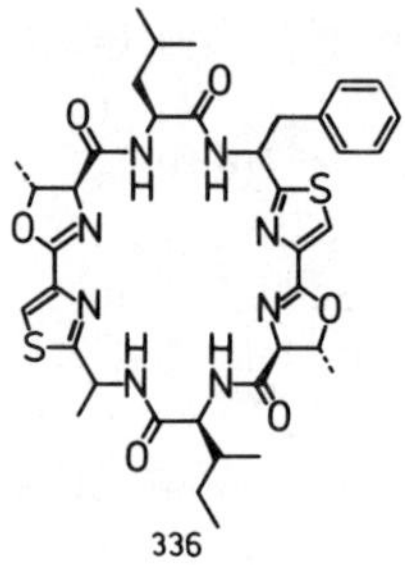

336

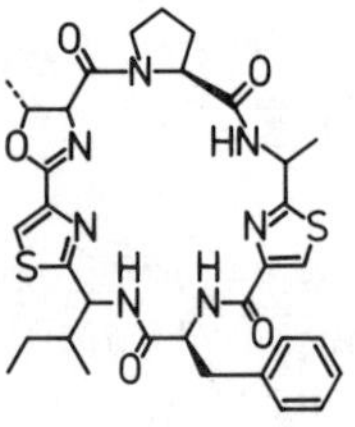

333

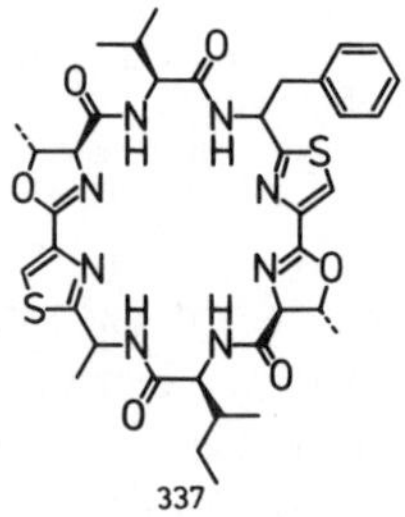

337

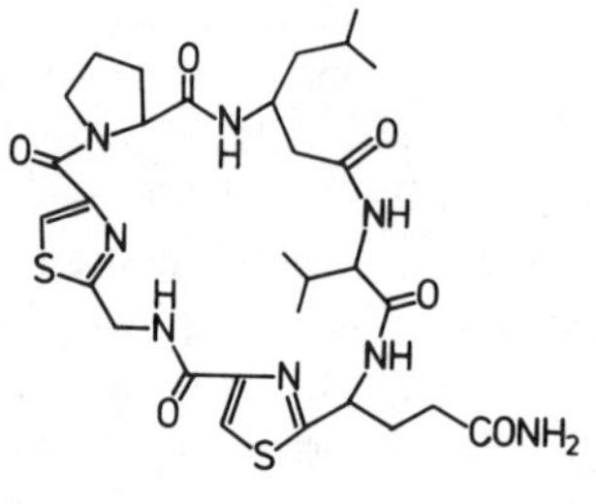

338

Macrolides containing 2-thiazolidone structural elements have been isolated from the red sponge *Latrunculia magnifica* Keller (*342*). Latrunculin-A {**339,** oil, $[\alpha]_D^{24}$ +152° ($CHCl_3$)} formed a crystalline acetal on methylation. X-ray structural analysis of this acetal gave the structure of latrunculin-A. Latrunculin-B {**340,** $[\alpha]_D^{24}$ +112° ($CHCl_3$)} was shown to be the 14-macrolide (**340**). Latrunculin-C is a stereoisomer of **339.** A biogenesis of the latrunculins is suggested. The latrunculins are strongly ichthyotoxic.

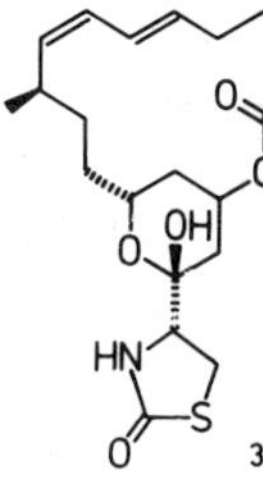

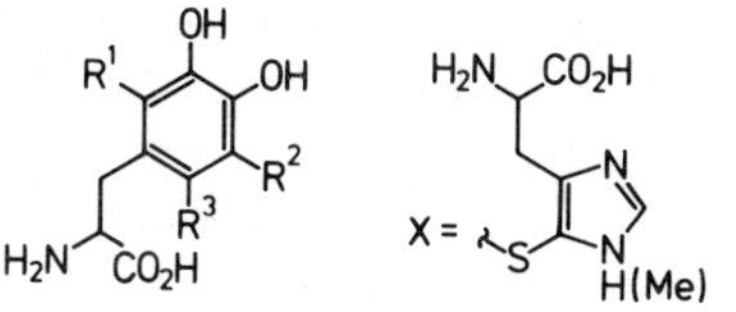

341 $R^1 = R^2 = X$, $R^3 = H$
342 $R^2 = R^3 = X$, $R^1 = H$
343 $R^1 = R^3 = X$, $R^2 = H$

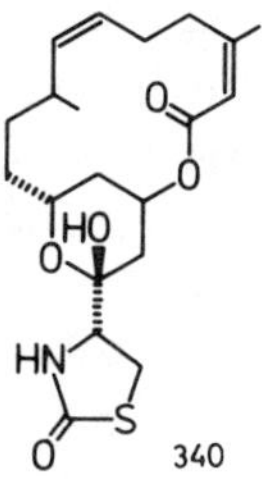

The iron-sequestering pigment adrenochrome of the branchial heart of the common octopus has been analyzed (*343*). Adrenochrome consists of a mixture of closely related peptides derived from glycine and the three amino acids adrenochromine A (**341**), B (**342**), and C (**343**). One of the building blocks of adrenochromines, namely, 1-methyl-5-mercapto-L-histidine has been isolated from the eggs of the sea urchin *Paracentrotus lividus* (*344*).

A unique sulfur-containing metabolite (**344,** mp 82–83°C) from the red tides

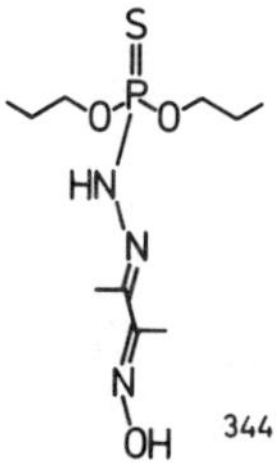

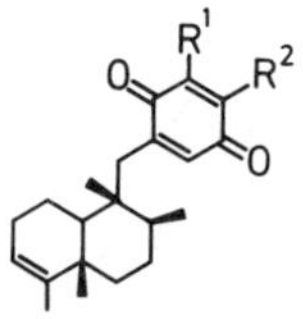

346 R^1=H, R^2=NHMe
347 R^1=NHMe, R^2=H

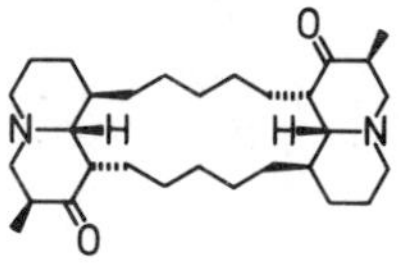

dinoflagellate *Gymnodinium breve* (*Ptychodiscus brevis*) yielded to X-ray crystallographic analysis (*345*). An ichthyotoxic novel type of bisquinolizidine alkaloid has been obtained from the sponge *Petrosia seriata* (*346*). Petrosin (**345**, mp 215–216°C) is devoid of optical activity. The structural assignment is the result of an X-ray diffraction analysis. The alkaloid was accompanied by at least three minor stereoisomers of unknown structures.

Another sponge, *Dysidea avara,* has yielded sesquiterpenoid aminoquinones (*347*). The structures of **346** (mp 160–163°C) and **347** (mp 153–155°C) were assigned on the basis of spectral analysis. The syntheses of **346** and **347** were accomplished from the corresponding quinone (**346,** $R^1 = R^2 = H$, avarone) obtained from the hydroquinol, avarol, present in the sponge in large amounts, by oxidation. Avarone on reaction with methylamine formed a mixture of **346** and **347.** The aminoquinones were found only as trace constituents in fresh material, while large amounts were recovered when the sponge was exposed to air and further extracted. These findings suggest that **346** and **347** may be artefacts formed during workup of the sponge material. The aminoquinones possess the interesting ability to induce developmental aberrations in sea-urchin eggs.

Many marine organisms have water-soluble substances with absorption maxima between 310 and 340 nm. In several cases the identities of such substances have been revealed as mycosporin derivatives. The structural diversity is exemplified by the following characteristic examples. Mytilin A (**348**) and mytilin B (**349**) were isolated as a 3:1 mixture from the edible mussel *Mytilus galloprovincialis* (*348*). These amino-acid conjugates (glycine, serine, and threonine)

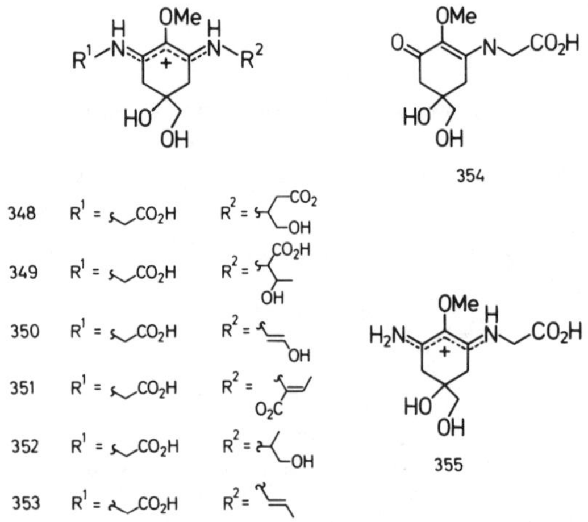

348 R^1 = CO_2H R^2 = CO_2 OH

349 R^1 = CO_2H R^2 = CO_2H OH

350 R^1 = CO_2H R^2 = OH

351 R^1 = CO_2H R^2 = O_2C

352 R^1 = CO_2H R^2 = OH

353 R^1 = CO_2H R^2 =

354

355

have λ_{max} 334 nm (H_2O). Asterina 330 (**350**) was obtained from the starfish *Asterina pectinifera* (*349*) and **351,** absorbing at 337 nm, from the ascidian *Halocynthia raretzi* (*350*). The red alga *Trichocarpus crinitus* contains **348** (*351*). The zoanthid *Palythoa tuberculosa* gave palythinol {**352,** $[\alpha]_D$ $-51.9°$ (H_2O), λ_{max} 332 nm (H_2O)} and palythene {**353,** $[\alpha]_D$ $-30.1°$ (H_2O), λ_{max} 360 nm (H_2O)} (*352*). The same organism had earlier yielded mycosporine-Gly [**354,** λ_{max} 310 nm (H_2O)] (*353*) and palythine [**355,** λ_{max} 320 nm (H_2O)] (*354*).

Certain members of the genus *Palythoa* have attracted the attention of natural product chemists for more than the last 10 years because they contain one of the most potent toxins known, palytoxin (*355*). The structure elucidation of palytoxin has been one of the greatest challenges in marine structural research. After many years of painstaking efforts, the collection of information about this unique toxin culminated in the publication of the total structure depicted in **356** for palytoxin from Hawaiian *Palythoa toxica* (*356*). Slightly different structures have been assigned to palytoxin from other *Palythoa* sp. (*356*). Taking advantage of the absolute stereochemistry determined by the Hirata group on degradation products of Okinawan palytoxin from *P. tuberculosa,* the Moore group assigned the absolute stereochemistry of 60 of the 64 chiral centers of the α anomer of palytoxin (*357*). A leading reference to the work of the Hirata group is given by Vemura *et al.* (*358, 359*).

356

Interestingly enough, a cultured *Vibrio* sp. from Hawaiian *P. toxica* seems to produce palytoxin (*357*).

A stereospecific synthesis of the C-30–C-43 segment of palytoxin has been published (*360*).

Polyamines have been isolated from marine organisms, e.g., the acarnidines (**75–77**), **357** and **358** from the soft coral *Sinularia brongersmai* (*361*), and **359** and **360** from a *Sinularia* sp. (*362*).

357

358

359

360

361 $R^1=R^2=H$
362 $R^1=H$, $R^2=OMe$
363 $R^1=Me$, $R^2=OH$

Another soft coral, *Sinularia flexibilis,* has yielded the tyramides **361–363** (*363*). A review of new chemistry of naturally occurring polyamines has appeared (*364*).

VI. Concluding Remarks

As evidenced by the number of structures determined, the field of marine alkaloids is firmly established and rapidly expanding. At present, a temporary segregation of this discipline seems justified on grounds of the uniqueness of most marine alkaloids as compared to alkaloids from terrestrial sources. Whether there are distinct features marking alkaloids as being of marine origin must await further research in this area. Certain structural types, e.g., the brominated alkaloids, are indicators of their origin. They may, however, represent only minor deviations in known metabolic pathways.

In those cases where the marine metabolites resemble terrestrial counterparts, the latter are nearly always microbial constituents. As repeatedly noted in this text, there is ample circumstantial evidence that many if not most of the structures treated here may be metabolites of an associated flora. Unfortunately, marine microorganisms are incompletely investigated. Marine microbiology is a rapidly growing discipline (*365*) and it is to be hoped that the knowledge now accumulating may prove valuable in interpreting the results of marine natural products chemistry.

Bearing in mind the uncertainty about the origin of the compounds, the attempts of construction of chemotaxonomical systems based on secondary metabolite analysis must be viewed with suspicion. The knowledge of marine ecology must be further advanced before a sound basis for this kind of chemotaxonomy may emerge.

An important offcast of classical natural products chemistry has been the identification of new pharmacologically valuable compounds and compounds with economic potential. In the case of marine products, the bioactive effects found have expanded greatly our knowledge of several basic phenomena in physiology, although few compounds have as yet proved economically important. As a whole, the impact of marine natural products research has occurred primarily on the intellectual level in widening the borders of imagination, especially in organic chemistry.

Many specific problems remain to be solved in marine alkaloid research ranging from the isolation and structure elucidation of the causative agent of ciguatera poisoning (*17*) to the determination of the structure of a paramagnetic violet pigment, $C_{21}H_{22}N_4O_5$, from the sea anemone *Calliactis parasitica* (*366*).

VII. Addendum

Since the conclusion of the present review a series of important investigations have appeared. The material is arranged according to the main sections of the text.

Section I. Introduction

A review on metabolites of marine algae and herbivorous marine molluscs covering the literature published between 1977 and October 1983 has appeared (*367*).

Section II. Guanidine Alkaloids

The search for the origin of tetrodotoxin (**6,** Section II,B) continues. For example it has been demonstrated that tetrodotoxin from the trumpet shell "boshubora" *Charonia sauliae* (*55*) originates from ingested starfish *Astropecten polyacanthus* (*368*). The origin of the starfish tetrodotoxin is still unaccounted for. A study of the local variation of toxicity of the puffer fish *Fugu niphobles* around Japan has appeared (*369*). Based on a marked local variation of toxicity, the authors are inclined to believe in an exogenous origin of the toxin and conclude that the Inland Sea of Seto may be a suitable field to search for the origin of the toxin.

Owing to the wide application of tetrodotoxin (**6**) and derivatives as neurophysiological tools, development of convenient synthetic routes to these compounds continues to be of interest, e.g., a series of highly functionalized pyrimidinones (*370*) and hydroquinazolines (*371*) have been prepared.

An investigation of the restibility of toxic and nontoxic crabs against tetrodotoxin and paralytic shellfish poison (Section II,C) has appeared (*372*) as has an investigation of the toxins of mussels infested with *Protogonyaulax catenella* from Senzaki Bay (*373*). Anatomical distribution and profiles of the toxins in highly infested scallops (*374*) and paralytic shellfish poison in sea scallops *Placopecten magellanicus* (*375*) have been subjects of study. Interestingly enough, a calcareous red alga has been identified as the primary source of paralytic shellfish toxins in coral reef crabs and gastropods (*376*).

Key intermediates in the total synthesis of saxitoxin (**7**) and gonyautoxins II (**9**) and III (**10**) have been prepared by an improved Blaise reaction (*377*). The structures of neosaxitoxin (**8**) and gonyautoxin II (**9**) were confirmed by ^{15}N-NMR studies of enriched samples obtained from ^{15}N-nitrate-fed *Gonyaulax tamarensis* (*378*).

The chemistry of imidazole alkaloids has recently been reviewed (*379*). The reader is referred to this review for a detailed discussion of the zoanthoxanthins (Section II,D) and oroidin and dibromophakellin (Section II,E). Compounds **49** and **50** have been reisolated from the Okinawan marine sponge *Hymeniacidon aldis* De Laubenfels in 2.1 and 0.3% yields, respectively (*380*). Hymenialdisine [**49,** yellow needles, mp 160–164°C(d)] was identified by X-ray crystallographic analysis and debromohymenialdisine [**50**, yellow needles, mp 171–180°C(d)] by spectroscopic analysis.

The revised structures of the polyandrocarpidines (**50–62,** Section II,E) have been confirmed, and hexahydropolyandrocarpidine I has been synthesized by condensation of 4-oxododecanoic acid and 1-amino-5-guanidinepentane, followed by reduction (H_2, 10% Pd–C) (*381*).

The total synthesis of (±)-ptilocaulin [(±)-**73,** section II,E] has been published (*382*). Scheme 12 summarizes the synthetic procedure, which starts with

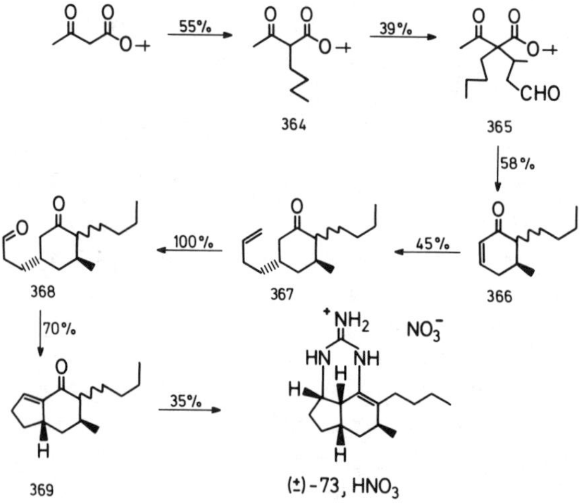

SCHEME 12

the alkylation of the anion of *tert*-butyl acetoacetate by butyl iodide in dioxane to yield the acetoacetate (**364**). Addition of crotonaldehyde to the sodium salt of **364** at −40°C in methanol gave a moderate yield of **365** as a mixture of stereoisomers. Cyclization and decarboxylation of **365** was effected by acid catalysis at room temperature, yielding a 1.7:1 trans–cis mixture of **366.** Reaction between **366** and the cuprate prepared from 3-butenylmagnesium bromide and $CuBr{\cdot}SMe_2$ left a 1.7:1 mixture of isomers (**367**), quantitatively transformed to **368** on ozonolysis. Cyclization to a 1:1 mixture of *cis-* and *trans-***369** was smoothly effected by hydrogen chloride in tetrahydrofuran. Azeotropic removal of water from a refluxing benzene solution of guanidine and *cis-***369** gave, after quenching with nitric acid, (±)-ptilocaulin nitrate identical to the natural product as shown by IR, ^{1}H-NMR, and MS comparison. Similar results were obtained with *trans-***369.**

An improved synthesis of (±)- and (−)-ptilocaulin (**73,** Section II,E) established the absolute stereochemistry of natural (+)-**73** (*383*). Another total synthesis of (−)-ptilocaulin was published simultaneously (*384*). A new addition to

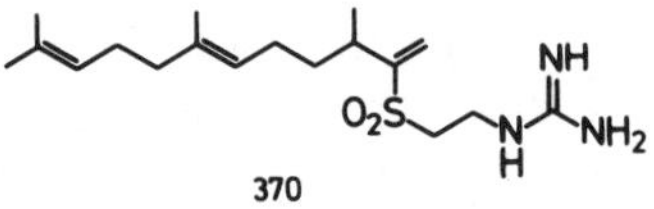

370

Section II,E is the antispasmodically active Agelasidine-A (**370**) isolated as the hydrochloride [mp 108–108.5°C, $[\alpha]_0^{25}$ + 19.1 (MeOH)]. An Okinawan sponge *Agelas* sp. yielded this unusual sesquiterpene taurocyamine derivative in 0.034% (wet weight) (*385*). The same compound (**370**) was later isolated (0.16% dry weight) from another *Agelas* sp. collected at Argulpelu Reef, Palau (*386*).

A general synthesis of the acarnidines (**75, 76,** and **77,** Section II,E) has been developed (*387*). Scheme 13 illustrates the preparation of naturally occuring 3,5-acarnidine (**75**). The amido alcohol (**371**) was prepared from 3-aminopropanol and 3,3-dimethylacryloyl chloride. Oxidation of **371** to the amido aldehyde (**372**) is difficult but proceeds in excellent yield, using dimethyl sulfoxide–oxalyl chloride at −70°C. Amido aldehyde **372** is unstable and was directly condensed with mono-BOC-protected 1,5-diaminopentane in chloroform over molecular sieves (4 Å) for 45 min, followed by sodium borohydride reduction of the aldimine (**373**), yielding 71% of the secondary amine (**374**). The mono-BOC-protected 1,5-diaminopentane was prepared in good yield through the use of (*S*)-(*tert*-butyloxycarbonyl)-4,6-dimethyl-2-mercaptopyridine. Lauroyl chloride

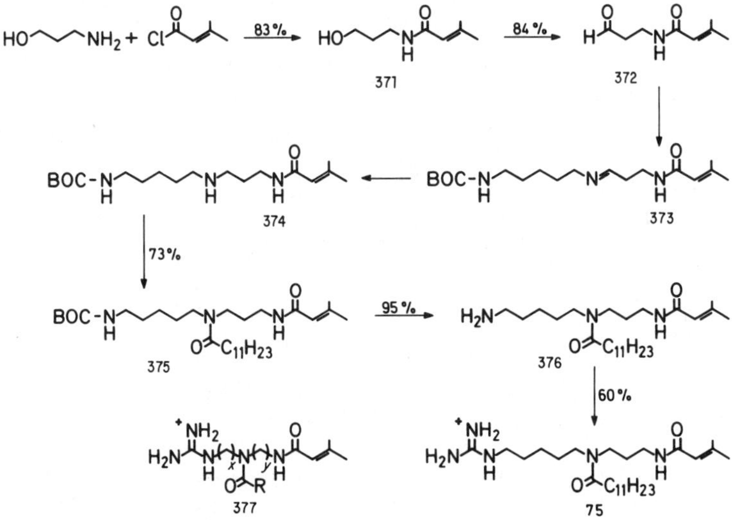

Scheme 13

acylated **374** to form **375,** which on treatment with trifluoroacetic acid generated the protonated amine (**376**). Neutralization of the ammonium salt of **376** gave the free amine (**376**), which was immediately treated with (*S*)-methylisothiouronium iodide to give the 3,5-acarnidine hydroiodide (**75**).

A number of synthetic acarnidines with varying chain lengths (**377,** $x = 2$–6, $y = 2$–5) were prepared by the same procedure.

Siphonodictidine (**378,** Section II,E), present in 1.06% of dry weight, is the major secondary metabolite of an undescribed Indo-Pacific sponge, *Siphonodictyon* sp. (*388*). The structure was determined from spectral data. This sponge

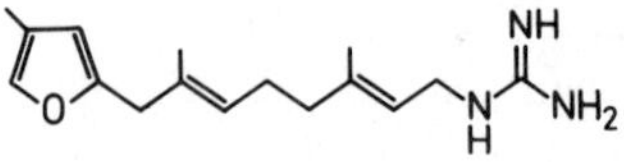

378

burrows deep into living coral heads, leaving only the oscular chimney exposed. The toxin (**378**) seems to serve to inhibit coral growth since there is always a 1–2-cm zone of dead coral polyps around the base of each oscular chimney. At a concentration of 100 ppm, acute toxicity was observed against the hard coral *Acropora formosa* in the course of 5–30 min. Toxin **378** inhibits the growth of *Staphylococcus aureus, Escherichia coli, Candida albicans, Pseudomonas aeruginosa,* and *Bacillus subtilis*. It did not inhibit the growth of two marine bacteria or cell division in fertilized sea-urchin eggs.

Section III. Indole Alkaloids

The sea hare *Aplysia dactylomela,* collected near La Parguera, Puerto Rico, has yielded 0.0004% of 2,3,5-tribromo-*N*-methylindole (**91,** mp 124–124.8°C, Section III,B). This compound was originally isolated from the red alga *Laurencia brongniartii* and may thus appear in the sea hare from its diet (*389*). Cytotoxicity testing of **91** showed no spectacular activity [ED_{50} 47 μg/ml for *in vitro* lymphocytic leukemia PS (P388) tissue culture].

Important additions to the simple indoles (Section III,B) are the four brominated methylthioindoles **379–382** secured from a Taiwanese collection of *Laurencia brongniartii* (**390**). These compounds were also found in the Okinawan red alga *Laurencia grevilleana,* which may be synonymous with *L. brogniartii* (*391*).

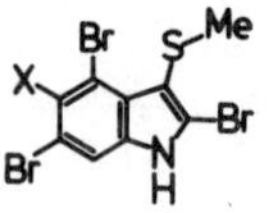

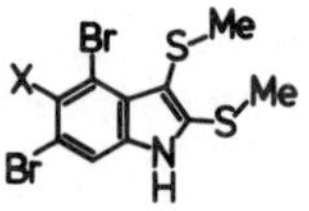

379 X=H
380 X=Br

381 X=H
382 X=Br

Estimates of biomass of the acorn worm *Ptychodera flava* Eschscholtz and excretion rate of metabolites have appeared (*392*).

New indole alkaloids (Section III,C) have been isolated from the red alga *Martensia fragilis* (family Delesseriaceae), namely, fragilamide {**383,** $[\alpha]_D$ $-32°$ (MeOH)}, martensine A {**384,** $[\alpha]_D$ $+42°$ (MeOH)}, and martensine B {**385,** mp 184–186°C, $[\alpha]_D$ $-18°$ (Me_2CO)}. Also, 10-epimartensine A (**386**) is

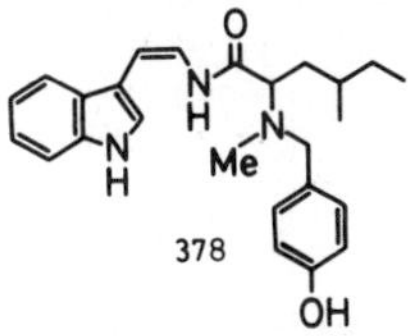

378

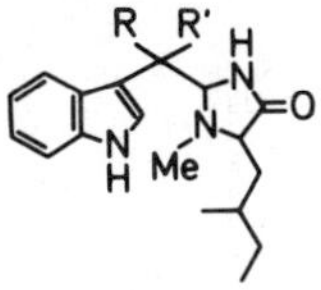

383 R = OH, R' = H
385 R = H, R' = OH

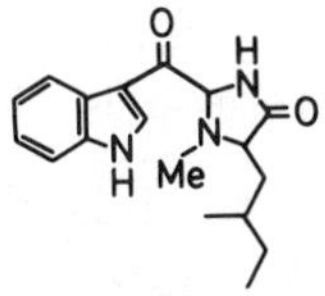

384

a minor alkaloid in the alga (*393*). The structures were solved by a combination of spectroscopic and chemical methods. Martensine A (**384**) shows antibiotic activity against *Bacillus subtilis, Staphylococcus aureus,* and *Mycobacterium smegmatis*.

A series of tryptophan related indoles (Section III,C) has been isolated and identified from the Carribean collonial tunicate *Eudistoma olivaceum* (*394, 395*). Eudistomins C (**387**), E (**388**), K (**389**), and L (**390**) all encompass the unprecedented oxathiazepine ring. Another member of this assembly, eudistomin F, with

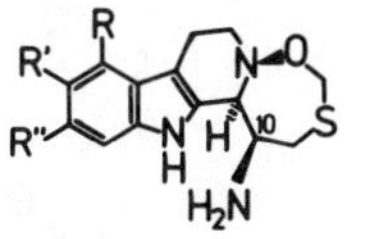

387	R = H	R' = OH	R'' = Br
388	R = Br	R' = OH	R'' = H
389	R = H	R' = H	R'' = Br
390	R = H	R' = Br	R'' = H

a substituent ($C_2H_3O_2$) at the amino group at position 10, was also isolated (*394*). Together with these interesting antiviral alkaloids an additional 11 β-carbolines, eudistomins A (**391**), D (**392**), G (**393**), H (**394**), I (**395**), J (**396**), M (**397**), N (**398**), O (**399**), P (**400**), and Q (**401**) were isolated and identified (*395*). The structure elucidations resulting from spectroscopic investigations. The β-carbolines showed modest antiviral activity.

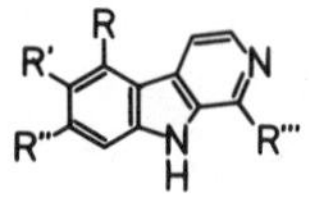

391	R=H	R'=OH	R''=Br	R'''=2-pyrrolyl
392	R=Br	R'=OH	R''=H	R'''=H
393	R=H	R'=H	R''=Br	R'''=1-pyrridin-2-yl
394	R=H	R'=Br	R''=H	R'''=1-pyrridin-2-yl
395	R=H	R'=H	R''=H	R'''=1-pyrridin-2-yl
396	R=H	R'=OH	R''=Br	R'''=H
397	R=H	R'=OH	R''=H	R''= 2-pyrrolyl
398	R=H	R'=Br	R''=H	R'''=H
399	R=H	R'=H	R''=Br	R'''=H
400	R=H	R'=OH	R''=Br	R'''=1-pyrridin-2-yl
401	R=H	R'=OH	R''=H	R'''=1-pyrridin-2-yl

The second example of the isolation of indole alkaloids from a marine bryozoan (Section III,E) have been reported (*396*). A collection of *Zoobotryon verticillatum* (Delle Chiaja, 1828) from San Diego Harbor yielded 0.0075% of dry weight 2,5,6-tribromo-*N*-methylgramine (**402**, mp 112–113°C) and 0.0114% of 2,5,6-tribromo-*N*-methylgramine *N*-oxide [**403**, mp 116–120°C(d)].

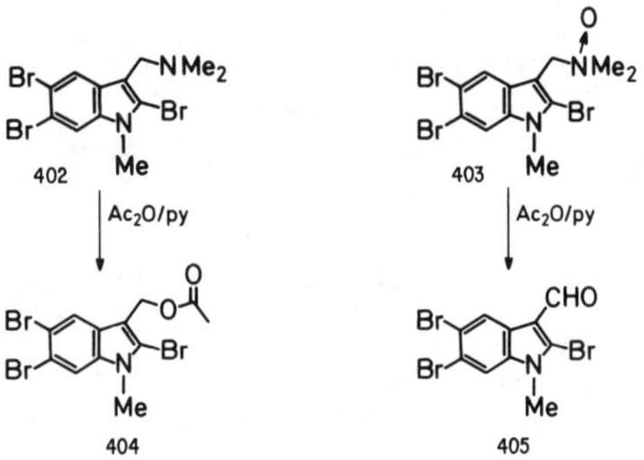

The structures were obtained by spectroscopic analysis. Oxidation of **402** with hydrogen peroxide generated **403,** which in turn on reduction gave **402.** The compounds were synthesized by N-methylation of gramine (75% yield), followed by bromination to 2,5,6-tribromo-*N*-methylgramine (**402**) in 15% yield. Acetylation of **402** resulted in displacement of the side-chain dimethylamino group to yield the acetate (**404**), while **403** under the same conditions gave the aldehyde (**405**).

Initial tests indicated **402** to inhibit cell division of the fertilized sea-urchin egg ($ED_{50} \simeq 16$ μg/ml).

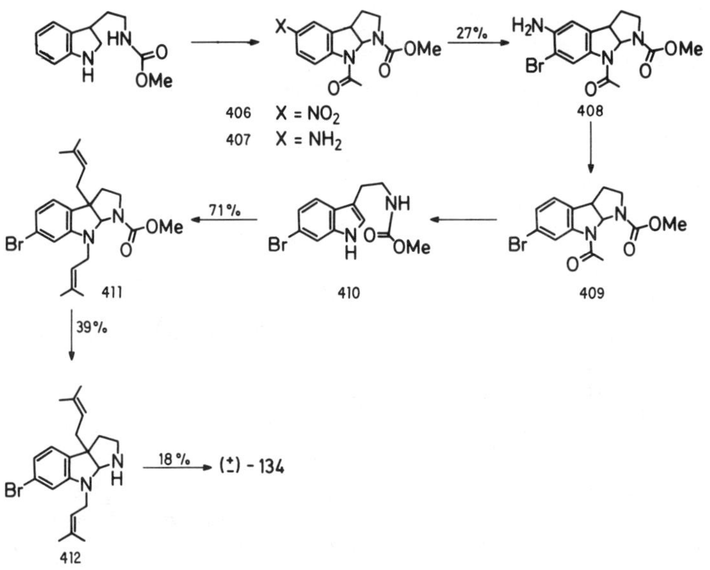

SCHEME 14

Synthesis of (±)-flustramine B [(±)-**134**] has been completed (*397*) according to Scheme 14 (Section III,E). Catalytic reduction of the 5-nitropyrroloindole (**406**) prepared from N_b-methoxycarbonyltryptamine gave the 5-amino derivative (**407**). Bromination with *N*-bromosuccinimide in dimethylformamide yielded the 5-amino-6-bromo derivative (**408**), which was deamineated with isoamyl nitrite in tetrahydrofuran, giving a combined yield of **409** from **406** of 60%. N_b-Carbomethoxy-6-bromotryptamine (**410**) was prepared in excellent yield from **409** by treatment with 10% sulfuric acid in methanol. Dimethylallyl bromide (10 equivalents) in an acetate buffer (pH 2.7) at room temperature transformed the tryptamine (**410**) to the diprenylated derivative (**411**), which with boiling 10% sodium hydroxide–ethanol (100 hr) was hydrolyzed to give **412.** Methylation with methyl iodide in potassium carbonate–acetone at room temperature gave (±)-**134,** while **412** was recovered in 57% yield. Reduction of **411** with lithium aluminum hydride in boiling dioxane, as expected, gave debromoflustramine B.

In connection with the discussion (Section III,E) of microorganisms in bryozoans, it may be noted that mycoplasma-like organisms have been found in larvae and adults of the marine bryozoan *Watersipora cucullata* (*398*). The consequences to the bryozoan of this association are at present unknown.

Section IV. Pyrrole Alkaloids

A series of bipyrroles (Section IV,B) have been obtained from the nembrothid nudibranchs *Roboastra tigris* Farmer 1978, *Tambje eliora* (Marcus and Marcus,

1967), and *Tambje abdere* Farmer 1978 (*399*). The same bipyrroles, tambjamine A (**413**), B (**414**), C (**415**), and D (**416**), all oils, were also isolated from a green bryozoan, *Sessibugula translucens* Osburn 1950.

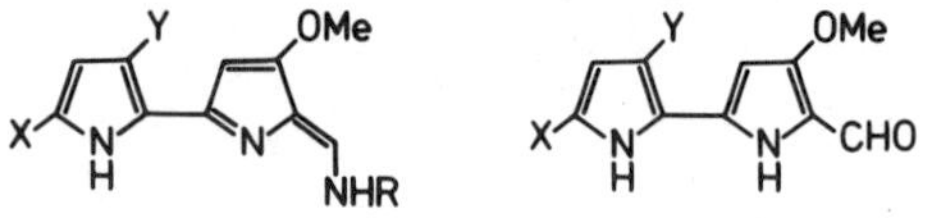

413 X = H , Y = H , R = H
414 X = Br, Y = H , R = H
415 X = H , Y = H , R = *i*-Bu
416 X = H , Y = Br, R = *i*-Bu

417 X = H , Y = H
418 X = Br, Y = H
419 X = H , Y = Br

During the isolation procedure, the tambjamines to some extent underwent hydrolysis, releasing the aldehydes **417–419** from the corresponding enamines. As a result of chemical and spectroscopic studies of **413–416** and aldehydes **417–419,** one of which (**417**) was a known compound, the structures of the tambjamines were determined. Seemingly, these metabolites exhibit complex ecological roles. The tambjamines are the major secondary metabolites of the bryozoan (0.45% of dry weight).

T. eliora and *T. abdere* presumably owe their content of tambjamines to a dietary source, namely, *S. translucens*. The contents in the two animals were 2.15 and 3.42% of dry weight, respectively. The *Tambje* species lay down a slime trail containing tambjamines, which can be followed by the large carnivorous predator *R. tigris*. This slime trail is presumably used to repel most potential predators; only the special predator *R. tigris* can detect the chemicals in the trail and use them to track the preferred prey. Attacked by *R. tigris, T. abdere* produces a yellow mucus from goblet cells in the skin. This defensive secretion often causes *R. tigris* to break off its attack and is presumably responsible for the preference of *T. eliora* rather than *T. abdere* as prey. *T. eliora* produces no defensive secretion.

Whether the tambjamines are metabolites of the bryozoan or some associated flora is at present unknown. It is, however, interesting to note that red marine bacteria of the genus *Benechea* (*400*) produce prodigiosin (**420**). Also, the red marine bacterium *Alteromonas rubra* produces prodigiosin (**420**) and cycloprodigiosin (*421*). The structure of the latter compound has recently been shown to be **421** on the basis of ^{1}H-NMR studies (*402*).

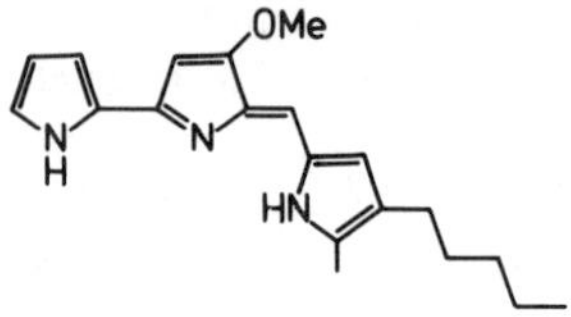

420

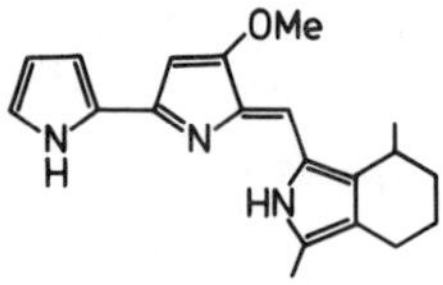

421

Since the tambjamines turn green on standing and during chromatography, the green pigments of the bryozoan are suspected to be dimers of the bipyrroles and thus related to the blue pigment **230** from a compound ascidian (*249*). The function of the tambjamines in the bryozoan is unknown; however, it is noteworthy that **413** and **414** inhibit cell division in the fertilized sea-urchin egg assay and show moderate antimicrobial activity against *Escherichia coli, Staphylococcus aureus, Bacillus subtilis,* and *Vibrio anguillarum.* The compounds **415** and **416** are more active than **413** and **414** in the above-mentioned assays and also show activity against *Candida albicans.* The tambjamines may thus have a function in controlling the symbionts in the bryozoan and also may act as antifouling agents.

An elegant synthesis of α-allokainic acid (**176**) has appeared (*403*) (Section IV,C). The mechanism of kainic acid neurotoxicity has been discussed (*404*).

From an intimate association between a soft coral, *Telesto* sp., and an unidentified sponge, 5-nonylpyrrole-2-carbaldehyde (**216,** $n = 8$, Section IV,F), has been identified and the structure confirmed by synthesis (*405*). The natural product was present in 0.11% of the dry weight of the associate organisms, but it was not possible to identify the source of the compound since it appeared to have perfused both organisms.

The flavonoidal alkaloid, phyllospadine (**422**), was isolated from the sea-grass *Phyllospadix iwatensis* collected in Yamagata Prefecture, Japan (*406*).

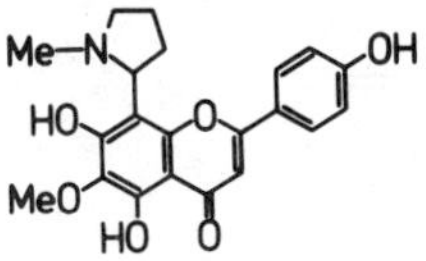

422

The complete structure determination of bonellin (**226,** Section IV,F) has been published (*407*). Bonellin [mp ~300°C(d), hydrochloride mp 236–240°C] was isolated as the dimethyl ester from female specimens collected in Marsaxlokk Bay, Malta. The structure was determined as depicted in **423** as a result of extensive spectroscopic studies coupled with chemical analysis and X-ray structural determination of anhydrobonellin methyl ester. Bonellin is accompanied by

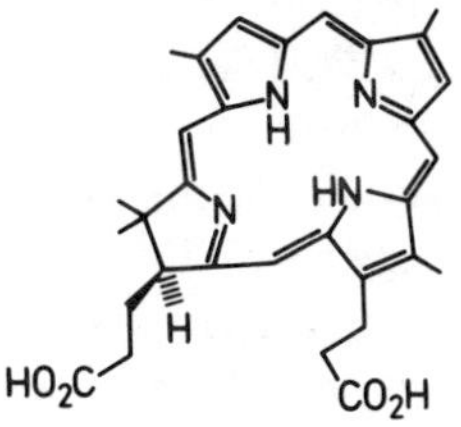

423

a series of mono-amino acid conjugate derivatives. A total synthesis of (±)-bonellin dimethyl ester [(±)-**423** dimethyl ester] has been reported (*408*).

A review of the biosynthesis of marine metabolites covers bile and *Aplysia* pigments (*409*).

Section V. Miscellaneous Alkaloids

A number of compounds have been reported from the sponge *Verongia aerophoba* collected in the bay of Naples (see Section V,B and ref. *38*). These reports rest on a misidentification, since direct comparison of the sponge with true *V. aerophoba,* collected from an area near Gallipoli, has revealed the previously examined sponge to be *Verongia cavernicola* (*410*). The species *Verongia aerophoba* (= *Aplysina aerophoba* Schmidt, 1862) and *Verongia cavernicola* Vacelet, 1959 are widespread along the Italian coasts.

The true *V. aerophoba* contains dienone **238,** aeroplysinin-1 (**254**) in addition to isofistularin-3 {$[\alpha]_D$ +108° (MeOH), 0.12%}, aerophobin-1 {$[\alpha]_D$ +187° (MeOH), 0.03%}, and aerophobin-2 {$[\alpha]_D$ +139° (MeOH), 0.08%}. Percentages are based on dry weight after extraction. Spectral studies of the metabolites, their acetylated derivatives, and hydrolysis products resulted in the assignment of

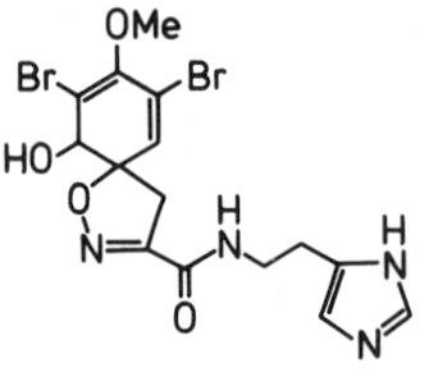

424

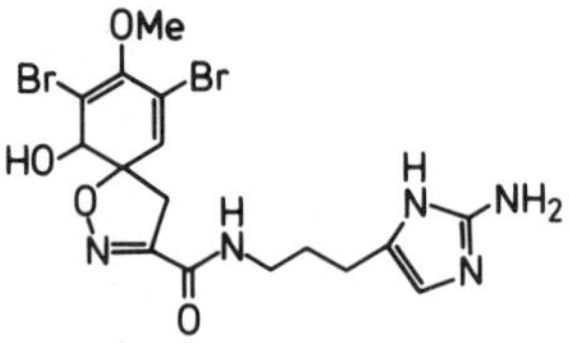

425

structure **424** to aerophobin-1 and **425** to aerophobin-2. Isofistularin-3 is isomeric with fistularin-3 (**263**) and can differ from this compound only in the configuration of one or more of the chiral centers. Isofistularin-3 is cytotoxic *in vitro* (KB cells), and the effective dose is 4 μg/ml (*410*).

Total syntheses of (±)-aerothionin [(±)-**259**] and (±)-homoaerothionin [(±)-**260**], summarized in Scheme 15, have been reported (*411*). The azlactone derived from 2-benzyloxy-3,5-dibromo-4-methoxybenzaldehyde was converted to **426** by successive treatments with potassium hydroxide, hydroxylamine, and benzyl chloride in 35% overall yield. Reaction with potassium carbonate followed by hydrogenolysis gave **427.** Dienone **428** was generated from **427** and thallium(III) trifluoroacetate in trifluoroacetic acid. Excess $Zn(BH_4)_2$ transformed **428** to **429** and 40% yield of the cis isomer. Reaction of **429** with diamines at room temperature gave (±)-**259** and (±)-**260,** respectively.

The full paper dealing with the structure elucidation of psammaplysin A (**264**) and psammaplysin B (**265,** Section V,B) have appeared (*412*). Apart from the detailed description of the structure elucidation of **264** and **265,** a total ^{13}C-NMR

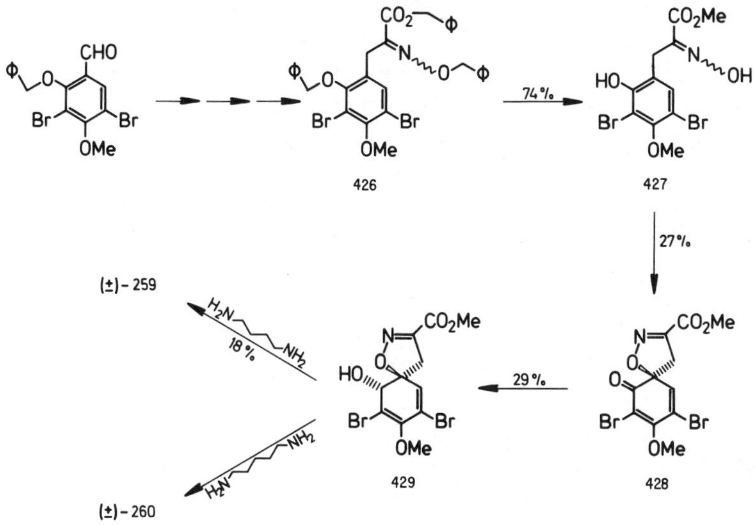

SCHEME 15

line assignment is presented for these alkaloids and for aerothionin (**259**), isolated from the Red Sea sponge *Denrilla praetensa,* as well. A biogenetic scheme for the syntheses of **264** and **265** starting from dibromotyrosine, is discussed.

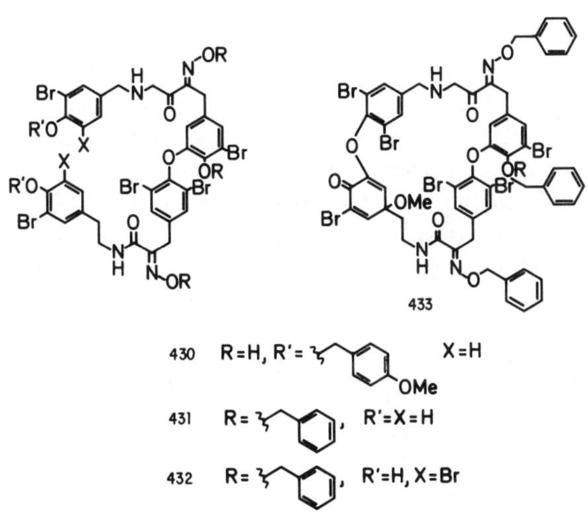

A synthesis of bastadin-6 (**271,** Section V,B) from bastadin-2 (**267**) has been described (*413*). Tribenzylbastadin-2 (**431**) was obtained in 82% yield by benzylation, followed by deprotection of **430** (*414*). Bromination of **431** afforded an 82% yield of dibromobastadin-2 tribenzyl ether (**432**), which on oxidation gave two macrocyclic dienones, one of which (**433,** 13% yield) gave the corresponding tribenzyl ether (tribenzyl-**271**) in 61% yield on reduction. Bastadin-6 (**271**) was obtained in 74% yield from the latter compound on hydrogenolysis.

Bastadin-6 trimethyl ether has been synthesized (*415*).

A report of the synthesis of 3-methylnavenone-B (**296**) has appeared (*416*).

The yellow zoochrome of the sponge *Verongia aerophoba* (Schmidt) has yielded to modern isolation and structure elucidation techniques (*417*). The unstable yellow pigment from which the sponge derives its name (*aerophoba* = air fearing) was shown to be 3,5,8-trihydroxy-4-quinolone (**434**). On exposure to air **434** rapidly turns navy blue, then red purple, and finally black. Since a nitrogen atmosphere prohibited the formation of the blue color, this is presumably due to formation of the quinone, which could be reduced to **434** with ascorbic acid.

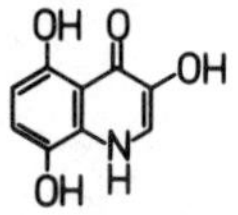

434

The isoquinolinequinone antibiotics from microorganisms keep expanding in numbers. A recent broad spectrum antimicrobial and antitumor agent is exemplified by cyanocycline A (**435**) from *Streptomyces flavogriseus* strain No. 491 (*418*).

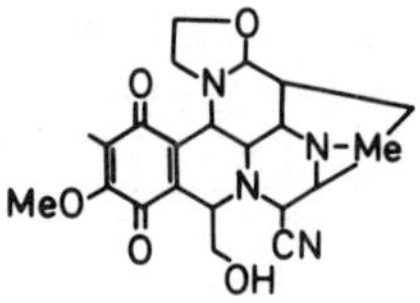

435

As mentioned in Section V,E the absolute configuration of dysidenin (**326**) and isodysidenin (**327**) was derived from an X-ray structural determination. Recently, however, it was shown by degradation of the thiazole nucleus with singlet oxygen followed by hydrolysis and GC analysis of the resulting alanin as the ethyl ester *N*-trifluoroacetate that dysidenin and isodysidenin actually had the opposite configurations of the ones represented by **326** and **327** (*419*). Dysidenin thus has the stereochemistry depicted in **436** and isodysidenin in **437.**

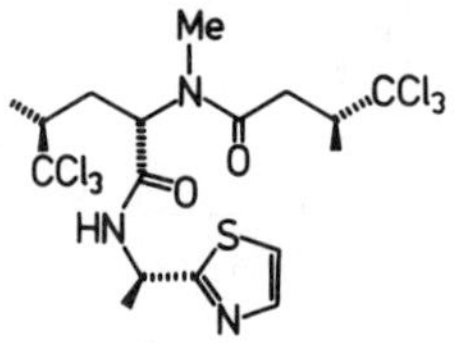

436 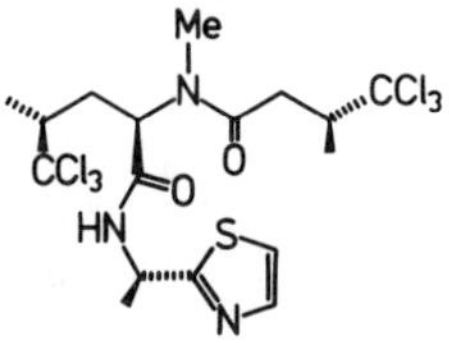 437

These examples should serve as a warning to the natural product chemist that although X-ray structure determination is still the single most reliable structure determination technique, the results can not always be relied on dogmatically. An excellent account of some of the pitfalls open for the noncrystallographer in this area has recently appeared (*420*).

Since the absolute configuration of 13-dimethyliso-dysidenin (**328**), 13-dimethyldysidenin (**329**), 11-monodechloro-13-dimethylisodysidenin (**330**), and 9-monodechloro-13-demethyl-isodysidenin (**331**) were all based on comparison with dysidenin and isodysidenin, the absolute configuration of the former derivatives is presumably the opposite of the ones depicted in **328, 329, 330,** and **331.**

Ascidiacyclamide {**438** (Section V,E), mp 139–139.5°C, $[\alpha]_D^{25}$ 164° ($CHCl_3$)} is a new number of the family of cytotoxic peptides isolated from ascidians (*421*). In this case the ascidian was an unidentified species collected from Rodda Reef, Queensland, Australia. The structure was solved by spectroscopic analysis.

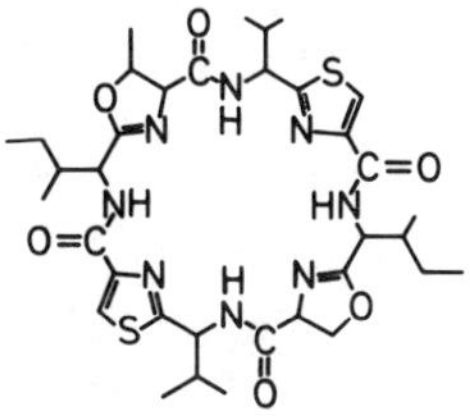

438

The compound is lethal to PV_4 cultured cells transformed with polyoma virus; 10 μg/ml caused *T/C* 100%. The animal also contained ulithiacyclamide (**334**).

The tunicate *Lissoclinum patella* has further yielded three cyclic peptides, **439** and the thiazolines **440** and **441** (*422*). The structures were assigned based on

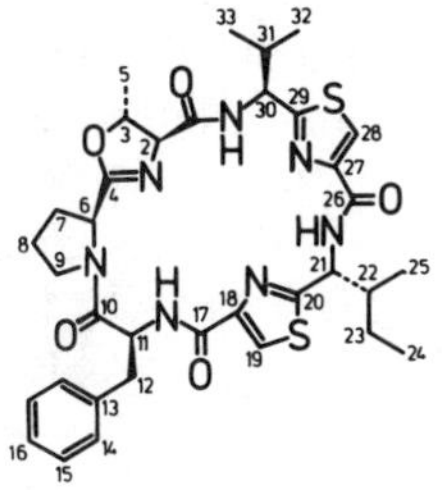

439

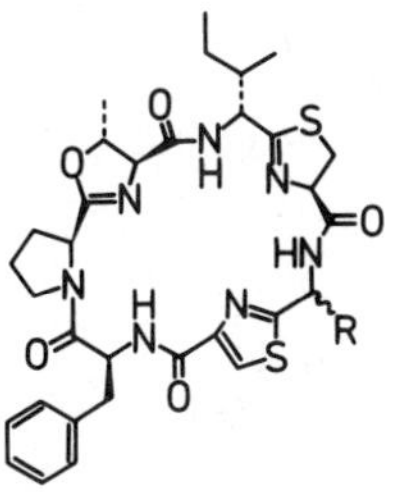

440 R=···CH_3
441 R=—CH_3

extensive chemical and spectroscopic studies, especially fast atom bombardment (FAB) mass spectrometry. The structure of ulicyclamide (**333**) was revised to the one depicted in **442.** The absolute configuration was assigned on basis of a chemical degradation of the 2-(1-aminoalkyl)thiazole-4-carboxylic acid unit as

442

described under the dysidenins (*423*). The peptides **439, 440,** and **441** displayed borderline cytotoxicity in L 1210 tissue culture assay.

An unidentified *Didemnum* sp. gave 0.39% (dry weight) of the iodinated thyramine **443** and smaller amounts of the urea **444** (*424*).

443 444

Two new ureido amino acids, namely lividine (**445**) [mp 225–226°C(d), $[\alpha]_D^{14}$ +20° (2 *N* HCl)] and grateloupine (**446**) [mp 178–179°C(d)] were secured from the red alga *Grateloupia livida* (*425*). The new amino acids were synthesized.

445 446

The full details of the structure elucidation of latrunculin A (**339**) and latrunculin B (**340**) have appeared (*426*). An interesting study of some of the physiological effects of these unique toxins has been published (*427*).

Four macrocyclic 1-ozaquinolizidines with vasodilative activity have been

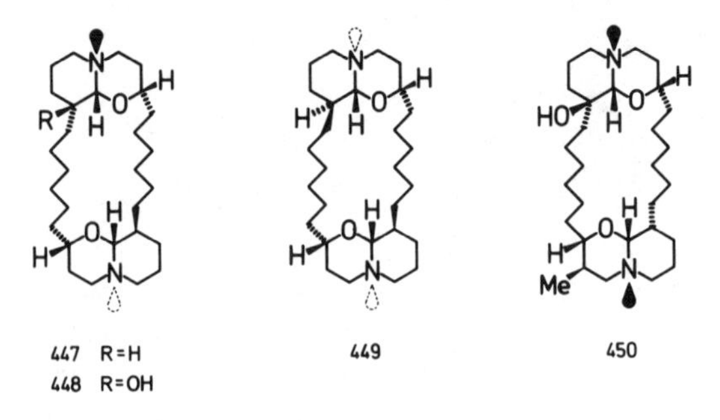

447 R=H
448 R=OH
449
450

secured from the Australian marine sponge *Xestospongia exigua* (*428*). The structure assignment of xestospongin C [(**447**) mp 149–150°C, [α] −2.4° ($CHCl_3$)] is the result of an X-ray structural investigation while the structures of xestospongin D (**448**) [mp 156–157°C, [α] +18.43° ($CHCl_3$)], xestospongin A (**449**) [mp 135–136°C, [α] +6.90° ($CHCl_3$)], and xestospongin B [(**450**) mp 178–181°C, [α] +7.10 ($CHCl_3$)] were assigned on the basis of spectroscopic studies.

A new phosphorus-containing toxin PB-1 (**451,** Section V,E) has recently been obtained from a 100 liter culture (1.2×10^9 cells) of the red tide dinoflagellate *Phytodiscus brevis* (*429*). Ichthyotoxicity against the common guppy

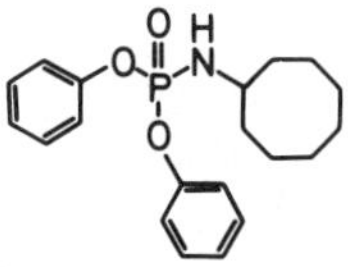

451

Lebestes reticulatus was determined as LD_{100} (1 hr) 1 ppm. The structure elucidation was carried out spectroscopically on 1 mg of toxin. Among other toxins, the sulfur-containing compound **344** was isolated in an amount of 2.5 mg from the same culture.

Synthesis of **451** was carried out by treating two equivalents of cycloocytylamine with *O,O*-diphenyl phosphochloridate in tetrahydrofuran for 30 min at room temperature, yielding **451** quantitatively. Attempts to incorporate ^{32}P into PB-1 gave ambiguous results since the activity of the PB-1 isolated was merely twice that of the background. These results lead the authors to conclude that PB-1 appears to be a metabolite of *P. brevis,* although the possibility that it may be derived from an unsuspected contaminant cannot be totally excluded.

Incidentally, two phosphorus-containing metabolites have recently been obtained from the culture filtrate of *Streptomyces lavendofolia* No. 630 (*430*). Fosfazinomycin A {**452,** mp 157–161°C(d), $[\alpha]_D^{25}$ 14.7° (H_2O)} and fosfazinomycin B {**453,** mp 148–150°C(d), $[\alpha]_D^{25}$ 17.2° (H_2O)} were identified from data on chemical degradation and spectroscopic studies. The fosfazinomycins are active against some filamentous fungi.

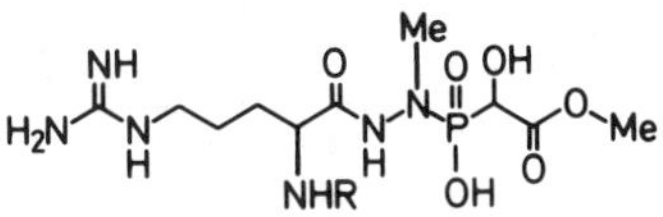

453 R = H

452 R =

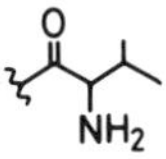

Aaptamine [**454** (Section V,E) mp 110–113°C] is a bright-yellow crystalline alkaloid isolated from the sponge *Aaptos aaptos,* collected at Okinawa (*431*). Aaptamine (**454**), obtained in 0.17% yield based on wet weight, is an example of the new skeleton 1*H*-benzo[*de*]-1,6-naphthyridine and possesses a remarkable α-

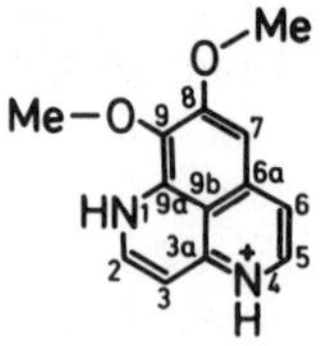

454

adrenoceptor blocking activity in the isolated rabbit-aorta test. The structure determination results from a combination of degradative experiments and spectroscopic studies.

Another new unique skeleton is represented in the fused pentacyclic aromatic alkaloid amphimedine (**455**, mp >360°C) isolated from a Pacific sponge, *Amphimedon* sp. (*432*). The structure elucidation was hampered by the paucity of

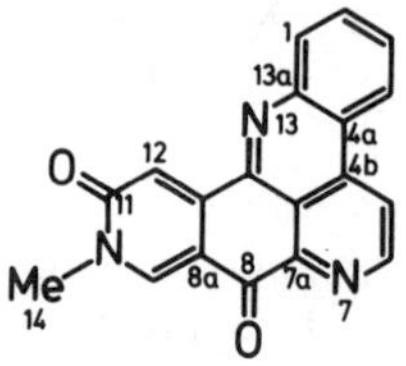

455

protons in the molecule ($C_{19}H_{11}N_3O_2$) and by the low solubility of the yellow solid. Proton homonuclear decoupling confirmed the contiguous positions of protons on carbons 1–4, that those on positions 5 and 6 constituted an isolated vicinal pair, and that the protons at postions 8 and 12 were each isolated. NOE measurements established the closeness of the protons at carbons 4 and 5 and 9 and those on the methyl group (C-14). UV, IR, and ^{13}C-NMR data confirmed the presence of amide carbonyl and a cross-conjugated ketone. The total structure was then established by three-bond proton–carbon couplings identified by single-frequency decoupling experiments and natural-abundance ^{13}C–^{13}C one-bond couplings obtained from a two-dimensional double-quantum coherence experiment (INADEQUATE experiment). The latter technique allowed the determination of the connectivity of the entire carbon network.

Studies of the chemical aspects of palytoxin (**356,** Section V,E) have appeared. The stereochemistry is dealt with in a series of papers (*433–436*). The stereocontrolled synthesis of the palytoxin C-101–C-115 fragment (*437*), the C-85–C-98 segment (*438*), and the C-23–C-37 segment (*439*) has been described.

A stereocontrolled synthesis of (2*R*,3*R*,5*R*,13*S*,14*R*)-(+)-aplidiasphingosine has been published (*440*).

A ciguatera-causing (Section VI) dinoflagellate, *Gambierdiscus toxicus,* was isolated from Hawaiian waters (*441*). This dinoflagellate seems to prefer epiphytic growth on the red alga *Spyridia filamentosa.* Two toxins were characterized physiologically, one of them exhibiting ciguatoxin-like activity. The outer wall of the viable cells had affinity to the fluorescence-labeled eel ciguatoxin antibody. The observations agree with those reported from French Polynesia.

A physiological study of purified ciguatoxin (*442*) concludes that ciguatoxin is a novel type of Na^+ channel toxin. Interestingly enough, the action of ciguatoxin is completely inhibited by tetrodotoxin.

Although nucleosides have not been treated in this review (see Section I), two new halogenated pyrrolo[2,3-*d*]pyrimidines (*443*) are so clearly marked as marine compounds that they will be mentioned here.

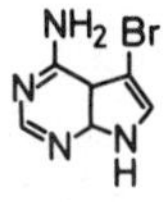

456

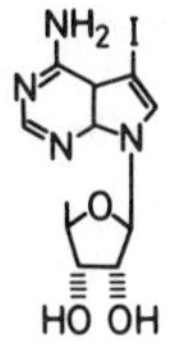

457

The first, 4-amino-5-bromopyrrolo[2,3-*d*]pyrimidine [**456,** mp 240–241°C(d)] was isolated from a sponge, *Echinodictyum* sp., in a yield of 0.0149% as the acetate based on dry weight. Compound **456** is synthetically available. The product exhibits strong activity of the isolated guinea pig trachea, an *in vitro* model for potential bronchodilators, and when administered intraperitoneally shows a 40% decrease in activity after amphetamine dosing. Intravenously administered **456** causes a lowering of blood pressure in deoxycorticosterone acetate-hypertensive rats.

The red alga *Hypnea valendiae,* collected at Quobba Lagoon, Western Australia, gave $3.18 \times 10^{-3}\%$, dry weight of 5′-deoxy-5-iodotubercidin {**457,** mp 227–228°C(d), $[\alpha]_D^{25}$ −55° (MeOH)}. The structure was established by spectroscopic investigations. Pharmacologically, 4-amino-7-(5′-deoxyribos-1′-yl)-5-iodopyrrolo[2,3-*d*]pyrimidine (**457**) produce pronounces muscle relaxation and hypothermia in mice and blocks polysynaptic and monosynaptic reflexes.

Reisolation of the active principle from another collection of the alga gave a much lower yield ($2.32 \times 10^{-4}\%$ of dry weight) of **457** and a $1.44 \times 10^{-3}\%$ yield of a new nucleoside, yellow gum {$[\alpha]_D^{22}$ −50° (MeOH)}, believed to be 4-amino-7-(5′-deoxyribos-1′α-yl)-5-iodopyrrolo[2,3-*d*]pyrimidine, namely, the 1′α isomer of **457.** The latter nucleoside is less active than **457.**

A purino-diterpene (**458**) was isolated from the marine sponge *Agelas mauri-*

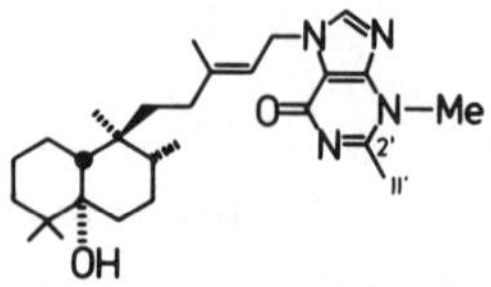

458

tiana collected from Enewetak. Seemingly the compound is an artifact since the carbons at position 2′ and 11′ are derived from an acetyl precursor (*444*). The structure of the true sponge metabolite is still unknown. The structural analysis of **458** was carried out by X-ray techniques.

An *Agelas* sp. collected at Argulpelu Reef, Palau, gave two new compounds, ageline A (**459**) [mp 175–176°C, $[\alpha]_D$ −8.4° ($CHCl_3$)] and ageline B (**460**) together with agelasidine A (**370**) (*445*).

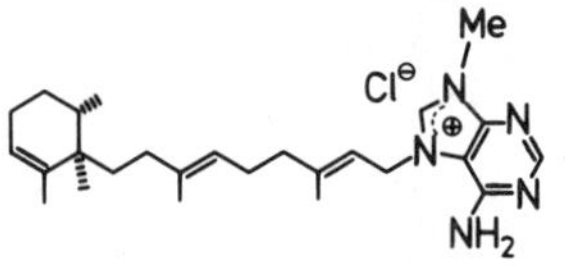

459 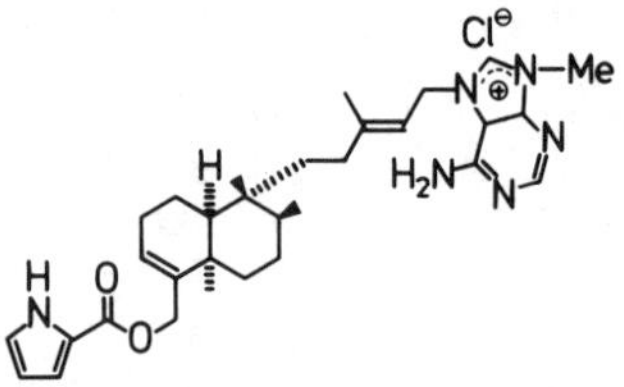 460

Ageline A (**459**) inhibited the grown of *Staphylococcus aureus, Bacillus subtilis,* and *Candida albicans* in the disk assay at 5 μg/disk.

References

1. D. B. Boylan and P. J. Scheuer, *Science* **155,** 52 (1967).
2. P. G. Nielsen, J. S. Carlé, and C. Christophersen, *Phytochemistry* **21,** 1643 (1982).
3. A. F. Cook, R. T. Bartlett, R. J. Quinn, and R. P. Gregson, *J. Org. Chem.* **45,** 4020 (1980).
4. M. R. Hagadone, B. J. Burreson, P. J. Scheuer, J. S. Finer, and J. Clardy, *Helv. Chim. Acta* **62** 2484 (1979); G. R. Schulte and P. J. Scheuer, *Tetrahedron* **38,** 1857 (1982).
5. M. Viscontini, W. Hummel, and A. Fisher, *Helv. Chim. Acta* **53,** 1207 (1970).
6. E. M. Low, *J. Mar. Res.* **8,** 97 (1949).
7. P. R. Bergquist, "Sponges." Hutchinson University Library, London, 1978.
8. L. D. Quin, *Top. Phosphorus Chem.* **4,** 23 (1967).
9. J. Vacelet, *J. Microsc. (Paris)* **9,** 333 (1970).
10. J. Vacelet, *J. Microsc. (Paris)* **12,** 363 (1971).
11. J. Vacelet, *J. Microsc. (Paris)* **23,** 271 (1975).
12. J. Vacelet and C. Donadey, *J. Exp. Mar. Biol. Ecol.* **30,** 301 (1977).
13. P. J. Burkholder, *in* "Biology and Geology of Coral Reefs" (O. A. Jones and R. Endean, eds.), Vol. 2, p. 117. Academic Press, New York, 1973.
14. G. M. Sharma and P. R. Burkholder, *J. Antibiot., Ser. A* **20,** 200 (1967).
15. P. J. Scheuer, "Chemistry of Marine Natural Products." Academic Press, New York, 1973.
16. M. H. Baslow, "Marine Pharmacology, A Study of Toxins and Other Biologically Active Substances of Marine Origin." Williams & Wilkins, Baltimore, Maryland, 1969.

17. B. W. Halstead, "Poisonous and Venomous Marine Animals of the World," rev. ed. Darwin Press, Princeton, New Jersey, 1978.
18. Y. Hashimoto, "Marine Toxins and Other Bioactive Marine Metabolites." Jpn. Sci. Soc. Press, Tokyo, 1979.
19. J. T. Baker and V. Murphy, "Handbook of Marine Science, Compounds from Marine Organisms," Vol. I. CRC Press, Cleveland, Ohio, 1976; Vol. II, 1981.
20. D. J. Faulkner and W. H. Fenical, eds., "Marine Natural Products Chemistry," Plenum, New York, 1977.
21. D. F. Martin and G. M. Padilla, eds., "Marine Pharmacognosy, Action of Marine Biotoxins at the Cellular Level." Academic Press, New York, 1973.
22. P. J. Scheuer, ed., "Marine Natural Products, Chemical and Biological Perspectives," Vol. 1. Academic Press, New York, 1978.
23. P. J. Scheuer, ed., "Marine Natural Products, Chemical and Biological Perspectives," Vol. 2. Academic Press, New York, 1978.
24. P. J. Scheuer, ed., "Marine Natural Products, Chemical and Biological Perspectives," Vol. 3. Academic Press, New York, 1980.
25. P. J. Scheuer, ed., "Marine Natural Products, Chemical and Biological Perspectives," Vol. 4. Academic Press, New York, 1981.
26. P. J. Scheuer, ed., "Marine Natural Products, Chemical and Biological Perspectives," Vol. 5. Academic Press, New York, 1983.
27. D. J. Faulkner and R. J. Andersen, *in* "The Sea" (E. D. Goldberg, ed.), Vol. 5, p. 679. Wiley (Interscience), New York, 1974.
28. R. H. Thomson, *Chem. Br.* **14,** 133 (1978).
29. W. Fenical, *Science* **215,** 923 (1982).
30. C. W. J. Chang, *J. Chem. Educ.* **55,** 684 (1978).
31. *Pure Appl. Chem.* **48,** 1–44 (1976).
32. *Pure Appl. Chem.* **51,** 1815–1911 (1979).
33. *Bull. Soc. Chim. Belg.* **89,** 1061–1106 (1980).
34. *Pure Appl. Chem.* **54,** 1907–2010 (1982).
35. P. J. Scheuer, *Isr. J. Chem.* **16,** 52 (1977).
36. P. J. Scheuer, *Naturwissenschaften* **58,** 549 (1971).
37. P. J. Scheuer, *Acc. Chem. Res.* **10,** 33 (1977).
38. L. Minale, G. Cimino, S. De Stefano, and G. Sodano, *Prog. Chem. Nat. Prod.* **33,** 1 (1976).
39. Y. Shimizu, *Recent Adv. Phytochem.* **13,** 199 (1979).
40. D. J. Faulkner, *Top. Antibiot. Chem.* **2,** 9 (1978).
41. R. E. Moore, *BioScience* **27,** 797 (1977).
42. H. W. Youngken, Jr. and Y. Shimizu, *in* "Chemical Oceanography" (J. P. Riley and G. Skirrow, eds.), 2nd ed., Vol. 4, p. 269. Academic Press, New York, 1975.
43. A. Der Marderosian, *J. Pharm. Sci.* **58,** 1 (1969).
44. G. D. Ruggieri, *Science* **194,** 491 (1976).
45. R. F. Nigrelli, M. F. Stempien, Jr., G. D. Ruggieri, V. R. Liguori, and J. T. Cecil, *Fed. Proc., Fed. Am. Soc. Exp. Biol.* **26,** 1197 (1967).
46. R. V. Southcott, *in* "Handbook of Clinical Neurology" (P. J. Vinken and G. W. Bruyn, eds.), Vol. 37, Part II, p. 27. Elsevier/North-Holland, New York, 1979.
47. B. W. Halstead, *Clin. Toxicol.* **18,** 1 (1981).
48. D. L. Fox, *Biochem. Biophys. Perspect. Mar. Biol.* **1,** 170 (1974).
49. G. Y. Kennedy, *Adv. Mar. Biol.* **16,** 352 (1979).
50. C. Christophersen and N. Jacobsen, *Annu. Rep. Prog. Chem., Sect. B* **76,** 433 (1979).
51. M. F. Grundon, senior reporter, "The Alkaloids," Vol. 11. Royal Society of Chemistry, London, 1981, and earlier volumes.

52. O. von Fürth, "Vergleichende chemische Physiologie der niederen Tiere." Fischer, Jena, 1903.
53. L. Chevolot, *in* "Marine Natural Products: Chemical and Biological Perspectives" (P. J. Scheuer, ed.), Vol. 4, p. 53. Academic Press, New York, 1981.
54. T. Goto, *in* "Marine Natural Products: Chemical and Biological Perspectives" (P. J. Scheuer, ed.), Vol. 3, p. 180. Academic Press, New York, 1980.
55. G. Prota, *in* "Marine Natural Products: Chemical and Biological Perspectives" (P. J. Scheuer, ed.), Vol. 3, p. 141. Academic Press, New York, 1980.
56. Y. Shimizu, *in* "Marine Natural Products: Chemical and Biological Perspectives" (P. J. Scheuer, ed.), Vol. 1, p. 1. Academic Press, New York, 1978.
57. Y. Shimizu, *Pure Appl. Chem.* **54,** 1973 (1982).
58. Y. Kishi, F. Nakatsubo, M. Aratani, T. Goto, S. Inoue, H. Kaboi, and S. Sugiura, *Tetrahedron Lett.* 5127, 5129 (1970); Y. Kishi, M. Aratani, T. Fukuyama, F. Nakatsubo, T. Goto, S. Inoue, H. Tanino, S. Sugiura, and H. Kakoi, *J. Am. Chem. Soc.* **94,** 9217 (1972); Y. Kishi, T. Fukuyama, M. Aratani, F. Nakatsubo, T. Goto, S. Inoue, H. Tanino, S. Sugiura, and H. Kakoi, *ibid.* 9219 (1972); H. Tanino and S. Inoue, *Tetrahedron Lett.* 335 (1974).
59. L. A. Pavelka, F. A. Fuhrman, and H. S. Mosher, *Heterocycles* **17,** 225 (1982).
60. T. Noguechi and Y. Hashimoto, *Toxicon* **11,** 305 (1973).
61. D. Schenmack, M. E. H. Howden, I. Spence, and R. J. Quinn, *Science* **199,** 188 (1978).
62. H. S. Mosher, F. A. Fuhrman, H. D. Buchwald, and H. G. Fisher, *Science* **144,** 1100 (1964).
63. L. A. Pavelka, Y. H. Kim, and H. S. Mosher, *Toxicon* **15,** 135 (1977).
64. T. Noguchi, J. Maruyama, Y. Ueda, K. Hashimoto, and T. Harada, *Bull. Jpn. Soc. Sci. Fish.* **47,** 909 (1981).
65. T. Yasumoto, Y. Oshima, N. Hosaka, and H. Miyakoshi, *Bull. Jpn. Soc. Sci. Fish.* **47,** 929 (1981).
66. H. Narita, T. Noguchi, J. Maruyama, Y. Ueda, K. Hashimoto, Y. Watanabe, and K. Hida, *Bull. Jpn. Soc. Sci. Fish.* **47,** 935 (1981).
67. T. Matsui, H. Sato, S. Hamada, and C. Shimizu, *Bull. Jpn. Soc. Sci. Fish.* **48,** 253 (1982).
68. E. J. Schantz, V. E. Ghazarossian, H. K. Schnoes, F. M. Strong, J. P. Springer, J. O. Pezzanite, and J. Clardy, *J. Am. Chem. Soc.* **97,** 1238 (1975).
69. J. Bordner, W. E. Thiessen, H. A. Bates, and H. Rapoport, *J. Am. Chem. Soc.* **97,** 6008 (1975).
70. Y. Shimizu, C. P. Hsu, W. E. Fallon, Y. Oshima, T. Miura, and K. Nakanishi, *J. Am. Chem. Soc.* **100,** 6791 (1978).
71. Y. Shimizu, L. J. Buckley, M. Alam, Y. Oshima, W. E. Fallon, H. Kasai, T. Miura, V. P. Gullo, and K. Nakanishi, *J. Am. Chem. Soc.* **98,** 5414 (1976).
72. G. L. Boyer, E. J. Schantz, and H. K. Schnoes, *J. Chem. Soc., Chem. Commun.* 889 (1978); T. Guchi, M. Kono, Y. Ueda, and K. Hashimoto, *J. Chem. Soc. Jpn., Chem. Ind. Chem.* 652 (1981).
73. Y. Shimizu, *Kagaku to Seibutsu* **18,** 792 (1980); Y. Shimuzu and C. P. Hsu, *J. Chem. Soc., Chem. Commun.* 314 (1981); C. F. Weichmann, G. L. Boyer, C. L. Divan, E. J. Schantz, and H. K. Schnoes, *Tetrahedron Lett.* **22,** 1941 (1981); M. Alam, Y. Oshima, and Y. Shimizu, *ibid.* **23,** 321 (1981); T. Noguchi, Y. Ueda, K. Hashimoto, and H. Seto, *Bull. Jpn. Soc. Sci. Fish.* **47,** 1227 (1981).
74. F. E. Koehn, S. Hall, C. Fix Wichmann, H. K. Schnoes, and P. B. Reichardt, *Tetrahedron Lett.* **23,** 2247 (1982).
75. S. Hall, P. B. Reichardt, and R. A. Neve, *Biochem. Biophys. Res. Commun.* **97** 649 (1980); M. Kobayashi and Y. Shimuzu, *J. Chem. Soc., Chem. Commun.* 827 (1981); C. F. Wichmann, W. P. Niemczura, H. K. Snoes, S. Hall, P. B. Reichardt, and S. D. Darling, *J. Am. Chem. Soc.* **103,** 6977 (1981).

76. R. S. Rogers and H. Rapoport, *J. Am. Chem. Soc.* **102,** 7335 (1980).
77. E. Jachim and J. Gentile, *Science* **162,** 915 (1968).
78. H. Tanino, T. Nakata, T. Kaneko, and Y. Kishi, *J. Am. Chem. Soc.* **99,** 2818 (1977).
79. Y. Kotaki, Y. Oshima, and T. Yasumoto, *Bull. Jpn. Soc. Sci. Fish.* **47,** 943 (1981).
80. T. Yasumoto, Y. Oshima, and T. Konta, *Bull. Jpn. Soc. Sci. Fish.* **47,** 957 (1981).
81. K. Koyama, T. Noguchi, Y. Ueda, and K. Hashimoto, *Bull. Jpn. Soc. Sci. Fish.* **47,** 965 (1981).
82. Y. Onove, T. Noguchi, J. Maruyania, K. Hashimoto, and T. Ikeda, *Bull. Jpn. Soc. Sci. Fish.* **47,** 1643 (1981).
83. Y. Onove, T. Noguchi, J. Maruyama, Y. Ueda, K. Hashimoto, and T. Ikeda, *Bull. Jpn. Soc. Sci. Fish.* **47,** 1347 (1981).
84. A. W. White, *Mar. Biol. (Berlin)* **65,** 255 (1981).
85. L. Cariello, S. Crescenzi, G. Prota, F. Giordana, and L. Mazzarella, *J. Chem. Soc., Chem. Commun.* 99 (1973); L. Cariello, S. Crescenzi, G. Prota, S. Capasso, F. Giordano, and L. Mazzarella, *Tetrahedron* **30,** 328 (1974).
86. L. Cariello, S. Crescenzi, G. Prota, and L. Zanetti, *Experientia* **30,** 849 (1974).
87. L. Cariello, S. Crescenzi, G. Prota, and L. Zanetti, *Tetrahedron* **30,** 3611 (1974).
88. L. Cariello, S. Crescenzi, G. Prota, and L. Zanetti, *Tetrahedron* **30,** 4191 (1974).
89. L. Cariello, S. Crescenzi, L. Zanetti, and G. Prota, *Comp. Biochem. Physiol. B* **63B,** 77 (1979).
90. R. E. Schwartz, M. B. Yunker, P. J. Scheuer, and T. Ottersen, *Tetrahedron Lett.* 2235 (1978).
91. R. E. Schwartz, M. B. Yunker, P. J. Scheuer, and T. Ottersen, *Can. J. Chem.* **57,** 1707 (1979).
92. Y. Komoda, S. Kaneko, M. Yamamoto, M. Ishikawa, A. Ital, and Y. Litaka, *Chem. Pharm. Bull.* **23,** 2464 (1975).
93. Y. Komodo, M. Shimizu, S. Kaneko, M. Yamamoto, and M. Ishikawa, *Chem. Pharm. Bull.* **30,** 502 (1982).
94. M. Braun and G. Büchi, *J. Am. Chem. Soc.* **98,** 3049 (1976).
95. R. P. Walker, D. J. Faulkner, D. van Engen, and J. Clardy, *J. Am. Chem. Soc.* **103,** 6772 (1981).
96. S. Forenza, L. Minale, R. Riccio, and E. Fattorusso, *J. Chem. Soc., Chem. Commun.* 1129 (1971).
97. E. E. Garcia, L. E. Benjamin, and R. I. Fryer, *J. Chem. Soc., Chem. Commun.* 78 (1973).
98. G. Cimino, S. De Rosa, S. De Stefano, L. Mazzarella, R. Puliti, and G. Sodano, *Tetrahedron Lett.* **23,** 767 (1982).
99. G. M. Sharma, J. S. Buyer, and M. W. Pomerantz, *J. Chem. Soc., Chem. Commun.* 435 (1980).
100. G. M. Sharma and P. R. Burkholder, *J. Chem. Soc., Chem. Commun.* 151 (1971).
101. G. M. Sharma and B. Maydoff-Fairchild, *J. Org. Chem.* **42,** 4118 (1977).
102. L. H. Foley and G. Büchi, *J. Am. Chem. Soc.* **104,** 1776 (1982).
103. G. Cimino, S. De Stefano, and L. Minale, *Comp. Biochem. Physiol. B* **47B,** 895 (1974).
104. L. Chevolot, S. Padua, B. N. Ravi, P. C. Blyth, and P. J. Scheuer, *Heterocycles* **7,** 891 (1977).
105. M. T. Cheng and L. K. Rinehart, Jr., *J. Am. Chem. Soc.* **100,** 7409 (1978).
106. B. Carté and D. J. Faulkner, *Tetrahedron Lett.* **23,** 3863 (1982).
107. B. Kazlauskas, P. T. Murphy, R. J. Quinn, and R. J. Wells, *Tetrahedron Lett.* 61 (1977).
108. K. H. Hollenbeak and F. J. Schmitz, *J. Nat. Prod.* **40,** 479 (1977).
109. P. Djura and D. J. Faulkner, *J. Org. Chem.* **45,** 735 (1980).
110. R. K. Oluda, D. Klain, R. B. Kinell, H. Li, and P. J. Scheuer, *Pure Appl. Chem.* **54,** 1907 (1982).

111. Y. Kishi, T. Goto, Y. Hirata, O. Shimomura, and F. H. Johnson, *Tetrahedron Lett.* 3427 (1966); Y. Kishi, T. Goto, S. Eguchi, Y. Hirata, E. Watanabe, and T. Aoyama, *ibid.* 3437 (1966).
112. Y. Kishi, T. Goto, S. Inoue, S. Sugiura, and H. Kishimoto, *Tetrahedron Lett.* 3445 (1966).
113. S. Inoue, H. Kaboi, M. Murata, T. Goto, and O. Shimomura, *Tetrahedron Lett.* 2685 (1977).
114. S. Inoue, H. Kakoi, and T. Goto, *J. Chem. Soc., Chem. Commun.* 1056 (1976).
115. S. Inoue, K. Okada, H. Kakoi, and T. Goto, *Chem. Lett.* 257 (1977).
116. S. Inoue, H. Kakoi, and T. Goto, *Tetrahedron Lett.* 2971 (1976).
117. G. C. Harbour, A. A. Tymiak, K. L. Rinehart, Jr., P. D. Shaw, R. G. Hughes, Jr., S. A. Mizsak, J. H. Coats, G. E. Zurenko, L. H. Li, and S. L. Kuentzel, *J. Am. Chem. Soc.* **103,** 5604 (1981).
118. G. T. Carter and K. L. Rinehart, Jr., *J. Am. Chem. Soc.* **100,** 4302 (1978).
119. K. Gustafson and R. J. Andersen, *J. Org. Chem.* **47,** 2167 (1982).
120. S. Heitz, M. Dugeat, M. Guyot, C. Brassy, and B. Bachet, *Tetrahedron Lett.* 1457 (1980).
121. J. M. Sieburth, "Microbial Seascapes." University Park Press, Baltimore, Maryland, 1975.
122. C. Christophersen, *in* "Marine Natural Products, Chemical and Biological Perspectives" (P. J. Scheuer, ed.), Vol. 5, p. 259. Academic Press, New York, 1983.
123. M. R. Brennan and K. L. Erickson, *Tetrahedron Lett.* 1637 (1978).
124. K. L. Erickson, M. R. Brennan, and P. A. Namnum, *Synth. Commun.* **11,** 253 (1981).
125. G. T. Carter, K. L. Rinehart, Jr., L. H. Li, S. L. Kuentzel, and J. L. Connor, *Tetrahedron Lett.* 4479 (1978).
126. H. G. Reinecke, H. W. Johnson, Jr., and J. F. Sebastian, *J. Am. Chem. Soc.* **91,** 3817 (1969).
127. T. Higa and P. J. Scheuer, *Naturwissenschaften* **62,** 395 (1975).
128. T. Higa and P. J. Scheuer, *Heterocycles* **4,** 231 (1976).
129. T. Higa and P. J. Scheuer, *Heterocycles* **3,** 227 (1976).
130. T. Higa and P. J. Scheuer, *in* "Marine Natural Products Chemistry" (D. J. Faulkner and W. H. Fenical, eds.), p. 35. Plenum, New York, 1977.
131. T. Higa, T. Fujiyama, and P. J. Scheuer, *Comp. Biochem. Physiol. B* **65,** 525 (1980).
132. M. F. Stempien, Jr., *Am. Zool.* **6,** 363 (1966).
133. S. J. Wratten, M. S. Wolfe, R. J. Andersen, and D. J. Faulkner, *Antimicrob. Agents Chemother.* **11,** 411 (1977).
134. H. Abe, M. Uchiyama, and R. Sato, *Agric. Biol. Chem.* **36,** 2259 (1972).
135. Y. Gopichand and F. J. Schmitz, *J. Org. Chem.* **44,** 4995 (1979).
136. R. S. Norton and R. J. Wells, *J. Am. Chem. Soc.* **104,** 3628 (1982).
137. J. F. Blount and R. J. Wells, *Aust. J. Chem.* (to be published).
138. G. Cimino and S. De Stefano, *Comp. Biochem. Physiol. C* **61C,** 361 (1978).
139. G. Mazzanti and D. Piccinelli, *Comp. Biochem. Physiol. C* **63C,** 215 (1979).
140. A. Schulman, M. I. B. Dick, and K. T. H. Farrer, *Nature (London)* **180,** 658 (1957).
141. T. Wieland, W. Motzel, and H. Herz, *Instus Liebigs Ann. Chem.* **581,** 10 (1953).
142. G. E. VanLear, G. O. Morton, and W. Fulmor, *Tetrahedron Lett.* 299 (1973).
143. P. Djura, B. B. Stierle, B. Sullivan, D. J. Faulkner, E. Arnold, and J. Clardy, *J. Org. Chem.* **45,** 1435 (1980).
144. K. L. Rinehart, Jr., P. D. Shaw, L. S. Shield, J. B. Gloer, G. C. Harbour, M. E. S. Koker, D. Samain, R. E. Schwarts, A. A. Tymiak, D. L. Weller, G. T. Carter, M. H. G. Munro, R. G. Hughes, Jr., H. E. Renis, E. B. Swynenberg, D. A. Stringfellow, J. J. Vavra, J. H. Coats, G. E. Zurenko, S. L. Kuentzel, L. H. Li, G. J. Bakus, R. C. Brusca, L. L. Craft, D. N. Young, and J. L. Connor, *Pure Appl. Chem.* **53,** 795 (1981).
145. E. R. Schultes and A. Hofmann, "The Botany and Chemistry of Hallucinogenes." Thomas, Springfield, Illinois, 1973.
146. G. Dellar, P. Djura, and M. V. Sargent, *J. Chem. Soc., Perkin Trans. 1* 1679 (1981).
147. W. D. Raverty, R. H. Thomson, and T. J. King, *J. Chem. Soc., Perkin Trans. 1* 1204 (1977).

148. G. A. Santos and M. S. Doty, *in* "Drugs from the Sea" (H. D. Freudenthal, ed.), p. 173. Marine Technology Society, Washington, D.C., 1968.
149. M. S. Doty and G. A. Santos, *Pac. Sci.* **24,** 351 (1970).
150. B. C. Maiti and R. H. Thomsen, *in* "Marine Natural Products Chemistry" (D. J. Faulkner and W. H. Fenical, eds.), p. 159. Plenum, New York, 1972.
151. S. C. Maiti, R. H. Thomsen, and M. Mahendran, *J. Chem. Res.* 1682 (1978).
152. T. Higa, *in* "Marine Natural Products: Chemical and Biological Perspectives" (P. J. Scheuer, ed.), Vol. 4, p. 93. Academic Press, New York, 1981.
153. R. J. Andersen and R. J. Stonard, *Can. J. Chem.* **57,** 2325 (1979).
154. R. J. Andersen, *Tetrahedron Lett.* 2541 (1978).
155. U. Schmidt, A. Lieberknecht, H. Griesser, and H. Bökens, *Tetrahedron Lett.* **23,** 4911 (1982).
156. R. J. Stonard and R. J. Andersen, *J. Org. Chem.* **45,** 3687 (1980).
157. R. J. Stonard and R. J. Andersen, *Can. J. Chem.* **58,** 2121 (1980).
158. J. Marchand, M. Païs, X. Monseur, and F.-X. Jarreau, *Tetrahedron* **25,** 937 (1969).
159. J. S. Ryland, "Bryozoans." Hutchinson University Library, London, 1970.
160. J. S. Carlé and C. Christophersen, *J. Am. Chem. Soc.* **102,** 5107 (1980).
161. C. Grøn and C. Christophersen, *Acta Chem. Scand.* **B38,** 709 (1984).
162. C. Christophersen and J. S. Carlé, *Naturwissenschaften* **65,** 440 (1978).
163. P. Wulff, J. S. Carlé, and C. Christophersen, *Comp. Biochem. Physiol.* **71B,** 523 (1982).
164. P. Wulff, J. S. Carlé, and C. Christophersen, *J. Chem. Soc., Perkin Trans. 1* 2895 (1981).
165. J. S. Carlé and C. Christophersen, *J. Am. Chem. Soc.* **101,** 4012 (1979).
166. J. S. Carlé and C. Christophersen, *J. Org. Chem.* **45,** 1586 (1980).
167. P. Muthusubramanian, J. S. Carlé, and C. Christophersen, *Acta Chem. Scand.* **B37,** 803 (1983).
168. J. S. Carlé and C. Christophersen, *J. Org. Chem.* **46,** 3440 (1981).
169. P. M. Scott, J. Polonsky, and H.-A. Merien, *J. Agric. Food Chem.* **27,** 201 (1979).
170. A. J. Birch and J. J. Wright, *Tetrahedron* **26,** 2329 (1970).
171. S. Inoue, J. Murata, N. Takamatsu, H. Nagano, and Y. Kishi, *Yakugaku Zasshi* **97,** 576 (1977).
172. G. Lutaud, *Cah. Biol. Mar.* **5,** 201 (1964).
173. G. Lutaud, *Cah. Biol. Mar.* **6,** 181 (1965).
174. G. Lutaud, *Arch. Zool. Exp. Gen.* **110,** 5 (1969).
175. R. M. Woollacott, *Mar. Biol.* **65,** 155 (1981).
176. J. H. Cardellina, II, F.-J. Marner, and R. E. Moore, *Science* **204,** 193 (1979).
177. R. E. Moore, *in* "Marine Natural Products: Chemical and Biological Perspectives" (P. J. Scheuer, ed.), Vol. 4, p. 1. Academic Press, New York, 1981.
178. M. Takashima, H. Sakai, and K. Arima, *Agric. Biol. Chem.* **26,** 660 (1962).
179. H. Nakata, H. Harada, and Y. Hirata, *Tetrahedron Lett.* 2515 (1966).
180. H. Fujiki, M. Mori, M. Nakayasu, M. Terada, T. Sugimura, and R. E. Moore, *Proc. Natl. Acad. Sci. U.S.A.* **78,** 3872 (1981).
181. H. Sakamoto, M. Terada, H. Fujiki, M. Mori, M. Nakayasu, T. Sugimura, and B. Weinstein, *Biochem. Biophys. Res. Commun.* **102,** 100 (1981).
182. M. Makayasu, H. Fujiki, M. Mori, T. Sugimura, and R. E. Moore, *Cancer Lett.* **12** 271 (1981); Y. Endo, K. Shudo, and T. Okamoto, *Chem. Pharm. Bull.* **30,** 3457 (1982).
183. N. Sakabe, H. Harada, Y. Hirata, Y. Tomite, and I. Nitta, *Tetrahedron Lett.* 2623 (1966).
184. T. Kosuge, H. Zenda, A. Ochiai, N. Masaki, M. Noguchi, S. Kimura, and H. Narita, *Tetrahedron Lett.* 2545 (1972).
185. T. Kosuge, K. Tsuji, K. Hirai, K. Yamaguchi, T. Okamoto, and Y. Iitaka, *Tetrahedron Lett.* 3417 (1981); T. Kosuge, K. Tsuji, and K. Hirai, *Chem. Pharm. Bull.* **30,** 3255 (1982).
186. P. Friedländer, *Chem. Ber.* **42,** 765 (1909).

187. J. T. Baker, *Endeavour* **32,** 11 (1974).
188. C. Christophersen, F. Wätjen, O. Buchardt, and U. Anthoni, *Tetrahedron* **34,** 2779 (1978).
189. P. Süsse and C. Krampe, *Naturwissenschaften* **66,** 110 (1979).
190. S. Larsen and F. Wätjen, *Acta Chem. Scand.* **A34,** 171 (1980).
191. C. Christophersen, F. Wätjen, O. Buchardt, and U. Anthoni, *Tetrahedron Lett.* 1747 (1977).
192. Y. Fujise, K. Miwa, and S. Ito, *Chem. Lett.* 631 (1980).
193. H. Fouquet and H. J. Bielig, *Angew. Chem.* **83,** 856 (1971).
194. J. H. Cardellina II, M. P. Kirkup, R. E. Moore, J. S. Mynderse, K. Seff, and C. J. Simmons, *Tetrahedron Lett.* 4915 (1979).
195. K. Sakano and S. Nakamura, *J. Antibiot.* **33,** 961 (1980).
196. M. Kaneda, K. Sakano, S. Nakamura, Y. Kushi, and Y. Iitaka, *Heterocycles* **15,** 993 (1981).
197. S. Kano, E. Sugino, and S. Hibino, *J. Chem. Soc., Chem. Commun.* 1241 (1980).
198. S. Takano, Y. Suzuki, and K. Ogasawara, *Heterocycles* **16,** 1479 (1981).
199. S. Kano, E. Sugino, S. Shibuya, and S. Hibino, *J. Org. Chem.* **46,** 3856 (1981).
200. S. Forenza, L. Minale, and R. Riccio, *J. Chem. Soc., Chem. Commun.* 1129 (1971).
201. I. J. Rinkes, *Recl. Trav. Chim. Pays-Bas* **60,** 303 (1941).
202. L. Chevolot, S. Padua, B. N. Ravi, P. C. Blyth, and P. J. Scheuer, *Heterocycles* **7,** 891 (1977).
203. P. R. Burkholder, R. M. Pfister, and F. M. Leitz, *Appl. Microbiol.* **14,** 649 (1966).
204. F. M. Lovell, *J. Am. Chem. Soc.* **88,** 4510 (1966).
205. S. Hanessian and J. S. Kaltenbronn, *J. Am. Chem. Soc.* **88,** 4509 (1966).
206. R. D. Hamilton and K. E. Austin, *Antonie van Leeuwenhoek* **33,** 257 (1967).
207. M. J. Gauthier, J. M. Shewan, D. M. Gibson, and J. V. Lee, *J. Gen. Microbiol.* **87,** 211 (1975).
208. R. J. Anderson, M. S. Wolfe, and D. J. Faulkner, *Mar. Biol. (Berlin)* **27,** 281 (1974).
209. M. F. Stempien, Jr., R. F. Nigrelli, and J. S. Chib, *164th Meet. Am. Chem. Soc.* Abstracts, MEDI 21 (1972), taken from reference *206*.
210. G. Impellizzeri, S. Mangiafico, G. Oriente, M. Piatelli, S. Sciuto, E. Fattorusso, S. Magno, C. Santacrose, and D. Sica, *Phytochemistry* **14,** 1549 (1975).
211. S. Sciuto, M. Piattelli, R. Chillemi, G. Furnari, and M. Cormaci, *Phycologia* **18,** 196 (1979).
212. F. Ardré, *Rev. Algol.* [N.S.] **12,** 73 (1977).
213. G. Impellizzeri, M. Piattelli, S. Sciuto, and E. Fattorusso, *Phytochemistry* **16,** 1601 (1977).
214. A. Welter, M. Marlier, and G. Dardenne, *Phytochemistry* **17,** 131 (1978).
215. E. Fattorusso and M. Piattelli, *in* "Marine Natural Products: Chemical and Biological Perspectives" (P. J. Scheuer, ed.), Vol. 3, p. 95. Academic Press, New York, 1980.
216. K. Daigo, *Yakugaku Zasshi* **79,** 350, 353, 356, 365 (1959); *Chem. Abstr.* **53,** 14218 (1959).
217. H. Watase and I. Nitta, *Bull. Chem. Soc. Jpn.* **30,** 899 (1957); I. Nitta, H. Watase, and Y. Tomie, *Nature (London)* **181,** 761 (1958); H. Watase, Y. Tomie, and I. Nitta, *Bull. Chem. Soc. Jpn.* **31,** 714 (1958).
218. H. Watase, *Bull. Chem. Soc. Jpn.* **31,** 932 (1958).
219. D. W. J. Cruickshank, *Acta Crystallogr.* **12,** 1952 (1959).
220. K. Nomoto, T. Iwashita, Y. Ohtune, T. Takemoto, and K. Daigo, to be published.
221. W. Oppolzer and K. Thirring, *J. Am. Chem. Soc.* **104,** 4978 (1982).
222. W. Oppolzer, C. Robbiani, and K. Bättig, *Helv. Chim. Acta* **63,** 2015 (1980).
223. Y. Ohfune and M. Tomita, *J. Am. Chem. Soc.* **104,** 3511 (1982).
224. W. Oppolzer and H. Andres, *Helv. Chim. Acta* **62,** 2282 (1979); *Tetrahedron Lett.* 3397 (1978).
225. J. Ueyanagi, H. Nawa, M. Honjo, R. Nakamori, K. Tanaka, Y. Ueno, and S. Tatsuoka, *Yakugaku Zasshi* **77,** 613, 618 (1957); *Chem. Abstr.* **51,** 16429, 16430 (1957); Y. Ueno, K. Tanaka, J. Ueyanagi, H. Nawa, Y. Sanno, M. Honjo, R. Nakamori, T. Sugawa, M. Uchibayashi, K. Usugi, and S. Tatsuoka, *Proc. Jpn. Acad.* **33,** 53 (1957).

226. M. Miyamoto, T. Sugawa, H. Morimoto, M. Uchibayashi, K. Tanaka, and S. Tatsuoka, *Yakugaku Zasshi* **77,** 580 (1957); M. Miyamoto, M. Honjo, Y. Sanno, M. Uchibayashi, K. Tanaka, and S. Tatsuoka, *ibid.* 568; *Chem. Abstr.* **51,** 16424, 16425 (1957); K. Tanaka, M. Miyamoto, M. Honjo, H. Morimoto, T. Sukawa, M. Uchibayashi, Y. Sanno, and S. Tatsuoka, *Proc. Jpn. Acad.* **33,** 47 (1957).
227. E. G. McGeer, J. W. Olney, and P. L. McBeer, "Kainic Acid as a Tool in Neurobiology." Raven Press, New York, 1978.
228. J. C. Watkins, *in* "Glutamate Transmitter in the Central Nervous System" (P. J. Roberts, J. Storm-Mathisen, and G. A. R. Johnston, eds.), p. 1. Wiley, New York, 1981.
229. J. H. Cardellina, II, F.-J. Marner, and R. E. Moore, *J. Am. Chem. Soc.* **101,** 240 (1979).
230. J. H. Cardellina, II, D. Dalietos, F.-J. Marner, J. S. Mynderse, and R. E. Moore, *Phytochemistry* **17,** 2091 (1978).
231. J. S. Mynderse and R. E. Moore, unpublished results (from reference *176*).
232. A. F. Rose, P. J. Scheuer, J. P. Springer, and J. Clardy, *J. Am. Chem. Soc.* **100,** 7665 (1978).
233. J. S. Mynderse and R. E. Moore, *J. Org. Chem.* **43,** 4359 (1978).
234. C. J. Simons, F.-J. Marner, J. H. Cardellina, II, R. E. Moore, and K. Seff, *Tetrahedron Lett.* 2003 (1979).
235. G. Cimino, S. De Stefano, and L. Minale, *Experientia* **31,** 1387 (1975).
236. D. B. Stierle and D. J. Faulkner, *J. Org. Chem.* **45,** 4980 (1980).
237. W. Hofheinz and W. E. Oberhänsli, *Helv. Chim. Acta* **60,** 660 (1977).
238. J. M. Frinke and D. J. Faulkner, *J. Am. Chem. Soc.* **104** 265 (1982); *ibid.* 5004.
239. T. Ohira, *J. Agric. Chem. Soc. Jpn.* **15,** 370 (1939); *Chem. Abstr.* **33,** 6245 (1939).
240. T. Ohira, *J. Agric. Chem. Soc. Jpn.* **16,** 1, 293 (1940); *Chem. Abstr.* **34,** 3683, 5826 (1940); T. Ohira, *ibid.* **18,** 915 (1942); *Chem. Abstr.* **45,** 4753 (1951); Y. Tasawa, *Acta Phytochim.* **15,** 141 (1949); T. Hira, H. Kido, and K. Imai, *Nippon Nogei Kagaku Kaishi* **29,** 520 (1955); *Chem. Abstr.* **53,** 1169 (1959).
241. T. Kaneko, T. Shiba, S. Wataria, S. Imai, T. Shimada, and K. Ueno, *Chem. Ind. (London)* 986 (1957); J. Rudinger and Z. Pravda, *Chem. Listy* **52,** 120 (1958); T. Shiba and S. Imai, *Nippon Kagaku Zasshi* **80,** 492 (1959); *Chem. Abstr.* **55,** 3452 (1961).
242. S. Konogaya, *Nippon Suisan Gakkaishi* **32,** 967 (1966).
243. S. Konagaya, *Nippon Suisan Gakkaishi* **33,** 417 (1967).
244. C. A. Dekker, D. Stone, and J. S. Fruton, *J. Biol. Chem.* **181,** 719 (1949).
245. L. Agius, J. A. Ballantine, V. Ferrito, V. Jaccarini, P. Murray-Rust, A. Petter, A. F. Psaila, and P. J. Schembri, *Pure Appl. Chem.* **51,** 1847 (1979).
246. W. Rüdiger, W. Klose, B. Rursch, N. Houvenaghel-Crevecoeur, and H. Budzikiewicz, *Justus Liebigs Ann. Chem.* **713,** 209 (1968).
247. W. Rüdiger, *Hoppe-Seyler's Z. Physiol. Chem.* **348,** 129 1554 (1967).
248. T. Ogata, N. Fusetani, and K. Yamaguchi, *Comp. Biochem. Physiol. B* **63B,** 239 (1979).
249. R. Kazlauskas, J. F. Marwood, P. T. Murphy, and R. J. Wells, *Aust. J. Chem.* **35,** 215 (1982).
250. Y. Hashimoto and J. Tsutsumi, *Bull. Jpn. Soc. Sci. Fish.* **27,** 859 (1961).
251. F. Wiedenmayer, "Shallow Water Sponges of the Western Bahamas," *Experientia, Suppl.* **28,** p. 287. Birkhaeuser, Basel, 1977.
252. C. J. Barwell and G. Blunden, *J. Nat. Prod.* **44,** 500 (1981).
253. K. C. Guven, A. Bora, and G. Sunam, *Phytochemistry* **9,** 1893 (1970).
254. H. Kawauchi and T. Sasaki, *Bull. Jpn. Soc. Sci. Fish.* **44,** 135 (1978).
255. K. Hori, T. Yamamoto, K. Miyazawa, and K. Ito, *J. Fac. Appl. Biol. Sci., Hiroshima Univ.* **18,** 65 (1979).
256. K. Hori, T. Yamamoto, K. Miyazawa, and K. Ito, *Bull. Jpn. Soc. Sci. Fish.* **46,** 559 (1980).
257. R. P. Gregson, R. R. Lohr, J. F. Marwood, and R. J. Quinn, *Experientia* **37,** 493 (1981).
258. E. Dreschel, *Z. Biol. (Munich)* **33,** 96 (1907).
259. C. T. Morner, *Hoppe Seyler's Z. Physiol. Chem.* **88,** 138 (1913).

260. D. Ackermann and C. Burchard, *Hoppe Seyler's Z. Physiol. Chem.* **271,** 183 (1941).
261. E. M. Low, *J. Mar. Res.* **10,** 239 (1951).
262. S. Hunt and S. W. Brever, *Biochim. Biophys. Acta* **252,** 401 (1971).
263. M. De Rosa, L. Minale, and G. Sodano, *Comp. Biochem. Physiol.* **45B,** 883 (1973).
264. A. A. Tymiak and K. L. Rinehart, Jr., *J. Am. Chem. Soc.* **103,** 6763 (1981).
265. G. Cimino, S. De Stefano, and L. Minale, *Experientia* **31,** 756 (1975).
266. A. Kelecom and G. J. Kannengiesser, *An. Acad. Bras. Cienc.* **51,** 633 (1979).
267. G. M. Sharma and P. R. Burkholder, *J. Antibiot., Ser. A* **20,** 200 (1967).
268. G. M. Sharma, B. Vig, and P. R. Burkholder, *J. Org. Chem.* **35,** 2823 (1970).
269. G. M. Sharma and P. R. Burkholder, *Tetrahedron Lett.* 4147 (1967).
270. J. Andersen and D. J. Faulkner, *Tetrahedron Lett.* 1175 (1973).
271. G. J. Kaspered, T. C. Bruice, H. Yagi, N. Kaubisch, and D. M. Jerina, *J. Am. Chem. Soc.* **94,** 7876 (1972).
272. A. Kelecom and G. J. Kannengiesser, *An. Acad. Bras. Cienc.* **51,** 639 (1979).
273. G. E. Krejcarek, R. H. White, L. P. Hager, W. O. McClure, R. D. Johnson, K. L. Rinehart, Jr., J. A. McMillan, I. C. Paul, P. D. Shaw, and R. C. Brusca, *Tetrahedron Lett.* 507 (1975).
274. K. Krohn, *Tetrahedron Lett.* 4667 (1975).
275. D. B. Borders, G. O. Morton, and E. R. Wetzel, *Tetrahedron Lett.* 2709 (1974).
276. T. N. Makarieva, V. A. Stonik, P. Alcolado, and Y. B. Elyakov, *Comp. Biochem. Physiol.* **68B,** 481 (1981).
277. T. N. Makarieva, V. A. Stonik, R. Gra, and Kh. Valukha, *Khim. Prir. Soedin.* 581 (1980).
278. E. Fattorusso, L. Minale, and G. Sodano, *J. Chem. Soc., Perkin Trans. 1* 16 (1972).
279. W. Fulmor, G. E. Van Lear, G. D. Morton, and R. O. Mills, *Tetrahedron Lett.* 4551 (1970).
280. L. Mazzarella and R. Puliti, *Gazz. Chim. Ital.* **102,** 391 (1972).
281. D. B. Cosulich and F. M. Lovell, *J. Chem. Soc., Chem. Commun.* 397 (1971).
282. E. Fattorusso, L. Minale, and G. Sodano, *J. Chem. Soc., Chem. Commun.* 751 (1970).
283. M. D'Ambrosio, A. Guerriero, P. Traldi, and F. Pietra, *Tetrahedron Lett.* **23,** 4403 (1982).
284. E. Fattorusso, L. Minale, G. Sodano, K. Moody, and R. H. Thomson, *J. Chem. Soc., Chem. Commun.* 752 (1970).
285. K. Moody, R. H. Thomson, E. Fattorusso, L. Minale, and G. Sodano, *J. Chem. Perkin Trans. 1* 18 (1972).
286. J. A. McMillan, I. C. Paul, Y. M. Goo, K. L. Rinehart, Jr., W. C. Krueger, and L. M. Pschigoda, *Tetrahedron Lett.* **22,** 39 (1981).
287. Y. Gopichand and F. J. Schmitz, *Tetrahedron Lett.* 3921 (1979).
288. Y. Kahsman, A. Groweiss, S. Carmely, Z. Kinamoni, D. Czarkie, and M. Rotem, *Pure Appl. Chem.* **54,** 1995 (1982).
289. R. Kazlauskas, R. O. Lidgard, P. T. Murphy, and R. J. Wells, *Tetrahedron Lett.* **21,** 2277 (1980).
290. R. Kazlauskas, R. O. Lidgard, P. T. Murphy, R. J. Wells, and J. F. Blount, *Aust. J. Chem.* **34,** 765 (1981).
291. S. Nishiyama and S. Yamamura, *Tetrahedron Lett.* **23,** 1281 (1982).
292. F. A. Hoppe-Seyler, *Hoppe-Seyler's Z. Physiol. Chem.* **222,** 105 (1933).
293. E. L. Gasteiger, P. C. Haake, and J. A. Gergen, *Ann. N. Y. Acad. Sci.* **90,** 622 (1960).
294. J. C. Netherton, III and S. Gurin, *J. Biol. Chem.* **257,** 11971 (1982).
295. P. H. List, *Planta Med.* **6,** 424 (1958).
296. D. Ackermann and R. Pant, *Hoppe-Seyler's Z. Physiol. Chem.* **326,** 197 (1971).
297. J. C. Madgwick and B. J. Ralph, *Bot. Mar.* **15,** 205 (1972).
298. G. Impellizzeri, S. Mangiafico, G. Oriente, M. Piattelli, S. Sciuto, E. Fattorusso, S. Magno, C. Santacrose, and D. Sica, *Phytochemistry* **14,** 1549 (1975).
299. P. R. Bergguist and W. P. Hartman, *Mar. Biol. (Berlin)* **3,** 247 (1969).

300. N. Takagi, H. Y. Hsu, and T. Takemoto, *Yakugaku Zasshi* **90,** 899 (1970).
301. K. Oka, S. Tanaka, H. Hasegawa, H. Imaio, I. Fujishiro, and R. Konishi, *Nagasaki Igakkai Zasshi* **35,** 564 (1960); *Chem. Abstr.* **54,** 17580 (1960).
302. H. L. Sleeper and W. Fenical, *J. Am. Chem. Soc.* **99,** 2367 (1977).
303. M. Sakakibara and M. Matsui, *Agric. Biol. Chem.* **43,** 117 (1979).
304. W. Fenical, H. L. Sleeper, V. J. Paul, M. O. Stallard, and H. H. Sun, *Pure Appl. Chem.* **51,** 1865 (1979).
305. R. D. Barnes, "Invertebrate Zoology," 2nd ed., p. 157. Saunders, Philadelphia, Pennsylvania, 1968.
306. Z. M. Bacq, *Arch. Int. Physiol.* **44,** 190 (1937).
307. W. R. Kem, R. M. Coates, and B. C. Abbot, *Fed. Proc., Fed. Am. Soc. Exp. Biol.* **28,** 610 (1969); W. R. Kem, B. C. Abbot, and R. M. Coates, *Toxicon* **9,** 15 (1971); W. R. Kem, *ibid.* 23.
308. E. Späth and L. Mamoli, *Chem. Ber.* **69,** 1082 (1936).
309. A. Orechoff and G. Menschikoff, *Chem. Ber.* **64,** 266 (1931).
310. W. R. Kem, K. N. Scott, and J. H. Duncan, *Experientia* **32,** 684 (1976).
311. W. R. Kem, *J. Biol. Chem.* **251** 4184 (1976); K. M. Blumenthal, *Biochemistry* **21,** 4229 (1982).
312. F. J. Schmitz, D. C. Campbell, K. Hollenbeak, D. J. Vanderah, L. S. Ciereszko, P. Stendler, J. D. Eckstrand, D. van der Helm, P. Kaul, and S. Kulkarni, *in* "Marine Natural Products Chemistry" (D. J. Faulkner and W. H. Fenical, eds.), p. 293. Plenum, New York, 1977.
313. F. J. Schmitz, K. H. Hollenbeak, and D. C. Campbell, *J. Org. Chem.* **43,** 3916 (1978).
314. K. Biemann, G. Büchi, and B. H. Walker, *J. Am. Chem. Soc.* **79,** 5558 (1957).
315. C. M. Wang, T. Narahashi, and T. J. Mende, *Toxicon* **11,** 499 (1973).
316. D. E. McIntyre, D. J. Faulkner, D. Van Engen, and J. Clardy, *Tetrahedron Lett.* 4163 (1979).
317. A. Kubo, S. Nakahara, R. Iwata, K. Takahashi, and T. Arai, *Tetrahedron Lett.* **21,** 3207 (1980).
318. S. Danishefsky, E. Berman, R. Cvetovich, and J. Minamikawa, *Tetrahedron Lett.* **21,** 4819 (1980); A. Kubo and S. Nakahra, *Chem. Pharm. Bull.* **29,** 595 (1981).
319. T. Arai and A. Kubo, *in* "The Alkaloids" (A. Brossi ed.), Vol. 21, pp. 55–100. Academic Press, New York, 1983.
320. H. Fukumi, H. Kurihara, T. Hata, C. Tamura, H. Mishima, A. Kubo, and T. Arai, *Tetrahedron Lett.* 3825 (1977); H. Fukumi, H. Kurihara, and H. Mishima, *Chem. Pharm. Bull.* **26,** 2175 (1978).
321. T. Arai, K. Takahashi, H. Nakahara, and A. Kubo, *Experientia* **36,** 1025 (1980).
322. T. Arai, K. Takahashi, A. Kubo, S. Nakahara, S. Sato, K. Aiba, and C. Tamura, *Tetrahedron Lett.* 2355 (1979).
323. T. Fukuyama and R. A. Sachleben, *J. Am. Chem. Soc.* **104,** 4957 (1982).
324. R. S. Jacobs, S. White, and L. Wilson, *Fed. Proc., Fed. Am. Soc. Exp. Biol.* **40,** 26 (1981).
325. E. Fattorusso, S. Forenza, L. Minale, and G. Sodano, *Gazz. Chim. Ital.* **101,** 104 (1971).
326. D. Coppini, *Gazz. Chim. Ital.* **80,** 36 (1950).
327. I. C. Wells, W. H. Elliot, S. A. Thayer, and E. A. Doisy, *J. BIol. Chem.* **196,** 331 (1952).
328. I. C. Wells, *J. Biol. Chem.* **196,** 321 (1952).
329. P. Wulff, J. S. Carlé, and C. Christophersen, *Comp. Biochem. Physiol.* **71B,** 525 (1982).
330. R. Kazlauskas, R. O. Lidgard, R. J. Wells, and W. Vetter, *Tetrahedron Lett.*, 3183 (1977).
331. C. Charles, J. C. Braekman, D. Daloze, and B. Tursch, *Tetrahedron* **36,** 2133 (1980).
332. C. Charles, J. C. Brackman, D. Daloze, and B. Tursch, *Tetrahedron Lett.* 1519 (1978).
333. K. L. Erickson and R. J. Wells, *Aust. J. Chem.* **35,** 31 (1982).
334. R. Kazlauskas, P. T. Murphy, and R. J. Wells, *Tetrahedron Lett.* 4945 (1978).
335. C. M. Ireland and P. J. Scheuer, *J. Am. Chem. Soc.* **102,** 5688 (1980).

336. C. M. Ireland, A. R. Durso, Jr., R. A. Newman, and M. P. Hacker, *J. Org. Chem.* **47,** 1807 (1982).
337. R. A. Lewin and N. W. Withers, *Nature (London)* **256,** 735 (1975).
338. E. H. Newcomb and T. D. Pugh, *Nature (London)* **253,** 533 (1975).
339. C. M. Ireland, A. R. Durso, Jr., and P. J. Scheuer, *J. Nat. Prod.* **44,** 360 (1981).
340. G. R. Pettit, Y. Kamano, Y. Fuji, C. L. Herald, M. Inoue, P. Brown, D. Gust, K. Kitahara, J. M. Schmidt, D. L. Doubek, and C. Michel, *J. Nat. Prod.* **44,** 482 (1981).
341. G. R. Pettit, Y. Kamano, P. Brown, D. Gust, M. Inoue, and C. L. Herald, *J. Am. Chem. Soc.* **104,** 905 (1982).
342. Y. Kashman, A. Groweiss, and U. Schmueli, *Tetrahedron Lett.* **21,** 3629 (1980).
343. G. Prota, S. Ito, and G. Nardi, *in* ''Marine Natural Products Chemistry'' (D. J. Faulkner and W. H. Fenical, eds.), p. 45. Plenum, New York, 1977.
344. A. Palumbo, M. Dishia, G. Misuraca, and G. Prota, *Tetrahedron Lett.* **23,** 3207 (1982).
345. M. Alam, R. Sanduja, M. Bilayet-Hossain, and O. van der Helm, *J. Am. Chem. Soc.* **104,** 5232 (1982).
346. J. C. Brackman, D. Dalose, P. Macedo de Abreu, C. Piccinni-Leopardi, G. Germain, and M. Van Meerssche, *Tetrahedron Lett.* **23,** 4277 (1982).
347. G. Cimino, S. De Rosa, S. De Stefano, L. Cariello, and L. Zanetti, *Experientia* **38,** 896 (1982).
348. F. Chioccara, G. Misuraca, E. Novellino, and G. Prota, *Tetrahedron Lett.* 3181 (1979).
349. H. Nakamura, J. Kobayashi, and Y. Hirata, *Chem. Lett.* 1413 (1981).
350. J. Kobayashi, J. Nakamura, and Y. Hirata, *Tetrahedron Lett.* **22,** 3001 (1981).
351. K. Yabe, I. Sekikawa, and I. Tsujino, *Hokkaido Kyoiku Daigaku Kiyo, Dai-2-bu, A* **30,** 271 (1980); *Chem. Abstr.* **95,** 2125 (1981).
352. S. Takano, D. Uemura, and Y. Hirata, *Tetrahedron Lett.* 4909 (1978).
353. S. Ito and Y. Hirata, *Tetrahedron Lett.* 2429 (1977).
354. S. Takano, D. Uemura, and Y. Hirata, *Tetrahedron Lett.* 2299 (1978).
355. R. E. Moore and P. J. Scheuer, *Science* **172,** 495 (1971).
356. R. E. Moore and G. Bartolini, *J. Am. Chem. Soc.* **103,** 2491 (1981).
357. R. E. Moore, G. Bartolini, J. Barchi, A. A. Bothner-By, J. Dadok, and J. Ford, *J. Am. Chem. Soc.* **104** 3776 (1982); *ibid.* 5572.
358. D. Uemura, K. Ueda, Y. Hirata, C. Katayama, and J. Tanaka, *Tetrahedron Lett.* **21,** 4857, 4861 (1980).
359. D. Uemura, K. Ueda, Y. Hirata, H. Naoki, and T. Iwashita, *Tetrahedron Lett.* **22,** 1909, 2781 (1981).
360. W. C. Still and I. Galynker, *J. Am. Chem. Soc.* **104,** 1774 (1982).
361. F. J. Schmitz, K. H. Hollenbeak, and R. S. Prasad, *Tetrahedron Lett.* 3387 (1979).
362. R. Kazlauskas, P. T. Murphy, B. N. Ravi, R. L. Sanders, and R. J. Wells, *Aust. J. Chem.* **35,** 69 (1982).
363. R. Kazlauskas, J. F. Marwood, and R. J. Wells, *Aust. J. Chem.* **33,** 1799 (1980).
364. B. Ganem, *Acc. Chem. Res.* **15,** 290 (1982).
365. J. M. Sieburth, ''Sea Microbes.'' Oxford Univ. Press, London and New York, 1979.
366. M. Barbier, *Naturwissenschaften* **69,** 341 (1982).
367. D. J. Faulkner *Nat. Prod. Rep. 1,* 251 (1984).
368. T. Noguchi, H. Narita, J. Maruyama, and K. Hashimoto *Bull. Japn. Soc. Sci. Fish.* **48,** 1173 (1982).
369. T. Matsui, S. Hamada, and K. Yamamori *Bull. Japn. Soc. Sci. Fish* **48,** 1179 (1982).
370. J. F. W. Keana, P. J. Boyle, M. Erion, R. Hartling, J. R. Husman, J. E. Richman, R. B. Roman, and F. M. Wah, *J. Org. Chem.* **48,** 3621 (1983).

371. J. F. W. Keana, J. S. Bland, P. J. Boyle, M. Erion, R. Hartling, J. R. Husman, R. B. Roman, G. Ferguson, and M. Parvez, *J. Org. Chem.* **48,** 3627 (1983).
372. K. Koyama, T. Noguchi, A. Uzu, and K. Hashimoto, *Bull. Japn. Soc. Sci. Fish.* **49,** 485 (1983).
373. T. Noguchi, J. Maruyama, Y. Onoue, K. Hashimoto, and T. Ikeda, *Bull. Japn. Soc. Sci. Fish.* **49,** 499 (1983).
374. J. Maruyama, T. Noguchi, Y. Onove, Y. Ueda, K. Hashimoto, and S. Kamimura *Bull. Japn. Soc. Sci. Fish.* **49,** 233 (1983).
375. G. S. Jamieson and R. A. Chandler, *Can. J. Fish. Aquat. Sci.* **40,** 313 (1983).
376. Y. Kotaki, M. Tajiri, Y. Oshima, and T. Yasumoto, *Bull. Japn. Soc. Sci. Fish.* **49,** 283 (1983).
377. S. M. Hannick and Y. Kishi, *J. Org. Chem.* **48,** 3833 (1983).
378. A. Hori and Y. Shimuzu, *J. Chem. Soc., Chem. Commun.* 790 (1983).
379. L. Maat and H. C. Beyerman, *In* "The Alkaloids" (A. Brossi, ed.), Vol. 22, Chapter 5, p. 281. Academic Press, New York, 1983.
380. I. Kitagawa, M. Kobayashi, K. Kitanaka, M. Kido, and Y. Kyogoku, *Chem. Pharm. Bull.* **31,** 2321 (1983).
381. K. L. Rinehart, Jr., G. C. Harbour, M. D. Graves, and M. T. Cheng, *Tetrahedron Lett.* **24,** 1593 (1983).
382. B. B. Snider and W. C. Faith, *Tetrahedron Lett.* **24,** 861 (1983).
383. B. B. Snider and W. C. Faith, *J. Am. Chem. Soc.* **106,** 1443 (1984).
384. W. R. Roush and A. E. Walts, *J. Am. Chem. Soc.* **106,** 721 (1984).
385. H. Nakamura, H. Wu, J. Kabayashi, Y. Ohizumi, Y. Hirata, T. Higashijima, and T. Miyazawa, *Tetrahedron Lett.* **24,** 4105 (1983).
386. R. J. Capon and D. J. Faulkner, *J. Am. Chem. Soc.* **106,** 1819 (1984).
387. J. W. Blunt, M. H. G. Munro, and S. C. Yorke, *Tetrahedron Lett.* **23,** 2793 (1982).
388. B. Sullivan, D. J. Faulkner, and L. Webb, *Science* **221,** 1175 (1983).
389. F. J. Schmitz, D. P. Michaud, and P. G. Schmidt, *J. Am. Chem. Soc.* **104,** 6415 (1982).
390. Reviewed by K. L. Erickson, ref. 26 p. 223.
391. T. Higa, personal communication.
392. T. Higa and S.-I. Sakemi, *J. Chem. Ecol.* **9,** 495 (1983).
393. M. P. Kirkup and R. E. Moore, *Tetrahederon Lett.* **24,** 2078 (1983).
394. K. L. Rinehart, Jr., and J. Kobayashi, G. C. Harbour, R. G. Hughes, Jr., S. A. Mizsak, and T. A. Scahill, *J. Am. Chem. Soc.* **106,** 1524 (1984).
395. J. Kobayashi, G. C. Harbour, J. Gilmore, and K. L. Rinehart, Jr., *J. Am. Chem. Soc.* **106,** 1526 (1984).
396. A. Sato and W. Fenical, *Tetrahedron Lett.* **24,** 481 (1983).
397. T. Hino, T. Tanaka, K. Matsuki, and M. Nakagawa, *Chem. Pharm. Bull.* **31,** 1806 (1983).
398. R. L. Zimmer and R. M. Woollacott, *Science* **220,** 208 (1983).
399. B. Carté and D. J. Faulkner, *J. Org. Chem.* **48,** 2314 (1983).
400. S. J. Giovannioni and L. Margulis, *Microbios* **30,** 47 (1981).
401. N. N. Gerber and M. J. Gauthier, *Appl. Environment Microbiol.* **37,** 1176 (1979).
402. H. Laatsch and R. H. Thomson, *Tetrahedron Lett.* **24,** 2701 (1983).
403. G. A. Kraus and J. O. Nagy, *Tetrahedron Lett.* **24,** 3427 (1983).
404. J. Garthwaite and G. Garthwaite, *Nature (London)* **305,** 138 (1983).
405. B. F. Bowden, P. S. Clezy, J. C. Coll, B. N. Ravi, and D. M. Tapiolas, *Aust. J. Chem.* **37,** 227 (1984).
406. M. Takagi, S. Funahashi, K. Ohta, and T. Nakabayashi, *Agric. Biol. Chem.* **44,** 3019 (1980).
407. J. A. Ballantine, A. F. Fsaila, A. Pelter, P. Murray-Rust, V. Ferrito, P. Schembri, and V. Jaccarine, *J. Chem. Soc., Perkin Trans. I* 1080 (1980).

408. C. J. Dutton, C. J. R. Fookes, and A. Battersby, *J. Chem. Soc., Chem. Commun.* 1237 (1983).
409. K. D. Barrow, *In* "Marine Natural Products, Chemial and Biological Perspectives" (D. J. Scheuer, ed.), Vol. 5, p. 51. Academic Press, New York, 1983.
410. G. Cimino, S. De Rosa, S. De Stefano, R. Self, and G. Sodano, *Tetrahedron Lett.* **24,** 3029 (1983).
411. S. Nishiyama and S. Yamamura, *Tetrahedron Lett.* **24,** 3351 (1983).
412. M. Rotem, S. Carmely, Y. Kashman, and Y. Loya, *Tetrahedron* **39,** 667 (1983).
413. S. Nishiyama, T. Suzuki, and S. Yamamura, *Chem. Lett.* 1851 (1981).
414. S. Nishiyama and S. Yamamura, *Tetrahedron Lett.* **23,** 1281 (1982).
415. S. Nishiyama, T. Suzuki, and S. Yamamura, *Tetrahedron Lett.* **23,** 3699 (1982).
416. J. Clelland and G. R. Knox, *J. Chem. Soc., Chem. Commun.* 1219 (1983).
417. G. Cimino, S. De Rosa, S. De Stefano, A. Spinella, and G. Sodano, *Tetrahedron Lett.* **25,** 2925 (1984).
418. T. Hayashi, T. Noto, T. Nawata, M. Okazaki, M. Sawada, and K. Ando, *J. Antibiot. (Tokyo)* **35,** 771 (1982).
419. J. E. Biskupiak and C. M. Ireland *Tetrahedron Lett.* **25,** 2935 (1984).
420. P. G. Jones, *Chem. Soc. Rev.* **13,** 157 (1984).
421. Y. Hamamoto, M. Endo, M. Nakagawa, T. Nakanishi, and K. Mizukawa, *J. Chem. Soc., Chem. Commun.* 313 (1983).
422. J. M. Wasylyk, J. E. Biskupiak, C. E. Costello, and C. M. Ireland, *J. Org. Chem.* **48,** 4445 (1983).
423. J. E. Biskupiak and C. M. Ireland, *J. Org. Chem.* **48,** 2302 (1983).
424. D. F. Sesin and C. M. Ireland, *Tetrahedron Lett.* **25,** 403 (1984).
425. T. Wakamiya, Y. Kobayashi, T. Shiba, K. Setogawa, and H. Matsutani, *Tetrahedron* **40,** 235 (1984).
426. A. Groweiss, U. Shmueli, and Y. Kashman, *J. Org. Chem.* **48,** 3512 (1983).
427. I. Spector, N. R. Shochet, Y. Kashman, and A. Groweiss, *Science* **4584,** 493 (1983).
428. M. Nakagawa, M. Endo, N. Tanaka, and L. Gen-Pei, *Tetrahedron Lett.* **25,** 3227 (1984).
429. M. DiNovi, D. A. Trainor, K. Nakanishi, R. Sanduja, and M. Alam, *Tetrahedron Lett.* **24,** 855 (1983).
430. T. Ogita, S. Gunji, Y. Fukazawa, A. Terahara, T. Kinoshita, H. Nagaki, and T. Beppu, *Tetrahedron Lett.* **24,** 2283 (1983).
431. H. Nakamura, J. Kobayashi, Y. Ohizumi, and Y. Hirata, *Tetrahedron Lett.* **23,** 5555 (1982).
432. F. J. Schmitz, S. K. Agarwal, S. P. Gunasekera, P. G. Schmidt, and J. N. Shoolery, *J. Am. Chem. Soc.* **105,** 4835 (1983).
433. L. L. Klein, W. W. McWhorter Jr., S. S. Ko, K.-P. Pfaff, Y. Kishi, D. Uemura, and Y. Hirata, *J. Am. Chem. Soc.* **104,** 7362 (1982).
434. S. S. Ko, J. M. Finan, M. Yonaga, Y. Kishi, D. Uemura, and Y. Hirata, *J. Am. Chem. Soc.* **104,** 7364 (1982).
435. H. Fujioka, W. J. Christ, J. K. Cha, J. Leder, Y. Kishi, D. Uemura, and Y. Hirata, *J. Am. Chem. Soc.* **104,** 7367 (1982).
436. J. K. Cha, W. J. Christ, J. M. Finan, H. Fujioka, Y. Kishi, L. L. Klein, S. S. Ko, J. Leder, W. W. McWhorter Jr., K.-P. Pfaff, M. Yonaga, D. Uemura, and Y. Hirata, *J. Am. Chem. Soc.* **104,** 7369 (1982).
437. S. S. Ko, L. L. Klein, K.-P. Pfaff, and Y. Kishi, *Tetrahedron Lett.* **23,** 4415 (1982).
438. W. W. McWhorter Jr., S. H. Kang, and Y. Kishi, *Tetrahedron Lett.* **24,** 2243 (1983).
439. J. Leder, H. Fujioka, and Y. Kishi, *Tetrahedron Lett.* **24,** 1463 (1983).
440. K. Mori, T. Umemura, *Tetrahedron Lett.* **23,** 3391 (1982).
441. Y. Shimizu, H. Shimizu, P. J. Scheuer, Y. Hokama, M. Oyama, and J. T. Miyahara, *Bull. Jpn. Soc. Sci. Fish.* **48,** 811 (1982).

442. J.-N. Bidard, M. P. M. Vijverberg, C. Frelin, E. Chungue, A.-M. Legrand, R. Bagnis, and M. Lazdunski, *J. Biol. Chem.* **259,** 8353 (1984).
443. R. Kazlauskas, P. T. Murphy, R. J. Wells, J. A. Baird-Lambert, and D. D. Jamieson, *Aust. J. Chem.* **36,** 165 (1983).
444. T. Nakatsu, D. J. Faulkner, G. K. Matsumoto, and J. Clardy, *Tetrahedron Lett.* **25,** 935 (1984).
445. R. J. Capon and D. J. Faulkner, *J. Am. Chem. Soc.* **106,** 1819 (1984).

——CHAPTER 3——

ARISTOTELIA ALKALOIDS

I. RALPH C. BICK AND MOHAMMAD A. HAI*

Chemistry Department, University of Tasmania
Hobart, Australia

I. Introduction

The *Aristotelia* spp. belong to the Elaeocarpaceae, a small family found mainly in tropical regions with about 350 species in seven genera. One of these, the

* Present address: Department of Chemistry, Jahangirnagar University, Savar, Dacca, Bangladesh.

THE ALKALOIDS, VOL. XXIV

ISBN 0-12-469524-8

genus *Elaeocarpus,* produces almost exclusively indolizidine alkaloids, the only exception being an indole base (*1*). A series of pyrrolidine alkaloids related structurally to these indolizidines has been isolated from a *Peripentadenia* sp. (*2*); of the remaining elaeocarpaceous genera, one *Aceratium* sp. has been found to contain alkaloids (*3*) that have, however, not been examined in detail.

The four *Aristotelia* spp. listed below from which alkaloids have so far been obtained are somewhat exceptional in that they grow in more temperate regions, in countries bordering the South Pacific. A fifth member of the genus, *A. australasica* F.v.M. from New South Wales, contains alkaloids that are at present under examination; several other *Aristotelia* spp. have been described, but they are less well defined botanically. The alkaloids so far isolated are all indolic, but otherwise have no structural relationship with the only indole alkaloid from *Elaeocarpus* spp. Like the great majority of known indole alkaloids, they appear to be constructed from tryptamine and a terpenoid unit, but they differ inasmuch as the latter has not undergone a preliminary rearrangement to an iridoid before incorporation into the alkaloid structure.

II. Occurrence

A. peduncularis (Labill) Hook. F. is a straggling shrub known as heart berry in its native Tasmania, where it grows in rain-forest areas (*4*). The following alkaloids have been isolated from it (in ppm of dried whole plant material, *5–10*): peduncularine (**1**, 15–30), isopeduncularine (**8**, 0–10), aristoteline (**11**, 1.5), aristoserratine (**12**, 1.5), sorelline (**13**, 0.5), tasmanine (**14**, 0.5), and hobartine (**15**, 0.07). The amount of isopeduncularine appears to vary considerably with different samples of plant material.

A. serrata (J. B. and G. Forst) W. R. B. Oliver is a small tree commonly called wineberry in New Zealand (Maori name: *makomako*), where it is endemic and widely distributed in lowlands and foothills (*11*). The alkaloid content of the dried whole plant material is considerably higher than that of *A. peduncularis,* and includes (in ppm, *8, 10, 12–16*) aristoteline (**11**, 700), isopeduncularine (**8**, 20), serratoline (**17**, 20), tasmanine (**14**, 15), makonine (**18**, 6), makomakine (**19**, 14), aristomakine (**20**, 11), aristotelinone (**21**, 27), aristoserratine (**12**, 10), aristoserratenine (**24**, 9), isohobartine (**25**, 6), serratenone (**26**, 7), aristomakinine (**27**, 6), and isosorelline (**28**, 11). This species was initially reported (*12*) to contain peduncularine (**1**), but a subsequent investigation (*10*) showed that the alkaloid concerned is isopeduncularine (**8**).

A. fruticosa Hook. f. (mountain wineberry) is a tall bush endemic in New Zealand, where it is found in upland country except north of Auckland. It is polymorphic, with true breeding as well as habit-induced forms, and it hybridizes freely with *A. serrata* (*11*). The dried whole plant material contains (in ppm, *10, 17*) isopeduncularine (**8,** 53), fruticosonine (**29,** 20), aristofruticosine (**32,** 18), isosorelline (**28,** 2), and isohobartine (**25,** 2).

A. chilensis (Mol.) Stunz is a small tree growing in moist and shady parts of central and southern Chile and western Argentina, where it is known as *maqui* or *koelón* (*18, 19*). The alkaloid content of the dried leaves and stems is low, and comprises (in ppm, *20, 21*) aristoteline (**11,** 17), aristotelone (**33,** 1), aristotelinine (**34,** 0.7), and aristone (**35,** 0.4).

III. Characterization and Structural Determination

A. Peduncularine (1) [$C_{20}H_{24}N_2$, small, fine needles from chloroform, mp 155–157°C, $[\alpha]_D^{19}$ −24° (MeOH), $[\alpha]_D^{21}$ −76° ($CHCl_3$) (*5, 6, 22*)]

Peduncularine was the first alkaloid to be isolated from *Aristotelia* spp., and it differs from others reported subsequently in being very sparingly soluble in cold chloroform: NMR spectra had to be recorded initially in $CDCl_3$–CD_3OD mixtures, but this solvent had the advantage of spreading out the spectra.

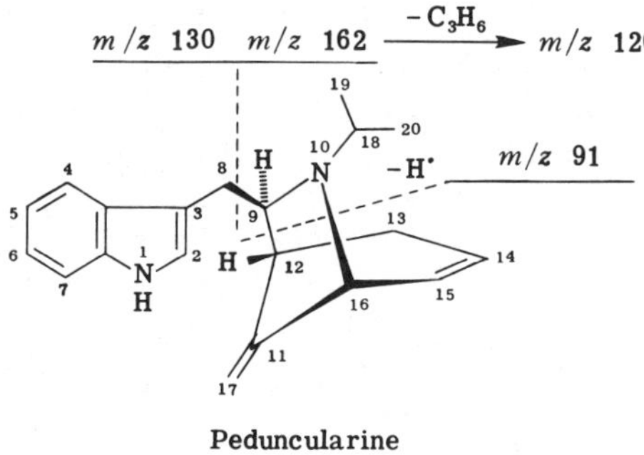

Peduncularine

1

The UV spectrum of peduncularine showed that it has an indole nucleus, and a positive Ehrlich test, together with the appearance in the ^{1}H-NMR spectrum of a broad aromatic proton signal weakly coupled to the N-1 proton, indicated an unsubstituted C-2 position; on the other hand, C-3 must bear a methylene group, in view of the strong *m*/*z* 130 ion in the mass spectrum and the signals attributable to a pair of geminally coupled protons in the allylic region of the ^{1}H-NMR spectrum. The geminal protons are coupled to a methine proton (H-9) that is not further coupled, and whose signal overlaps that of another methine proton (H-18); the chemical shifts of the latter two (δ2.85–3.1) suggested that they are located on carbons α to the basic nitrogen (N-10). The H-18 proton is coupled to the protons of two methyl groups, which thus form part of an isopropyl group.

However, it was considered initially (*5*) that this group was unlikely to be attached to nitrogen since the M − 15 peak in the mass spectrum is extremely weak, and a Herzig–Meyer determination on peduncularine gave a negligible value for an *N*-alkyl group; apart from this, an isopropyl group attached to nitrogen had not been reported in a natural product up to that time. On the other hand, the nonindolic nitrogen must be tertiary, since there is only one proton exchangeable with D_2O in the ^1H-NMR spectrum. The latter spectrum, furthermore, shows signals for four olefinic protons, and hydrogenation experiments confirmed the presence of two double bonds in peduncularine, one of which must be in a vinylidene group since two of the olefinic protons give sharp singlets that disappear on hydrogenation, being replaced by a doublet from a methyl group.

The other two olefinic protons (H-14 and H-15) are vicinally coupled, and one is further coupled to a proton (H-16) that resonates around δ3.9 (in $CDCl_3$), corresponding to a doubly allylic location adjacent to nitrogen. Of the remaining protons, a geminal pair (attached to C-13) are also allylic from their chemical shifts; their very large mutual coupling constant (18 Hz) indicated that they each make equal dihedral angles with the plane of an adjacent double bond. The methylene group to which they belong must thus be coplanar with an olefinic group, and this suggested that both groups might be in a five-membered ring (*5*). One of the same methylene protons is further coupled to a methine proton (H-12) whose signal also overlaps those of H-9 and H-18, and which was assigned initially (*5*) a location α to the basic nitrogen. These observations suggested a pyrrolizidine nucleus for peduncularine (*5*) with the isopropyl group attached to an α-methine carbon. The available evidence at this stage fitted structure **2** except that the methine proton of the isopropyl group resonates at lower field

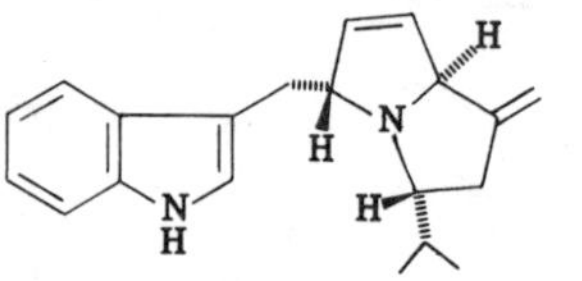

2 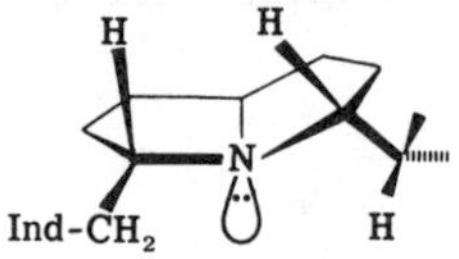 **3**

than expected from this formulation, and shows no coupling with the proton of the methine group to which the isopropyl is attached. It was suggested that this might be due to mutual restriction of rotation by the bulky skatolyl and isopropyl groups attached in peri positions and cis to one another (cf. **3**), which could result in a dihedral angle of 90° between the two methine protons in question and the deshielding of the one in the isopropyl group by the lone pair electrons on the nitrogen (*5*).

The molecular formula of peduncularine suggested that it was composed of a tryptamine and a terpenoid unit, and an alternative structural formula (**4**) was put forward (*23*) based on the same data as before, in which the terpenoid residue conforms to the isoprene rule. However, this formula agreed less well with the

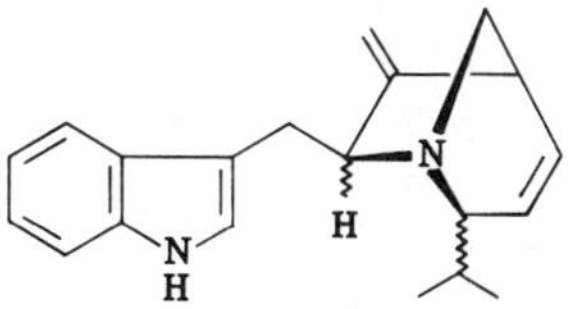

4

spectroscopic evidence: in particular, protons of the methylene group placed next to a nitrogen instead of a planar olefinic group would have little more than half the observed coupling constant, and structural formula **4** no longer explains the low chemical shift of the methine proton in the isopropyl group, nor why this proton shows no coupling with the vicinal methine proton.

Further evidence from high-resolution ^{1}H- and ^{13}C-NMR spectroscopy made structural formulas **2** and **4** both untenable for peduncularine: a weak coupling was detected between the vicinally coupled olefinic proton H-14 and one of the methylene protons with the very large geminal coupling which are attached to C-13. Furthermore, heteronuclear decoupling experiments identified the methine carbon (C-18) of the isopropyl group; its chemical shift (δ50.9) showed that it must be attached to nitrogen, and this was confirmed by experiments with a lanthanide shift reagent. The amended structural formula **1** was put forward (*22*), which fitted the previous data and the new evidence. No coupling could be detected between the protons H-9 and H-12, but a molecular model showed that if the skatolyl group is attached exo to C-9, the dihedral angle between these methine protons is 90°. The model also showed that the cyclohexene ring is tilted sharply at the corner (C-11) bearing the exocyclic double bond, but the rest of the ring is planar, resulting in an extra large geminal coupling for the allylic methylene protons on C-13. The mass spectrum accords well with structure **1,** and in particular shows a distinct m/z 91 ion arising from the cyclohexene ring and the sp^2 methylene carbon attached to it. The base peak at m/z 162 is complementary to the m/z 130 ion referred to previously, and is formed by an α cleavage; it gives rise to another prominent ion at m/z 120 through loss of the isopropyl group by an onium reaction (*6*).

Structural formula **1** was confirmed by degradative experiments (*6*): Hofmann degradation gave product **5** whose UV spectrum showed it to be a 3-vinylindole,

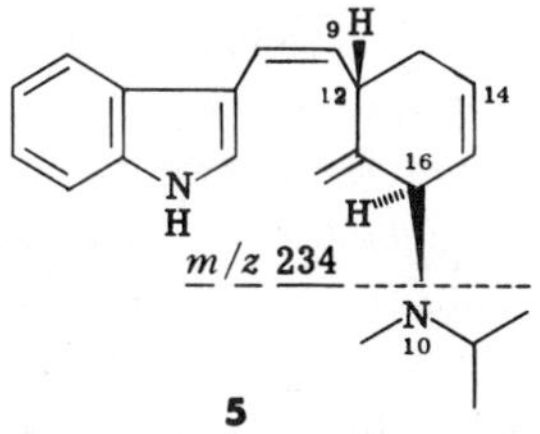

5

and whose ^{1}H-NMR spectrum indicated the presence of a doubly allylic proton coupled to one of the protons of the conjugated vinyl group. This observation is of especial interest since in peduncularine itself no coupling could be detected between the corresponding protons H-9 and H-12 (cf. **1**). The original diallylic proton H-16, α to the nitrogen N-10, is still present in **5**, indicating that no shift in double bonds had taken place during the degradation. The spectrum also provided clear evidence for the presence of the *N*-isopropyl group, and of a new *N*-methyl group. These features were confirmed by the mass spectrum of **5**, the base peak of which corresponds to the loss of N-10 together with both of its alkyl substituents (*6*).

Catalytic hydrogenolysis of peduncularine (*6, 10*) gave the two diastereomeric secopeduncularines **6** and **7** in which, apart from reduction of the olefinic groups, the bond between N-10 and the doubly allylic C-16 has been broken according to evidence from the ^{1}H-NMR and mass spectra. Compounds **6** and **7** each have two secondary nitrogens: there are now two exchangeable hydrogens in both products, and the major one gave an *N*-acetyl derivative. Structure **1** established by these experimental data represents the relative stereochemistry, and the absolute configuration of peduncularine is still unknown.

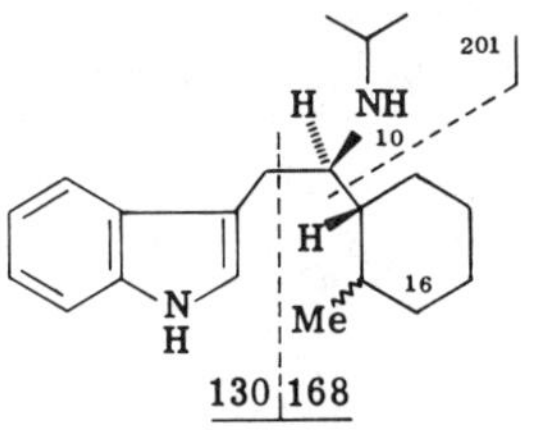

6 Me *exo*
7 Me *endo*

B. Isopeduncularine (8) [$C_{20}H_{24}N_2$, small needles, mp 113–114°C $[\alpha]_D^{19}$ −45° (MeOH), $[\alpha]_D^{19}$ −40° ($CHCl_3$) (*10*)]

The UV, mass, and ^{13}C-NMR spectra of this alkaloid are practically identical with those of peduncularine (**1**); there are, however, small but distinct dissimilarities in the IR, ^{1}H-NMR, and CD spectra, and considerable differences in specific rotation and in melting point, which is depressed for a mixture of the two bases. There is a large solubility difference: peduncularine is sparingly soluble and isopeduncularine readily soluble in chloroform.

Spectroscopic evidence confirmed the presence of the same structural features in **8** as in **1**, and the two alkaloids are evidently diastereomeric. Of the three chiral centers in structures **1** and **8**, two are at bridgeheads of interlocking rings, and are tentatively assumed to be the same in both peduncularine and isopedun-

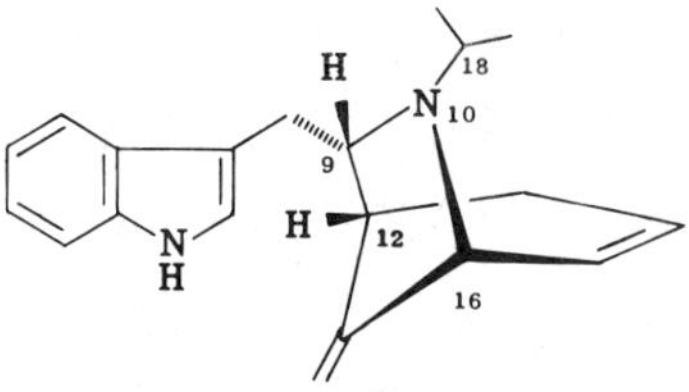

Isopeduncularine

8

cularine, which must then differ in configuration at the remaining center C-9. The ^{1}H-NMR signals for the H-9 protons in both alkaloids are obscured by overlapping signals from protons attached to C-12 and C-18. In the case of peduncularine (**1**), no coupling could be detected between the H-9 and H-12 protons, which evidently make a 90° dihedral angle, but with isopeduncularine (**8**), decoupling experiments showed that H-9 and H-12 are weakly coupled, and in consequence the skatolyl group in the latter has been assigned an endo configuration at C-9. A Hofmann degradation, by which the asymmetry at this center is destroyed, gave the same product **5** as for peduncularine (**1**), but on the other hand, hydrogenolysis of isopeduncularine (**8**) furnished two diastereomeric products **9** and **10,** which differ in properties from the corresponding products **6** and **7**

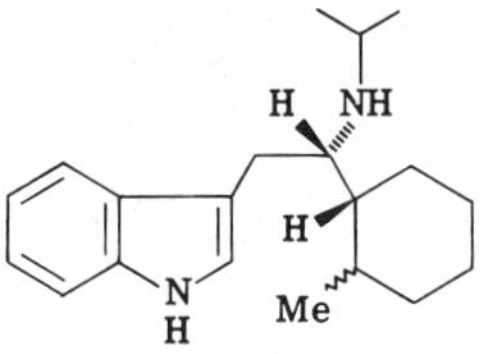

9 Me *exo*
10 Me *endo*

from peduncularine (**1**): in particular, there are considerable differences in specific rotation for all four compounds. The more polar product from each alkaloid crystallized; the two crystalline substances have the same R_f values, but differ from one another in IR spectra and in melting point, which is depressed for a mixture of the two (*10*).

The difference in configuration at C-9 in the case of peduncularine (**1**) and isopeduncularine (**8**) may account for the large difference in solubility between the two alkaloids in chloroform: an exo configuration would, from molecular models, cause obstruction of the basic nitrogen (N-10) by the bulky skatolyl group and impede its solvation by the comparatively large molecules of chloroform. The same factor may be responsible for the considerable difference in specific rotations of **1** and **8** in chloroform.

C. Aristoteline (**11**) {$C_{20}H_{26}N_2$, orthorhombic crystals from methanol, mp 164°C, $[\alpha]_D^{22}$ +23° ($CHCl_3$) [from *A. peduncularis* (*9*) and *A. serrata* (*12*)]; **11a**: $[\alpha]_D^{25}$ −23°($CHCl_3$) [from *A. chilensis* (*20*)]}

Spectroscopic evidence, together with a negative Ehrlich test, showed that aristoteline has an indole nucleus in which C-2 and C-3 are substituted. The nonindolic nitrogen (N-10) is secondary since it can be acetylated; the ^{1}H-NMR spectrum showed the presence of three quaternary methyl groups attached to

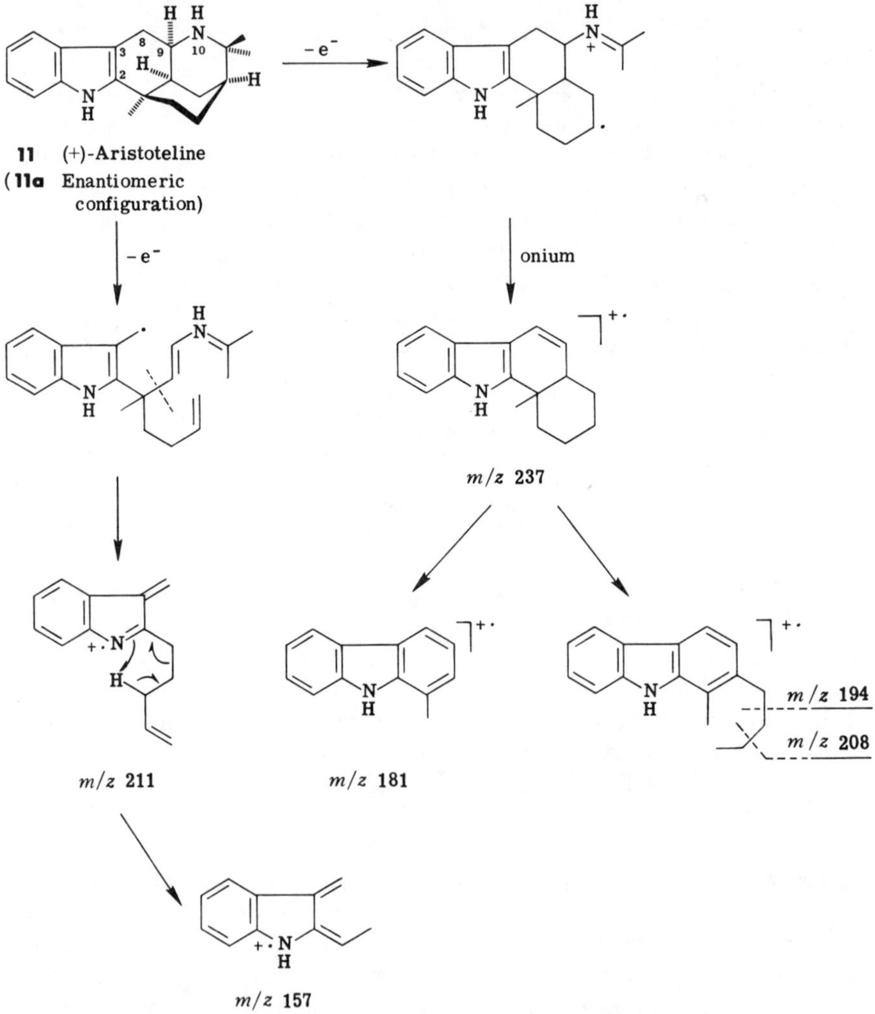

Scheme 1. Principal MS peaks of aristoteline (**11**).

saturated carbons, two of which evidently form part of a *gem*-dimethyl group adjacent to nitrogen as indicated by the strong M − CH_3 and M − C_3H_7N peaks in the mass spectrum (Scheme 1, *9, 12, 20*) and the fact that their proton signals underwent a large and equal downfield shift in the ^{1}H-NMR spectrum on addition of a lanthanide shift reagent (*12, 20*). At the same time, this addition produced a downfield shift in signal of an α-proton (H-9) resonating at δ3.6 and coupled to a pair of geminal protons, which as indicated by their chemical shifts and large coupling constants must be in a methylene group (C-8) attached to the indole nucleus (*12, 20*). The remainder of the structure of aristoteline was deduced from mass spectrometry (Scheme 1, *20*), and the complete structure and absolute stereochemistry were confirmed by X-ray crystallography. The hydrobromide of the dextrorotatory alkaloid formed orthorhombic crystals with the unit cell dimensions a = 19.462(3), b = 10.443(1), and c = 9.889(1), and the 2895 unique data obtained enabled the absolute configuration of the molecule to be determined with an R factor of .036 (*12*). The free base **11a** formed a methanol adduct, which likewise gave orthorhombic crystals; these belong to the space group $P2_12_12_1$ with a = 13.158(2), b = 21.816(4), and c = 6.563(2)Å. Least-squares refinements based on 1917 reflections gave an R factor of .051 for the structure and relative configuration shown in **11a** (*24*). The structure and absolute stereochemistry have also been confirmed by a synthesis that starts from (−)-β-pinene (*25*) (See Section IV).

D. ARISTOSERRATINE (**12**) [$C_{20}H_{24}N_2O_2$, square prisms from methanol, mp 199°C, $[\alpha]_D^{19}$ +22.5°($CHCl_3$) (*8, 26*)]

The spectroscopic data for aristoserratine resemble closely those for aristoteline (**11**): it has an indole nucleus with substituents at C-2 and C-3, a basic secondary amino group with a *gem*-dimethyl group adjacent to it, and a methyl group attached to a quaternary carbon. The ^{1}H-NMR spectrum is quite similar to that of aristoteline (**11**), and decoupling experiments enabled a chain of connectivities to be established, which largely corresponds to that in **11.** However, aristoserratine has a ketone group replacing one of the methylene groups of aristoteline, as shown by the molecular formula and by the IR and NMR spectra. Since the UV spectrum shows that the carbonyl group is not conjugated with the indole nucleus, this leaves three possible locations for it; that shown in **12** is strongly favored by both NMR spectra, and also by the mass spectrum, in which the ions at m/z 237, 211, and 208 in the spectrum of aristoteline (**11,** Scheme 1) are replaced by ions at m/z 251, 225, and 222, respectively (*8*). However, attempts to confirm this structure by direct correlation with aristoteline proved unsuccessful: aristoserratine is too unstable for direct reductive methods, and the ketone group is too hindered to be derivatized as a preliminary to reduction. The corresponding dihydroaristoserratine is readily obtained by borohydride reduc-

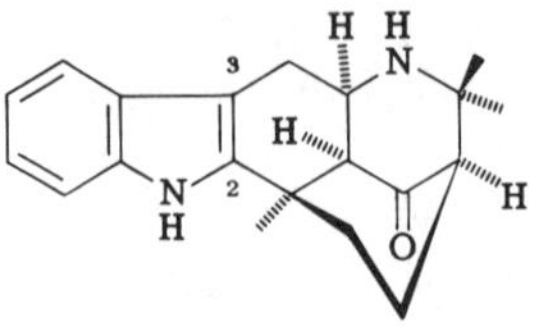

Aristoserratine
12

tion, but attempts at dehydration of this secondary alcohol followed by reduction were likewise unsuccessful (*8, 26*). The structure and relative stereochemistry shown in **12** were finally confirmed by X-ray crystallography (*26*). A monoclinic crystal, space group $P2_1$, a = 14.836(5), b = 8.568(3), c = 6.633(3) Å, β 98.05(3)°, Z2, yielded 1107 reflections, from which least-squares refinements led to solution of the structure with an R factor of .034.

A CD study on **11** and **12** had previously shown (*8*) that the two bases belong to the same stereochemical series, and since the absolute configuration of (+)-aristoteline is known, that for aristoserratine is thereby fixed as in **12**; this stereochemistry is also in accord with the octant rule and the positive Cotton effect at 302 nm produced by the ketone group (*8*).

E. Sorelline (13) [$C_{20}H_{24}N_2$, crystals, mp 165–168°C, $[\alpha]_D^{22}$ +157°($CHCl_3$) (*7*)]

The UV and NMR spectra indicated the presence of a 3-substituted indole nucleus, and a strong m/z 130 ion in the mass spectrum showed that the substituent is a methylene group. Furthermore, the ^{1}H-NMR spectrum shows that the geminal protons of this group (on C-8) are coupled to the proton (H-9) of a methine group α to a nitrogen (N-10), which in turn is in a secondary amino group since the ^{13}C-NMR spectrum indicated that only two carbons, one quaternary (C-18) and one methine (C-9), are attached to it (*7*). The former is the only quaternary sp^3 carbon in sorelline and must bear the two methyl groups whose unsplit signals appear in the ^{1}H-NMR spectrum.

The ^{13}C-NMR spectrum shows, moreover, the presence of two double bonds, which must be conjugated, as indicated by the UV spectrum; one is in a vinylidine group according to both NMR spectra, while the other has protons at-

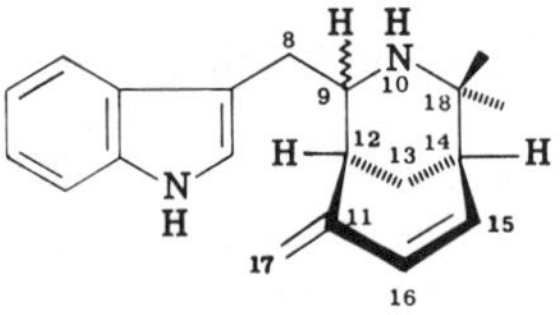

Sorelline
13

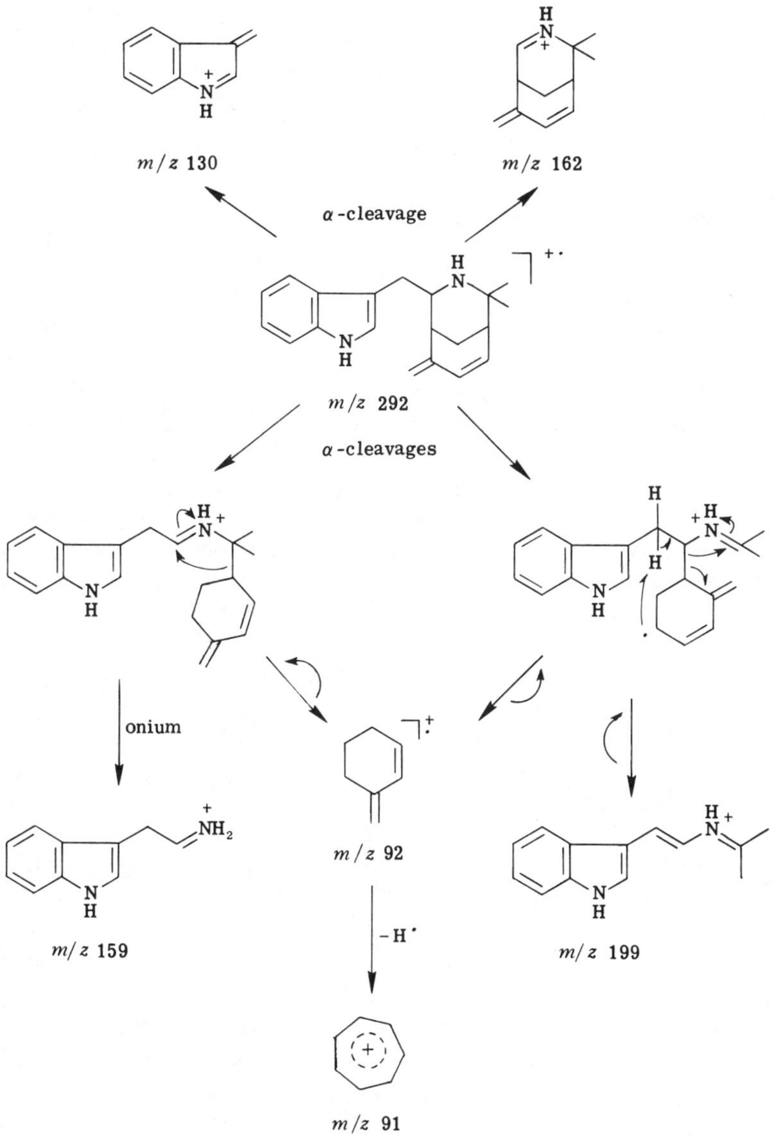

m/z 130

m/z 162

α-cleavage

m/z 292

α-cleavages

onium

m/z 159

m/z 92

$-H^{\cdot}$

m/z 199

m/z 91

SCHEME 2. Principal MS peaks of sorelline (**13**).

tached in a *Z* configuration. By proton-decoupling experiments, a sequence could be traced extending from these vicinally coupled olefinic protons H-16 and H-15 through H-14 and the geminal protons on C-13 to the allylic H-12, and finally to the above-mentioned H-9. Structure **13** established by these observations is in concordance with the mass spectrum (Scheme 2), but the spectroscopic data are not able to define unequivocally the configuration of H-9 relative to that of H-12; the dihedral angles made by these protons are the same whether the skatolyl group is attached exo or endo. The absolute stereochemistry at the remaining centers C-14 and C-12 has, however, been established as shown in **13** by CD measurements (*27*).

F. Tasmanine (**14**) [$C_{20}H_{26}N_2O$, crystals from methanol, mp 249–250°C, $[\alpha]_D$ −150°(MeOH), $[\alpha]_D^{20}$ −132°($CHCl_3$) (*9, 28*)]

Tasmanine as compared to **11** has an additional oxygen, which is present in a lactam group as indicated in the IR spectrum of the base, and the lactam in turn forms part of an oxindole nucleus as shown by its UV spectrum. There is a general resemblance in the chemical shifts and multiplicities of the sp^3 carbons in the ^{13}C-NMR spectra of tasmanine (**14**) and aristoteline (**11**), but the former has an extra quaternary carbon signal at δ53.7, which replaces an sp^2 singlet at δ104 in the spectrum of **11,** and which is ascribed to the spiro carbon (C-3) of the oxindole nucleus (*9, 28*). The ^{1}H-NMR spectra of **14** and **11** also show many similarities in the aliphatic proton region, and a series of decoupling experiments (*9*) established that the same functional groups joined together in the same fashion are present in the two alkaloids. A structure of type **14** is indicated for tasmanine by these observations, which is supported by the mass spectrum (Scheme 3, *9*). This structure was confirmed by LAH reduction (*28*) of tasmanine, which gave (+)-aristoteline (**11**) in 30% yield. This experiment at the same time fixed the absolute stereochemistry except for the spiro center (C-3).

To gain an insight into the stereochemistry at C-3, a comparison was made between the CD spectra of tasmanine (**14,** *9*) and aristoserratenine (**24,** *27*), an indolenine alkaloid with a structure similar to that of **14,** whose absolute stereochemistry has been fixed (*28*). Both alkaloids can be transformed to (+)-aristoteline (**11**), and in consequence they must have the same configurations at all chiral centers apart from the spiro carbon. The CD spectra of **14** and **24** reveal intense Cotton effects of opposite sign around 250 nm, and also around 225 nm; the specific rotations of the two alkaloids (at 589 nm) are likewise of opposite sign (*9,27*). These observations suggest that the two bases differ in stereochemistry at their spiro carbons, and that the absolute configuration of tasmanine is to be represented by **14.** However, an assignment on this basis is open to question in view of the difference in nature of the chromophores of tasmanine and aristoserratine.

Further evidence was sought from a study of the anisotropic effects caused by the aromatic ring and the carbonyl group of tasmanine on protons in the immediate vicinity (*9*). As formulated in **14,** the H-12 proton lies in the deshielding zone of the carbonyl group and resonates at distinctly lower field (δ2.56) than H-14

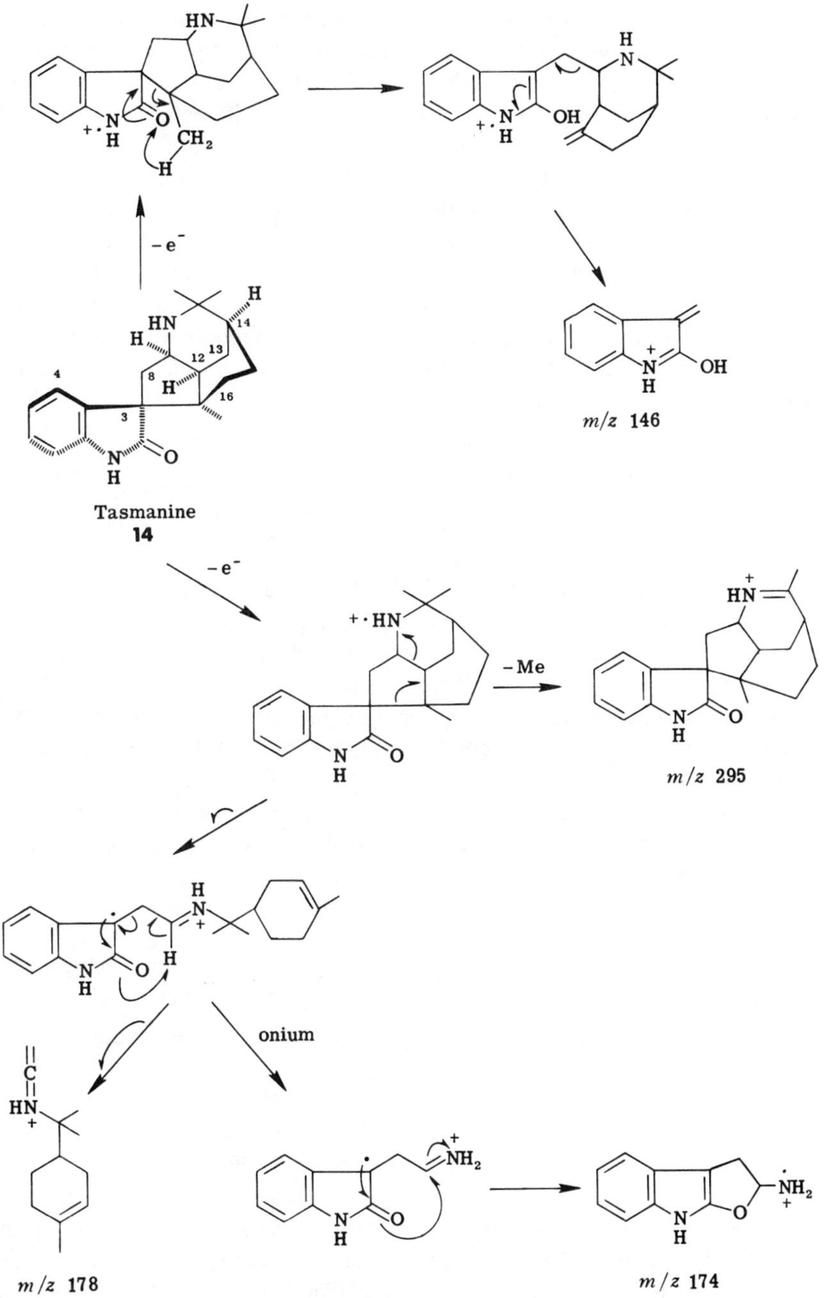

Tasmanine
14

m/z 146

m/z 295

onium

m/z 178

m/z 174

SCHEME 3. Principal MS peaks of tasmanine (**14**).

(δ1.31), which is remote from the carbonyl. Similarly, H_{endo}-16 as formulated in **14** lies close to the plane of the aromatic ring, and is distinctly deshielded (δ3.00) as compared to H_{exo}-16 (δ0.75). However, this approach is likewise of somewhat doubtful validity since similar effects might be expected for the stereoisomer with the alternative configuration at the spiro atom, in which the locations of the aromatic ring and the carbonyl group are reversed. The ambiguity was finally resolved (*28*) by NOE experiments on tasmanine: irradiation of the aromatic proton H-4 produced a distinct enhancement in the signal due to H_{endo}-16 and vice versa. In a molecular model corresponding to **14,** the two protons in question are seen to be close to one another in space; but with the epimeric configuration at the spiro atom, they would be too remote for an NOE effect to take place, and structure **14** thus represents the absolute configuration for all chiral centers in tasmanine.

G. HOBARTINE (15) [$C_{20}H_{26}N_2$, colorless crystals, mp 149–150.5°C, $[\alpha]_D^{22}$ −20°($CHCl_3$) (*7*)]

Spectroscopic evidence showed that hobartine, like sorelline (**13**), has a 3-substituted indole nucleus, and in particular NMR spectroscopy enabled a chain of connectivities to be traced, which largely corresponds to that of **13.** However, hobartine has two more hydrogens than sorelline and only one double bond. On the other hand, it has an extra methylene and an extra methyl group, both of which must be attached to the olefinic group since the methyl protons form a broad singlet at δ1.81 and show allylic coupling with the only olefinic proton in the ^{1}H-NMR spectrum; this proton is further coupled to the geminal protons of the methylene group, which in turn form part of the above-mentioned chain. Structure **15,** which is indicated by these observations, has been confirmed by the synthesis of racemic hobartine (see Section IV).

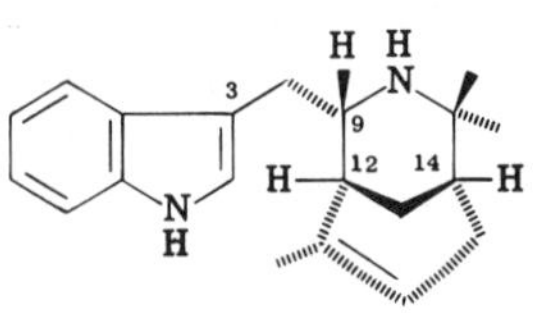

Hobartine
15

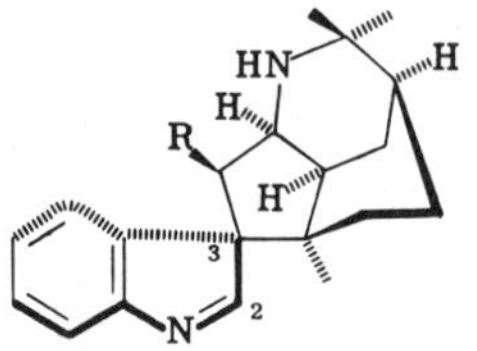

16 R = OH
24 Aristoserratenine: R = H

The absolute configuration of (−)-hobartine as shown in **15** has recently been determined by synthesis (*59*) and by conversion into (+)-aristoteline (**11**), whose absolute stereochemistry is known (see Addendum).

H. SERRATOLINE (**17**) [$C_{20}H_{26}N_2O$, rhombic crystals from methanol, $[\alpha]_D^{19}$ −64.25°($CHCl_3$) (*13,16*)]

As compared to aristoteline (**11**), serratoline has an extra oxygen, which is part of a hydroxyl group as indicated by the IR and mass spectra. Apart from this, many features in common with **11** can be detected by ^{1}H-NMR and ^{13}C-NMR

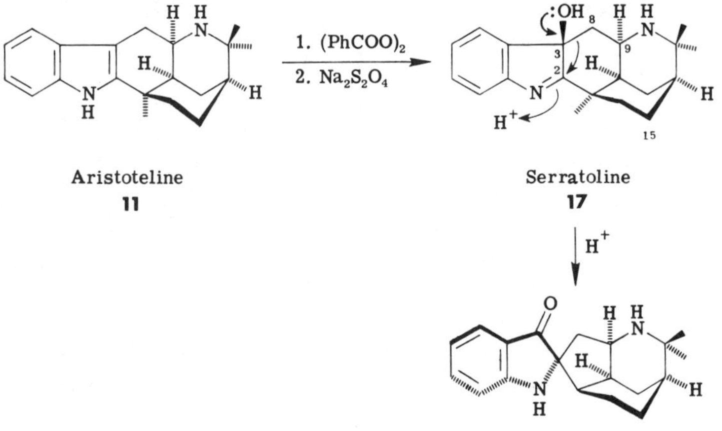

1. $(PhCOO)_2$
2. $Na_2S_2O_4$

Aristoteline
11

Serratoline
17

Aristotelone
33

spectroscopy, but the UV spectrum of serratoline shows that it has an indolenine nucleus. On borohydride reduction, a dihydro product was obtained whose UV and ^{1}H-NMR spectra show it to be an indoline derivative. When refluxed in dilute acid, the indoline was dehydrated and converted to (+)-aristoteline (**11**), but on the other hand, serratoline was resistant to rearrangement when refluxed with alkali (*13*). The latter observation suggested that the hydroxyl group was not attached to C-3, and structural formula **16** was tentatively assigned to serratoline on the basis of the close similarity of its ^{1}H-NMR spectrum with that of the dihydroaristotelinone (**23**); a possible mechanism for the acid-catalyzed conversion of dihydroserratoline to aristoteline (**11**) was put forward (*13*). Structural formula **16** has now been revised to **17** in the light of further spectroscopic evidence (*16*): a comparison of the ^{13}C-NMR spectrum of serratoline with that of the indolenine alkaloid aristoserratenine (**24**) revealed many similarities, but the former spectrum contains only four doublets in the sp^2 region as compared to five in the latter; on the other hand, the serratoline spectrum has an extra low-field quaternary carbon signal at δ189.0 instead of the doublet at δ178.7 ascribed to C-2 in aristoserratenine (**24**). Moreover, the quaternary signal at δ83.9 in the serratoline spectrum is distinctly downfield from the corresponding signal at δ70.8, which is attributed to C-3 in the case of aristoserratenine. This suggests that the hydroxyl group in serratoline is attached to C-3, and points to the revised formula **17** for this alkaloid (*16*).

Structure **17** is supported by the fact that when refluxed in mineral acid, serratoline rearranges to a new crystalline base whose UV, IR, and mass spectra show that it is a ψ-indoxyl derivative (*16*). The melting point, mass, and in particular IR spectra of this subtance correspond with those of the alkaloid aristotelone (**33**), and although a rigorous comparison could not be made for lack of material, the two bases are in all probability identical (*16*).

Finally, (+)-aristoteline (**11**) was converted to serratoline by oxidation with benzoyl peroxide to a hydroperoxide, followed by reduction with dithionite. These experiments confirmed the structure and absolute stereochemistry represented by **17** for serratoline, except for the configuration at C-3. It is evident from a molecular model of **17** that the proton attached to C-9 forms equal dihedral angles with those on C-8, whereas in a model with the epimeric configuration for the C-3 hydroxyl, the corresponding dihedral angles are considerably different. The coupling constants determined experimentally for H-9 with H_{exo}-8 and with H_{endo}-8 were identical (2.9 Hz), and this observation completes the evidence for the assignment to serratoline of the absolute stereochemistry shown in **17.**

I. Makonine (18) [$C_{20}H_{22}N_2O$, hexagonal crystals from methanol, mp 310–312°C(d), $[\alpha]_D^{19}$ +431°(MeOH + $CHCl_3$, 1:2) (*14*)]

Makonine has two fewer hydrogens than has aristotelinone (**21**), to which it shows many points of resemblance in other respects: from spectroscopic evi-

Hg(OAc)$_2$

Makonine
18

Aristotelinone
21

NaBH$_4$

NaBH$_4$

23

dence it is shown to have an indole nucleus substituted at C-2 and C-3, and three quaternary methyl groups, of which two are in a *gem*-dimethyl group. It also has a conjugated carbonyl group, as shown by the UV, IR, and ^{13}C-NMR spectra, but the latter spectrum shows an olefinic carbon instead of the sp^3 methine carbon (C-9) located α to the nonindolic nitrogen in **21,** and the ^{1}H-NMR spectrum indicates that there is no exchangeable hydrogen attached to this nitrogen (N-10). These observations suggested a structure of type **18** for makonine, which has been confirmed by the conversion of aristotelinone (**21**) to makonine in 25% yield by mercuric acetate oxidation; furthermore, both **18** and **21** gave the same secondary alcohol **23** on borohydride reduction. These experiments at the same time fix the absolute stereochemistry of makonine.

J. MAKOMAKINE (**19**) [$C_{20}H_{26}N_2$, crystals from chloroform, mp 99–100°C, $[\alpha]_D^{19}$ +131°($CHCl_3$) (*14*)]

Makomakine is isomeric with hobartine (**15**) and isohobartine (**25**), and spectroscopic evidence shows that, like them, it has a 3-substituted indole nucleus, one double bond, a *gem*-dimethyl group, and two secondary nitrogens. The UV

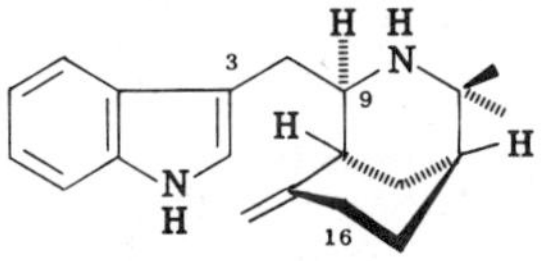

H H N 3 9 H H N H 16

Makomakine
19

and mass spectra of the three bases are almost identical; there are also considerable similarities in their ^{1}H- and ^{13}C-NMR spectra, and decoupling experiments enabled a chain of connectivities to be traced in makomakine similar to those of **15** and **25.** However, the NMR spectra show that one fewer methyl group is present in makomakine, and that the double bond is in a vinylidine group, the protons of which are coupled allylically to the endo proton on C-16; this in turn is linked to the above-mentioned chain. These observations point to structure **19** for makomakine, which has been confirmed by an acid-catalyzed cyclization at room temperature to (+)-aristoteline (**11**). This experiment at the same time fixes the absolute stereochemistry of makomakine, including the configuration at C-9 (*14*). The structure and stereochemistry in structure **19** have also been confirmed by a synthesis starting from (−)-β-pinene (*25*) (see Section IV).

K. ARISTOMAKINE (**20**) [$C_{20}H_{26}N_2$, amorphous, $[\alpha]_D^{22}$ −79.1°($CHCl_3$) (*15*)]

Aristomakine is isomeric with aristoteline (**11**), with which it has a number of features in common: the UV spectrum shows the presence of an indole nucleus, which must be substituted at C-2 and C-3 as indicated by the negative Ehrlich test and the NMR spectra; the latter also give evidence of a quaternary C-methyl group. However, the spectra in addition show the presence of an olefinic group, and aristomakine thus has one ring fewer than aristoteline (**11**). As in the latter case, both nitrogens are secondary as shown by the appearance of two broad singlets exchangeable with D_2O in the ^{1}H-NMR spectrum; further evidence comes from signals at δ51.5 and 46.6 in the ^{13}C-NMR spectrum, corresponding to two sp^3 methine carbons adjacent to nitrogen. A proton attached to one of these carbons produces a septet in the ^{1}H-NMR spectrum, and it is coupled to two sets of methyl protons: aristomakine thus has an *N*-isopropyl group. The other methine proton forms part of a sequence that was elucidated by spin decoupling and by a comparison of the NMR spectra with those of aristoteline (**11**). Structure **20,** suggested by this evidence, is in full accord with the mass spectrum (Scheme 4). A comparison of the CD spectrum with those of various stereoisomeric yohimbanes (*29*) suggests that the carbocyclic rings have a trans-fused hexalin structure with the absolute stereochemistry shown in **20** (*27*); and since H-9 and H-12 have a coupling constant of only 3.5 Hz, they cannot have a

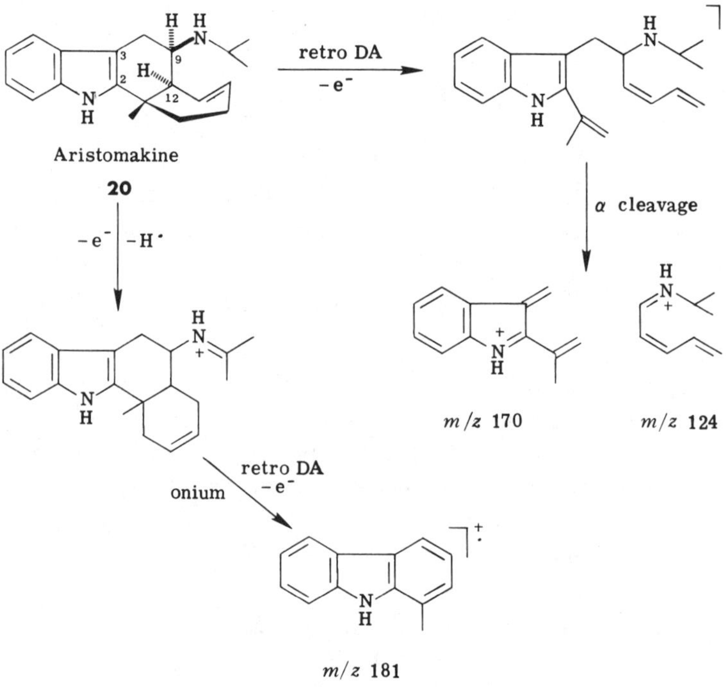

Aristomakine

20

retro DA
$-e^-$

$-e^-$ | $-H^\cdot$

α cleavage

onium

retro DA
$-e^-$

m/z 170

m/z 124

m/z 181

SCHEME 4. Principal MS peaks of aristomakine (**20**).

trans–diaxial relationship. Structure **20** with a cis configuration of these protons thus represents the absolute stereochemistry of the molecule.

L. Aristotelinone (21) [$C_{20}H_{24}N_2O$, fine needles from methanol, changing around 255°C to longer needles which remain unaltered up to 320°, $[\alpha]_D^{19}$ +122.7° (MeOH + $CHCl_3$, 1:1) (*13*)]

Aristotelinone is isomeric with aristoserratine (**12**), which it resembles in having a carbonyl group replacing one of the methylene groups of aristoteline (**11**). The UV and IR spectra show that this carbonyl group is conjugated with the indole nucleus, and a comparison of the NMR spectra of **21** and **11** suggests that the carbonyl replaces the methylene group at C-8 in aristoteline (**11**). The geminal protons H_{exo}-8 and H_{endo}-8 of **11** are coupled to a proton (H-9) vicinal to N-10; this coupling is absent in the ^{1}H-NMR spectrum of aristotelinone (**21**), and the corresponding H-9 proton resonates at lower field. The remainder of the NMR spectra of aristoteline and aristotelinone correspond with one another, and the evidence indicates structure **21** for the latter alkaloid.

The structure and absolute stereochemistry of aristotelinone were confirmed by LAH reduction, which gave (+)-aristoteline (**11**) together with a pair of diastereomeric alcohols (**22** and **23**). One of the alcohols has a strong M − 18 peak in its mass spectrum and was assigned structure **23** in which the hydroxyl group and the proton on C-9 are trans–diaxial. This alcohol is the sole product on borohydride reduction of aristotelinone, and it can be further reduced to aristoteline with LAH.

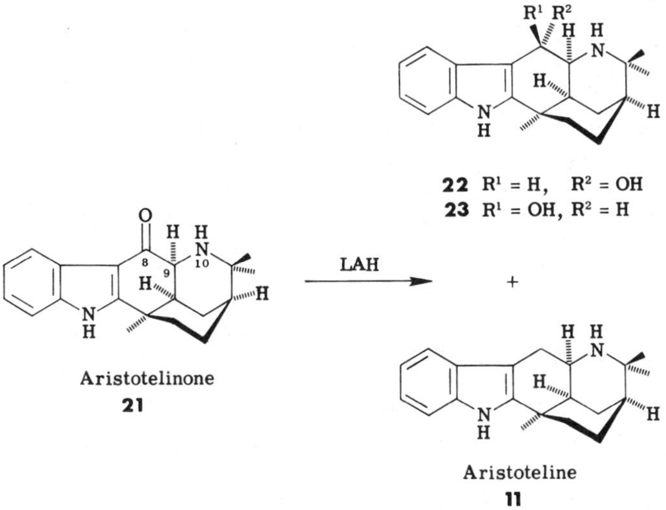

Aristotelinone
21

LAH

22 R¹ = H, R² = OH
23 R¹ = OH, R² = H

+

Aristoteline
11

M. Aristoserratenine (24) [$C_{20}H_{26}N_2$, amorphous, $[\alpha]_D^{19}$ +58°($CHCl_3$) (*28*)]

Aristoserratenine is isomeric with aristoteline (**11**), but its UV spectrum demonstrated that it is an indolenine. The 2 position is unsubstituted, as shown by a sharp singlet at δ8.00 and by a doublet at δ178.7 in the NMR spectra. The remainder of the ^{13}C spectrum for **24** closely corresponds with that of **11**, except

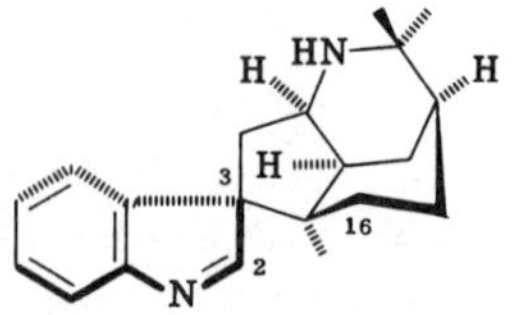

Aristoserratenine

24

for a singlet attributed to C-3 that appears at δ104 in the spectrum of aristoteline. This singlet is shifted to δ70.8 in the aristoserratenine spectrum, and is consistent with the spiro carbon of an indolenine nucleus. The rest of the ^1H-NMR spectrum of aristoserratenine likewise corresponds in general terms with that for aristoteline (**11**), and decoupling experiments permit a sequence of protons to be traced in the spectrum of **24**, which parallels that for **11**. The mass spectra of the two alkaloids are also very similar, and the evidence thus points to an indolenine structure of type **24** for aristoserratenine. This conclusion has been confirmed by an acid-catalyzed rearrangement of aristoserratenine to (+)-aristoteline in good yield, which at the same time fixes the absolute stereochemistry of structure **24** except for the configuration at the spiro carbon (C-3).

In order to clarify the stereochemistry at this point, a series of NOE experiments were undertaken on aristoserratenine (**24**), in the course of which a 27% decrease in signal intensity of the endo proton attached to C-16 was observed on irradiation of the C-2 proton (*28, 30*): these two protons must thus be located relatively close to one another in space, and no comparable effect was observed on protons in other positions. Conversely, irradiation of the C-16 endo proton produced a similar, though predictably smaller, negative enhancement of the C-2 proton. These data support the configuration at C-3 shown in structure **24**, which thus represents the absolute stereochemistry of aristoserratenine.

N. Isohobartine (25) [$C_{20}H_{26}N_2$, crystals from chloroform, mp 134–135°C, $[\alpha]_D^{19}$ −30°($CHCl_3$) (*10*)]

The UV, mass, and CD spectra of isohobartine closely resemble those of hobartine (**15**), but the melting points and specific rotations differ; and while the IR spectra of the two alkaloids are in general alike, there are dissimilarities in the fingerprint region. The ^1H-NMR spectra of the two bases are almost identical

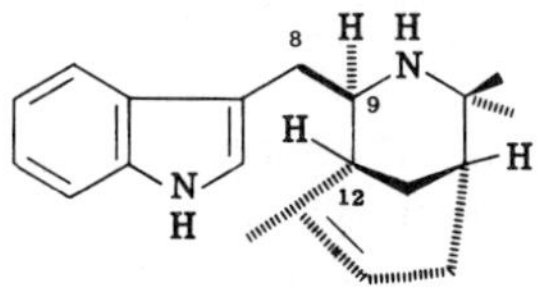

Isohobartine
25

with one another, but there is a significant difference: whereas the methylene protons attached to C-8 in hobartine (**15**) resonate at δ2.84 and 2.69 and show a geminal coupling of 15.5 Hz, as well as further couplings of 6.4 and 7.7 Hz, respectively, to the C-9 proton, the corresponding C-8 protons of isohobartine produce a two-proton doublet at δ2.85, being split by the methine proton on C-9 ($J = 7.0$ Hz); they show no geminal coupling. These data indicate that hobartine (**15**) and isohobartine (**25**) are diastereomers with the same structure and stereochemistry except for their configurations at C-9. ^{1}H-NMR spectroscopy is of little assistance in assigning the correct configurations at this center for the two alkaloids, since the couplings between H-9 and H-12 are 2.5 and 2.7 Hz for **15** and **25,** respectively, as expected from molecular models, which indicate approximately equal dihedral angles in the two cases.

However, the absolute stereochemistry of hobartine has now been fixed by synthesis (*59*), and isohobartine is in consequence assigned the corresponding structure with the epimeric configuration at C-9. The structure and the stereochemistry at C-12 and C-14 have been confirmed by the fact that natural (−)-isohobartine could be cyclized at room temperature under conditions of acid catalysis to (+)-aristoteline (**11,** *10*).

O. SERRATENONE (26) [$C_{20}H_{24}N_2O$, amorphous, $[\alpha]_D^{19}$ −45°($CHCl_3$) (*16*)]

The spectroscopic data for serratenone show many similarities with those of sorelline (**13**), hobartine (**15**), and makomakine (**19**): thus serratenone must have an indole nucleus with a C-3 substituent that consists of a methylene group attached to a heterocyclic system similar to those of **13, 15, 19,** and **25,** in which is included a secondary amino group linked to a *gem*-dimethyl group. However,

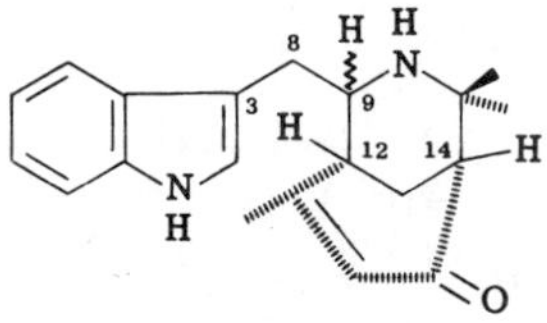

Serratenone
26

serratenone differs from the other bases in that it has an α,β-unsaturated ketone group, as shown by the IR, UV, and NMR spectra; in particular, the UV spectrum is practically identical with that of an equimolar mixture of tryptamine and mesityl oxide. The only olefinic proton in the ^{1}H-NMR spectrum is coupled allylically to the protons of a methyl group, and also to a methine proton (H-12), which in turn, as shown in decoupling experiments, is part of a whole sequence of protons attached to a chain of carbons that extends from the above-mentioned methylene group (C-8) up to a carbon (C-14) located between the carbonyl and the *gem*-dimethyl group. Consideration of these data led to structure **26** for serratenone, and the absolute stereochemistry at C-14 and C-12 has been fixed by the fact that its CD spectrum shows a negative *R* band (*27*); however, the configuration at C-9 remains in doubt, as in the case of sorelline (**13**) and isosorelline (**28**).

P. Aristomakinine (27) [$C_{17}H_{20}N_2$, amorphous, $[\alpha]_D^{19}$ −72°($CHCl_3$) (*10*)]

Aristomakinine differs from all the other *Aristotelia* bases in having 17 carbon atoms instead of 20. In many respects its UV, IR, and NMR spectra show a close resemblance to those of aristomakine (**20**): it has an indole nucleus substituted at C-2 and C-3, also an olefinic and quaternary methyl group, but the *N*-isopropyl group present in **20** is lacking; furthermore, it has three exchangeable hydrogens and only one carbon, a methine, adjacent to the basic amino group (N-10), which is thus primary. The evidence suggests that aristomakinine (**27**) has the same

1. Me_2CO
2. $NaBH_4$

Aristomakinine **27** → Aristomakine **20**

skeleton as aristomakine (**20**), but lacks the *N*-isopropyl group, and this inference was confirmed by treating aristomakinine with acetone, followed by reduction with sodium borohydride to produce aristomakine in 50% yield. The structure and absolute stereochemistry of aristomakinine are thus represented by **27.**

Q. Isosorelline (28) [$C_{20}H_{24}N_2$, crystals from methanol, mp 160–162°C $[\alpha]_D^{21.5}$ +120°($CHCl_3$) (*10*)]

The UV, mass, CD, and ^{1}H-NMR spectra of isosorelline resemble very closely those of sorelline (**13**), but their specific rotations are distinctly different, and although their IR spectra show a general resemblance to one another, there are

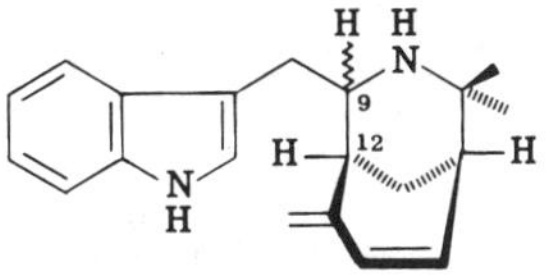

Isosorelline
28

significant dissimilarities; furthermore, mixing of the two bases produced a depression of the melting point. These observations suggested that they have the same skeleton and differ only in stereochemistry at C-9, at which position a difference in configuration would not be expected to produce an observable difference in coupling constant between H-9 and H-12: as in the case of hobartine (**15**) and isohobartine (**25**), the dihedral angles are virtually the same in both cases. Isosorelline and sorelline are thus diastereomers with epimeric but at present unknown configurations at C-9.

R. Fruticosonine (29) [$C_{20}H_{28}N_2O$, square prisms from ether, mp 120.1°C, $[\alpha]_D^{20}$ +45.7°($CHCl_3$) (*17*)]

Evidence from the IR and ^{1}H-NMR spectra suggested the presence of two secondary amino groups; the UV and IR spectra indicated a carbonyl group and an indole nucleus; and a positive Ehrlich test, together with a doublet at δ7.02 in the ^{1}H-NMR spectrum, showed that C-2 was unsubstituted. On the other hand, C-3 evidently bears an ethylene chain from the strong *m*/*z* 130 and *m*/*z* 144 peaks in the mass spectrum, and from the four-proton multiplet between δ2.96 and 2.86, which further suggests that the chain is attached to the aliphatic nitrogen N-10. This conclusion was supported by the fact that the base peak *m*/*z* 182 in the mass spectrum is complementary to the *m*/*z* 130 ion, and could be formed by an α cleavage.

The ^{1}H-NMR spectrum showed signals for three methyl groups, two of which form a six-proton singlet and must be in a *gem*-dimethyl group attached to N-10 from the intense peak in the mass spectrum at *m*/*z* 201, which could arise from an

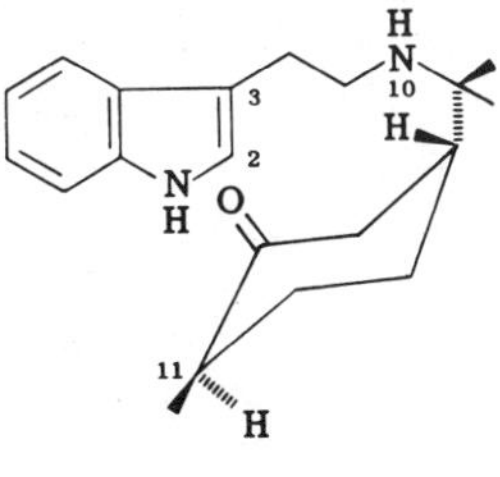

Fruticosonine
29

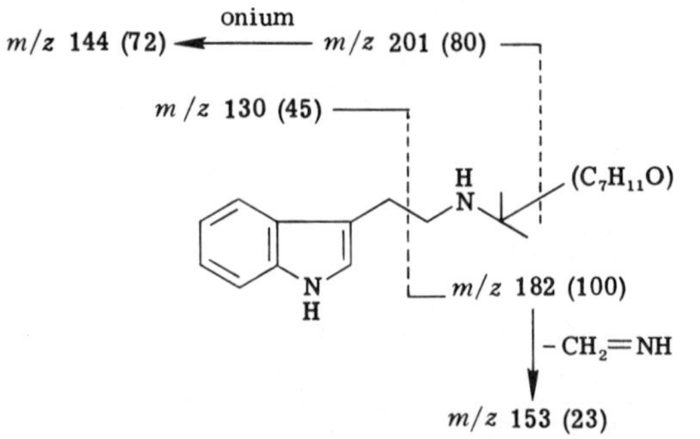

m/z 144 (72) ← onium — m/z 201 (80)
m/z 130 (45)
(C$_7$H$_{11}$O)
m/z 182 (100)
− CH$_2$=NH
m/z 153 (23)

30 Principal MS peaks of fruticosonine (**29**) (intensities in %)

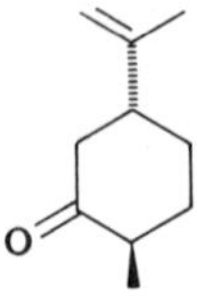

31 (+)-Dihydrocarvone

alternative α cleavage (cf. **30**). ^{1}H-NMR evidence showed that the remaining methyl is attached to a methine group (C-11), which in turn could be linked to the carbonyl from the chemical shift of its proton (δ2.19). The formula of fruticosonine and the absence of olefinic groups showed that an aliphatic ring must be present. A series of decoupling experiments identified this as 2-methylcyclohexanone and enabled the tentative structural formula **29** to be put forward. This structure has been fully confirmed by synthesis (see Section IV) and by X-ray crystallography. The crystals measured were tetragonal and belonged to the space group $P4_32_12$ with unit-cell dimensions a = 8.847(2) and c = 47.857(9) Å. The R factor on refinement of the 1428 data was .062 and enabled the relative stereochemistry as well as the structure to be fixed. The absolute stereochemistry followed from the virtual identity of the CD curves of fruticosonine and (+)-dihydrocarvone (**31**) (*27*).

S. Aristofruticosine (32) [$C_{20}H_{24}N_2$, amorphous, $[\alpha]_D^{15}$ +50.5°($CHCl_3$) (*10*)]

Spectroscopic evidence and a positive Ehrlich test showed that this alkaloid has an indole nucleus substituted at C-3 but not at C-2. The ^{1}H-NMR and mass spectra indicated that the substituent is a methylene attached to a methine group (C-9), which is located α to the nonindolic nitrogen N-10; the latter must be tertiary, since there is only one proton exchangeable with D_2O in the ^{1}H-NMR

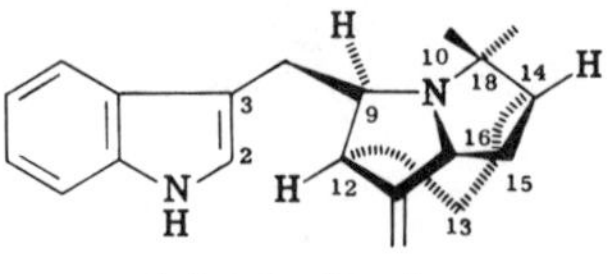

Aristofruticosine
32

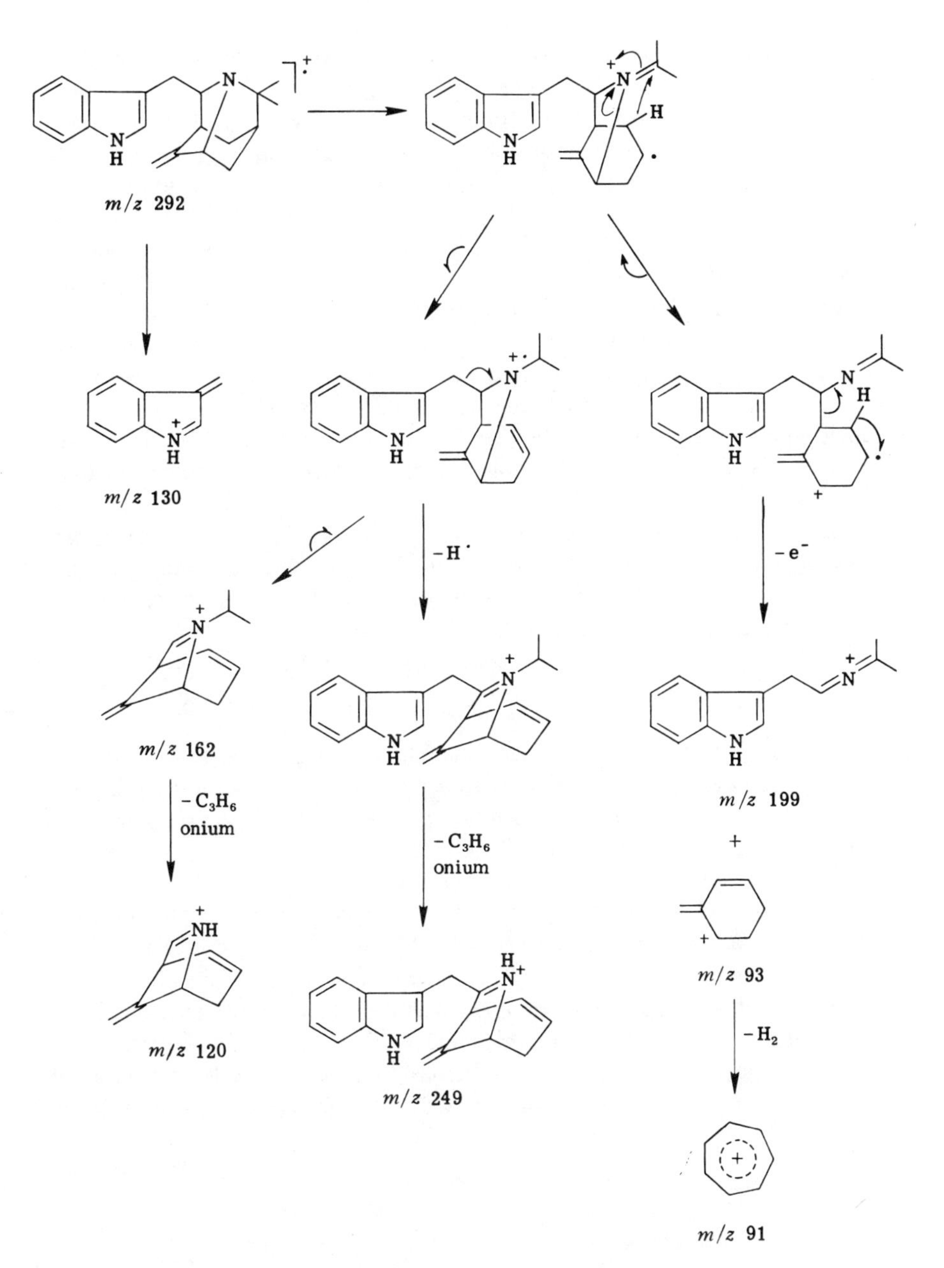

SCHEME 5. Principal MS peaks of aristofruticosine (**32**).

spectrum, and there are three carbon signals between δ62 and 66 in the ^{13}C-NMR spectrum. Two of these signals, including the one from C-9, are from methine carbons, while the third comes from the only sp^3 quaternary carbon (C-18); this must in consequence form part of the *gem*-dimethyl group whose presence is attested by both NMR spectra. The remaining methine group (C-16) linked to N-10 has a proton coupled allylically to a pair of olefinic protons, which must be in a vinylidine group from NMR evidence. The same methine proton H-16 is also coupled to one of a geminal pair of protons (on C-15), which is further coupled to another methine proton, H-14. This in turn is coupled to another pair of geminal protons (on C-13), each of which is again coupled to another methine proton resonating at δ2.26, and in consequence assigned a location on a carbon (C-12) adjacent to the vinylidine group. These data, together with the mass spectrum (Scheme 5) and biogenetic considerations, suggest that aristofruticosine has the unique 3-azatricyclo[3.3.1.$0^{3,7}$]nonane structure **32.**

The methine protons on C-9 and C-12 show no coupling, and a model indicates that they form a dihedral angle close to 90°, provided the skatolyl group is attached exo to C-9. The tentative structural formula **32** thus represents the relative stereochemistry of aristofruticosine (*10*).

T. Aristotelone (33) [$C_{20}H_{26}N_2O$, mp 218–222°C (*20*)]

Aristotelone is isomeric with tasmanine (**14**), but its UV and IR spectra showed it to be a ψ-indoxyl derivative. Paucity of material precluded a detailed NMR examination, but the mass spectroscopic fragmentation pattern (Scheme 6) corresponds to the proposed structural formula **33.**

The indolenine alkaloid serratoline (**17**), which has a C-3 hydroxyl group, undergoes an acid-catalyzed rearrangement to a ψ-indoxyl whose structure and absolute stereochemistry is represented by **33.** This compound corresponds to aristotelone in melting point and MS and, in particular, in the details of its IR spectrum. Although it was not possible to confirm their identity by direct comparison for lack of a sample of aristotelone, the two compounds are in all probability the same, and the likely relationship to serratoline lends support to the tentative structure and stereochemistry for aristotelone represented in **33.**

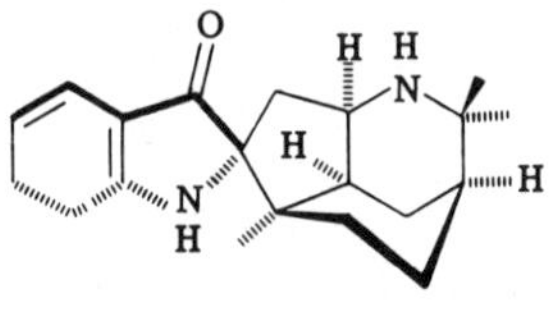

Aristotelone
33

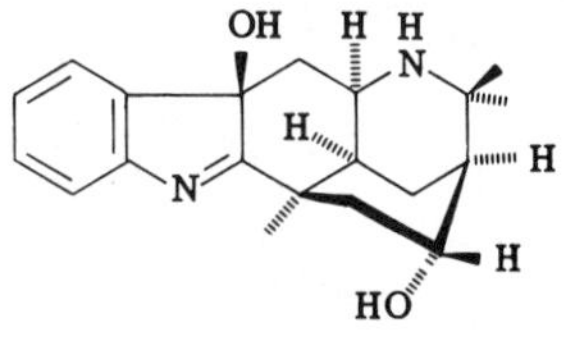

Aristotelinine
34

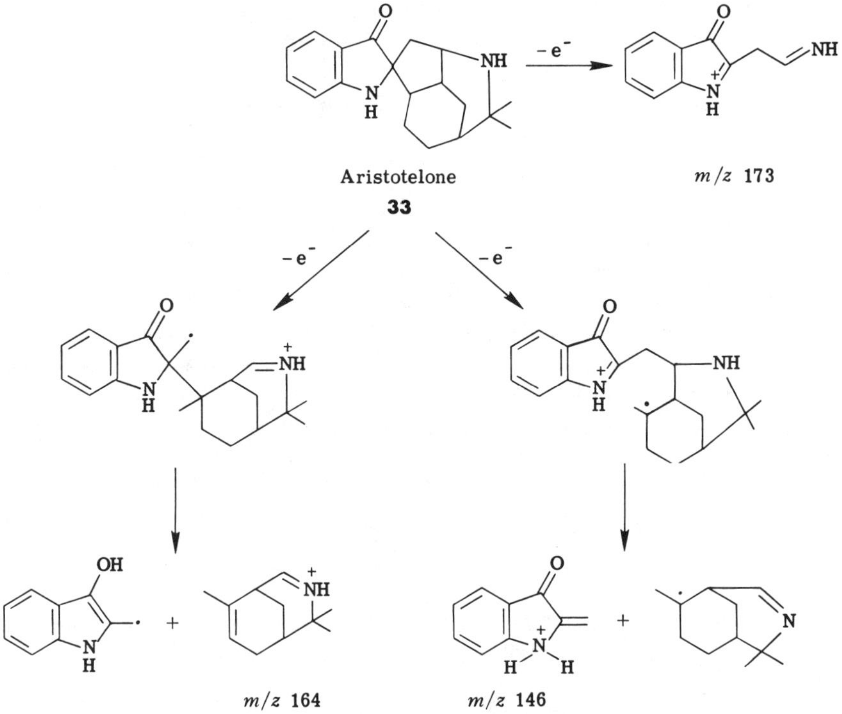

Aristotelone

33

m/z 173

m/z 164

m/z 146

SCHEME 6. Principal MS peaks of aristotelone (**33**).

U. ARISTOTELININE (**34**) [$C_{20}H_{26}N_2O$, long needles, mp 246–250°C (*21, 31*)]

The UV spectrum of aristotelinine is characteristic of an indolenine, and its IR spectrum indicated the presence of hydroxyl groups. The complete structure and stereochemistry was established by X-ray crystallographic analysis. An orthorhombic crystal belonging to the space group $P2,2,2$, with unit cell dimensions $a = 12.710(4)$, $b = 10.115(1)$, and $c = 13.035(4)$ Å provided 1532 reflections, and refinement of the data yielded a final R value of .045. The X-ray crystallographic evidence led to the structure and stereochemistry shown in **34,** and aristotelinine thus proves to be the 15-*endo*-hydroxy derivative of serratoline (**17**).

V. ARISTONE (**35**) [$C_{20}H_{24}N_2O$, orthorhombic crystals from benzene–ethyl acetate, mp 240–242°C (*21, 32*)]

The IR spectrum of aristone showed the presence of a ketone group, and the UV spectrum is characteristic of a substituted aniline. Lack of material precluded

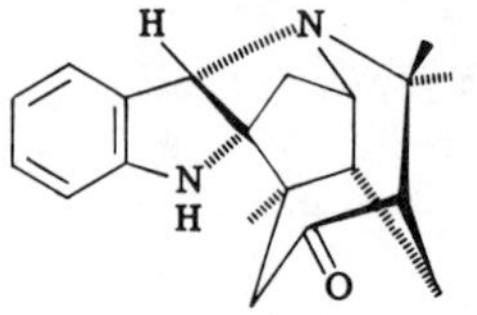

Aristone

35

a detailed spectroscopic examination, and the structure was solved by X-ray crystallography. The crystals belong to the space group $P2,2,2$, with cell dimensions $a = 11.075(3)$, $b = 18.004(4)$, and $c = 7.7973(1)$ Å. The measurements established both structure and stereochemistry and demonstrated that the molecule has a basket-like cavity. This novel indoline structure **35** with six interlocking rings and six chiral centers represents the most complex *Aristotelia* alkaloid so far described.

IV. Synthesis

Fruticosonine (**29**), the simplest *Aristotelia* alkaloid from a structural point of view, was the first to be synthesized (*10,17*). *o*-Toluidine was converted by

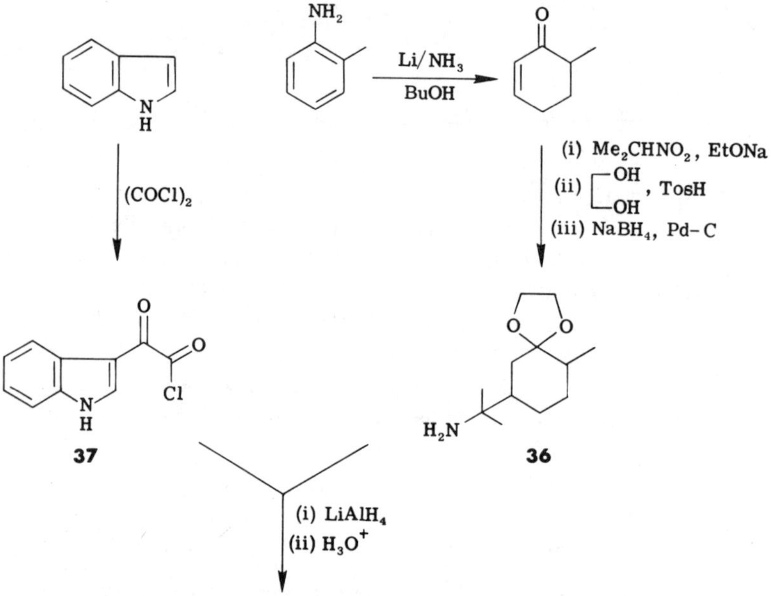

SCHEME 7. Synthesis of fruticosonine (**29**).

Birch reduction to 6-methylcyclohex-2-enone (*33*); a Michael condensation with 2-nitropropane, followed by protection of the carbonyl and reduction of the nitro group, gave the primary amine **36,** which was then treated with the acid chloride **37** formed from oxalyl chloride and indole (*34*). The resulting ketoamide was reduced with LAH to give racemic fruticosonine in 30% overall yield (Scheme 7).

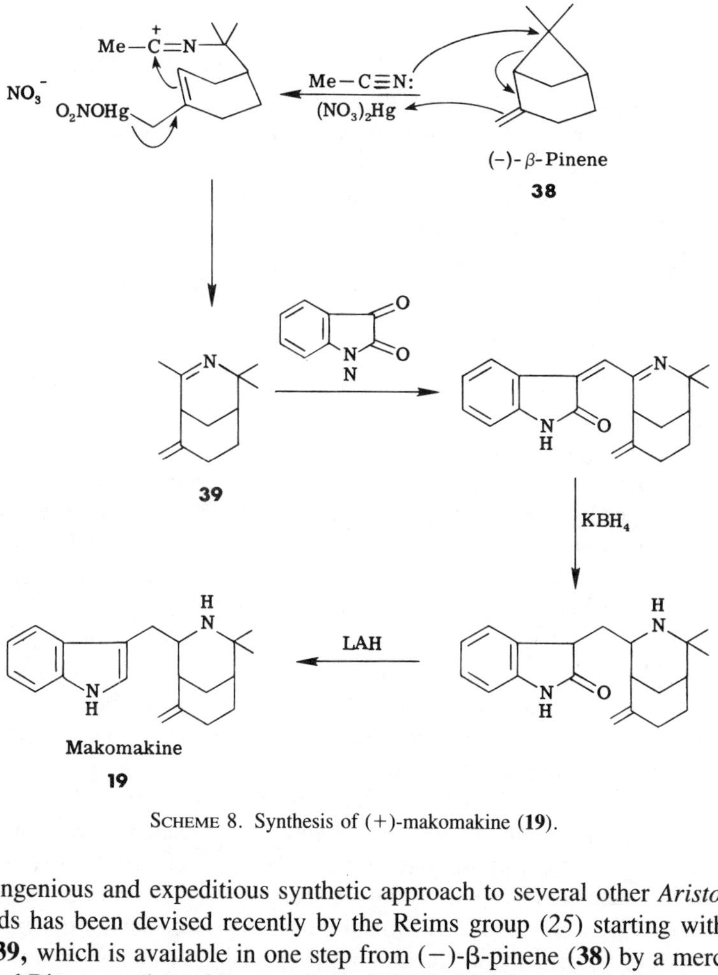

SCHEME 8. Synthesis of (+)-makomakine (**19**).

An ingenious and expeditious synthetic approach to several other *Aristotelia* alkaloids has been devised recently by the Reims group (*25*) starting with the imine **39,** which is available in one step from (−)-β-pinene (**38**) by a mercury-mediated Ritter reaction with acetonitrile (*35,36*). Condensation of **39** with isatin, followed by a two-stage reduction, gave in 40% yield (+)-makomakine (**19**), which had previously been cyclized to (+)-aristoteline (**11**) under conditions of acid catalysis (*14*) (Scheme 8).

A similar synthesis starting with (−)-α-pinene (**40**) led to hobartine (**15**), but the product was racemic owing to the rapid equilibration of the intermediate

MeCN, $Hg(NO_3)_2$

$Me-\overset{+}{C}=N$ NO_3^- $HgONO_2$ $\rightleftharpoons$ $Me-\overset{+}{C}=N$ O_2NOHg NO_3^-

40 **41**

organomercurial **41**; the (±)-hobartine was likewise cyclized to (±)-aristoteline (*25*). Since (+)-aristoteline has been converted to serratoline (**17,** *13,16*) which in turn has been transformed to aristotelone (**33,** *16*), the French route to **11** also constitutes formally a synthesis of **17** and **33,** and by a further extension, aristotelinone (**21**) should be accessible through mild oxidation of **11**; aristotelinone has already been transformed to makonine (**18,** *14*).

In a modification of the synthetic procedure shown in Scheme 8, acetonitrile was replaced with indole-3-ylacetonitrile in the Ritter reaction to give improved yields of **11**, (±)-**15,** and **19** after borohydride reduction (*55*).

V. Biogenesis

No labeling experiments have so far been reported on any *Aristotelia* sp., but preliminary incorporation experiments on *A. serrata* plants have been carried out (*37*), and suggestions have been put forward (*9, 21, 38*) concerning the possible biogenetic pathways to the major *Aristotelia* alkaloids. Some of the hypothetical intermediates in the proposed schemes have now been shown to exist and have been isolated and characterized, and an extension to the suggested pathways that includes these and other more recently discovered alkaloids is shown in Scheme 9. The transformations involved, such as oxidation at an α position to nitrogen, or formation of a new C—N bond through nucleophilic attack by a nitrogen, are well-known naturally occurring reactions for which there are many analogies in alkaloid biosynthesis. Some of the transformations have been carried out in the laboratory under mild conditions, and are described in Section III.

Scheme 9 is capable of a number of variations that would accord equally well with the present state of knowledge of these alkaloids; one such variation indicated in the scheme involves the biosynthesis of the 2,3-disubstituted indole aristoteline (**11**), which could be formed directly by electrophilic cyclization at C-2 of a 3-substituted indole such as makomakine (**19**) or hobartine (**15**), or alternatively by rearrangement of the indolenine aristoserratenine (**24**). For the sake of simplicity, no indication of stereochemistry is given in the scheme, and although it might be considered of little advantage to do so until more detailed knowledge is available, some significant points are nevertheless obscured as, for instance, in the final stage of the pathway to aristone (**35**) involving the transfer

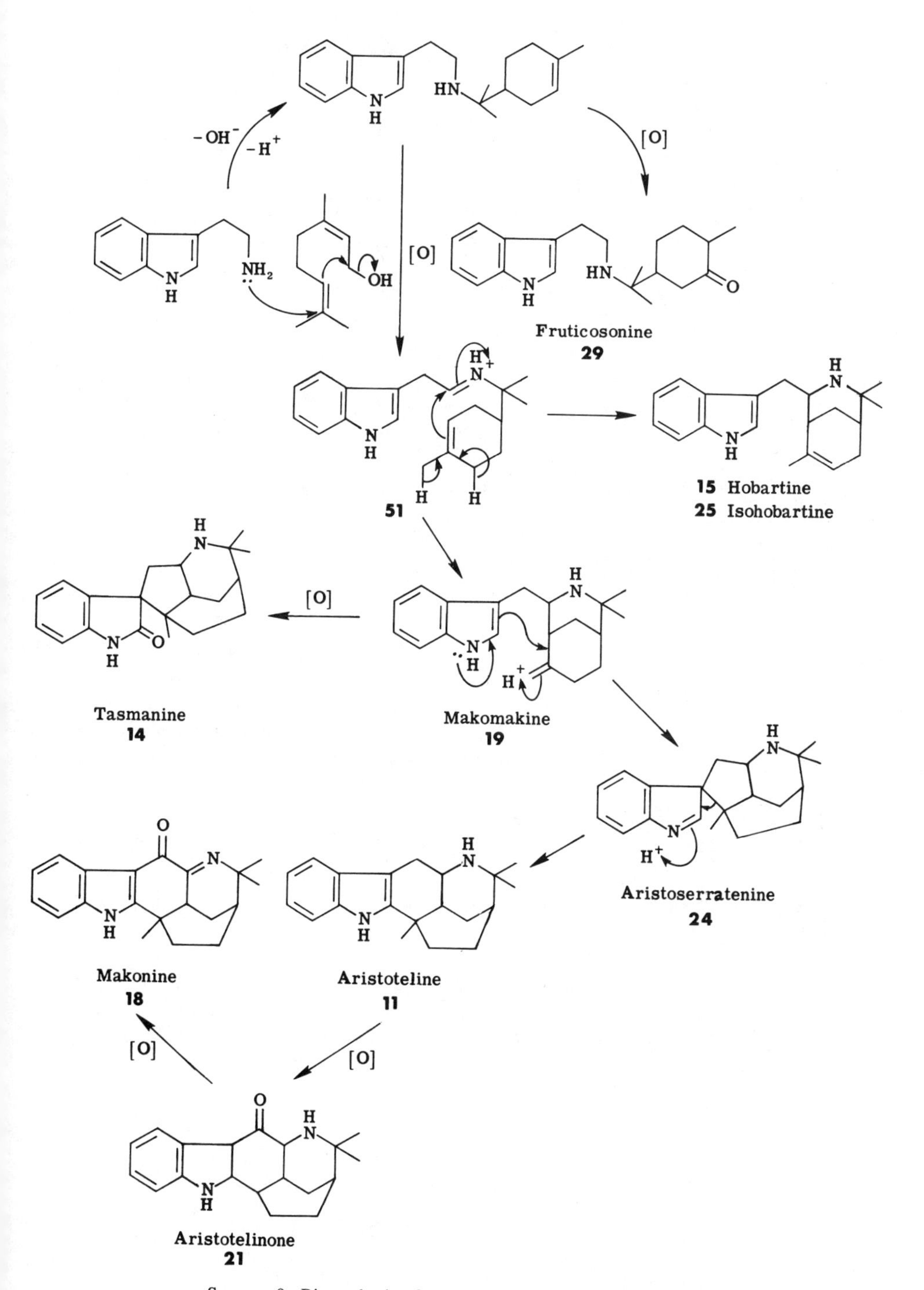

SCHEME 9. Biosynthesis of *Aristotelia* alkaloids. (*Continued*.)

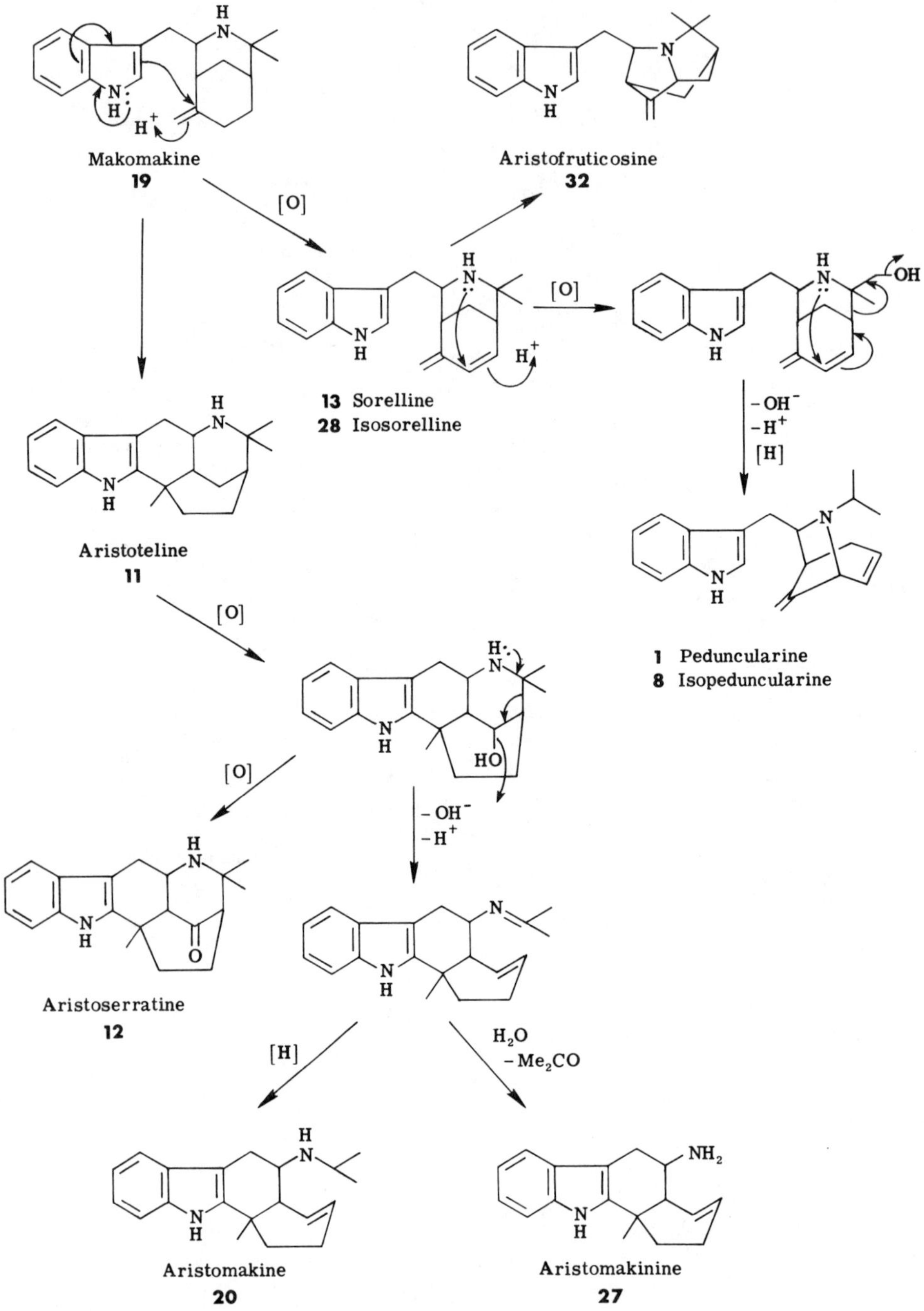

Makomakine
19

Aristofruticosine
32

[O]

[O]

13 Sorelline
28 Isosorelline

- OH⁻
- H⁺
[H]

Aristoteline
11

1 Peduncularine
8 Isopeduncularine

[O]

[O]

- OH⁻
- H⁺

Aristoserratine
12

[H]

H_2O
$-Me_2CO$

Aristomakine
20

Aristomakinine
27

SCHEME 9 (*continued*)

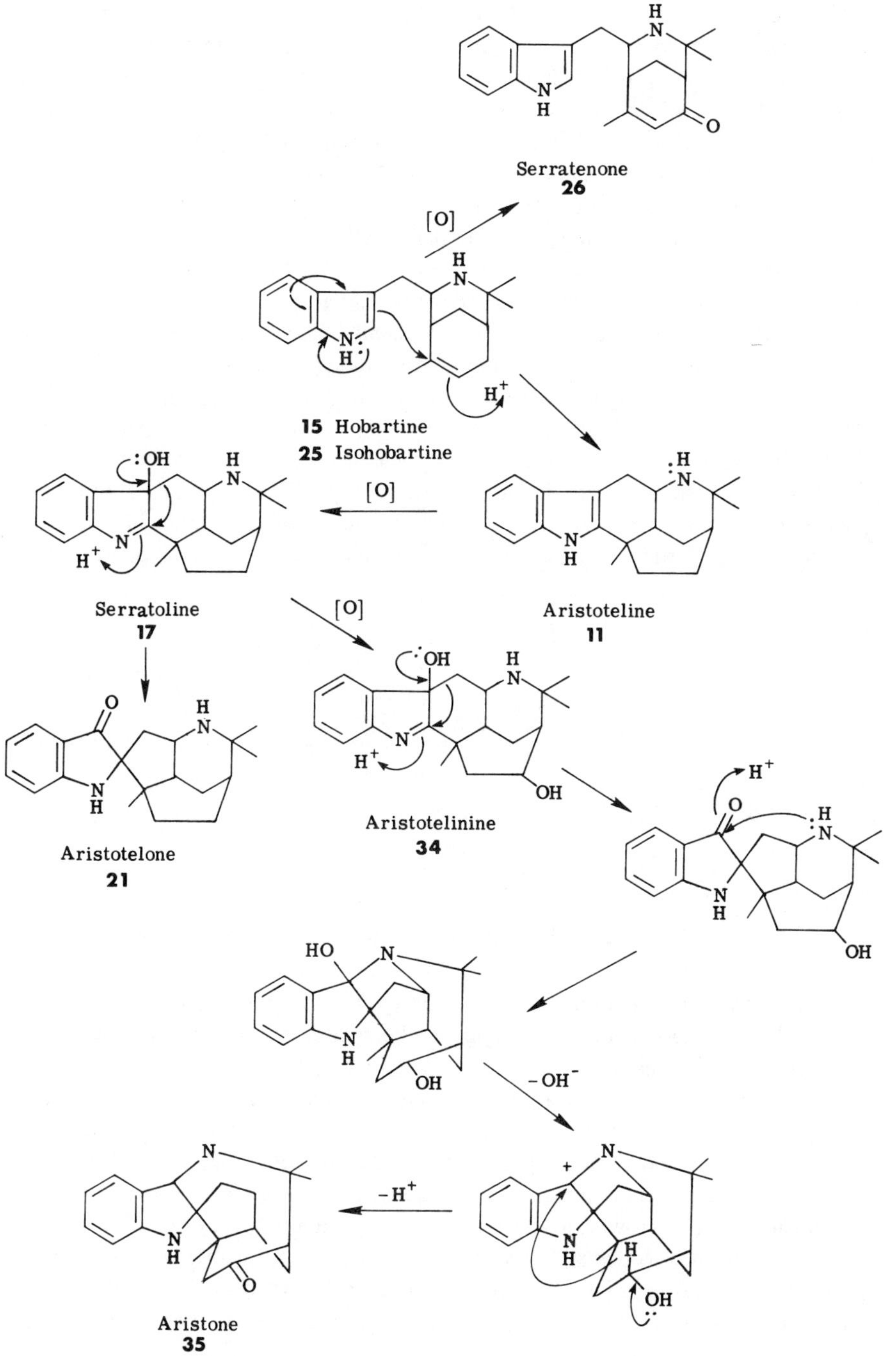

Serratenone
26

[O]

15 Hobartine
25 Isohobartine

[O]

Serratoline
17

Aristoteline
11

[O]

Aristotelone
21

Aristotelinine
34

$-OH^-$

$-H^+$

Aristone
35

SCHEME 9 (*continued*)

of a hydride ion: molecular models show that the two centers between which the proposed transfer takes place are quite close to one another in space.

It is already clear that different species, or even the one species, can produce alkaloids that are closely related structurally but belong to different stereochemical series: thus *A. serrata* elaborates serratenone (**26**) on the one hand, and on the other makomakine (**19**), isohobartine (**25**), and isosorelline (**28**). Several cases of diastereomers have been reported and one of enantiomers, and others are implicit in Scheme 9, as, for instance, in the transformation of hobartine to serratenone: one of these would have to be an antipode of the base known at present.

At the time of their discovery, the *Aristotelia* alkaloids appeared to constitute a unique series of indole bases in which an unchanged terpenoid unit has been incorporated into the molecule without preliminary conversion to an iridoid, although various rearrangements might take place subsequently. Recent studies on *Borreria* spp. (*39*) suggest that plants of this genus are also capable of the biosynthesis of indole alkaloids in which an unrearranged terpenoid unit has been built in, but the alkaloids are, however, structurally different from the *Aristotelia* bases. The first *Borreria* alkaloid in which the terpenoid nature of the nontryptamine portion of the molecule was clearly recognizable is borrecapine (**42,** *39*).

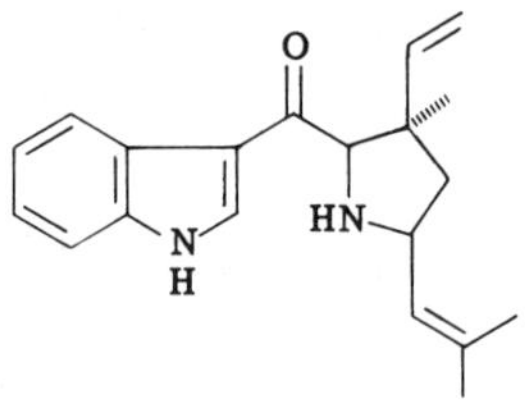

O
HN
N
H

Borrecapine
42

VI. Pharmacology and Applications in Folk Medicine

Few *Aristotelia* alkaloids occur in nature in sufficient quantity to permit a comprehensive pharmacological study to be made, but several have recently become readily accessible by synthesis (see Section IV). Studies on the physiological effects of a few that are available have been initiated, and although the complete results have not yet been reported, preliminary tests show that (+)-aristoteline (**11**), and certain others, decrease blood pressure and pulse rate (*40*), but it was found to be inactive in inducing interferon production in mice (*41*). Peduncularine (**1**) showed low activity against human breast cancer cells (*40*).

At least two *Aristotelia* spp. have been used in native remedies. *A. serrata* was employed in Maori medicine for the treatment of burns, sores, eye troubles, and rheumatism (*42–47*). The leaves were warmed on hot coals, then bandaged on burns (*44–46*); alternatively, leaves were first boiled with water, and the extracts were then used in the treatment of burns, boils, and sore eyes (*42–44, 46*). Extracts made by soaking the bark in cold water were also used for treating sore eyes (*46*). Similar infusions of bark (*43, 44, 46, 47*) or leaves (*42, 44*) made with hot water were used in the form of a warm bath in cases of rheumatism, both by the Maoris and by early settlers in New Zealand. The fruit is edible and has been used for making preserves (*43*).

The Maori name makomako for *A. serrata* resembles maki, one of the words used for the South American species *A. chilensis* by the Mapuche Indians (*48*), and some of the applications of the latter plant in the folk medicine of various Araucanian tribes are also similar. Crushed fresh leaves are used as a poultice for the treatment of boils and tumors (*19, 49–51*), and decoctions of leaves and soft bark likewise for pimples (*48*). Poultices made from leaves are also applied to the back region of the kidneys to counteract fevers (*19, 49–52*); infusions prepared from leaves are taken internally for the same purpose, and in addition for treating intestinal tumors (*19, 52*). Similar extracts of leaf or leaf sap are also used as a gargle in alleviating mouth sores (*19*) and throat complaints (*19, 49–52*). Dried and powdered leaves (*50–52*) or fruit (*48*) are applied to wounds and cuts, either directly or in the form of an ointment (*19, 51*). A warm extract of leaves and soft bark is also employed as a bath for treating hemorrhoids (*48*), and the sap from the bark, mixed with water, is administered to women to facilitate childbirth (*48*).

The fruit is edible (*19, 51*), and well-known in Chile; there is evidence for its use as a foodstuff in prehistoric times from its occurrence, together with other local and imported foods, in the pre-Columbian necropolis at Ancón, Peru (*53*). The berries are eaten fresh or made into preserves (*51, 53*); they are also used for making *teku,* a wine with tonic properties (*19, 48, 51, 54*), and a form of spirits called *tuen* (*54*). Infusions made from the fruit, or the fruit itself, are administered in the treatment of fevers, stomach disorders, enteritis, and diarrhea (*19, 49–52, 54*).

VII. Addendum

A review of alkaloids of *Aristotelia* spp. has appeared (*56*).

A further three minor alkaloids have been reported from *A. peduncularis* (*57*):

1. Peduncularistine (**43**) [0.18 ppm from dried plant material, amorphous]. Its structure and stereochemistry were deduced from a comparison of its spectroscopic data with those of other alkaloids from the same plant. In particular, similarities in CD spectra with aristoteline (**11**) indicated that the two bases had the same configuration. The structure and stereochemistry of **43** were confirmed by its conversion to aristoserratine (**12**) by hydrogenation of the olefinic group.

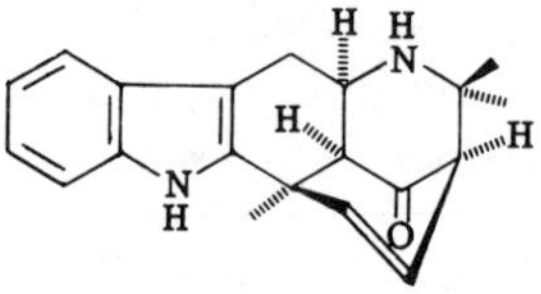

Peduncularistine
43

2. Triabunnine (**44**) [0.05 ppm from dried plant material, amorphous]. Its UV spectrum showed it to be an indolenine; ^{1}H-NMR data indicated quaternary carbons at C-2 and C-3, and further evidence from proton decoupling experiments pointed to structure **44** with a β-substituted hydroxyl group at C-3. The structure and absolute stereochemistry shown in **44** were confirmed by photochemical oxidation of aristoserratine (**12**) to triabunnine.

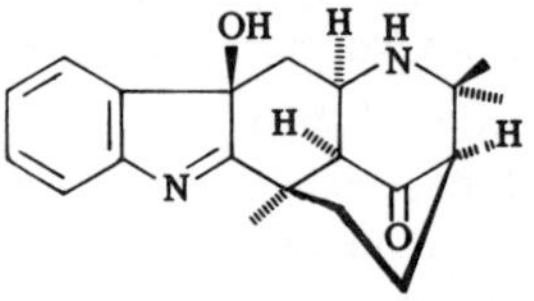

Triabunnine
44

3. Aristolarine (**45**) [0.11 ppm from dried plant material, yellow crystals, mp 181–181.5°C]. The electronic spectrum showed that aristolarine is a ψ-indoxyl derivative; it has from NMR evidence a secondary alcohol group and only two methyls. Structure **45** was suggested on the basis of spectroscopic data and biogenetic considerations.

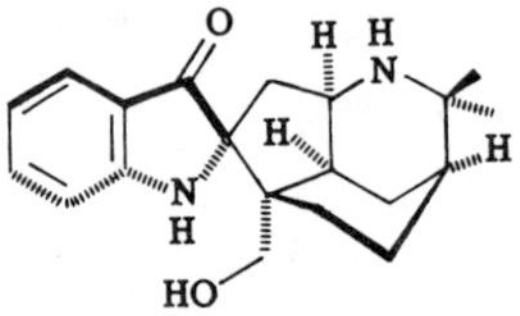

Aristolarine
45

Preliminary results from an examination of *A. australasica* have established the presence of the following alkaloids (*58*):

1. Bisaristone [mp 303–305°C] the major base, a dimer formed from two aristone units, with two carbonyl groups; very sparingly soluble.
2. Aristone (**35**) [mp 268–270°C, $[\alpha]_D$ −130°].

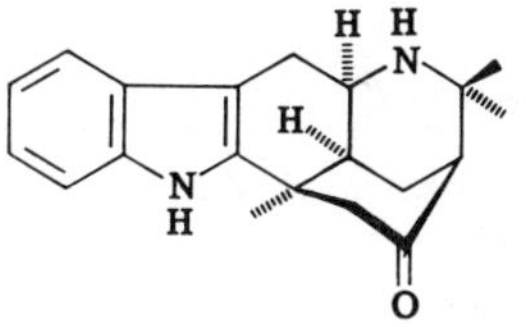

Aristolasicone
46

3. Aristoteline (**11**).
4. Aristolasicone (**46**) [mp 205–210°C, $[\alpha]_D^{20}$ −161°] whose structure has been established from spectroscopic data.

The plant contains at least one other previously undescribed alkaloid.

An expedient synthesis of naturally occurring (−)-hobartine (**15**) in five steps with an overall yield of 30% has fixed the previously unknown absolute stereochemistry of this base (*59*). The synthesis (Scheme 10) starts from (*S*)-α-terpineol (**47**), and proceeds *via* the intermediate **48,** which is one of the stages postulated in the biosynthesis of hobartine (Scheme 9). Synthetic (−)-hobartine (**15**) was transformed into (+)-aristoteline (**11**) to confirm the stereochemical relationship between these two bases.

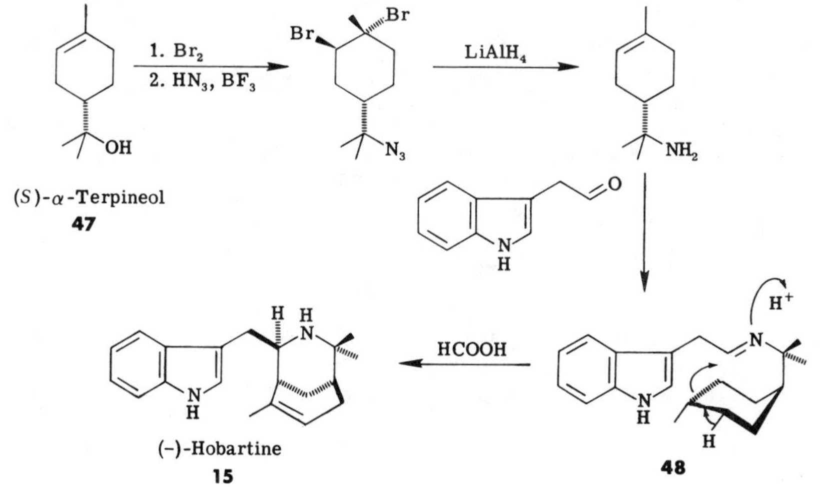

1. Br_2
2. HN_3, BF_3

$LiAlH_4$

HCOOH

(*S*)-α-Terpineol
47

(-)-Hobartine
15

48

SCHEME 10. Synthesis of (−)-hobartine.

REFERENCES

1. S. R. Johns and J. A. Lamberton, *in* "The Alkaloids" (R. H. F. Manske, ed.), Vol. 14, p. 326. Academic Press, New York, 1973.
2. J. A. Lamberton, Y. A. G. P. Gunawardana, and I. R. C. Bick, *J. Nat. Prod.* **46,** 235 (1983).

3. J. W. Loder, personal communication.
4. W. M. Curtis, "The Student's Flora of Tasmania," Part 1, p. 86. Government Printer, Hobart, 1956.
5. I. R. C. Bick, J. B. Bremner, N. W. Preston, and I. C. Calder, *J. Chem. Soc., Chem. Commun.* 1155 (1971).
6. H.-P. Ros, R. Kyburz, N. W. Preston, R. T. Gallagher, I. R. C. Bick, and M. Hesse, *Helv. Chim. Acta* **62,** 481 (1979).
7. R. Kyburz, E. Schöpp, I. R. C. Bick, and M. Hesse, *Helv. Chim. Acta* **62,** 2539 (1979).
8. M. A. Hai, N. W. Preston, R. Kyburz, E. Schöpp, I. R. C. Bick, and M. Hesse, *Helv. Chim. Acta* **63,** 2130 (1980).
9. R. Kyburz, E. Schöpp, I. R. C. Bick, and M. Hesse, *Helv. Chim. Acta* **64,** 2555 (1981).
10. M. A. Hai, Ph.D. Thesis, University of Tasmania, Hobart (1981).
11. H. H. Allan, "The Flora of New Zealand," p. 333. Govt. Printing Office, Wellington, 1961.
12. B. F. Anderson, G. B. Robertson, H. P. Avey, W. F. Donovan, I. R. C. Bick, J. B. Bremner, A. J. T. Finney, N. W. Preston, R. T. Gallagher, and G. B. Russell, *J. Chem. Soc., Chem. Commun.* 511 (1975).
13. I. R. C. Bick, M. A. Hai, N. W. Preston, and R. T. Gallagher, *Tetrahedron Lett.* **21,** 545 (1980).
14. I. R. C. Bick and M. A. Hai, *Heterocycles* **16,** 1301 (1981).
15. I. R. C. Bick, and M. A. Hai, *Tetrahedron Lett.* **22,** 3275 (1981).
16. I. R. C. Bick, M. A. Hai, and N. W. Preston, *Heterocycles* **20,** 667 (1983).
17. N. Chaichit, B. M. Gatehouse, I. R. C. Bick, M. A. Hai, and N. W. Preston, *J. Chem. Soc., Chem. Commun.* 874 (1979).
18. C. Muñoz Pizarro, "Sinopsis de la flora chilena," p. 126. Universidad de Chile, Santiago de Chile, 1959.
19. M. Muñoz, E. Barrera, and T. Mesa, "El uso medicinal y alimentico de plantas nativas y naturalizadas en Chile," Publ. Ocas. No. 33, p. 37. Museo Nacional de Historia Natural, Santiago de Chile, 1981.
20. D. S. Bhakuni, M. Silva, S. A. Matlin, and P. G. Sammes, *Phytochemistry* **15,** 574 (1976).
21. M. Bittner, M. Silva, E. M. Gopalakrishna, W. H. Watson, V. Zabel, S. A. Matlin, and P. G. Sammes, *J. Chem. Soc., Chem. Commun.* 79 (1978).
22. I. R. C. Bick and R. T. Gallagher, *10th IUPAC Symp. Chem. Nat. Prod., 1976* Abstract E8 (1976).
23. J. A. Joule, *Alkaloids (London)* **3,** 193 (1973).
24. W. H. Watson, V. Zabel, M. Silva, and M. Bittner, *Cryst. Struct. Commun.* **11,** 141 (1982).
25. C. Mirand, G. Massiot, and J. Lévy, *J. Org. Chem.* **47,** 4169 (1982).
26. I. R. C. Bick, M. A. Hai, V. A. Patrick, and A. H. White, *Aust. J. Chem.* **36,** 1037 (1983).
27. G. Snatzke, M. A. Hai, and I. R. C. Bick, unpublished results.
28. M. A. Hai, N. W. Preston, H.-P. Husson, C. Kan-Fan, and I. R. C. Bick, *Tetrahedron,* in press.
29. L. Bartlett, N. J. Dastoor, J. Hrbek, Jr., W. Klyne, H. Schmid, and G. Snatzke, *Helv. Chim. Acta* **54,** 1238 (1971).
30. R. A. Bell and J. K. Saunders, *Can. J. Chem.* **46,** 3421 (1968).
31. E. M. Gopalakrishna, W. H. Watson, M. Silva, and M. Bittner, *Acta Crystallogr., Sect. B* **B34,** 3378 (1978).
32. V. Zabel, W. H. Watson, M. Bittner, and M. Silva, *J. Chem. Soc., Perkin Trans. 1* 2842 (1980).
33. G. Stork and W. N. White, *J. Am. Chem. Soc.* **78,** 4604 (1956).
34. K. N. F. Shaw, A. McMillan, A. G. Gudmundson, and M. D. Armstrong, *J. Org. Chem.* **23,** 1171 (1958).

35. B. Delpech and Q. Khuong-Huu, *J. Org. Chem.* **43,** 4898 (1978).
36. A. Pancrazi, I. Kabore, B. Delpech, and Q. Khuong-Huu, *Tetrahedron Lett.* 3729 (1979).
37. I. R. C. Bick and R. K. Crowden, unpublished experiments.
38. I. R. C. Bick, M. A. Hai, and N. W. Preston, *Heterocycles* **12,** 1563 (1979).
39. A. Jössang, J.-L. Pousset, H. Jacquemin, and A. Cavé, *Tetrahedron Lett.* 4317 (1977).
40. H.-P. Ros, Farmos-Medipolar, (personal communication).
41. R. L. Buchanan, Bristol-Myers (personal communication).
42. G. Usher, "A Dictonary of Plants used by Man," p. 59. Constable, London, 1974.
43. C. Macdonald, "Medicines of the Maori," pp. 50–51. Collins, Auckland, 1973.
44. S. G. Brooker and R. C. Cooper, *Econ. Bot.* **15,** 1 (1961).
45. O. L. G. Adams, "Maori Medicinal Plants," Bull. No. 2. Auckland Botanical Society, Auckland, 1945.
46. S. G. Brooker, R. C. Cambie, and R. C. Cooper, *in* "New Zealand Medicinal Plants," p. 50. Heinemann, Auckland, 1981.
47. "Cookery Calendar," Spec. Maori Sect. Poverty Bay Federation of Womens' Institutes, Gisborne, ca. 1935.
48. M. Gusinde, *Anthropos* **31,** 555, 851 (1971).
49. P. Pacheco, M. Chiang, C. Marticorena, and M. Silva, *in* "Química de las plantas chilenas usadas en medicina popular I," p. 96. Laboratorio de Química de Productos Naturales, Departamento de Botánica, Universidad de Concepción, Concepción, 1977.
50. J. Zin, "La salud por medio de las plantas medicinales, especialmente de Chile," p. 246. Universidad de Chile, Santiago de Chile, 1919.
51. A. Murillo, "Plantes médicinales du Chili," pp. 28 and 29. Exposition Universelle de Paris, Section Chilienne, Paris, 1889.
52. M. Montes and T. Wilkomirsky, "Plantas chilenas en medicina popular, ciencia, y folklore," pp. 64 and 66. Facultad de Farmacia, Universidad de Concepción, Concepción, Chile, 1978.
53. M. A. Towle, "The Ethno-botany of Pre-Columbian Peru," pp. 63, 98, and 125. Aldine, Chicago, Illinois, 1961.
54. Dujardin-Beaumetz and E. Égasse, "Les plantes médicinales indigènes et exotiques," p. 792. Doin, Paris, 1889.
55. R. V. Stevens and P. M. Kenney, *J. Chem. Soc., Chem. Commun.* 384 (1983).
56. J. E. Saxton, *in* "Indoles: Monoterpenoid indole alkaloids" (J. E. Saxton, ed.), chapter II, p. 47. Wiley, New York, 1983.
57. R. Kyburz, E. Schöpp, and M. Hesse, *Helv. Chim. Acta* **67,** 804 (1984).
58. J. C. Quirion, C. Kan-Fan, H.-P. Husson, and I. R. C. Bick, unpublished results.
59. T. Darbre, C. Nussbaumer, and H.-J. Borschberg, *Helv. Chim. Acta* **67,** 1040 (1984).

—CHAPTER 4—

APORPHINE ALKALOIDS

TETSUJI KAMETANI AND TOSHIO HONDA

Institute of Medicinal Chemistry
Hoshi University
Tokyo, Japan

I. Introduction

Since the aporphine and oxoaporphine alkaloids were reviewed in Volumes IX and XIV of this treatise (*1*), a variety of important developments have taken place in the area of this group of alkaloids. Concerning the isolation and characterization of new aporphine alkaloids, more than 200 alkaloids have appeared to date, and this achievement can mostly be attributed to the application of modern physical methods. Biogenesis of aporphine alkaloids was also well investigated, and tracer work showed that the proerythrinadienones could be another important intermediate for the conversion of tetrahydroisoquinolines to aporphines. In the syntheses of aporphine alkaloids, phenolic oxidative coupling, the Pschorr reaction, the photo-Pschorr reaction, photocyclization, the benzyne-mediated reac-

THE ALKALOIDS, VOL. XXIV

ISBN 0-12-469524-8

tion, and the quinone-mediated reaction have played an important role. Moreover, enzymatic model syntheses were applied to the synthesis of this class of alkaloids. Pharmacologies and properties of aporphine alkaloids are also discussed in this review.

The homoaporphines will be dealt with in the chapter on phenethylisoquinolines, therefore they are omitted from this review. The numbering system used throughout this chapter is as follows:

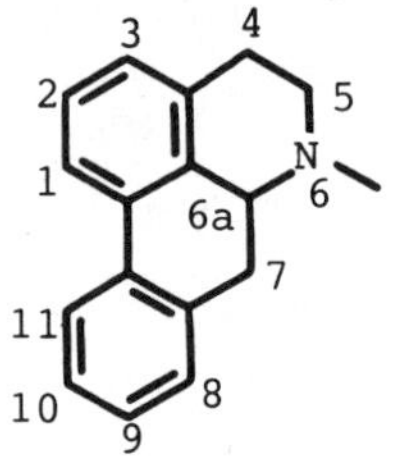

II. New Aporphine Alkaloids

More than 200 aporphine alkaloids including oxoaporphines and bisisoquinoline alkaloids were newly isolated and characterized since the last reviews in Volumes IX and XIV of this treatise (*1*) were published. These new alkaloids are listed in the tables together with their botanical sources, and some of those that were isolated in recent years are further explained by showing their structure.

A. Monomeric Aporphine Alkaloids

Newly isolated monomeric aporphine alkaloids are listed in Table I (*2–131*), although a number of reviews concerning aporphine alkaloids have appeared (*132, 133*). Proaporphines, phenanthrene alkaloids, and azafluoranthene alkaloids were excluded even though they were of interest to chemists in this field from biological and structural points of view.

1. (−)-Norliridinine and (−)-3-Hydroxynornuciferine

(−)-Norliridinine (**1**) and (−)-3-hydroxynornuciferine (**2**) were isolated from *Polyalthia acuminata* Thw. (Annonaceae), and their mass spectra show molecular ion peaks at *m*/*z* 297, which corresponds to $C_{18}H_{19}NO_3$, and a base peak at *m*/*z* 196 (*53*).

The NMR spectra reveal two methoxyl singlets at 3.63 and 3.93 ppm for **1,** and at 3.72 and 3.98 ppm for **2.** The 11-H signal for **1** appears as a doublet at 8.24 ppm with J = 8.2 Hz, and the corresponding doublet is found at 8.21 ppm

TABLE I

NEWLY ISOLATED APORPHINES

Alkaloid	Structure	Plant source	Reference
O-Methylbulbocapnine	$R^1 + R^2 = OCH_2O$, $R^3 = R^6 = R^7 = H$, $R^4 = R^5 = OMe$, $R^8 = Me$	*Lindera oldhamii* Hemsl.	2
Norcorydine	$R^1 = OH$, $R^2 = R^4 = R^5 = OMe$, $R^3 = R^6 = R^7 = R^8 = H$	*Popowia* cf *cyanocarpa* Laut. and K. Schurum.	3
N-Methylhernovine	$R^1 = R^4 = OMe$, $R^2 = R^5 = OH$, $R^3 = R^6 = R^7 = H$, $R^8 = Me$	*Croton wilsonii* Griseb.	4
10-*O*-Methylhernovine	$R^1 = R^4 = R^5 = OMe$, $R^2 = OH$, $R^3 = R^6 = R^7 = R^8 = H$	*Croton wilsonii* Griseb.	4
N-Methyl-10-*O*-methylhernovine	$R^1 = R^4 = R^5 = OMe$, $R^2 = OH$, $R^3 = R^6 = R^7 = H$, $R^8 = Me$	*Croton wilsonii* Griseb.	4
Lindcarpine	$R^1 = R^5 = OMe$, $R^2 = R^4 = PH$, $R^3 = R^6 = R^7 = R^8 = H$	*Lindera pipericarpa* Boerl	5
N-Methyllindcarpine	$R^1 = R^5 = OMe$, $R^2 = R^4 = OH$, $R^3 = R^6 = R^7 = H$, $R^8 = Me$	*Phoebe clemensii* Allen	6
Norglaucine	$R^1 = R^2 = R^5 = R^6 = OMe$, $R^3 = R^4 = R^7 = R^8 = H$	*Duguetia* species	.7

(*continued*)

TABLE I (*Continued*)

Alkaloid	Structure	Plant source	Reference
Wilsonirine	$R^1 = OH, R^2 = R^5 = R^6 = OMe, R^3 = R^4 = R^7 = R^8 = H$	*Popowia* cf *cyanocarpa* Laut. and K. Schurum	*3*
Predicentrine	$R^1 = R^5 = R^6 = OMe, R^2 = OH, R^3 = R^4 = R^7 = H, R^8 = Me$	*Beilschmiedia podagrica* Kostermans	*8*
Norpredicentrine	$R^1 = R^5 = R^6 = OMe, R^2 = OH, R^3 = R^4 = R^7 = R^8 = H$	*Beilschmiedia podagrica* Kostermans	*8*
Bracteoline	$R^1 = R^5 = OH, R^2 = R^6 = OMe, R^3 = R^4 = R^7 = H, R^8 = Me$	*Papaver bracteatum* Lindl., *P. orientale*	*9*
Nornantenine	$R^1 = R^2 = OMe, R^3 = R^4 = R^7 = R^8 = H, R^5 + R^6 = OCH_2O$	*Cassytha racemosa* Nees	*10*
Nordomesticine	$R^1 = OH, R^2 = OMe, R^3 = R^4 = R^7 = R^8 = H, R^5 + R^6 = OCH_2O$	*Cassytha pubescens* R. Br.	*11*
Hernandine	$R^1 + R^2 = OCH_2O, R^3 = R^4 = OMe, R^5 = OH, R^6 = R^7 = R^8 = H$	*Hernandia bivalvis* Benth.	*12*
Ocokryptine	$R^1 = R^5 = OMe, R^2 + R^3 = OCH_2O, R^4 = OH, R^6 = R^7 = H, R^8 = Me$	*Ocotea* species	*13*
Oconovine	$R^1 = R^2 = R^3 = R^5 = OMe, R^4 = OH, R^6 = R^7 = H, R^8 = Me$	*Ocotea* species	*13*
Thalicsimidine (Purpureine)	$R^1 = R^2 = R^3 = R^5 = R^6 = OMe, R^4 = R^7 = H, R^8 = Me$	*Thalictrum simplex*, *Annona purpurea* L.	*14,15*
Norpurpureine	$R^1 = R^2 = R^3 = R^5 = R^6 = OMe, R^4 = R^7 = R^8 = H$	*Annona purpurea* L.	*15*
O-Demethylpurpureine	$R^1 = R^5 = R^6 = OMe, R^2 + R^3 = OH + OMe, R^4 = R^7 = H, R^8 = Me$	*Annona purpurea* L.	*15*
Ocopodine	$R^1 + R^2 = OCH_2O, R^3 = R^4 = H, R^5 = R^6 = R^7 = OMe, R^8 = Me$	*Ocotea macropoda*	*13*
Nuciferoline	$R^1 = R^2 = OMe, R^3 = R^4 = R^6 = R^7 = H, R^5 = OH, R^8 = Me$	*Papaver caucasicum* Marsch.–Bieb.	*16*

(+)-Apoglaziovine	$R^1 = R^5 = OH, R^2 = OMe, R^3 = R^4 = R^6 = R^7 = H, R^8 = Me$	*Ocotea variabilis*	*17*
(+)-Nuciferine	$R^1 = R^2 = OMe, R^3 = R^4 = R^5 = R^6 = R^7 = H, R^8 = Me$	*Papaver* species	*18*
(+)-Litsedine	$R^1 + R^2 = OCH_2O, R^3 = R^6 = R^7 = R^8 = H, R^4 = R^5 = OMe$	*Litsea nitida* Roxb.	*19*
(+)-Caaverine	$R^1 = OH, R^2 = OMe, R^3 = R^4 = R^5 = R^6 = R^7 = R^8 = H$	*Liriodendron tulipifera*	*20*
(+)-Isolaureline	$R^1 + R^2 = OCH_2O, R^3 = R^4 = R^5 = R^7 = H, R^6 = OMe, R^8 = Me$	*Liriodendron tulipifera*	*21*
(+)-Cryptodorine	$R^1 + R^2 = R^5 + R^6 = OCH_2O, R^3 = R^4 = R^7 = R^8 = H$	*Cryptocarya odorata* (Panch. and Seb.) Guillaum	22
(+)-*N*-Acetylnornantenine	$R^1 = R^2 = OMe, R^3 = R^4 = R^7 = H, R^5 + R^6 = OCH_2O, R^8 = Ac$	*Liriodendron tulipifera* L.	*23,24*
(+)-*N*-Acetyl-3-methoxynornantenine	$R^1 = R^2 = R^3 = OMe, R^4 = R^7 = H, R^5 + R^6 = OCH_2O, R^8 = Ac$	*Liriodendron tulipifera* L.	*23,24*
(+)-*O,O*-Dimethylcorytuberine	$R^1 = R^2 = R^4 = R^5 = OMe, R^3 = R^6 = R^7 = H, R^8 = Me$	*Hernandia jamaicensis*	*25*
(+)-*O*-Methylcassyfiline	$R^1 = R^2 = OCH_2O, R^3 = R^5 = R^6 = OMe, R^4 = R^7 = R^8 = H$	*Cassytha americana*	*26*
Litseferine	$R^1 + R^2 = OCH_2O, R^3 = R^4 = R^7 = R^8 = H, R^5 = OH, R^6 = OMe$	*Litsea sebifera*	*27*
(+)-Lirioferine	$R^1 = R^2 = R^6 = OMe, R^3 = R^4 = R^7 = H, R^5 = OH, R^8 = Me$	*Liriodendron tulipifera*	*28*
(+)-Liriotulipiferine	$R^1 = R^6 = OMe, R^2 = R^5 = OH, R^3 = R^4 = R^7 = H, R^8 = Me$	*Liriodendron tulipifera*	*28*
(+)-Leucoxylopine	$R^1 + R^2 = OCH_2O, R^3 = R^5 = R^6 = R^7 = OMe, R^4 = H, R^8 = Me$	*Ocotea leucoxylon*	*29*
(+)-Ocoxylopine	$R^1 + R^2 = OCH_2O, R^3 = R^5 = R^6 = OMe, R^4 = H, R^7 = OH, R^8 = Me$	*Ocotea leucoxylon*	*29*
(+)-Leucoxine	$R^1 + R^2 = OCH_2O, R^3 = R^4 = H, R^5 = R^6 = OMe, R^7 = OH, R^8 = Me$	*Ocotea leucoxylon*	*29*

(*continued*)

TABLE I (*Continued*)

Alkaloid	Structure	Plant source	Reference
(+)-Delporphine	$R^1 = R^2 = R^5 = OMe$, $R^3 = R^6 = OH$, $R^4 = R^7 = H$, $R^8 = Me$	*Delphinium dictyocarpum*	*30*
(+)-Glaufine	$R^1 = OMe$, $R^2 = R^4 = R^5 = OH$, $R^3 = R^6 = R^7 = H$, $R^8 = Me$	*Glaucium fimbrilligerum*	*31*
(+)-Laetanine	$R^1 = R^6 = OMe$, $R^2 = R^5 = OH$, $R^3 = R^4 = R^7 = R^8 = H$	*Litsea laeta* Benth. & Hook. f.	*32*
(+)-Thalisopynine	$R^1 = R^2 = R^3 = R^5 = OMe$, $R^4 = R^7 = H$, $R^6 = OH$, $R^8 = Me$	*Thalictrum isopyroides* C. A. Mey.	*33*
(+)-Isothebaidine	$R^1 = R^4 = OH$, $R^2 = OMe$, $R^3 = R^5 = R^6 = R^7 = H$, $R^8 = Me$	*Papaver orientale* L.	*34*
Ocotominarine	$R^1 + R^2 = R^6 + R^7 = OCH_2O$, $R^3 = R^5 = OMe$, $R^4 = H$, $R^8 = Me$	*Ocotea minarum*	*35*
Ocominarine	$R^1 + R^2 = R^6 + R^7 = OCH_2O$, $R^3 = R^4 = H$, $R^5 = OMe$, $R^8 = Me$	*Ocotea minarum*	*35*
Norleucoxylonine	$R^1 + R^2 = OCH_2O$, $R^3 = R^5 = R^6 = R^7 = OMe$, $R^4 = H$, $R^8 = Me$	*Ocotea minarum*	*35*
Isooconovine	$R^1 = OH$, $R^2 = R^3 = R^4 = R^5 = OMe$, $R^6 = R^7 = H$, $R^8 = Me$	*Ocotea minarum*	*35*
(+)-Hernagine	$R^1 = R^2 = R^4 = OMe$, $R^3 = R^6 = R^7 = R^8 = H$, $R^5 = OH$	*Hernandia nymphaefolia*	*36*
Isocorytuberine	$R^1 = R^5 = OH$, $R^2 = R^4 = OMe$, $R^3 = R^6 = R^7 = H$, $R^8 = Me$	*Glaucium fimbrilligerum*	*37*
Laetine	$R^1 = OMe$, $R^2 = OH$, $R^3 = R^6 = R^7 = R^8 = H$, $R^4 + R^5 = OCH_2O$	*Litsea laeta*	*38*
N,O,O-Trimethylsparsiflorine	$R^1 = R^2 = R^5 = OMe$, $R^3 = R^4 = R^6 = R^7 = H$, $R^8 = Me$	*Thalictrum foliolosum*	*39*

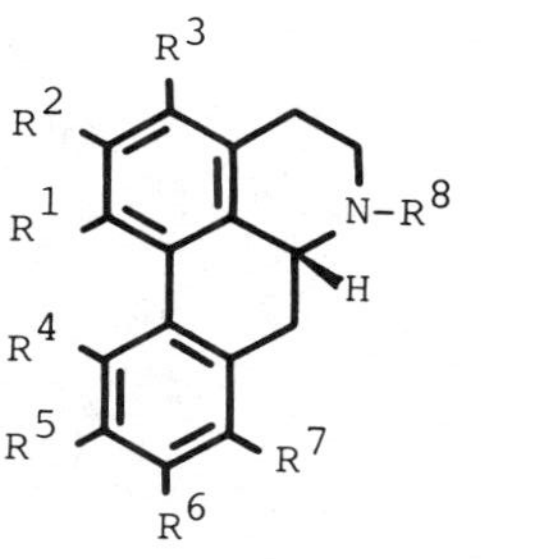

(−)-Dicentrine	$R^1 + R^2 = OCH_2O$, $R^3 = R^4 = R^7 = H$, $R^5 = R^6 = OMe$, $R^8 = Me$	*Duguetia* species	*7*
O-Methylpukateine	$R^1 + R^2 = OCH_2O$, $R^3 = R^5 = R^6 = R^7 = H$, $R^4 = OMe$, $R^8 = Me$	*Laurelia novaezelandiae* A. Cunn.	*40*
(−)-Mecambroline	$R^1 = R^2 = OCH_2O$, $R^3 = R^4 = R^6 = R^7 = H$, $R^5 = OH$, $R^8 = Me$	*Phoebe clemensii* Allen, *Laurelia novaezelandiae* A. Cunn.	*6,40*
(−)-*O*-Methyllirinine	$R^1 = R^2 = R^3 = OMe$, $R^4 = R^5 = R^6 = R^7 = H$, $R^8 = Me$	*Liriodendron tulipifera*	*41*
(−)-Obovanine	$R^1 + R^2 = OCH_2O$, $R^3 = R^5 = R^6 = R^7 = R^8 = H$, $R^4 = OH$	*Magnolia obovata* Thunb.	*42*
(−)-Stephalagine	$R^1 + R^2 = OCH_2O$, $R^3 = OMe$, $R^4 = R^5 = R^6 = R^7 = H$, $R^8 = Me$	*Stephania dinklagei* Diels	*43,44*
(−)-Elmerrillicine	$R^1 + R^2 = OCH_2O$, $R^3 = OMe$, $R^4 = OH$, $R^5 = R^6 = R^7 = R^8 = H$	*Elmerillia papuana* (Schltr.) Dandy	*45*
Floripavidine	$R^1 = OMe$, $R^2 =$ L-Rhamnoside, $R^3 = R^4 = R^5 = R^6 = R^7 = H$, $R^8 = Me$	*Papaver floribundum*	*46*
(−)-Tuliferoline	$R^1 = R^2 = R^3 = OMe$, $R^4 = R^5 = R^6 = R^7 = H$, $R^8 = Ac$	*Liriodendron tulipifera*	*47*
(−)-Norlaureline	$R^1 + R^2 = OCH_2O$, $R^3 = R^4 = R^6 = R^7 = R^8 = H$, $R^5 = OMe$	*Guatteria elata*	*48*
(−)-Puterine	$R^1 + R^2 = OCH_2O$, $R^3 = R^4 = R^5 = R^7 = R^8 = H$, $R^6 = OMe$	*Guatteria elata*	*48*

(*continued*)

TABLE I (*Continued*)

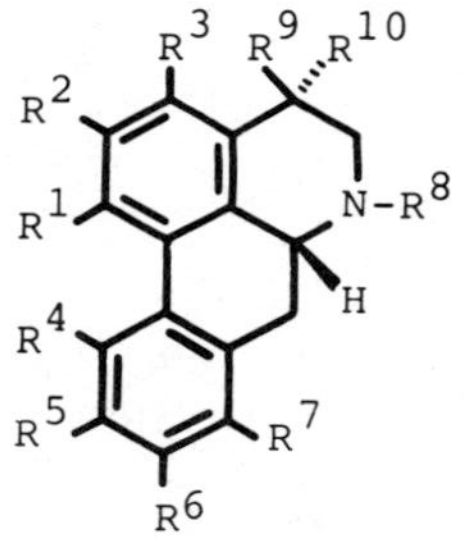

Alkaloid	Structure	Plant source	Reference
(−)-Zenkerine	$R^1 = R^5 = OMe, R^2 = OH, R^3 = R^4 = R^6 = R^7 = R^8 = H$	*Isolona pilosa*	*49*
Stesakine	$R^1 + R^2 = OCH_2O, R^3 = R^4 = R^5 = H, R^6 = OH, R^7 = OMe, R^8 = Me$	*Stephania sasakii*	*50*
(−)-Roemeroline	$R^1 + R^2 = OCH_2O, R^3 = R^4 = R^5 = R^7 = H, R^6 = OH, R^8 = Me$	*Stephania sasakii* *Roemeria refracta* (Stev.) DC.	*51,52*
(−)-Norliridinine	$R^1 = R^3 = OMe, R^2 = OH, R^4 = R^5 = R^6 = R^7 = R^8 = H$	*Polyalthia acuminata*	*53,54*
(−)-3-Hydroxynornuciferine	$R^1 = R^2 = OMe, R^3 = OH, R^4 = R^5 = R^6 = R^7 = R^8 = H$	*Polyalthia acuminata* (*Guatteria melosma* Diels)	*53*
(−)-Norannuradhapurine	$R^1 + R^2 = OCH_2O, R^3 = R^4 = R^5 = R^8 = H, R^6 = OMe, R^7 = OH$	*Polyalthia acuminata*	*53*
Steporphine	$R^1 + R^2 = OCH_2O, R^3 = R^4 = R^5 = R^6 = R^7 = H, R^8 = Me, R^9$ or $R^{10} = OH$	*Stephania sasakii* Hayata	*55*
4-Hydroxyanonaine	$R^1 = R^2 = OCH_2O, R^3 = R^4 = R^5 = R^6 = R^7 = R^8 = R^{10} = H, R^9 = OH$	*Laurelia philippiana*	*56*

(−)-4-Hydroxycrebanine	$R^1 = R^2 = R^6 = R^7 = OMe, R^3 = R^4 = R^5 = R^9 = H, R^8 = Me, R^{10} = OH$	*Stephania sasakii*	*51*
	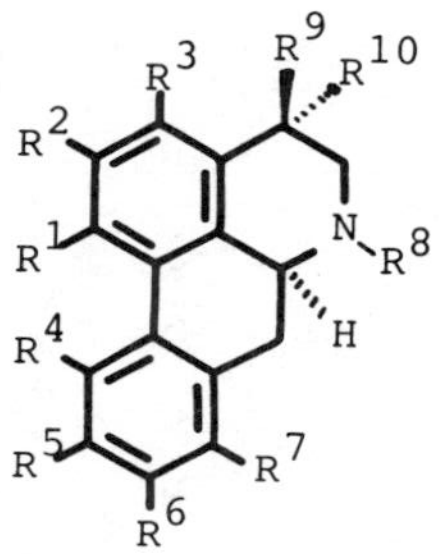		
Cataline	$R^1 = R^2 = R^5 = R^6 = OMe, R^3 = R^4 = R^7 = R^{10} = H, R^8 = Me, R^9 = OH$	*Glaucium flavum* Cr. var. vestitum	*57*
(+)-4-Hydroxybulbocapnine	$R^1 + R^2 = OCH_2O, R^3 = R^6 = R^7 = R^{10} = H, R^4 = R^9 = OH, R^5 = OMe, R^8 = Me$	*Glaucium vitellinum* Boiss. and Buhse	*58*
(+)-4-Hydroxynornantenine	$R^1 = R^2 = OMe, R^3 = R^4 = R^7 = R^8 = R^{10} = H, R^5 + R^6 = OCH_2O, R^9 = OH$	*Laurelia philippiana* Looser	*59*
(+)-Srilankine	$R^1 = R^5 = R^6 = OMe, R^2 = R^9 = OH, R^3 = R^4 = R^7 = R^{10} = H, R^8 = Me$	*Alseodaphne semicarpifolia* Nees	*60*
4-Hydroxydicentrine	$R^1 + R^2 = OCH_2O, R^3 = R^4 = R^7 = R^9 = H, R^5 = R^6 = OMe, R^8 = Me, R^{10} = OH$	*Ocotea minarum*	*35*
Glaufidine	$R^1 = R^{10} = OH, R^2 = R^4 = R^5 = OMe, R^3 = R^6 = R^7 = R^9 = H, R^8 = Me$	*Glaucium fimbrilligerum*	*61*

(*continued*)

TABLE I (*Continued*)

Alkaloid	Structure	Plant source	Reference
	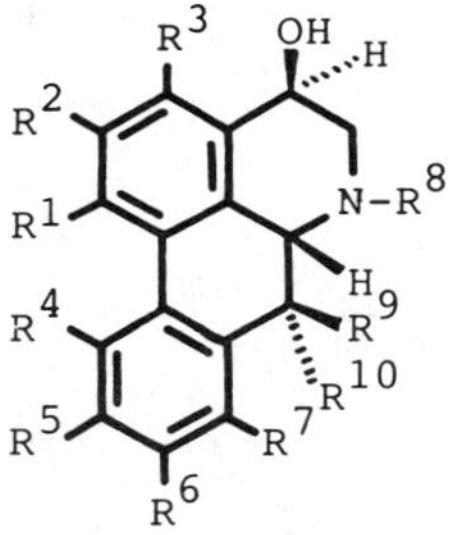		
(+)-Pachystaudine	$R^1 + R^2 = OCH_2O$, $R^3 = R^4 = R^5 = R^6 = R^7 = R^{10} = H$, $R^8 = Me$, $R^9 = OMe$	*Pachypodanthium* species	*62*
(+)-Norpachystaudine	$R^1 + R^2 = OCH_2O$, $R^3 = R^4 = R^5 = R^6 = R^7 = R^8 = R^{10} = H$, $R^9 = OMe$	*Pachypodanthium* species	*62*
	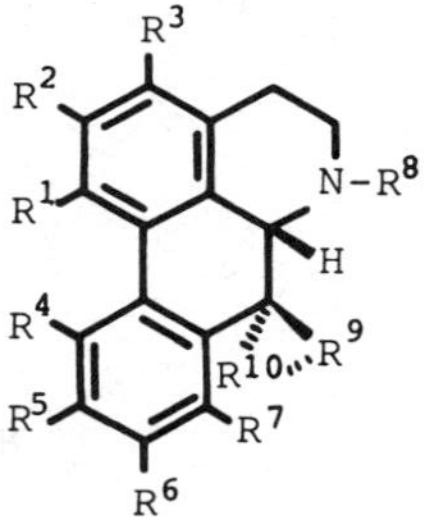		

(−)-Duguetine	$R^1 + R^2 = OCH_2O$, $R^3 = R^4 = R^7 = R^{10} = H$, $R^5 = R^6 = OMe$, $R^8 = Me$, $R^9 = OH$	*Duguetia* species	*7*
(−)-Michelanugine	$R^1 + R^2 = OCH_2O$, $R^3 = R^4 = R^5 = R^7 = R^8 = R^9 = H$, $R^6 = OMe$, $R^{10} = OH$	*Michelia lanuginosa* Wall.	*63,64*
(−)-Oliveridine	$R^1 + R^2 = OCH_2O$, $R^3 = R^4 = R^5 = R^7 = R^{10} = H$, $R^6 = OMe$, $R^8 = Me$, $R^9 = OH$	*Polyalthia oliveri* Engl.	*64*
(+)-Oliverine	$R^1 + R^2 = OCH_2O$, $R^3 = R^4 = R^5 = R^7 = R^{10} = H$, $R^6 = R^9 = OMe$, $R^8 = Me$	*Polyalthia oliveri* Engl.	*64*
(+)-Pachypodanthine	$R^1 + R^2 = OCH_2O$, $R^3 = R^4 = R^5 = R^6 = R^7 = R^8 = R^{10} = H$, $R^9 = OMe$	*Pachypodanthium staudtii* Engl. and Diels, *P. oliveri* Engl.	*65*
Oliveroline	$R^1 + R^2 = OCH_2O$, $R^3 = R^4 = R^5 = R^6 = R^7 = R^{10} = H$, $R^8 = Me$, $R^9 = OH$	*Pachypodanthium oliveri* Engl.	*66*
Noroliveridine	$R^1 + R^2 = OCH_2O$, $R^3 = R^4 = R^5 = R^7 = R^8 = R^{10} = H$, $R^6 = OMe$, $R^9 = OH$	*Pachypodanthium oliveri* Engl.	*66*
Pachyconfine	$R^1 = OMe$, $R^2 = R^9 = OH$, $R^3 = R^4 = R^5 = R^6 = R^7 = R^{10} = H$, $R^8 = Me$	*Pachypodanthium confine* Engl. and Diels	*67*
(+)-Polysuavine	$R^1 + R^2 = OCH_2O$, $R^3 = R^4 = R^5 = R^7 = R^{10} = H$, $R^6 = OH$, $R^8 = Me$, $R^9 = OMe$	*Polyalthia species*	*68*
(+)-Noroliverine	$R^1 + R^2 = OCH_2O$, $R^3 = R^4 = R^5 = R^7 = R^8 = R^{10} = H$, $R^6 = R^9 = OMe$	*Polyalthia species*	*68*
(+)-Polyalthine	$R^1 + R^2 = OCH_2O$, $R^3 = R^6 = OMe$, $R^4 = R^5 = R^7 = R^{10} = H$, $R^8 = Me$, $R^9 = OH$	*Polyalthia species*	*68*
Anaxagoreine	$R^1 = OMe$, $R^2 = R^{10} = OH$, $R^3 = R^4 = R^5 = R^6 = R^7 = R^8 = R^9 = H$	*Anaxagorea dolichocarpa, A. prinoides*	*69*
(−)-Ayuthianine	$R^1 + R^2 = OCH_2O$, $R^3 = R^4 = R^5 = R^6 = R^9 = H$, $R^7 = OMe$, $R^8 = Me$, $R^{10} = OH$	*Stephania cepharantha*	*70*
(−)-Sukuhodianine	$R^1 + R^2 = OCH_2O$, $R^3 = R^4 = R^5 = R^9 = H$, $R^6 = R^7 = OMe$, $R^8 = Me$, $R^{10} = OH$	*Stephania cepharantha*	*70*
(−)-Noroliveroline	$R^1 + R^2 = OCH_2O$, $R^3 = R^4 = R^5 = R^6 = R^7 = R^8 = R^{10} = H$, $R^9 = OH$	*Polyalthia acuminata*	*53*

(*continued*)

TABLE I (*Continued*)

Alkaloid	Structure	Plant source	Reference
(±)-Isoboldine	$R^1 = R^6 = OH$, $R^2 = R^5 = OMe$, $R^3 = R^4 = R^7 = H$, $R^8 = Me$	*Glaucium flavum*	*71*
O-Demethylnuciferine	$R^1 = OH$, $R^2 = OMe$, $R^3 = R^4 = R^5 = R^6 = R^7 = H$, $R^8 = Me$	*Papaver persicum* Lindl.	*72*
(±)-Variabiline	$R^1 = OH$, $R^2 = OMe$, $R^3 = R^4 = R^6 = R^7 = H$, $R^5 = N(CH_2Ph)_2$, $R^8 = Me$	*Ocotea variabilis*	*17*
Lirinine	$R^1 = R^2 = OMe$, $R^3 = OH$, $R^4 = R^5 = R^6 = R^7 = H$, $R^8 = Me$	*Liriodendron tulipifera* L.	*41*
Steporphine	$R^1 + R^2 = OCH_2O$, $R^3 = OH$, $R^4 = R^5 = R^6 = R^7 = H$, $R^8 = Me$	*Stephania sasakii* Hayata	*55*
Duguentine	$R^1 = R^2 = R^5 = R^6 = OMe$, $R^3 = R^4 = R^7 = H$, $R^8 = Me$	*Duguentia* species	*7*
1-Hydroxy-2,9,10Trimethoxy-aporphine	$R^1 = OH$, $R^2 = R^5 = R^6 = OMe$, $R^3 = R^4 = R^7 = H$, $R^8 = Me$	*Glaucium flavum*	*73*
Liridinine	$R^1 + R^3 = OMe$, $R^2 = OH$, $R^4 = R^5 = R^6 = R^7 = H$, $R^8 = Me$	*Ocotea glaziovii*	*74,75*
N-Acetylnornuciferine	$R^1 = R^2 = OMe$, $R^3 = R^4 = R^5 = R^6 = R^7 = H$, $R^8 = Ac$	*Liriodendron tulipifera*	*76*

N-Acetylasimilobine	$R^1 = OMe, R^2 = OH, R^3 = R^4 = R^5 = R^6 = R^7 = H, R^8 = Ac$	*Liriodendron tulipifera*	*76*
Isopiline	$R^1 = OH, R^2 = R^3 = OMe, R^4 = R^5 = R^6 = R^7 = R^8 = H$	*Isolona pilosa*	*49*
O-Methylnorlirinine	$R^1 = R^2 = R^3 = OMe, R^4 = R^5 = R^6 = R^7 = R^8 = H$	*Liriodendron tulipifera*	*77*
Polygospermine	$R^1 = R^2 = R^3 = OMe, R^4 + R^5 = OCH_2O, R^6 = R^7 = R^8 = H$	*Polyalthia oligosperma* (Danguy) Diels	*78*
Noroconovine	$R^1 = R^2 = R^3 = R^5 = OMe, R^4 = OH, R^6 = R^7 = R^8 = H$	*Polyalthia oligosperma* (Danguy) Diels	*78*
Xyloguyelline	$R^1 = R^3 = OMe, R^2 = OH, R^4 = R^7 = R^8 = H, R^5 + R^6 = OCH_2O$	*Xylopia danguyella*	*79*
Danguyelline	$R^1 = R^2 = R^5 = OMe, R^3 = R^4 = OH, R^6 = R^7 = R^8 = H$	*Xylopia danguyella*	*79*
Norstephalagin	$R^1 + R^2 = OCH_2O, R^3 = OMe, R^4 = R^5 = R^6 = R^7 = R^8 = H$	*Xylopia buxifolia*	*79*
Buxifoline	$R^1 + R^2 = OCH_2O, R^3 = R^6 = OMe, R^4 = R^5 = R^7 = R^8 = H$	*Xylopia buxifolia*	*79*
Hernangerine	$R^1 + R^2 = OCH_2O, R^3 = R^6 = R^7 = R^8 = H, R^4 = OMe, R^5 = OH$	*Hernandia papuana, H. ovigera*	*80,81*
Calycinine	$R^1 + R^2 = OCH_2O, R^3 = R^5 = R^7 = R^8 = H, R^4 = OH, R^6 = OMe$	*Duguetia calycina* Benoist	*82*
Norisodomesticine	$R^1 = OMe\ R^2 = OH, R^3 = R^4 = R^7 = R^8 = H, R^5 + R^6 = OCH_2O$	*Xylopia danguyella*	*79*

(*continued*)

TABLE I (*Continued*)

Alkaloid	Structure	Plant source	Reference
	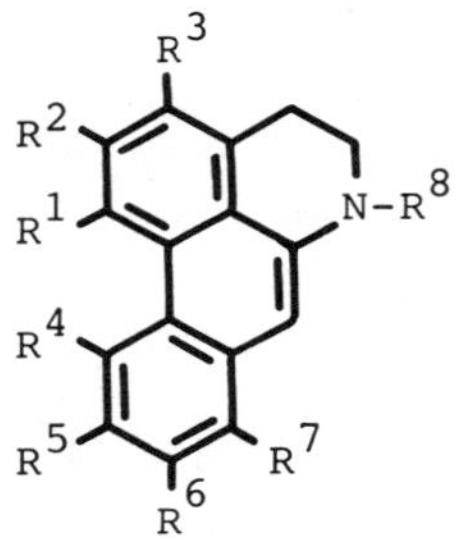		
Dehydrodicentrine	$R^1 + R^2 = OCH_2O$, $R^3 = R^4 = R^7 = H$, $R^5 = R^6 = OMe$, $R^8 = Me$	*Ocotea macropoda*	*13*
Dehydroglaucine	$R^1 = R^2 = R^5 = R^6 = OMe$, $R^3 = R^4 = R^7 = H$, $R^8 = Me$	*Glaucium flavum* Cranz, *G. flavum* var. leocarpum	*83*
Dehydroocoteine	$R^1 + R^2 = OCH_2O$, $R^3 = R^5 = R^6 = OMe$, $R^4 = R^7 = H$, $R^8 = Me$	*Ocotea puberula* (Nees et Mart.) Nees	*84*
Dehydroocopodine	$R^1 + R^2 = OCH_2O$, $R^3 = R^4 = H$, $R^5 = R^6 = R^7 = OMe$, $R^8 = Me$	*Ocotea macropoda*	*25*
Dehydronantenine	$R^1 = R^2 = OMe$, $R^3 = R^4 = R^7 = H$, $R^5 + R^6 = OCH_2O$, $R^8 = Me$	*Nandina domestica* Thunb., *Ocotea macrophylla*	*85,86*
Dehydronorglaucine	$R^1 = R^2 = R^5 = R^6 = OMe$, $R^3 = R^4 = R^7 = R^8 = H$	*Glaucium flavum*	*87*
Cabudine	$R^1 + R^2 = OCH_2O$, $R^3 = CH_2OH$, $R^4 = R^5 = R^6 = R^7 = H$, $R^8 = Me$	*Thalictrum isopyroides*	*88*
Dehydroisolaureline	$R^1 + R^2 = OCH_2O$, $R^3 = R^4 = R^5 = R^7 = H$, $R^6 = OMe$, $R^8 = Me$	*Liriodendron tulipifera*	*89*
Dehydrostesakine	$R^1 + R^2 = OCH_2O$, $R^3 = R^4 = R^5 = H$, $R^6 = OH$, $R^7 = OMe$, $R^8 = Me$	*Stephania sasakii*	*90*

Dehydrocrebanine	$R^1 + R^2 = OCH_2O$, $R^3 = R^4 = R^5 = H$, $R^6 = R^7 = OMe$, $R^8 = Me$	*Stephania sasakii*	*90*
Dehydrophanostenine	$R^1 + R^2 = OCH_2O$, $R^3 = R^4 = R^7 = H$, $R^5 = OH$, $R^6 = OMe$, $R^8 = Me$	*Stephania sasakii*	*50*
Dehydrostephanine	$R^1 + R^2 = OCH_2O$, $R^3 = R^4 = R^5 = R^6 = H$, $R^7 = OMe$, $R^8 = Me$	*Stephania kwangsiensis, S. cepharantha*	*91,92*
Dehydrocorydine	$R^1 = OH$, $R^2 = R^4 = R^5 = OMe$, $R^3 = R^6 = R^7 = H$, $R^8 = Me$	*Glaucium fimbrilligerum*	*37*
Subsessiline	$R^1 = R^2 = R^3 = OMe$, $R^4 = R^5 = R^7 = R^8 = H$, $R^6 = OH$	*Guatteria subsessilis*	*93,94*
Oxostephanine	$R^1 + R^2 = OCH_2O$, $R^3 = R^4 = R^5 = R^6 = R^8 = H$, $R^7 = OMe$	*Stephania japonica*	*95*
Oxolaureline	$R^1 + R^2 = OCH_2O$, $R^3 = R^4 = R^6 = R^7 = R^8 = H$, $R^5 = OMe$	*Magnolia soulangeana* Soul–Bod, *Laurelia novaezeladiae* A. Cunn.	*96,97*
		Guatteria elata R. E. Fries	*98*
Oxoputerine	$R^1 + R^2 = OCH_2O$, $R^3 = R^5 = R^6 = R^7 = R^8 = H$, $R^4 = OMe$	*Guatteria elata*	*98*
Liriodendronine	$R^1 = R^2 = OH$, $R^3 = R^4 = R^5 = R^6 = R^7 = R^8 = H$	*Liriodendron tulipifera*	*99*
Oxopukateine	$R^1 + R^2 = OCH_2O$, $R^3 = R^5 = R^6 = R^7 = R^8 = H$, $R^4 = OH$	*Duguetia eximia*	*100*

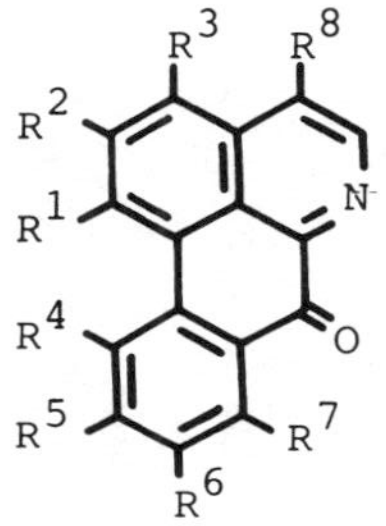

(*continued*)

TABLE I (*Continued*)

Alkaloid	Structure	Plant source	Reference
Splendidine	$R^1 = R^2 = R^8 = OMe$, $R^3 = R^4 = R^5 = R^6 = R^7 = H$	*Abuta rufescens* Aublet	*101*
Glaunine	$R^1 = R^2 = R^5 = OMe$, $R^3 = R^6 = R^7 = R^8 = H$, $R^4 = OH$	*Glaucium fimbrilligerum*	*102*
Oxocrebanine	$R^1 + R^2 = OCH_2O$, $R^3 = R^4 = R^5 = R^8 = H$, $R^6 = R^7 = OMe$	*Stephania sasakii*	*50*
Oxoanolobine	$R^1 + R^2 = OCH_2O$, $R^3 = R^4 = R^5 = R^7 = R^8 = H$, $R^6 = OH$	*Guatteria melosma*, *Telitoxicum peruvianum*	*103,104*
Ocominarone	$R^1 + R^2 = OCH_2O$, $R^3 = R^4 = R^8 = H$, $R^5 = R^6 = R^7 = OMe$	*Ocotea minarum* Nees (Mez.)	*35*
Peruvianine	$R^1 = R^2 = OMe$, $R^3 = R^4 = R^5 = R^7 = R^8 = H$, $R^6 = OH$	*Telitoxicum peruvianum*	*104*
Isomoschatoline	$R^1 = R^2 = OMe$, $R^3 = OH$, $R^4 = R^5 = R^6 = R^7 = R^8 = H$	*Guatteria melosma* Diels *Cleistopholis patens* Engl. and Diels	*54*
	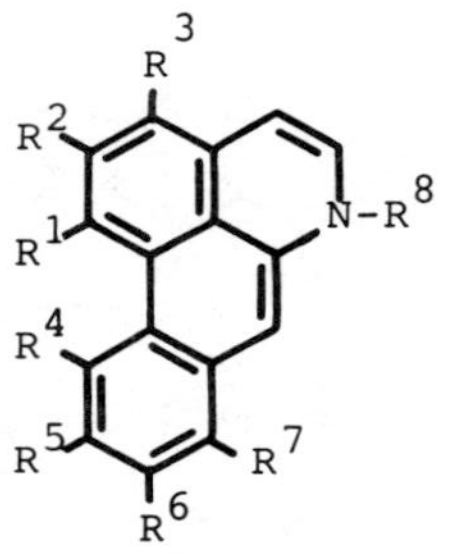		
Didehydroocoteine	$R^1 + R^2 = OCH_2O$, $R^3 = R^5 = R^6 = OMe$, $R^4 = R^7 = H$, $R^8 = Me$	*Ocotea puberula* (Nees and Mart.) Nees	*84*

	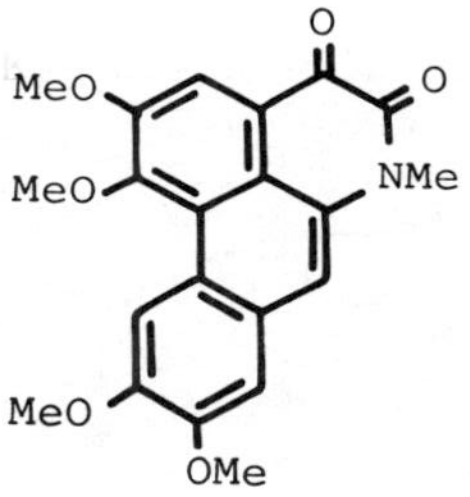		
Pontevedrine		*Glaucium flavum* Cr. var. *vestitum*	*57,105, 106*
	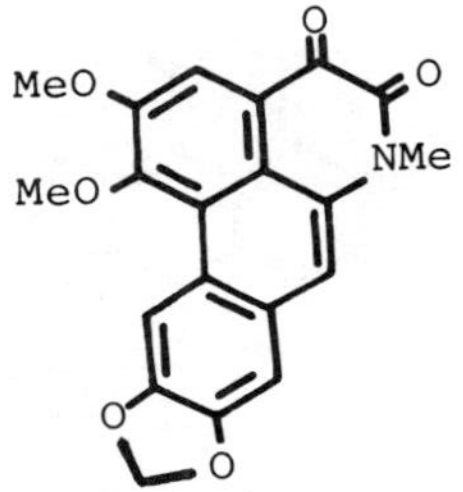		
4,5-Dioxodehydronantenine		*Nandina domestica*	*107*

(*continued*)

TABLE I (*Continued*)

Alkaloid	Structure	Plant source	Reference
4,5-Dioxodehydrocrebanine	O, O, NMe, OMe, OMe	*Stephania sasakii*	*90*
Tuberosinone	O, O, NH, HO	*Aristolochia tuberosa*	*108*
Tuberosinone *N*-β-glucoside	O, O, N, HO, H, HO, H, OH, H, OH, O, OH, H, H, H	*Aristolochia tuberosa*	*108*

	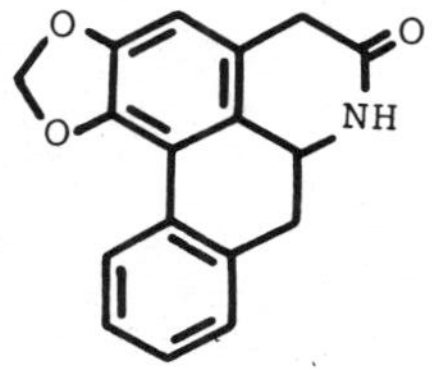		
Fuseine		*Fusea longifolia* (Aubl.) Safford	*109*
	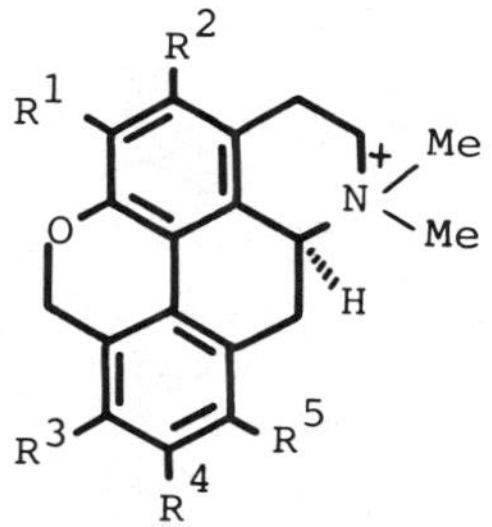		
Thalphenine	R^1 = OMe, R^2 = R^5 = H, R^3 + R^4 = OCH_2O	*Thalictrum polygamum* Muhl.	*110*

(*continued*)

TABLE I (*Continued*)

Alkaloid	Structure	Plant source	Reference
	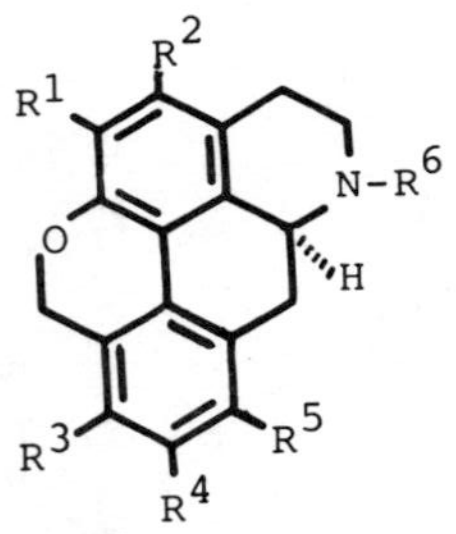 R1, R2, R3, R4, R5, N–R6, O, H		
(+)-Bisnorthalphenine	R^1 = OMe, R^2 = R^5 = R^6 = H, R^3 + R^4 = OCH_2O	*Thalictrum polygamum* Muhl.	*111*
(+)-*N*-Demethylthalphenine	R^1 = OMe, R^2 = R^5 = H, R^3 + R^4 = OCH_2O, R^6 = Me	*Thalictrum revolutum* DC.	*112*
	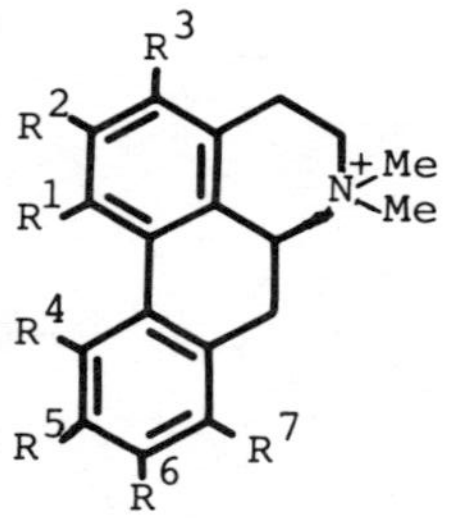 R1, R2, R3, R4, R5, R6, R7, N+ Me Me		
Remrefidine	R^1 + R^2 = OCH_2O, R^3 = R^4 = R^5 = R^6 = R^7 = H	*Roemeria refracta* DC.	*113*

Nantenine methochloride	$R^1 = R^2 = OMe, R^3 = R^4 = R^7 = H, R^5 + R^6 = OCH_2O$	*Thalictrum polygamum* Muhl.	*114*
(+)-*N*,*N*-Dimethyllindcarpine	$R^1 = R^5 = OMe, R^2 = R^4 = OH, R^3 = R^6 = R^7 = H$	*Menispermum canadense* L.	*115*
Zanthoxyphylline	$R^1 = R^2 = R^4 = OMe, R^3 = R^5 = R^6 = R^7 = H$	*Zanthoxylum oxyphyllum* Edgew.	*116*
(+)-Bulbocapnine *N*-metho-salt	$R^1 + R^2 = OCH_2O, R^3 = R^6 = R^7 = H, R^4 = OH, R^5 = OMe$	*Corydalis cava*	*117*
N-Oxy-*N*-methylpachypondathine	$R^1 + R^2 = OCH_2O, R^3 = R^4 = R^5 = R^6 = R^7 = R^9 = H, R^8 = OMe$	*Polyalthia oliveri* Engl.	*66*
N-Oxyoliveroline	$R^1 + R^2 = OCH_2O, R^3 = R^4 = R^5 = R^6 = R^7 = R^9 = H, R^8 = OH$	*Polyalthia oliveri* Engl.	*66*
N-Oxyguatterine	$R^1 + R^2 = OCH_2O, R^3 = OMe, R^4 = R^5 = R^6 = R^7 = R^9 = H, R^8 = OH$	*Pachypodanthium confine* Engl. and Diels	*67*
N-Oxyoliveridine	$R^1 + R^2 = OCH_2O, R^3 = R^4 = R^5 = R^7 = R^9 = H, R^6 = OMe, R^8 = OH$	*Enantia pilosa* Exell	*118*
N-Oxyoliverine	$R^1 + R^2 = OCH_2O, R^3 = R^4 = R^5 = R^7 = R^9 = H, R^6 = R^8 = OMe$	*Enantia pilosa* Exell	*118*

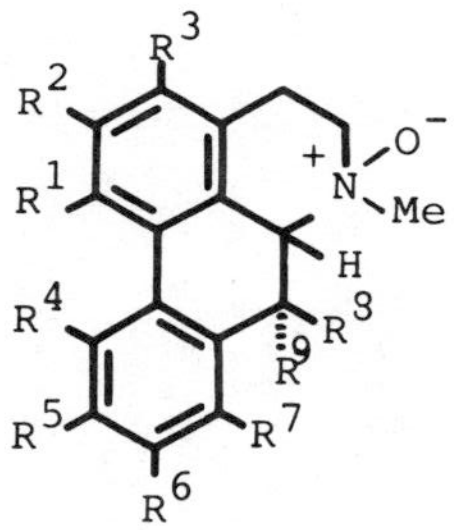

(*continued*)

TABLE I (*Continued*)

Alkaloid	Structure 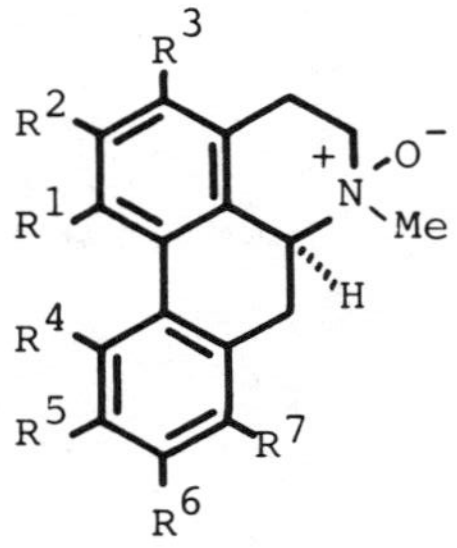	Plant source	Reference
Thalicmidine *N*-oxide	$R^1 = OH, R^2 = R^5 = R^6 = OMe, R^3 = R^4 = R^7 = H$	*Thalictrum minus*	*119*
Preocoteine *N*-oxide	$R^1 = OH, R^2 = R^3 = R^5 = R^6 = OMe, R^4 = R^7 = H$	*Thalictrum minus*	*119*
(+)-Isocorydine *N*-oxide	$R^1 = R^2 = R^5 = OMe, R^3 = R^6 = R^7 = H, R^4 = OH$	*Berberis integerrima* Bge.	*120*
Corydine *N*-oxide	$R^1 = OH, R^2 = R^4 = R^5 = OMe, R^3 = R^6 = R^7 = H$ 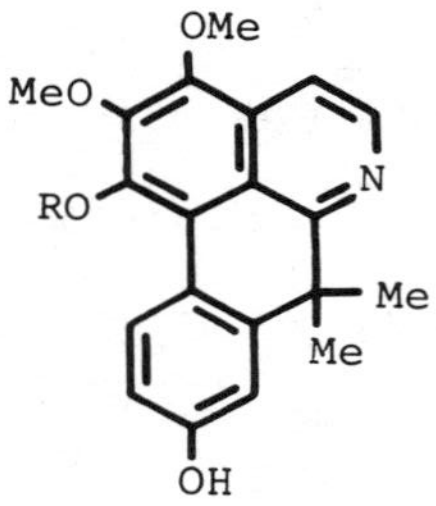	*Glaucium fimbrilligerum*	*37*

Melosmine	R = H	*Guatteria melosma*	*121*
Melosmidine	R = Me	*Guatteria melosma*	*121*
Guadiscine	R = H	*Guatteria discolor*	*122*
Guadiscoline	R = OMe	*Guatteria discolor*	*122*
(+)-Guattescine		*Guatteria scandens*	*123*

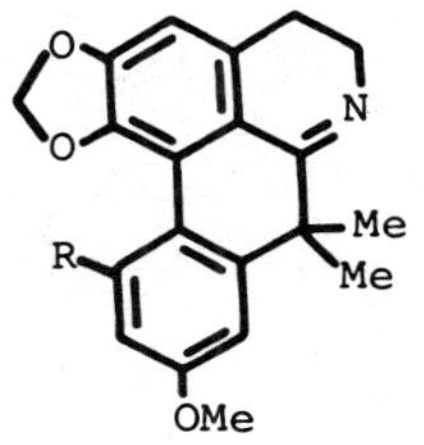

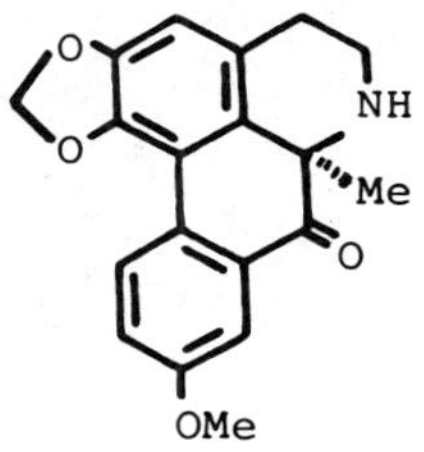

(*continued*)

TABLE I (*Continued*)

Alkaloid	Structure	Plant source	Reference
(−)-Guattescidine	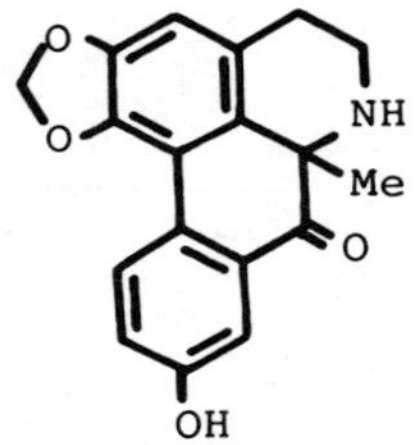 NH, Me, O, OH	*Guatteria scandens*	*123*
Duguecalyne	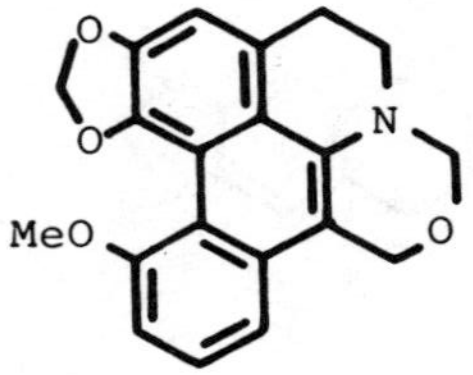 N, O, MeO	*Duguetia calycins*	*124*
Duguenaine	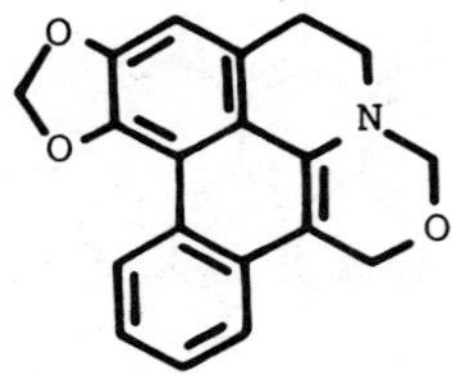 N, O	*Duguetia calycins*	*124*

Corunnine (Glauvine)	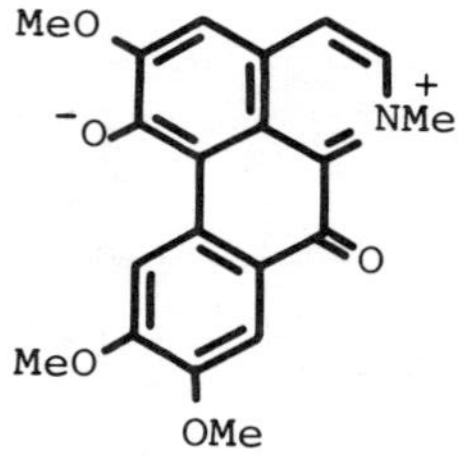	*Glaucium flavum, G. grandiflorum, G. serpieri*	*125–128*
Nandazurine	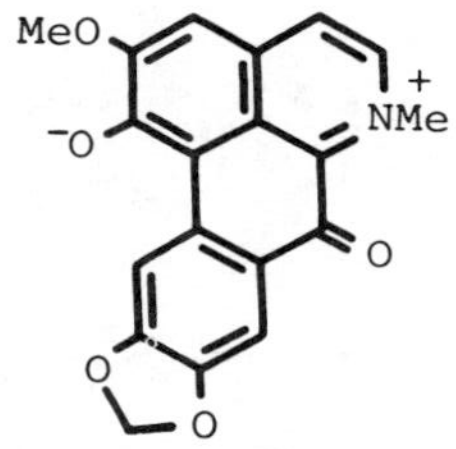	*Nandina domestica* Thunb.	*129*
Glaunidine (Arosine)	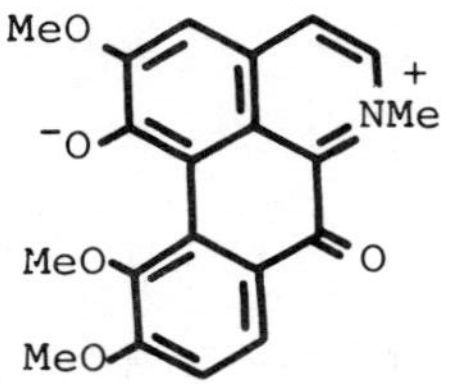	*Glaucium fimbrilligerum* *G. flavum* Cr. var. *Vestitum*	*102* *130*

(*continued*)

TABLE I (*Continued*)

Alkaloid	Structure	Plant source	Reference
Arosinine	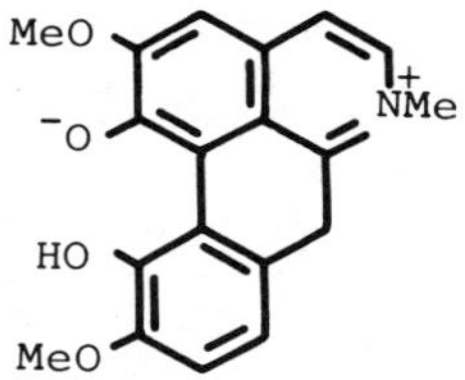 MeO, $^{-}$O, HO, MeO, $\overset{+}{N}$Me	*Glaucium flavum* Cr. var. *vestitum*	*130*
Uthongine	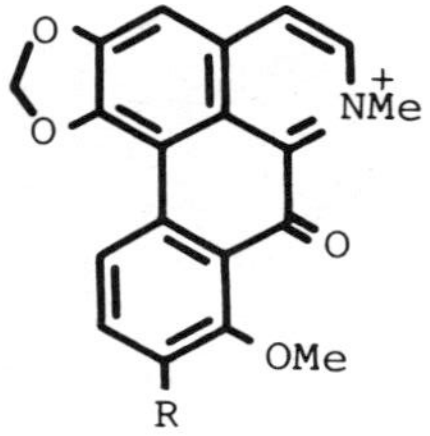 R = OMe	*Stephania venosa*	*131*
Thailandine	R = H	*Stephania venosa*	*131*

with J = 7.5 Hz in the case of **2.** Treatment of both compounds with diazomethane afforded the known aporphine (−)-*O*-methylisopiline (**3**). The location of a phenolic function in **1** and **2** was established by comparison of their UV spectra. Both compounds exhibit a bathochromic shift in their spectra by addition of alkali owing to the presence of phenolic groups. But **2** also shows a hyperchromic effect at 315 nm, while **1** does not. It is known that monophenolic

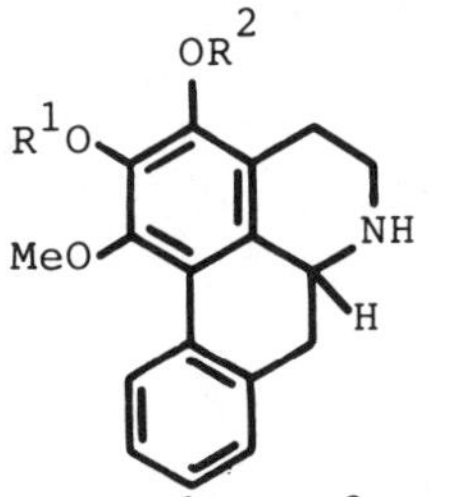

1 R^1=H, R^2=Me
2 R^1=Me, R^2=H
3 R^1=R^2=Me

aporphines having the hydroxyl group at the C-9 position show a bathochromic shift as well as a hyperchromic shift in their UV spectra upon addition of base. Observation of a similar effect in **2** is, therefore, rationalized by assuming that **2** has a phenolic function at the corresponding position in ring A, i.e., at C-3. Thus the structures of **1** and **2** were determined. To complete the characterization of **1** and **2,** both alkaloids were N-methylated with formaldehyde and formic acid to give the known (−)-liridinine and (−)-3-hydroxynuciferine, respectively.

2. Ayuthianine and Sukuhodianine

Ayuthianine (**4**) was isolated from the dried tuberous root powder of *Stephania venosa* Spreng. (*S. rotunda* Lour.) (Menispermaceae), and its mass spectrum shows the molecular ion peak at *m/z* 325, which corresponds to $C_{19}H_{19}NO_4$. Its NMR spectrum exhibits 7-H at 5.52 ppm as a doublet with J = 2.4 Hz, indicat-

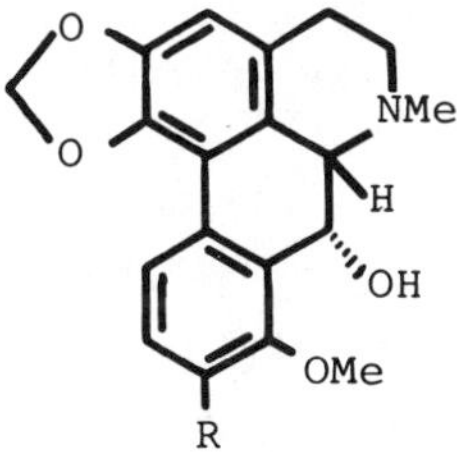

4 R=H
5 R=OMe

ing a cis relationship between 6a-H and 7-H. Sukuhodianine (**5**) is also present in *S. venosa* together with **4**, and its molecular formula was derived from its mass spectrum as $C_{20}H_{21}NO_5$ (*m/z* 353). Its NMR spectrum also shows 7-H at 5.47 ppm as a doublet with J = 2.7 Hz (*70*). On the basis of the survey of the literature (*69, 132*), the following conclusions concerning the C-7 oxygenated aporphines are drawn: the occurrence of C-7 oxygenated aporphines is limited to the four families Annonaceae, Menispermaceae, Magnoliaceae, and Lauraceae; C-7 oxygenated aporphines inevitably belong to the C-6a *R* configuration; the trans configuration between 6a-H and 7-H is found only among alkaloids of the Annonaceae for C-7 oxygenated aporphines; it is rare among the Annonaceae alkaloids that C-7 methoxylated aporphines occur. In these alkaloids, 6a-H and 7-H are trans to each other.

3. Stesakine

Stesakine (**6**) was obtained from *Stephania sasakii* Hayata (Menispermaceae) as colorless, columnar crystals [mp 188–189°C, $[\alpha]_D$ −78.7°($CHCl_3$)] (*50*). Its mass spectrum established the molecular formula as $C_{19}H_{19}NO_4$. Its UV system is consistent with the formula, and it exhibits a bathochromic shift in alkaline solution. The NMR spectrum shows the presence of a methoxyl group, an *N*-methyl group, one methylenedioxy function, and three aromatic protons in addition to a hydroxyl proton at 4.55 ppm. Its IR spectrum shows the presence of a phenolic OH group. Diazomethane treatment of **6** gave crebanine (**7**). For deter-

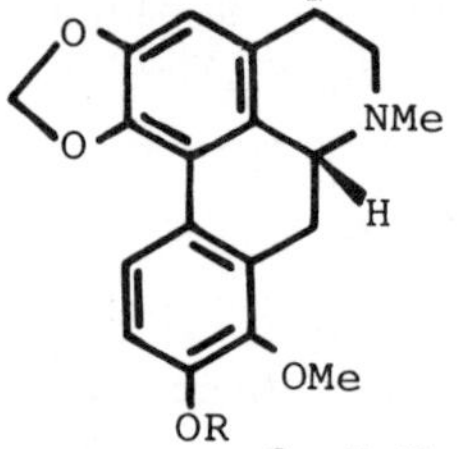

NMe
H
OMe
OR
6 R=H
7 R=Me

mination of the position of the phenolic function, deuterium exchange of the aromatic proton was carried out using 5% NaOD–D_2O at 120°C in a sealed tube for 20 hr. The doublet signal of the C-11 aromatic proton at 7.81 ppm changed to a singlet, and the signal of the C-10 aromatic proton at 6.93 ppm disappeared. Therefore the phenolic position was determined to be at C-9.

4. Dehydrostesakine and Dehydrocrebanine

Dehydrostesakine (mp 201–203°C) was isolated from *Stephania sasakii* Hayata as colorless, columnar crystals (*50*). Its IR spectrum shows a phenolic OH at

3500 cm^{-1}. Elemental analysis and the mass spectrum established the formula as $C_{19}H_{17}NO_4$. Its NMR spectrum shows the presence of a methoxyl, an *N*-methyl, one methylenedioxy group, and four aromatic protons in addition to a hydroxyl proton at 5.88 ppm. O-Methylation of **8** with diazomethane afforded dehydrocrebanine (**9**). In order to determine the position of the OMe group, nuclear Overhauser effect (NOE) and internuclear double resonance (INDOR) were applied. In the INDOR determination of **8,** no peak was observed in reference to the aromatic proton of C-10 at 6.99 ppm by monitoring the signal of

8 R=H
9 R=Me

the methoxyl at 3.96 ppm, but it was observed in regard to the hydroxyl group at 5.88 ppm. These facts suggest that the position of the methoxyl group is at C-8. This observation indicates the presence of NOE, that is, the signal area of the aromatic proton at 6.83 ppm increases 9.8 or 15.8% by monitoring the signal of the OMe at 3.96 ppm or NMe at 3.10 ppm, respectively.

Dehydrocrebanine (**9**) was also isolated from *Stephania sasakii* Hayata (Menispermaceae) as pale yellow needles, mp 150–152°C, and its composition of $C_{20}H_{19}NO_4$ was determined by mass spectrum and elemental analysis (*50*). The UV and mass spectra show that it is a dehydroaporphine alkaloid. ^{1}H-NMR showed the presence of two methoxyl, an *N*-methyl, and one methylenedioxy group together with four aromatic protons. Dehydrogenation of crebanine with iodine in dioxane gave dehydrocrebanine (**9**). Its UV spectrum [λ_{max}^{EtOH} nm (logϵ): 248 (sh, 4.36), 272 (4.77), 296 (sh, 4.17), 337 (4.15), 385 (3.49)] is consistent with its structure.

5. Dehydrocorydine

Dehydrocorydine was isolated from the epigeal part and roots of *Glaucium fimbrilligerum* as an amorphous powder. Its UV spectrum showed maxima at 220 (log ϵ 4.33), 310 (4.27), and 340 nm (4.10). Its NMR spectrum showed the presence of an *N*-methyl group at 2.96 ppm, three methoxyl groups at 3.65, 3.89, and 3.93 ppm, and four aromatic protons at 6.32 and 6.97 ppm as each singlet and 7.10 and 7.34 ppm as each doublet with J = 8 Hz. The Adams hydrogenation of dehydrocorydine led to corydine (**11**). Thus the structure of dehydrocorydine was deduced to be $\Delta^{6a,7}$-dehydroaporphine (**10**) (*37*).

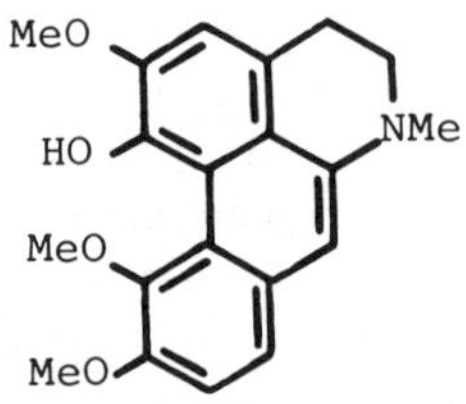

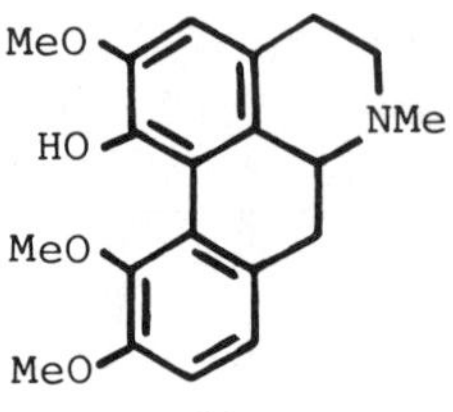

6. Glaunine

Glaunine (**12**) was isolated from *Glaucium fimbrilligerum* as a green substance. Its IR spectrum shows absorption bands at 1590, 1660, and 3410 cm^{-1}. Its UV spectrum [λ_{max}^{EtOH} nm (log ϵ): 250 (4.40), 272 (4.22), 310 (inflection) (3.97), 348 (3.87), 406 (2.75), and 600 (2.68)] changes on acidification [λ_{max}^{EtOH} nm 248 (log ϵ): (4.46), 263 (inflection) (4.41), 385 (4.32), 320 (inflection) (3.88), 375 (2.94), and 470 (inflection) (2.60)]. Its NMR ($CDCl_3$) spectrum exhibits the presence of three methoxyl groups at 3.78, 4.03, and 4.04 ppm, and in the aromatic region there were doublets at 7.18 and 8.50 ppm (J = 8 Hz) and 7.72 and 8.77 ppm (J = 5 Hz) and also a singlet at 7.17 ppm. When glaunine (**12**) was reduced with zinc in sulfuric acid, norisocorydine was obtained, and the methylation of the latter by the Craigs method afforded isocorydine (**13**). Hence the structure of glaunine was assigned as **12** (*102*).

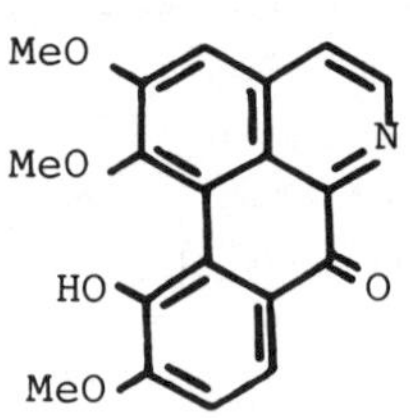

7. Glaunidine (Arosine)

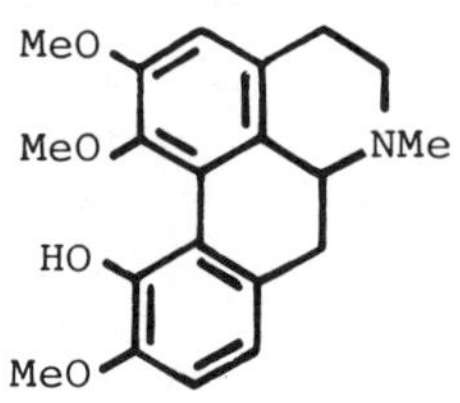

Glaunidine (**14**) was obtained from *Glaucium fimbrilligerum* and *G. flavum* as dark-green prisms [mp 245–248°C(d)] (*102, 103*). It is green in neutral or basic solution and yellow on addition of acid. High resolution MS and elemental analysis established its molecular formula as $C_{20}H_{17}NO_5$. Its IR spectrum shows the presence of a highly conjugated carbonyl system and a phenolic function. Its UV spectrum in ethanol or basic solution exhibits the characteristic quaternary

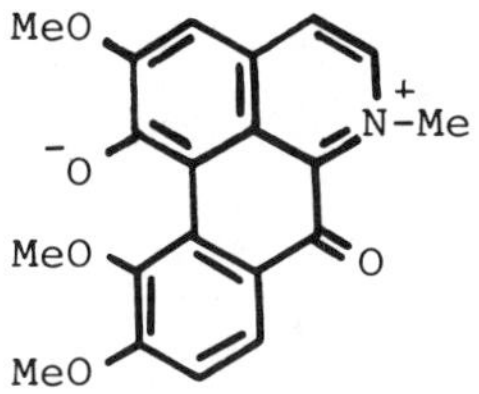

14

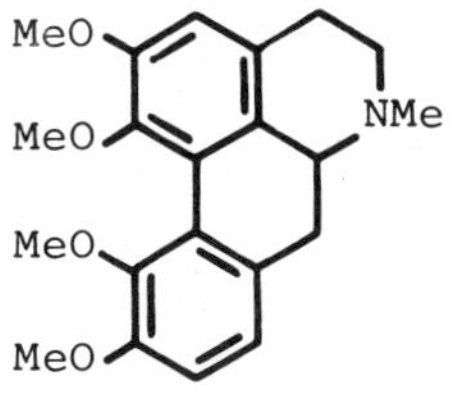

15

oxoaporphine system. On addition of acid, a strong hypsochromic shift of the absorption bands was observed, which is very similar to that of an oxoaporphine and its methiodide by direct comparison. Its NMR spectrum exhibits three methoxyl signals at 3.87, 4.11, and 4.21 ppm, one quaternary *N*-methyl group at 4.71 ppm, one aromatic proton at 7.46 ppm, and two AB quartets. Reduction of **14** with zinc in acetic and hydrochloric acids under reflux afforded corydine (**11**). The unambiguous establishment of its structure was proved by its synthesis. Photolysis (eosine, O_2) of *O,O*-dimethylcorytuberine (**15**) gave the oxoaporphine whose treatment with an excess of methyl iodide in acetone at room temperature for several hours produced the methiodide as red prisms (mp 155–157°C), the subsequent refluxing of which in dry acetone gave arosine (**14**).

8. Arosinine

Arosinine (**16**) was isolated from *Glaucium flavum* Cr. var. *vestitum* and crystallized as very dark-green needles [mp 302–305°C(d)] (*130*). It is green in neutral or basic solution and red in acid solution. High resolution MS and microanalysis established its molecular formula as $C_{19}H_{15}NO_5$ (M^+, 337). Its IR spectrum exhibits the presence of a phenolic function and a highly conjugated carbonyl system. A strong hypsochromic shift was again observed in its UV spectrum on addition of acid. Its NMR spectrum showed a quaternary *N*-methyl signal at 4.71 ppm in addition to two methoxyl groups at 4.12 and 4.20 ppm and five aromatic protons. Reduction of **16** with zinc in acetic and hydrochloric acids under reflux afforded corytuberine (**17**). When *O*-methylglaunine (**18**) was heated at 150°C for 16 hr, arosine (**14**) and arosinine (**16**) were formed in 15 and 45%

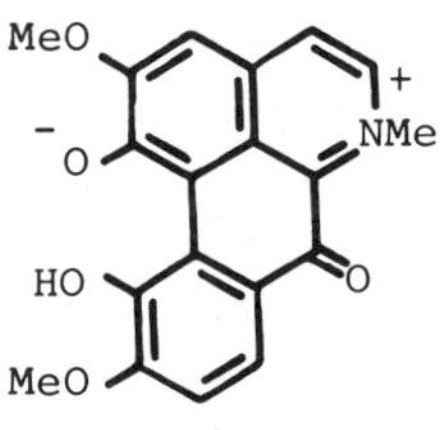

16 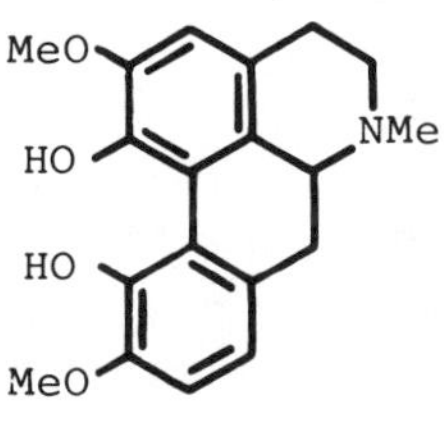 17 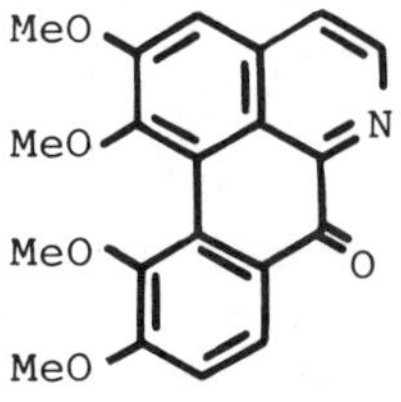 18

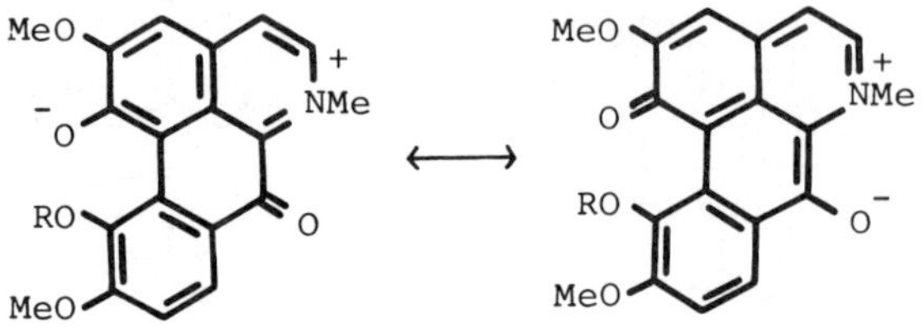

yields, respectively. Furthermore, when arosine was treated in the same way, arosinine was obtained. Arosine and arosinine can be represented by the corresponding forms **19** and **22,** which are also prevalent in acid solution.

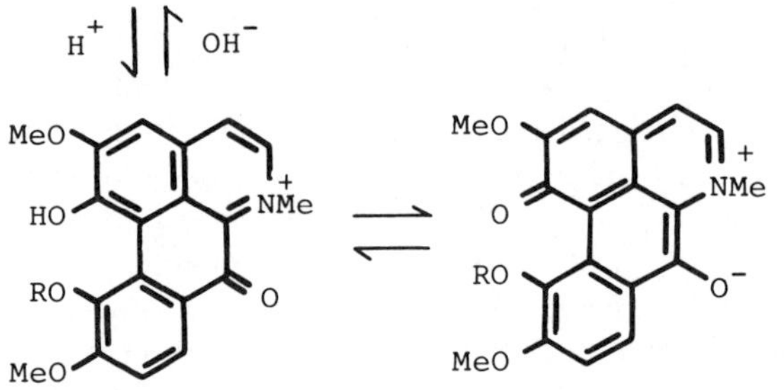

9. Uthongine and Thailandine

Uthongine (**23**) was isolated from the roots of *Stephania venosa* Spreng. (*S. rotunda* Lour.) (Menispermaceae), and its spectral data provided the basis for suggesting formula **23** as the structure for uthongine (*131*). Of particular importance in its NMR spectrum is the downfield *N*-methyl singlet at 4.83 ppm. Uthongine is partially decomposed by chromatography on silica gel to 7-oxocrebanine (**25**). The N-methylation of 7-oxocrebanine with methyl iodide on refluxing in acetonitrile afforded uthongine iodide in 80% yield. Thailandine (**24**) is also present in the same plant as uthongine, and its NMR spectrum exhibits a characteristic downfield *N*-methyl singlet at 4.86 ppm. As with uthongine, chromatography of thailandine resulted in some N-demethylation to furnish 7-oxostephanine (**26**). Additionally, N-methylation of 7-oxostephanine with methyl iodide in acetonitrile afforded amorphous thailandine iodide in 75% yield. Both of the semisynthetic alkaloids are identical with the natural products by comparison of their NMR spectra in CF_3CO_2D (*131*).

An interesting conclusion is that some or part of the oxoaporphine isolated from plant sources may exist in nature as the corresponding *N*-metho salts. Such salts would then undergo easy N-demethylation upon chromatographic purification.

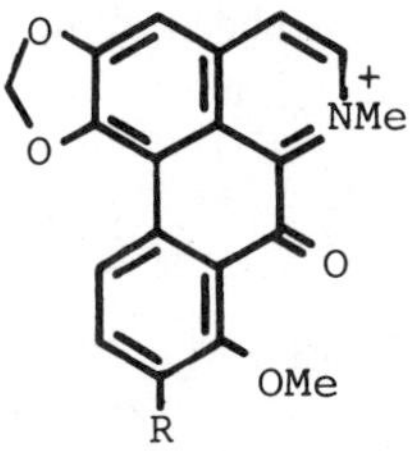

23 R=OMe
24 R=H

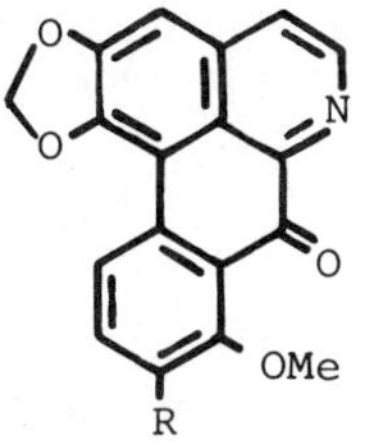

25 R=OMe
26 R=H

10. Melosmine and Melosmidine

Melosmine (**27**) was isolated from the stembark of *Guatteria melosma* Diels (Annonaceae) as pale-yellow cuboidal crystals (mp 104°C; $[\alpha]_D$ 0°) (*121*). Its mass spectrum exhibits the molecular ion peak at m/z 337 as $C_{20}H_{19}NO_4$. The UV spectrum [λ_{max}^{EtOH} nm (log ϵ): 218sh (4.34), 204sh (4.41), 278 (4.40), and 412 (3.85) is not similar to that of a simple aporphine or oxoaporphine. The bathochromic shift, upon the addition of base, indicated the phenolic nature of the alkaloid. The NMR spectrum partially resembled that of a 1,2,3,9-tetrasubstituted oxoaporphine, and showed the presence of four aromatic protons at 8.50 and 7.71 ppm each as a doublet with J = 6 Hz, and 8.91 (d, J = 9 Hz), 6.89 (1H, dd, J = 3 and 9 Hz) and 7.22 ppm (d, J = 3 Hz), two methoxyl signals at 4.00 and 4.17 ppm and two *gem*-dimethyl groups at 1.73 ppm. Its IR spectrum indicated the absence of a carbonyl group at C-7, therefore the position of the *gem*-dimethyl group was assigned to be at C-7. Based on the spectral data, the structure of **27** was suggested to be 2,3-dimethoxy-7,7-dimethyl-7*H*-dibenzo[*de,q*]quinoline-1,9-diol, and this was confirmed by X-ray crystallographic analysis (*121*).

Melosmidine (**28**) is also present in *G. melosma* and is isolated as a yellow amorphous solid (mp 170–171°C, $[\alpha]_D$ 0°). Its spectral data were very similar to those of melosmine, and treatment of **28** with ethereal diazomethane gave a

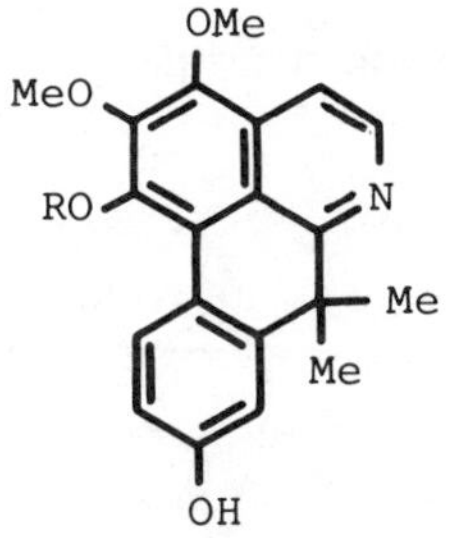

27 R=H
28 R=Me

product identical with *O,O*-dimethylmelosmine. The position of the phenolic function was determined by its NMR spectrum.

11. Telazoline

Telazoline (**29**) was obtained as reddish-brown prisms (mp 240–243°C) from *Telitoxicum peruvianum* (Menispermaceae) (*104*). Its mass spectrum afforded the molecular formula $C_{17}H_{12}N_2O_2$. The UV spectrum exhibits maxima at 242sh (log ε 4.52), 251 (4.53), 283 (4.47), 317sh (3.93), and 470 nm (4.08), and after

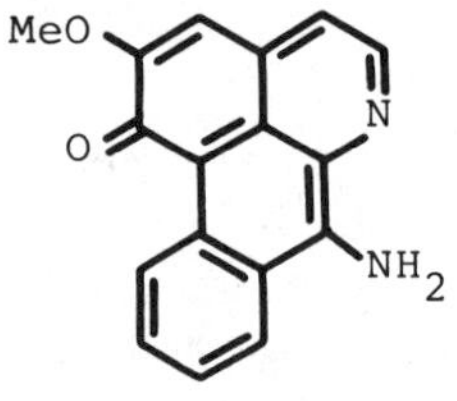

MeO O N NH2

29

addition of NaOH, those at λ_{max} 241sh (4.54), 250 (4.55), 273 (4.47), 324sh (3.85), 370 (3.30), and 471 nm (4.62). Its NMR spectrum showed the presence of a methoxyl group at 4.09 ppm and the rest of the spectrum consisted of a seven-aromatic-proton signal as a singlet at 7.00 (1H), a pair of doublets at 7.66 and 8.75 (1H each, J = 5 Hz), a pair of multiplets at 7.44 and 7.69 ppm (1H each), and a pair of doublets at 8.51 and 8.57 ppm (1H each, J = 8 Hz). Based on these spectral data, tentative structural formula **29** was proposed for talazoline. Further study will be needed for its confirmation.

12. 4,5-Dioxodehydrocrebanine

4,5-Dioxodehydrocrebanine (**30**) was obtained from *Stephanina sasakii* as orange needles, mp 278–280°C, and its mass spectrum and elemental analysis established its formula as $C_{20}H_{15}NO_6$. The UV spectrum [λ_{max}^{EtOH} nm (log ε): 220 (4.52), 244.5 (4.62), 308 (4.21), 321 (4.25), and 435 (4.21)] indicated a highly conjugated system in comparison with that of dehydroaporphine, and its IR

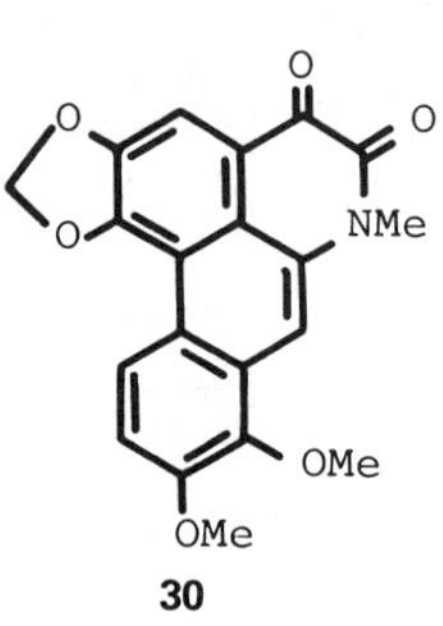

O O O O NMe OMe OMe

30

spectrum showed a conjugated ketone or a six-membered lactam at 1660 and 1593 cm^{-1}. Its ^{1}H-NMR spectrum showed the presence of two methoxyl groups, a methylenedioxy function, an *N*-methyl group and four aromatic protons. Moreover, the signals of the *N*-methyl group and two aromatic protons at 7.55 and 7.83 ppm are unusually low shielded. This fact suggested that the B ring of the alkaloid was highly strained as expected for 4,5-dioxodehydroaporphine. The structure was confirmed by the conversion of dehydrocrebanine (**9**) to 4,5-dioxodehydrocrebanine by air oxidation with an alkaline catalyst (*90*).

B. Dimeric Aporphine Alkaloids

Since the first isolation of the aporphine–benzylisoquinoline alkaloid [thalicarpine was reported by Kupchan in 1963 (*134*)] more than 30 dimeric aporphine alkaloids were found in nature. Concerning proaporphine– and aporphine–benzylisoquinoline dimers and aporphine–pavine dimers, a complete list including their physical and spectral data as well as their botanical sources was published in 1979 (*135*). Therefore, only the dimeric aporphine alkaloids isolated after 1979 are mentioned in this review. The numbering system used throughout in this review for dimeric aporphine alkaloids is as follows.

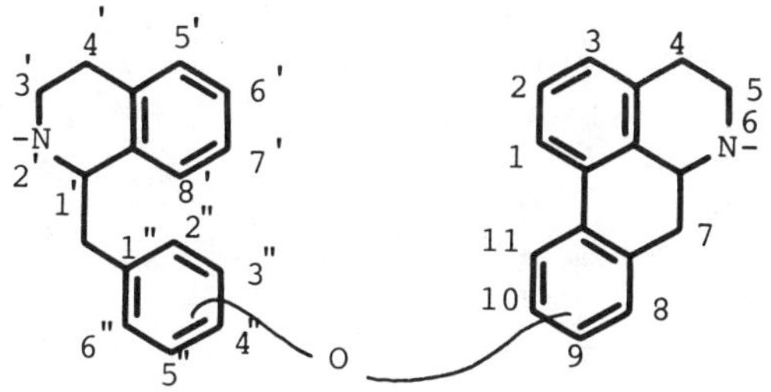

1. Chitraline and 1-*O*-Methylpakistanine

Chitraline (**31**) was isolated as an amorphous powder from *Berberis orthobotrys* Bienert ex Aitch. (*136*). It developed a greenish coloration on TLC and its

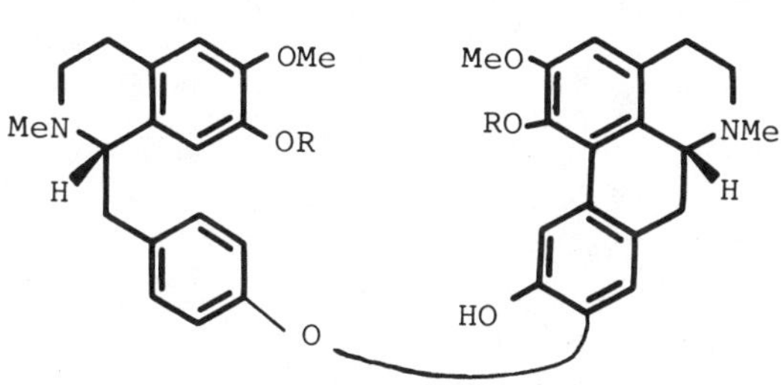

31 R=H
32 R=Me

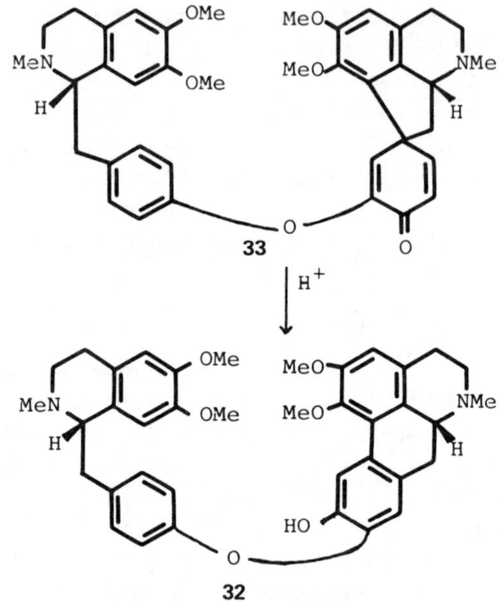

acetylation yielded 1,10,7′-tri-*O*-acetylchitraline. Chitraline has the molecular formula $C_{36}H_{38}N_2O_6$ and $[\alpha]_D$ +136° (c = 0.172, MeOH). Its mass spectral breakdown pattern with m/z 593 (M^+ − 1), 401, 192, and 177 is strongly suggestive of an aporphine–benzylisoquinoline of the pakistanine series, possessing three phenolic functions, one of which is located in ring A of the benzylisoquinoline moiety. The UV spectrum of **31** [λ_{max}^{MeOH} 220sh (4.51), 268sh (4.03), 278 (4.10), 292sh (3.94), and 304 nm (3.96)] is closely related to that of pakistanine. Its NMR spectrum showed the presence of two methoxyl signals instead of the three signals found in pakistanine. Thus the structure of chitraline was determined to be **31.** 1-*O*-Methylpakistanine (**32**) had previously not been known as a natural product, but as a semisynthetic compound from the dienone–phenol rearrangement of pakistanine (**33**) in 3 *N* sulfuric acid at 70°C.

2. Kalashine and Khyberine

Kalashine (**34**) was isolated from the roots of *Berberis orthobotrys* Bienert ex Aitch. (Berberidaceae) as a minor component (*136, 137*), whose mass spectral fragmentation pattern is characteristic of an apoprhine–benzylisoquinoline dimer with peak m/z 607 (M^+ − 1) and base peak m/z 206. The other intense fragments (m/z 403, 311, 296, and 107) suggested that this alkaloid was bonded at the lower portion of the benzylisoquinoline residue through a diphenyl ether linkage to the aporphine ring. Since its NMR spectrum shows the absence of a

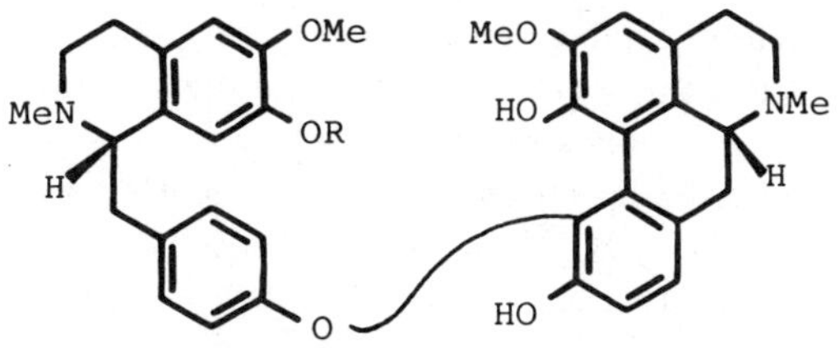

34 R=Me
35 R=H

downfield proton signal at 11-H, it is assumed to be substituted at the C-11 position. The UV spectrum of **34** [λ_{max}^{MeOH} (log ϵ): 220 (4.54), 272 (4.04), 290sh (3.74), and 304 nm (3.70)] is consonant with substitution at the C-11 position, since it lacks the strong absorbance near 280 nm characteristic of 1,2,9,10-substituted aporphines. Its absolute configuration was also determined by the CD spectrum. Kalashine is the first aporphine–benzylisoquinoline alkaloid known to be substituted at C-11 position.

Khyberine (**35**) is present in *Berberis calliobotrys* Aitch. ex Bienert in about 1 ppm, and has mp 145–147°C ($CHCl_3$–MeOH). Its mass spectrum shows *m/z* 693 ($M^+ - 1$) (denoting the molecular formula $C_{36}H_{38}N_2O$ and loss of a proton) 403, 402, 296, 192, and 107. The UV spectrum [λ_{max}^{MeOH} (log ϵ): 220sh (4.53), 264sh (3.97), 272 (4.02), 292sh (3.80), and 304 nm (3.70)] is very close to that for kalashine, pointing to 1,2,10,11-substitution around the aporphine nucleus. The absence of a downfield proton peak around at 8.0 ppm suggested that this alkaloid has a substituent at C-11, and the lack of a three-proton signal near at 3.40 ppm indicates the presence of a phenol rather than a methoxyl at C-7′. Based on the comparison of spectral data for **35** with those of kalashine, the structure of khyberine was deduced to be 7′-demethylkalashine (*137*).

3. Northalicarpine

Northalicarpine (**36**) was isolated from the roots of *Thalictrum revolutum* DC (Ranunculaceae) as an amorphous solid with $[\alpha]_D$ +108° (c = 0.25, MeOH)

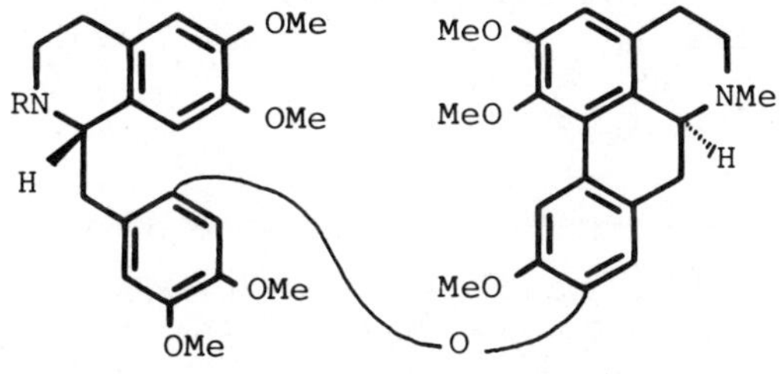

36 R=H
37 R=Me

(*138*). Its NMR spectrum exhibits the presence of seven methoxyls and a downfield one-aromatic proton as a singlet and six additional aromatic protons. The UV spectrum with absorption greater than 300 nm was suggestive of an aporphine moiety. The mass spectrum shows a molecular ion peak at *m/z* 682 corresponding to formula $C_{40}H_{46}N_2O_8$, in which one of the oxygens was present as a diphenyl ether. The structure of northalicarpine including its absolute stereochemistry was unambiguously determined by its conversion to thalicarpine (**37**) by N-methylation with formaldehyde and sodium borohydride.

4. (+)-Istanbulamine

Istanbulamine (**38**) was found in the roots and rhizomes of *Thalictrum minus* L. var. *microphyllum* Boiss. (Ranunculaceae) and its mass spectrum indicated

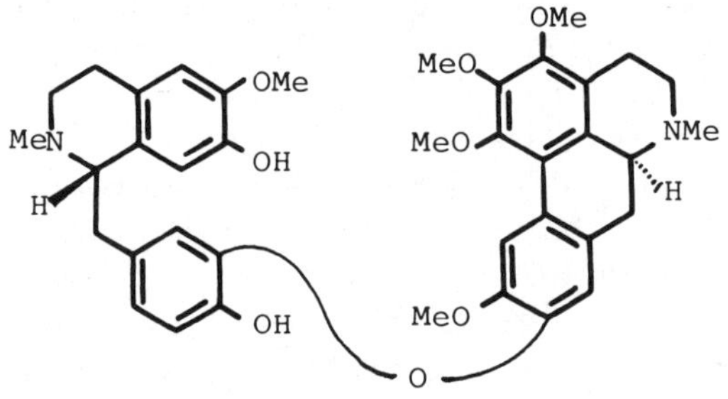

38

the molecular formula as $C_{39}H_{44}N_2O_8$ with a small molecular ion *m/z* 668 and a base peak *m/z* 192 owing to the facile formation of the dihydroisoquinolinium cation through cleavage of the C-1′ to C-2′ bond. The UV spectrum suggested the presence of a 1,2,9,10- or 1,2,3,9,10-substituted aporphine system. The presence of a characteristic aromatic proton at 8.03 ppm in its NMR spectrum indicated that C-11 is unsubstituted. A nuclear Overhauser enhancement study was carried out to confirm the chemical shift assignment. Structure assignment was done based on the spectral data, and its absolute configuration was determined by its CD pattern. Istanbulamine is the first aporphine–benzylisoquinoline alkaloid formed from one (+)-reticuline-type unit linked to a (+)-*N*-methylcoclaurine moiety (*139*).

5. (+)-Bursanine and (+)-Iznikine

Bursanine (**39**) and iznikine (**40**) were isolated together with istanbulamine (**38**) from the roots and rhizomes of *Thalictrum minus,* and both alkaloids are structural isomers having $C_{40}H_{46}N_2O_9$ as the common molecular formula. The NMR spectra coupled with NOE studies as well as mass and UV spectra led to

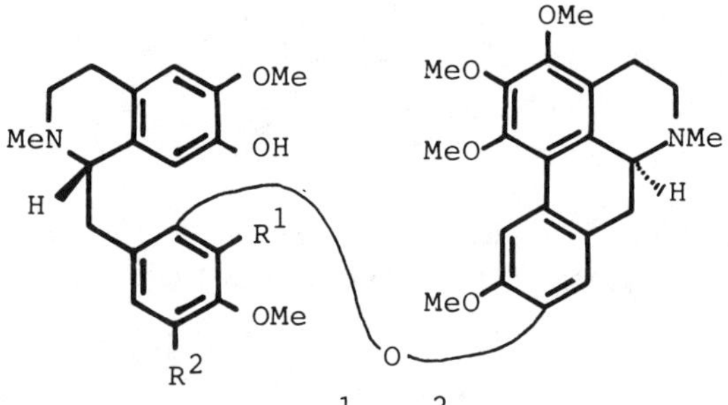

39 R^1=H R^2=OH

40 R^1=OH, R^2=H

the structural assignments. Again the absolute configuration of both alkaloids were derived from their CD spectra (*139*).

6. Beccapoline and Beccapolinium

Beccapoline (**41**) was isolated from *Polyalthia beccarii* (Annonaceae) (*140*) [mp >280°C(d) and $[\alpha]_D$ 0°]. Its mass spectrum shows a molecular ion at *m/z* 608 (M^+) which corresponds to $C_{37}H_{24}N_2O_7$, and the fragmentation pattern suggests the dimeric aporphine for beccapoline. The UV spectrum [λ_{max}^{MeOH} (log ϵ): 218 (4.52), 231 (4.62), 250 (4.53), 279 (4.40), 310 (4.01), 366 (4.07), 428 (4.22), and 440 nm (4.21)] indicates a highly conjugated system, which exhibits a bathochromic shift in acidic solution. Its IR spectrum shows the presence of an oxoaporphine system at 1650 cm^{-1}. Its NMR spectrum indicates the presence of two methylenedioxy groups, two methoxyl groups, and one *N*-methyl group. Reduction of **41** with zinc in hydrochloric acid solution afforded the deoxydecahydrobeccapoline. Thus the structure of beccapoline was determined to be a bisaporphine having an oxostephanine linked to tetrahydrostephanine.

Beccapolinium (**42**) is the quaternary salt [mp 250°C(d) and $[\alpha]_D$ 0°]. Its mass

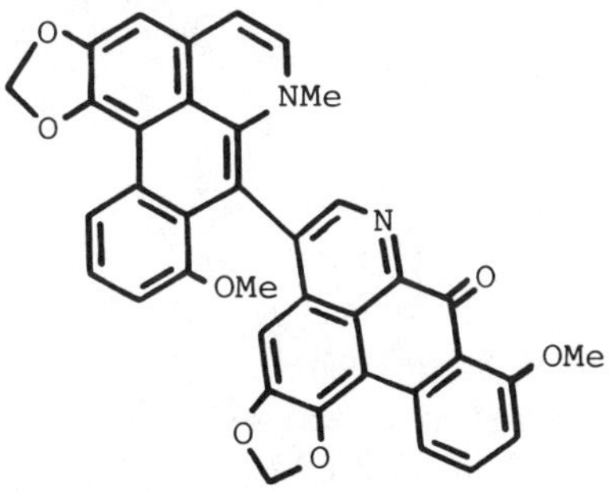

41

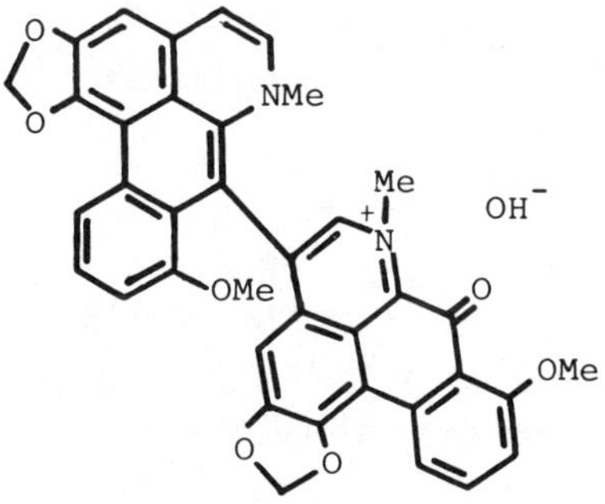

42

spectrum shows its molecular formula as $C_{38}H_{27}N_2O_7$. Its IR, NMR, and UV spectra are consistent with structure **42** (*140*).

7. (+)-Uskudaramine

(+)-Uskudaramine (**43**) was isolated from *Thalictrum minus* L. var *microphyllum* (Ranunculaceae), and its molecular formula was determined to be $C_{39}H_{44}N_2O_8$. This base is structurally isomeric with the diphenolic bisaporphine (+)-istanbulamine (**38**). Its NMR spectrum exhibits the presence of five methoxyl groups, two *N*-methyl groups, and six aromatic protons. The UV spectrum exhibits an absorption at 312 nm, diagnostic of an aporphine system and shows the expected bathochromic shift in alkaline solution attributable to the phenolic function as well as a hyperchromic shift, which is associated with the presence of a phenolic function at either C-3 or C-9 of the aporphine moiety. The presence of a phenolic function at C-9 was confirmed by the NMR spectrum of its triacetate, which displays a downfield shift of 11-H from 7.99 to 8.10 ppm. The relative position of each of the six aromatic protons was determined in relation to each of the methoxyl substituents from its NOE study. Separate irradiations of the aromatic protons substantiated the fact that all of these protons belong to the benzylisoquinoline moiety except for 11-H. The CD spectrum of **43,** with a Cotton effect maximum at 244 nm and a negative through a 212 nm, is very close to that of (+)-istanbulamine and hence the absolute stereochemistry was determined (*141*).

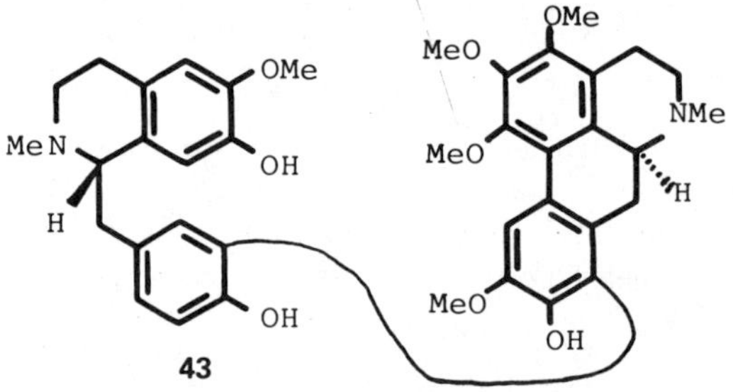

8. (+)-*N*-2′-Noradiantifoline

(+)-*N*-2′-Noradiantifoline (**44**) was isolated from *Thalictrum minus* L. var. *microphyllum* Boiss. Its molecular formula was determined as $C_{41}H_{48}N_2O_9$ and it has $[\alpha]_D$ +39°. Its NMR spectrum shows the presence of eight methoxyl groups, one *N*-methyl group, and six aromatic protons as all singlets. The 11-H proton appears characteristically downfield at 8.05 ppm. The position of the *N*-methyl group was determined to be present on the aporphine rather than the

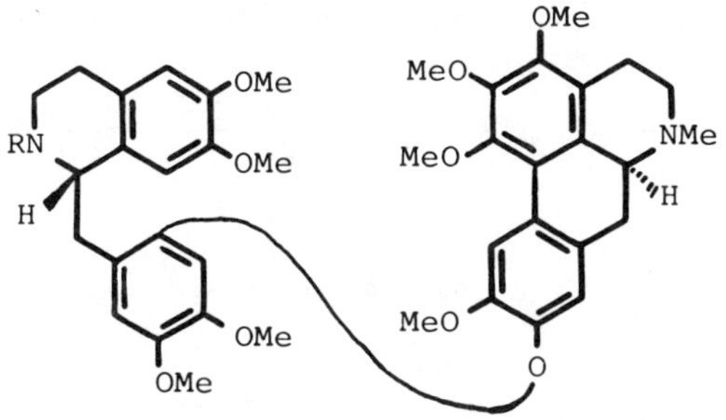

44 R=H
45 R=Me

tetrahydroisoquinoline by its mass spectrum, which shows the base peak at *m/z* 192 attributable to the dihydroisoquinolinium cation formed from cleavage of the C-1′ to C-α bond. The UV spectrum of **44** with maxima at 280, 302sh, and 314 nm is similar to that of the known adiantifoline (**45**). The absolute configuration was determined by its CD spectrum (*142*).

III. Occurrence of Known Aporphines in Plants

Table II summarizes the known aporphine alkaloids recently reisolated from plant sources (*143–299*).

TABLE II
THE KNOWN APORPHINE ALKALOIDS RECENTLY REISOLATED FROM PLANT SOURCES

Alkaloid	Plant source	Reference
Asimilobine	*Uvaria chamae*	*143*
	Anaxagorea prinoides	*69*
	A. dolichocarpa	*69*
	Zizyphus sativa	*144*
	Liriodendron tulipifera	*28,76,89*
	Ocotea glaziovii	*145*
	Magnolia obovata	*42*
	Mitrella kentii	*146*
	Schefferomitra subaequalis	*147*
	Laurelia philippiana	*56*
Actinodaphnine	*Litsea laurifolia*	*148*
	L. nitida	*149*
	L. sebifera	*150*
	Hernandia cordigera	*151*
N-Acetylnornuciferine	*Liriodendron tulipifera*	*28*
(±)-Apoglaziovine	*Ocotea glaziovii*	*145*

(*continued*)

TABLE II (*Continued*)

Alkaloid	Plant source	Reference
(+)-Apoglaziovine	*O. variabilis*	*17*
Atheroline	*Nemuaron vieillardii*	*152*
	Monimia rotundifolia	*153*
	Laurelia sempervirens	*154*
	L. philippiana	*154*
Anolobine	*Schefferomitra subaequalis*	*147*
N-Acetylanonaine	*Zanthoxylum bungeanum*	*155*
N-Acetylasimilobine	*Liriodendron tulipifera*	*28*
Bisnorthalphenine	*Thalictrum revolutum*	*112*
Anonaine	*Polyalthia emarginata*	*78*
	Isolona pilosa	*49*
	Annona cherimolia	*156*
	Cananga odorata	*157*
	Doryphora sassafras	*158*
	Mitrella kentii	*146*
	Nelumbo nucifera	*159*
	Xylopia buxifolia	*79*
	Laurelia philippiana	*56*
Bracteoline	*Papaver orientale*	*34*
	P. bracteatum	*160*
	Corydalis gortschakovii	*161*
Bulbocapnine	*Glaucium vitellinum*	*58,162*
	G. flavum	*163*
	G. pulchrum	*162*
	Corydalis vaginans	*164*
	C. marschalliana	*164–166*
	C. cava	*167,168*
	C. bulbosa	*169*
	C. ledebouriana	*170*
	C. slivenensis	*171*
Caaverine	*Liriodendron tulipifera*	*89*
	Isolona pilosa	*49*
	Ocotea glaziovii	*145*
Cassythicine	*Ocotea brachybotra*	*172*
Corytuberine	*Corydalis pallida* var. *tenuis*	*173*
	C. suaveolens	*174*
	Papaver lecoquii	*175*
	P. albiflorum ssp. *austromoravicum*	*175*
	P. albiflorum ssp. *albiflorum*	*175*
	P. litwinowii	*176*
Crebanine	*Stephania cepharantha*	*92*
Cassameridine	*Litsea kawakamii*	*177*
Corunnine	*Thalictrum foetidum*	*178*
Cassamedine	*Siparuna guianensis*	*109*
Boldine	*Litsea laurifolia*	*148*
	L. turfosa	*179*

TABLE II (*Continued*)

Alkaloid	Plant source	Reference
Boldine	*L. leefeana*	*180*
	L. sebifera	*150*
	L. wightiana	*150*
	Machilus duthei	*181*
	Sassafras albidum	*182*
	Monimia rotundifolia	*153*
Glaucine	*Mahonia repens*	*183*
	Litsea laeta	*38*
	L. triflora	*184*
	Uvaria chaemae	*143*
	Glaucium vitellinum	*58,162*
	G. oxylobum	*185*
	G. elegans	*186*
	G. grandiflorum	*125*
	G. leiocarpum	*187*
	G. flavum	*188,189*
	Croton draconoides	*190*
	Liriodendron tulipifera	*28,76,89*
	Alphonsea ventricosa	*191*
	Ocotea macrophylla	*86*
	Magnolia obovata	*42*
	Aconitum yezoense	*192*
	Thalictrum foetidum	*178*
	Corydalis bulbosa	*169*
	C. ambigua	*193*
	Papaver strictum	*194*
	P. spicatum ssp. *luschanii*	*194*
	P. spicatum ssp. *spicatum*	*194*
Dehydroroemerine	*Stephania sasakii*	*50*
	S. kwangsiensis	*91*
	S. micrantha	*195*
	Liriodendron tulipifera	*89*
	Papaver spicatum ssp. *luschanii*	*194*
	P. spicatum ssp. *spicatum*	*194*
Corydine	*Guatteria moralessi*	*196*
	G. cubensis	*196*
	Dicranostigma leptopodium	*197*
	Glaucium fimbrilligerum	*37*
	G. vitellinum	*58*
	G. pulchrum	*162*
	G. elegans	*186*
	G. flavum	*198*
	G. leiocarpum	*187*
	Decentra spectrabilis	*199*

(*continued*)

TABLE II (*Continued*)

Alkaloid	Plant source	Reference
Corydine	*Litsea triflora*	*184*
	Mahonia aquifolium	*200*
	Zanthoxylum oxyphyllum	*116*
	Thalictrum dioicum	*201,202*
	Corydalis gortschakovii	*161*
	C. intermedia	*161*
	C. merschalliana	*165,166*
	C. bulbosa	*169*
	C. slivenensis	*171*
	Stephania dinklagei	*44*
	Aconitum leucostomum	*203*
	Xylopia danguyella	*79*
	Papaver lecoquii	*175*
	P. albiflorum ssp. *austromoravicum*	*175*
	P. litwinowii	*176*
	P. oreophilum	*204*
	Mahonia repens	*183*
Dicentrine	*Ocotea brachybotra*	*172*
	O. minarum	*35*
	Glaucium vitellinum	*58,162*
	G. flavum	*163*
	Cissampelos pareira	*205*
	Litsea nitida	*149*
	Lindera oldhamii	*2*
Dicentrinone	*Litsea salicifolia*	*38*
	Ocotea minarum	*35*
	Glaucium vitellinum	*58*
Dehydrodicentrine	*Glaucium vitellinum*	*58*
	Cissampelos pareira	*205*
Dehydroocoteine	*Thalictrum isopyroides*	*206*
Dehydrocrebanine	*Stephania cepharantha*	*92*
Dehydroglaucine	*Liriodendron tulipifera*	*23,24,28,76*
Dehydrostephanine	*Stephania micrantha*	*195*
Dehydronantenine	*Corydalis bulbosa*	*169*
	C. slivenensis	*171*
	C. marschalliana	*166*
Domesticine	*Corydalis gortschakovii*	*161*
	C. marschalliana	*165,166*
	C. cava	*168*
	C. bulbosa	*169*
	C. slivenensis	*171*
	C. suaveolens	*174*
Guatterine	*Pachypodanthium confine*	*67*
Glaunine	*Glaucium fimbrilligerum*	*37*
Glaunidine	*Glaucium fimbrilligerum*	*37*
Hernovine	*Neolitsea variabillima*	*207*

TABLE II (*Continued*)

Alkaloid	Plant source	Reference
Homomoschatoline	*Annona acuminata*	*208*
	Abuta rufescens	*101*
	Liriodendron tulipifera	*76,209*
	Triclisia patens	*210*
	T. gillettii	*211*
	Duguetia eximia	*212*
	Guatteria subsessilis	*93*
Imenine	*Abuta rufescens*	*101*
Isothebaine	*Papaver bracteatum*	*160,213*
	P. orientale	*34,214,215*
	P. pseudo-orientale	*215*
Isoboldine	*Hernandia cordigera*	*151*
	Neolitsea fuscata	*216*
	Xylopia danguyella	*79*
	Guatteria melosma	*121*
	Schefferomitra subaequalis	*147*
	Uvaria chamae	*143*
	Thalictrum alpinum	*217*
	T. foetidum	*178*
	Glaucium fimbrilligerum	*37*
	G. elegans	*186*
	G. grandiflorum	*125*
	Machilus duthei	*181*
	Litsea triflora	*184*
	L. laurifolia	*148*
	Fumaria parviflora	*218*
	F. vaillantii	*219*
	Cryptocarya longifolia	*220*
	Mahonia aquifolium	*200*
	Berberis integerrima	*120*
	Delphinium dictyocarpum	*221*
	Sassafras albidum	*182*
	Corydalis gortschakovii	*161*
	C. cava	*168*
	C. marschalliana	*165,166*
	C. bulbosa	*169*
	C. slivenensis	*171*
	Ocotea glaziovii	*145*
	Erythrina abyssinica	*222*
	Monodora angolensis	*223*
Laurotetanine	*Machilus duthei*	*179*
	Cryptocarya longifolia	*220*
	C. odorata	*22*
	Litsea laurifolia	*148*

(*continued*)

TABLE II (*Continued*)

Alkaloid	Plant source	Reference
Laurotetanine	*L. sebifera*	*150*
	Hernandia cordigera	*151,224*
	Laurelia sempervirens	*154*
	L. philippiana	*154*
	Actinodaphne obovata	*150*
	Nemuaron vieillardii	*152*
	Xylopia danguyella	*79*
	Monimia rotundifolia	*153*
Isocorydine	*Glaucium fimbrilligerum*	*37*
	G. flavum	*198*
	G. leiocarpum	*187*
	G. pulchrum	*162*
	G. vitellinum	*58,162*
	G. elegans	*186*
	Papaver rhoeas	*225*
	P. fugax	*226*
	P. lisae	*227*
	P. comutatum	*228*
	P. lecoquii	*175*
	P. albiflorum ssp. *austromoravicum*	*175*
	P. litwinowii	*176*
	P. oreophilum	*204*
	Litsea triflora	*184*
	Hernandia cordigera	*224*
	Stephania sasakii	*50*
	S. kwangsiensis	*91*
	S. micrantha	*195*
	Mahonia aquifolium	*200*
	M. repens	*183*
	Ocotea macrophylla	*86*
	Pteridophyllum racemosum	*229*
	Doryphora sassafras Endlicher	*158*
	Cryptocarya odorata	*22*
	Enantia polycarpa	*230*
	Corydalis ledebouriana	*170*
	Dicranostigma leptopodium	*197*
(−)-Isocorydine	*Corydalis slivenensis*	*171*
Laurifoline	*Cocculus laurifolius*	*231*
	Zanthoxylum species	*232*
	Legnephora moorei	*233*
Laurolitsine	*Litsea laurifolia*	*148*
	L. kawakamii	*177*
	L. akoensis	*177*
	L. turfosa	*179*
	L. leefeana	*180*
	Machilus duthei	*181*

TABLE II (*Continued*)

Alkaloid	Plant source	Reference
Laurolitsine	*Phoebe formosana*	*234*
	Monimia rotundifolia	*153*
Launobine (Norbulbocapnine)	*Illigera luzonensis*	*235*
Lirinidine	*Ocotea glaziovii*	*145*
	Liriodendron tulipifera	*20*
Lanuginosine	*Stephania sasakii*	*90*
	S. japonica	*95*
	Xylopia buxifolia	*79*
	Annona squamosa	*236*
	Polyalthia emarginata	*78*
	P. oliveri	*66*
	Enantia pilosa	*118*
	Uvariopsis guineensis	*237*
	Magnolia campbelli	*64,238*
Liridinine	*Papaver armeniacum*	*226*
	P. tauricola	*226*
Laurelliptine	*Nectandra rigida*	*239*
	Litsea laurifolia	*148*
	L. triflora	*184*
	Monodora tenuifolia	*240*
Liriodenine	*Liriodendron turipifera*	*23,28,76,89,209*
	Annona cherimolia	*156*
	A. acuminata	*208*
	A. cristalensis	*241*
	Phoebe formosana	*234*
	Mitrella kentii	*146*
	Stephania sasakii	*90*
	Guatteria cubensis	*196*
	Xylopia buxifolis	*79*
	Siparuna gilgiana	*242*
	S. guianensis	*109*
	Laurelia novae-zelandiae	*97*
	Rhigiocarya racemifera	*243*
	Pachygone ovata	*244*
	Polyalthia emarginata	*78*
	P. oliveri	*66*
	Xanthoxylum cuspidatum	*245*
	Enantia pilosa	*118*
	Fusae longifolia	*109*
	Eupomatia laurina	*246*
	Cananga odorata	*157*
	Talauma mexicana	*247*
	Uvariopsis guineensis	*237*
	Magnolia campbelli	*238*

(*continued*)

TABLE II (*Continued*)

Alkaloid	Plant source	Reference
Liriodenine	*M. obovata*	*42*
	M. mutabilis	*238*
	Doryphora sassafras	*158*
Leucoxylonine	*Ocotea minarum*	*35*
Leucoxine	*Ocotea minarum*	*35*
Lysicamine	*Telitoxicum peruvianum*	*104*
	Stephania sasakii	*50*
	Annona acuminata	*208*
	Abuta rufescens	*101*
Mecambrine	*Papaver urbanianum*	*234*
Menisperine	*Magnolia grandiflora*	*248*
	Papaver oreophilum	*204*
N-Methyllaurotetanine	*Cryptocarya longifolia*	*220*
	C. odorata	*22*
	Litsea laurifolia	*148*
	L. triflora	*184*
	L. sebifera	*150*
	Thalictrum revolutum	*138*
	T. dioicum	*201*
	Glaucium vitellinum	*58*
	Delphinium dictyocarpum	*221*
	Liriodendron tulipifera	*28*
	Actinodaphne obovata	*150*
	Nemuaron vieillardii	*152*
	Monimia rotundifolia	*153*
	Hernandia cordigera	*151*
	Papaver strictum	*194*
N-Methyllindcarpine	*Glaucium fimbrilligerum*	*31,37*
	G. vitellinum	*58*
	G. cherimolia	*162*
N-Methylglaucine	*Stephania dinklagei*	*243*
N,*N*-Dimethyllindcarpine	*Coscinium fenestratum*	*249*
N-Methylhernovine	*Neolitsea variabillima*	*207*
	Lindera oldhamii	*2*
N-Methylisocorydine	*Cocculus laurifolius*	*231*
	Zanthoxylum coriaceum	*250*
	Z. culantrillo	*250*
	Rhigiocarya racemifera	*243*
	Fumaria parviflora	*243*
N-Methylnandigerine	*Litsea laurifolia*	*148*
	Lindera oldhamii	*2*
N-Methylushinsunine	*Elmerillia papuana*	*45*
N-Methylasimilobine	*Papaver lacerum*	*251*
	P. rhoeas	*225*
	P. urbanianum	*252*
	Stephania cepharantha	*92*

TABLE II (*Continued*)

Alkaloid	Plant source	Reference
N-Methylasimilobine	*Xylopia buxifolia*	*79*
N-Methylactinodaphnine	*Litsea kawakamii*	*177*
	L. laurifolia	*148*
	Hernandia cordigera	*151*
N-Methylcorydine	*Kolobopetalum auriculatum*	*253*
	Stephania dinklagei	*243*
	Polyalthia oliveri	*66*
Nuciferine	*Papaver tauricola*	*226*
	P. oreophilum	*204*
	Liriodendron tulipifera	*28*
	Ziziphus amphibia A. Cheval.	*254*
	Nelumbo nucifera	*159,255*
Nornantenine	*Hernandia cordigera*	*151,224*
	Laurelia sempervirens	*154*
	L. philippiana	*56*
	Xylopia danguyella	*79*
Nantenine	*Papaver tauricola*	*226*
	Laurelia sempervirens	*154*
	Ocotea macrophylla	*86*
	O. variabilis	*17*
	Cassytha filiformis	*256*
	Corydalis bulbosa	*169*
	C. slivenensis	*171*
	C. marschalliana	*166*
Nordicentrine	*Litsea salicifolia*	*38*
Magnoflorine	*Pachygone ovata*	*257*
	Thalictrum fauriei	*258*
	T. alpinum	*217*
	T. revolutum	*259*
	T. podocarpum	*260*
	T. minus	*261–263*
	T. polyganum	*264*
	T. foetidum	*178*
	T. lucidum	*265*
	T. longistylum	*266*
	Cocculus laurifolius	*231*
	C. carolinus	*267,268*
	Zanthoxylum microcarpum	*269*
	Z. species	*232*
	Z. culantrillo	*250*
	Kolobopetalum auriculatum	*253*
	Tinospora cordifolia	*270*
	Papaver rupifragum	*271*
	P. somniferum	*272*

(*continued*)

TABLE II (*Continued*)

Alkaloid	Plant source	Reference
Magnoflorine	*P. oreophilum*	*204*
	P. albiflorum ssp. *austromoravicum*	*175*
	P. lecoquii	*175*
	Anamirta cocculus	*273*
	Meconopsis cambrica	*274*
	Aristolochia macedonica	*275*
	Croton turumiquirensis	*276*
	Pycnarrhena longifolia	*277*
	Rhigiocarya racemifera	*243*
	Stephania japonica var. *australis*	*278*
	S. elegans	*279*
	Fagara mayu Engler	*206*
	F. leprieurii	*280*
	F. rubescens	*280*
	F. viridis	*280*
	Papaveraceae callus tissues	*281*
	Glaucium flavum	*282*
	Eschscholtzia californica	*282a*
	Legnephora moorei	*233*
	Cyclea barbata	*283*
	Mahonia repens	*183*
	M. aquifolium	*284*
	Heptacyclum zenkeri	*285*
	Berberis koreana	*286*
Norboldine	*Sassafras albidum*	*182*
	Litsea wightiana	*150*
Norglaucine	*Liriodendron tulipifera*	*28*
	Alphonsea ventricosa	*191*
	Monimia rotundifolia	*153*
Nornuciferine	*Liriodendron tulipifera*	*28*
	Nelumbo nucifera	*159,255*
	Xylopia buxifolia	*79*
	Zizyphus sativa	*287*
	Isolona pilosa	*49*
(±)-Nornuciferine	*Liriodendron tulipifera*	*288*
Nandazurine	*Corydalis bulbosa*	*169*
Neolitsine	*Glaucium vitellinum*	*58*
	Hernandia cordigera	*151*
Norushinsunine	*Annona cherimolia*	*156*
(Michelalbine)	*Elmerillia papuana*	*45*
	Liriodendron tulipifera	*28,76*
Norcorydine	*Glaucium fimbrilligerum*	*37*
	Stephania dinklagei	*44*
	Xylopia danguyella	*79*
	Laurelia philippiana	*56*
Norisocorydine	*Hernandia cordigera*	*151,224*

TABLE II (*Continued*)

Alkaloid	Plant source	Reference
Norisocorydine	*Cryptocarya longifolia*	*220*
	Glaucium fimbrilligerum	*37*
	Nemuaron vieillardii	*152*
	Xylopia danguyella	*79*
Nandigerine	*Hernandia cordigera*	*224*
	Litsea laurifolia	*148*
	Neolitsea variabillima	*207*
Ovigerine	*Hernandia cordigera*	*224*
Oxoputerine	*Laurelia novae-zelandiae*	*56*
	Duguetia calycina	*82*
	D. eximia	*212*
Puterine	*Duguetia calycina*	*82*
(+)-Remrefidine	*Papaver fugax*	*289*
Roemrefidine	*Papaver litwinowii*	*176,290*
Ocoteine	*Ocotea minarum*	*35*
	O. leucoxylon	*29*
	Thalictrum strictum	*291*
Ocopodine	*Ocotea minarum*	*35*
	O. brachybotra	*172*
Oxonantenine	*Corydalis bulbosa*	*169*
	C. marschalliana	*171*
	Siparuna gilgiana	*242*
Oxoglaucine	*Glaucium grandiflorum*	*125*
(*O*)-Methylatheroline	*G. serpieri*	*125*
	Liriodendron tulipifera	*28,76*
O-Methylpukateine	*Duguetia calycina*	*82*
Obovanine	*Duguetia calycina*	*82*
	Laurelia novae-zelandiae	*56*
Oxolaureline	*Laurelia novae-zelandiae*	*56*
O-Methylcassyfiline	*Thalictrum strictum*	*291*
O-Methyldomesticine	*Nandina domestica*	*292*
Roemerine	*Stephania kwangsiensis*	*91*
	Papaver tauricola	*226*
	P. lacerum	*251*
	P. spicatum ssp. *luschanii*	*194*
	P. spicatum ssp. *spicatum*	*194*
	P. albiflorum ssp. *albiflorum*	*175*
	P. albiflorum ssp. *austromoravicum*	*175*
	P. strictum	*194*
	P. litwinowii	*176*
	Isolona pilosa	*49*
	Cananga odorata	*157*
	Phoebe formosana	*234*
	Nelumbo nucifera	*159*

(*continued*)

TABLE II (*Continued*)

Alkaloid	Plant source	Reference
(±)-Roemerine	*Liriodendron tulipifera*	*288*
Predicentrine	*Litsea triflora*	*184*
	Ocotea min rum	*35*
	O. brachybotra	*172*
	Liriodendron tulipifera	*28,29*
	Glaucium oxylobum	*185*
	Corydalis cava	*168*
	C. bulbosa	*169*
	C. slivenensis	*171*
Preocoteine *N*-oxide	*Thalictrum minus*	*293*
Sparsifoline	*Croton sparsiflorus*	*294*
Stesakine	*Stephania cepharantha*	*92*
Subsessiline	*Telitoxicum peruvianum*	*104*
Stepharine	*Legnephora moorei*	*233*
	Stephania dinklagei	*44*
Steporphine	*Stephania dinklagei*	*44*
Stephanine	*Stephania micrantha*	*195*
	S. cepharantha	*92*
	S. kwansiensis	*295*
Thaliporphine	*Uvaria chamae*	*143*
	Thalictrum alpinum	*217*
	T. foetidum	*178*
	Croton draconoides	*190*
	Liriodendron tulipifera	*28*
	Glaucium flavum	*87*
	Corydalis bulbosa	*169*
	Mahonia repens	*183*
Thaliporphine *N*-oxide	*Berberis integerrima*	*120*
Thalphenine	*Thalictrum minus* race B	*296*
	T. revolutum	*259*
	T. rugosum	*297*
Thalicminine	*Thalictrum strictum*	*291*
	T. minus	*293*
	T. dioicum	*298*
	Ocotea puberula	*299*
	O. minarum	*35*
Thalicmidine	*Thalictrum minus*	*293*
Thalicmidine *N*-oxide	*Thalictrum minus*	*293*
Thalicmine	*Thalictrum minus*	*293*
	Ocotea puberula	*299*
Xylopine	*Annona squamosa*	*236*
	Xylopia buxifolia	*79*
Ushinsunine	*Cananga odorata*	*157*
	Phoebe formosana	*234*
	Stephania venosa	*70*

IV. Biogenesis

Aporphine alkaloids are known to be derived from the corresponding phenolic tetrahydrobenzylisoquinolines by direct oxidative coupling or by the formation of dienone derivatives (proaporphines or proerythrinadienones), which then rearrange into aporphines through dienone–phenol or dienol–benzene rearrangements. Actually, the feeding experiment of labeled reticuline in *Papaver somniferm* has shown that isoboldine (**47**) was derived from reticuline (**46**) by direct ortho–para oxidative coupling (*300*). Moreover, (±)-4′-*O*-methylnorlaudanosine, (±)-reticuline, and (±)-norreticuline were shown to be effective precursors for boldine (**48**) in *Litsea glutinosa* (*301*). Since (+)-reticuline but not (−)-

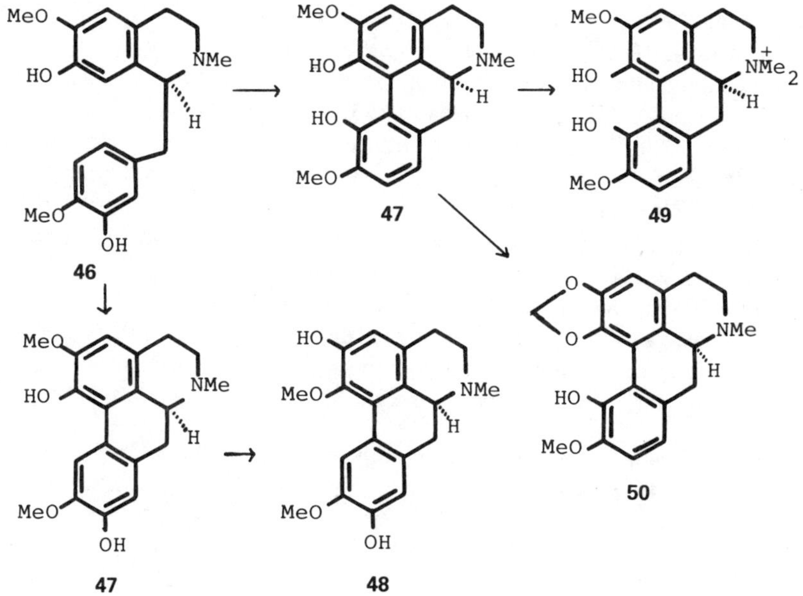

MeO HO NMe H MeO OH **46** → MeO HO NMe H HO MeO **47** → MeO HO $\overset{+}{N}Me_2$ H HO MeO **49**

47 → O O NMe H HO MeO **50**

46 ↓ MeO HO NMe H MeO OH **47** → HO MeO NMe H MeO OH **48**

reticuline is a possible precursor for boldine and (+)-isoboldine was specifically incorporated into boldine, the postulated biosynthesis of boldine is believed to be as follows:

(+)-Norreticuline (**51**) → (+)-Reticuline (**46**) → (+)-Isoboldine (**47**) → (+)-Boldine (**48**)

However, bulbocapnine (**50**) has been shown to be derived directly from reticuline in *Corydalis cava,* by ortho–ortho phenolic oxidative coupling (*302*). In the biosynthesis of isocorydine (**13**), labeled nororientaline, norprotosinomenine, norlaudanidine, and reticuline were fed to *Annona squamosa,* and it was shown that only reticuline was significantly incorporated. This result leads to the conclusion that isocorydine was biosynthetically formed from reticuline by direct

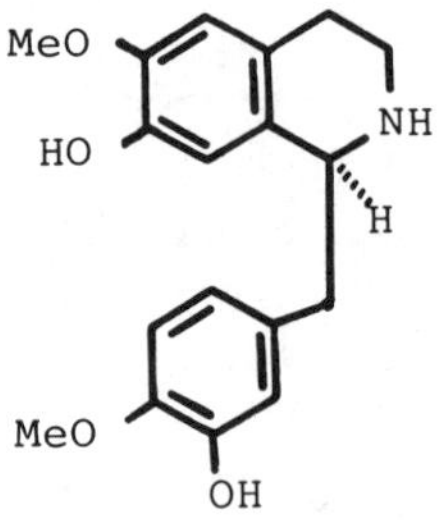

MeO
HO
NH
H
MeO
OH
51

ortho–ortho phenolic oxidative coupling (*303*). Magnoflorine (**49**) was also shown to be derived by ortho–ortho phenolic coupling of reticuline in *Aquilegia* species and *Cocculus laurifolius,* respectively, and corytuberine (**17**) must be the intermediate (*304, 305*). Another possible biogenetic route for aporphines involves dienone intermediates. This pathway was postulated by the incorporation of (+)-orientaline (**52**) into isothebaine (**55**) via orientalinone (**53**) and orientalinol (**54**). The dienol–benzene rearrangement involved in the above biosynthesis resulted in the loss of one oxygen in its D ring (*306*). Proerythrinadienone has also been shown to be a key intermediate for the biosynthesis of aporphines. Thus corydine (**11**), boldine (**48**), glaucine (**57**) and dicentrine (**58**) were biosynthesized from protosinomenine (**56**) via the corresponding dienone (**59**) and (**60**) in *Dicentra eximia* (Fumariaceace) (*307*).

It is of interest to note that the methylenedioxy bridge of (+)-thalpenine (**61**) may arise from nanthenine isolated from the same plant as thalpenine by oxidation of one of the methyl groups via the plausible intermediate (**63**) (*111, 114*).

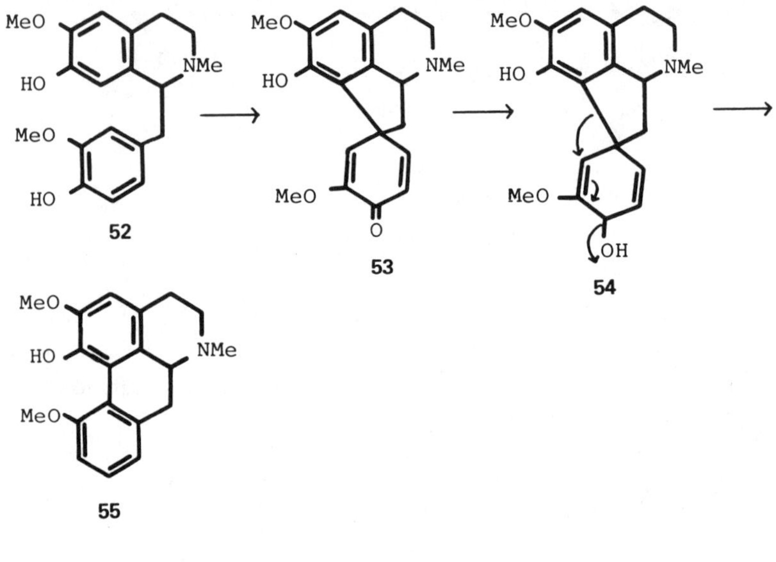

MeO
HO
NMe
MeO
HO
52
→
MeO
HO
NMe
MeO
O
53
→
MeO
HO
NMe
MeO
OH
54
→

MeO
HO
NMe
MeO
55

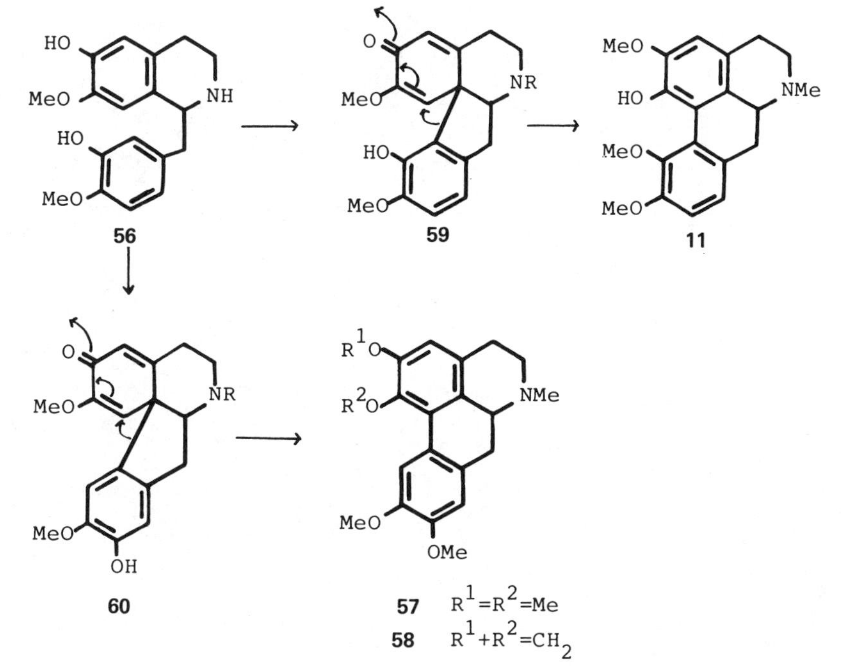

56 59 11

60 57 $R^1=R^2=Me$
58 $R^1+R^2=CH_2$

Recently two different biogenetic schemes were proposed, in which morphinandienones and protoberberinium salts were suggested as possible precursors, respectively *(308, 309)*.

Oxidation of laudanosine (**64**) with VOF_3–TFA afforded glaucine in 43% yield via the morphinandienone (**65**) and the proerythrinadienone (**66**). Furthermore, the treatment of (±)-*O*-methylflavinantine (**67**) with concentrated hydrochloric acid yielded 1,2-dihydroxy-9,10-dimethoxyaporphine (**68**) in 98% yield, whose methylation leads to the synthesis of glaucine (**57**). A similar result was obtained from the mixture of corydine (**11**) and *O*-methylcorydine (**70**). Based on the consideration of the above results, it is strongly suggested that the morphinandienones could be *in vivo* precursors of aporphine alkaloids *(308)*.

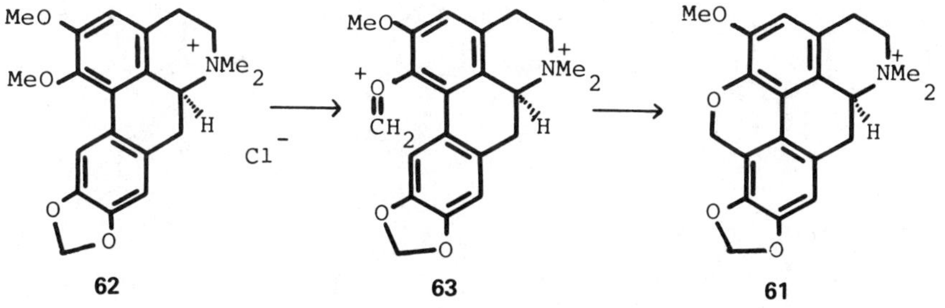

62 63 61

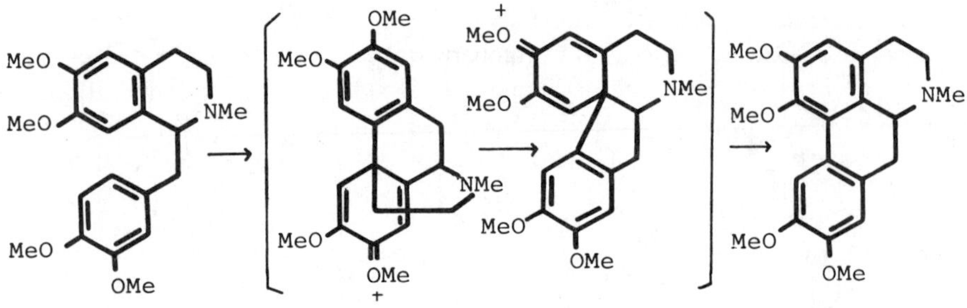

64 → [**65**] → [**66**] → **57**

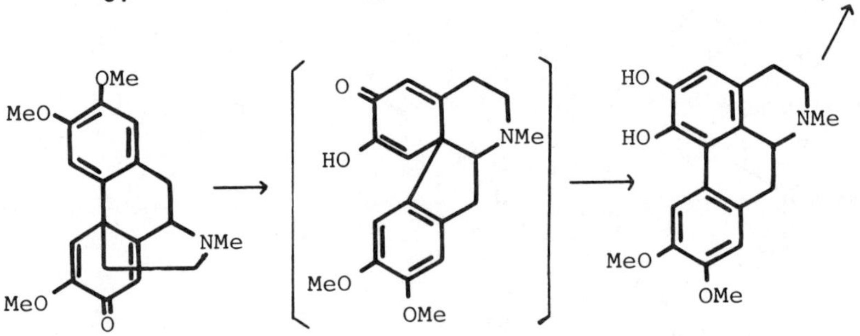

67 → [ ] → **68** → **57**

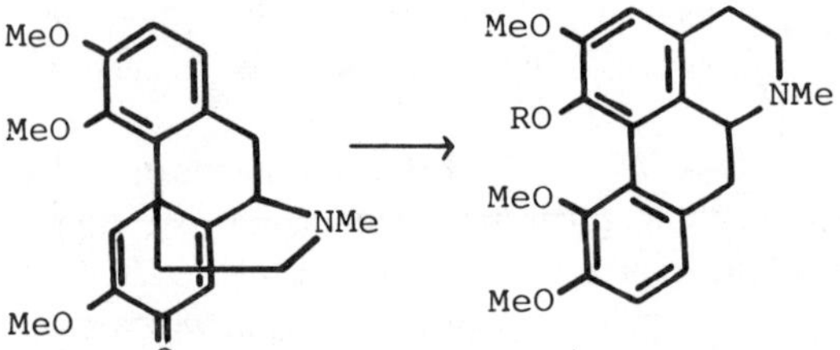

69 → **11** R=H, **70** R=Me

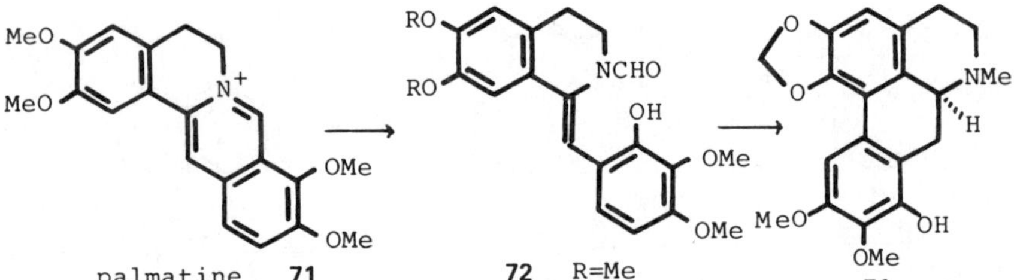

palmatine **71** → **72** R=Me, **73** R+R=CH_2 → **74**

An alternative biogenetic route proposed for aporphines does not involve phenolic oxidative coupling but protoberberinium salts. The oxidation of palmatine (**71**) with *m*-chloroperbenzoic acid, followed by hydrolysis, afforded polycarpine (**72**). Since the formation of aporphines from enamides by photocyclization was already documented, similar conversion might occur in the plant. Thus the transformation of protoberberinium salts to aporphine alkaloids via benzylisoquinoline enamides would constitute an alternative plausible biogenetic pathway (*309*).

V. Synthesis

In the synthesis of aporphine alkaloids, four major reactions, the Pschorr reaction, phenolic or nonphenolic oxidative coupling, benzyne-mediated reaction, and photocyclization have widely been utilized as key reactions.

This review summarizes the extensive applications of these reactions as well as others, such as lead tetraacetate oxidation, for the synthesis of aporphines.

A. Pschorr Reaction

The Pschorr reaction is well known for the synthesis of phenanthrene derivatives from diazotized compounds by intramolecular arylation. Application of the Pschorr reaction for aporphine synthesis, therefore, involves a nucleophilic reaction of the 8-position of the isoquinoline ring with the aromatic cation derived from the corresponding diazonium salt as follows:

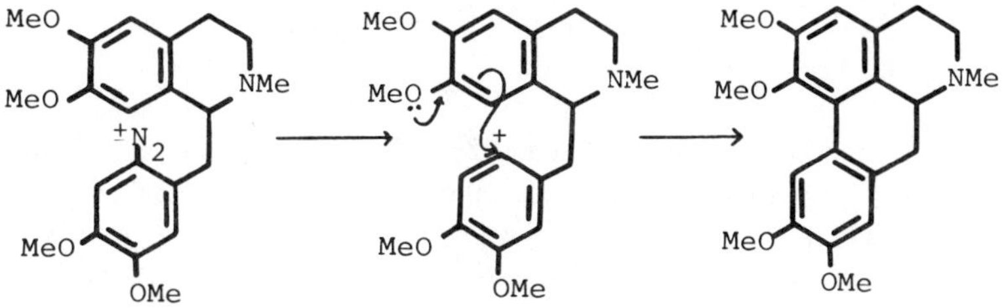

MeO, MeO, NMe, $\overset{+}{N}_2$, MeO, OMe → MeO, MeO, NMe, +, MeO, OMe → MeO, MeO, NMe, MeO, OMe

The requisite nitro-1-benzylisoquinolines for the Pschorr reaction are usually synthesized from the corresponding amides by the Bischler–Napieralski cyclization or the Reissert reaction of *o*-nitrotoluene derivatives with 3,4-dihydroisoquinolines. The yield of the Pschorr reaction varies and depends on the substituents in the precursor. Syntheses of a number of aporphine alkaloids have been accomplished by employing the Pschorr cyclization as a key step. The aporphine alkaloids so synthesized since 1972 were as follows: thaliporphine (*310, 311*), bracteoline (*312*), nuciferine (*310*), nantenine (*313*), glaucine (*310, 311*), coninnine (*314*), apomorphine (*315*), apocodeine (*315*), predicentrine (*316*), *N*-methylhernagine (*224*), stephanine (*317*), isocorytuberine (*318*), nornantenine (*319*), *N*-acetylnornantenine (*319*), thalicarpine (*310*), corunnine (*320, 321*), liriodenine (*321*), cassameridine (*321*), and others.

B. Photo-Pschorr Reaction

As previously described, the Pschorr reaction involves aromatic nucleophilic reaction. Since this type of reaction usually afforded the desired aporphines in low yield, the more effective method, which involves a radical intermediate

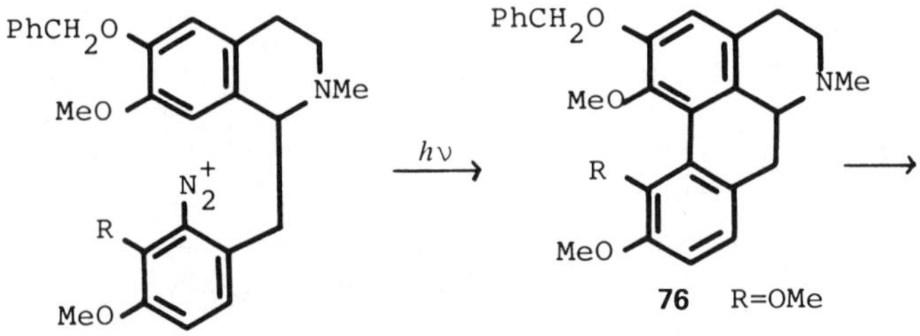

75 R=OMe
80 R=OCH_2Ph

76 R=OMe

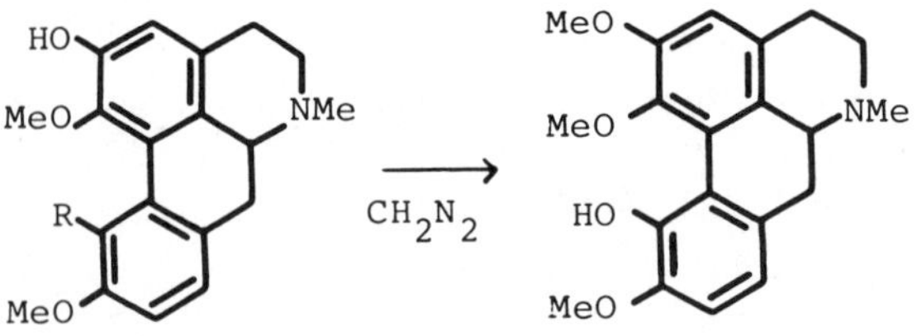

77 R=OMe
78 R=OH

CH_2N_2

79

instead of a cation, was investigated (*322, 323*). Thus photolysis of the diazonium salt brings about a homolysis of the carbon–nitrogen bond of the diazo group and leads to the formation of the aporphine skeleton. This reaction was

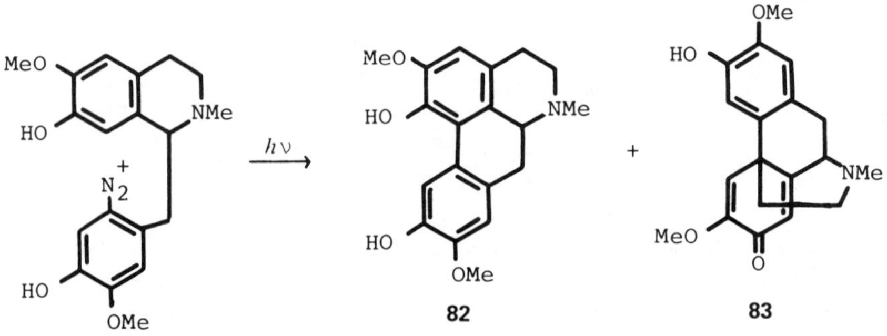

81 → **82** + **83**

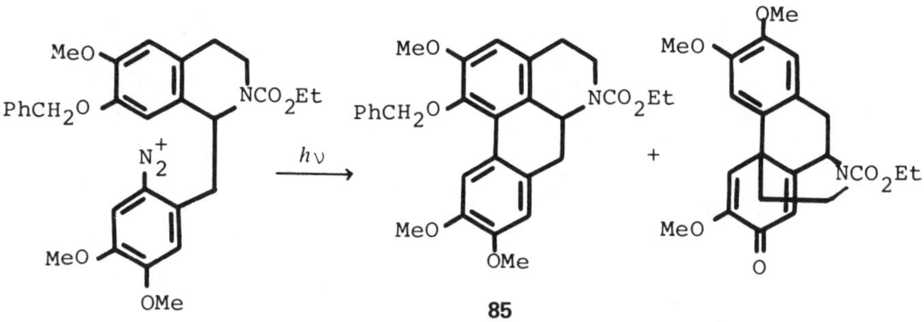

named the photo-Pschorr reaction, and yields are usually higher than those of the Pschorr reaction. When the isoquinoline (**75**) was irradiated by a UV lamp equipped with a pyrex filter, the aporphine (**76**) was isolated in 17% yield (*324*). On the other hand, the usual Pschorr reaction for **75** afforded the same aporphine in only 3% yield, which was then converted to *N,O*-dimethylhernovine (**77**). Similar results were obtained in the synthesis of *N*-methyllindecarpine (**78**) and isocorydine (**79**) from the diazotized compound (**80**). It is worth noting that the side reactions, such as diazo coupling in the phenolic isoquinoline and loss of a protecting group, which usually were observed in the Pschorr reaction (*325*), were eliminated in the photo-Pschorr reaction. Similarly, when the diazonium isoquinoline **81** was subjected to the photo-Pschorr reaction, the aporphine alkaloid bracteoline **82** was formed together with the morphinandienone flavinantine **83** (*326*). Moreover, irradiation of the N-protected diazoisoquinoline **84** afforded the aporphine **85** instead of the protoberberine (*327*). *N*-Methyllindecarpine (**78**) was also synthesized directly by irradiation of the phenolic diazotized compound (**86**) (*325*).

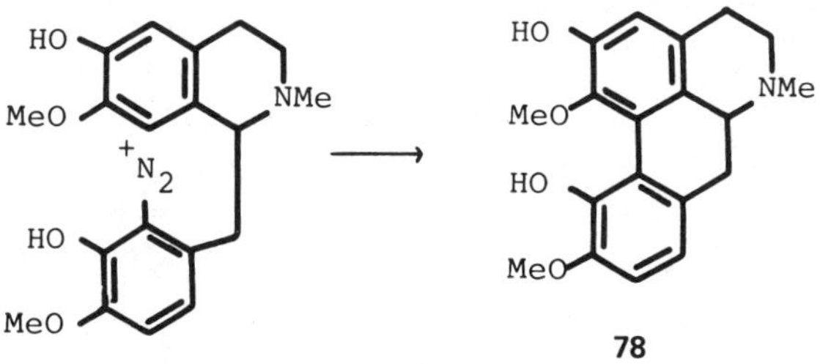

C. Photocyclization

Syntheses of aporphines by means of photolysis of stilbene derivatives were reported by Cava and Lenz in 1966 (*328–330*). Ultraviolet irradiation of 1-benzylidene-2-ethoxycarbonylisoquinoline (**87**) in the presence of iodine af-

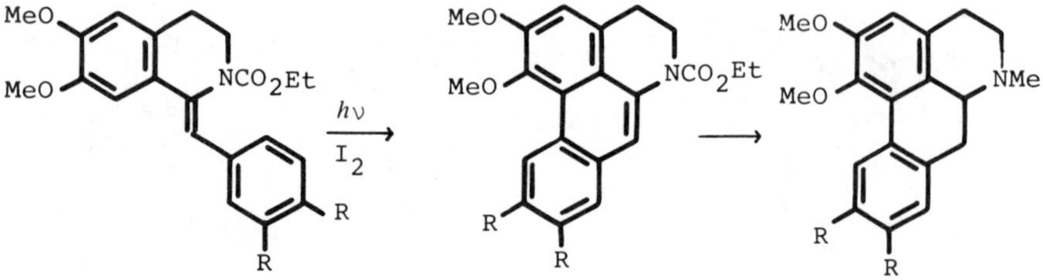

MeO MeO NCO₂Et R R **87** R=H **93** R=OMe $\xrightarrow[I_2]{h\nu}$ MeO MeO NCO₂Et R R **88** R=H **94** R=OMe ⟶ MeO MeO NMe R R **89** R=H **57** R=OMe

forded the dehydronuciferine **88,** which on treatment with lithium aluminum hydride was converted to nuciferine (**89**) in the usual manner (*328*).

Though the photolysis of 1-benzylidene-2-methylisoquinoline (**90**), either in the presence or absence of oxidizing agents, gave none of the cyclized product,

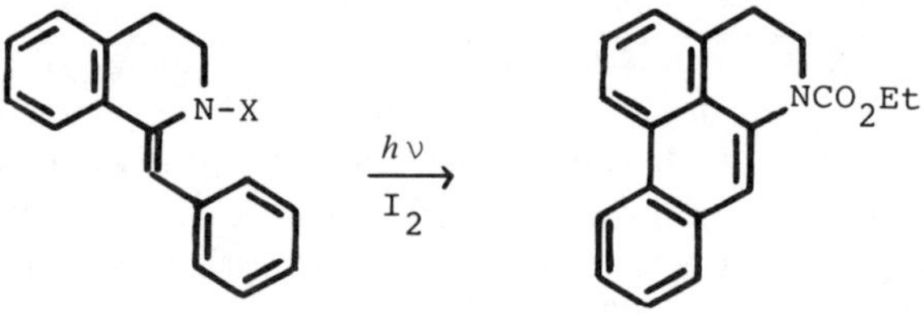

N-X **90** X=Me **91** $X=CO_2Et$ $\xrightarrow[I_2]{h\nu}$ NCO_2Et **92**

the corresponding *N*-ethoxycarbonyl derivative (**91**) led to the formation of the aporphine (**92**) by the same treatment, and the interpretation of the above results is well discussed by Lenz (*329*). Similar reaction for the *N*-ethoxycarbonylstilbene **93** gave rise to the dehydroaporphine **94**, whose reduction afforded glaucine (**57**) (*328*). When the photocyclization of halogenated stilbene was carried out without an oxidizing agent, the dehydroaporphine system was also formed (*328*).

For instance, photolysis of halogenated stilbenes (**95** and **96**) in the presence of calcium carbonate as an acid scavenger, afforded the corresponding dehydroap-

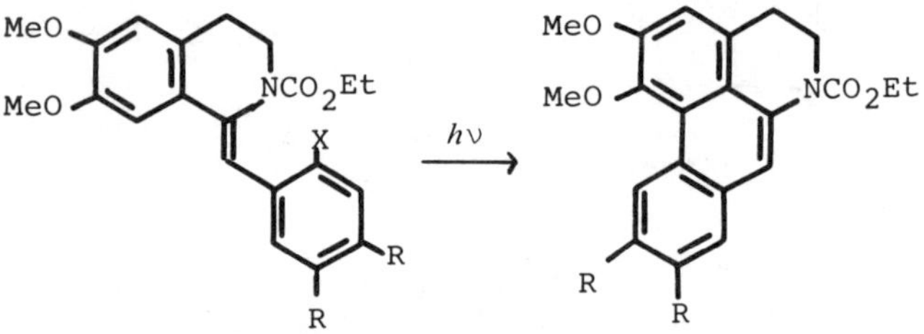

MeO MeO NCO_2Et X R R **95** R=H, X=Cl **96** R=OMe, X=Br $\xrightarrow{h\nu}$ MeO MeO NCO_2Et R R **97** R=H **98** R=OMe

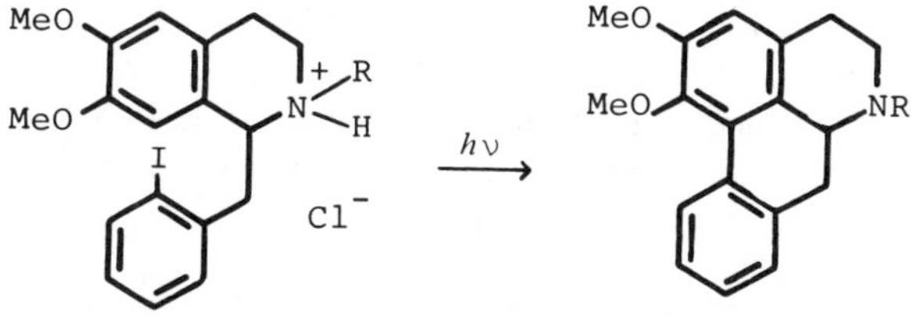

100 R=H
101 R=Me

99 R=H
89 R=Me

orphines **97** and **98**, respectively. This reaction was applied to the synthesis of dicentrine (*331*), cassameridine (*331*), cassamedine (*332*), and neolitsine (*333*).

It has also been documented by Kupchan (*334*) that the photocyclization of 1-benzyl-1,2,3,4-tetrahydroisoquinolinium salts was an effective method for the synthesis of aporphine alkaloids. Thus nornuciferine (**99**) and nuciferine (**89**) were successfully synthesized from **100** and **101**, respectively, by means of an intramolecular photocyclization. Thus nandazurine (*335*) and corunnine (*335*) were synthesized from the corresponding iodoisoquinolines by photolysis. Kupchan also indicated that the presence of a nitrogen lone pair obstructs this cyclization. Actually, photolysis of the free base of **100** did not give the expected aporphine, whereas the N-acylated derivative of **100** yielded the cyclized aporphine under similar reaction conditions. More recently, Spangler and Kametani have reported the syntheses of aporphines by the irradiation of the phenolic bromoisoquinolines in the presence of alkali, both methods of which generate the phenoxide anions independently (*336, 337*). Irradiation of 6′-bromoorientaline (**102**), 6′-bromoreticuline (**103**), and its methylenedioxy analog (**104**) gave bracteoline (**105**) (*337*), isoboldine (**47**) (*338*), and domesticine (**106**)

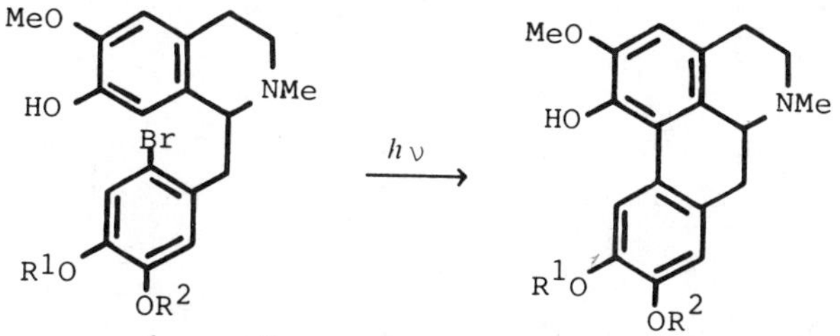

102 R^1=H, R^2=Me
103 R^1=Me, R^2=H
104 R^1+R^2=CH_2

105 R^1=H, R^2= Me
47 R^1=Me, R^2=H
106 R^1+ R^2=CH_2

(*337*), respectively, in addition to the morphinandienone alkaloids. Thus *N*-methyllaurotetanine (*339*), cassythicine (*339, 340*), pukateine (*339*), liriodenine (*341, 342*), norpredicentrine (*343*), atheroline (*344*), cryptodorine (*345*), 7-

methylcryptodorine (*345*), roemeroline (*340*), anolobine (*346*), isoboldine (*347*), norglaucine (*347*), oliveroline (*342*), ushinsunine (*342*), norushinsunine (*342*), noroliveroline (*342*), and steporphine (*348*) were synthesized from the corresponding phenolic or nonphenolic halogenated isoquinoline derivatives by the application of nonoxidative photocyclization.

Furthermore, the irradiation of the phenolic bromoisoquinoline **107** under basic conditions afforded the pentacyclic compound **109** via the quinone intermediate **108** in one step. Treatment of **109** with methyl iodide gave the quaternary aporphine thalphenine (**61**) (*349*).

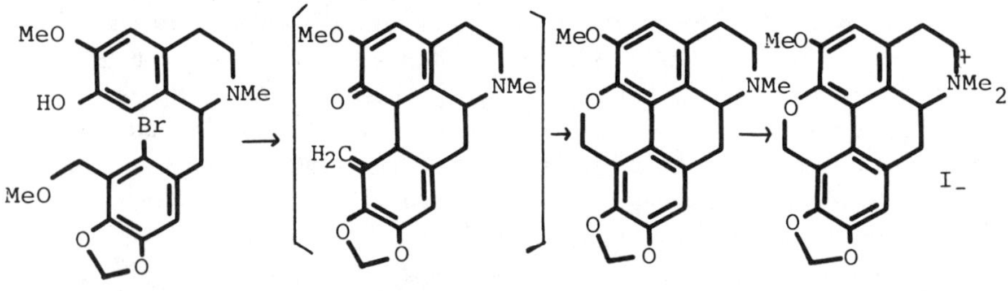

MeO HO NMe Br MeO O O → MeO O NMe H_2C O O → MeO O NMe O O → MeO O $\overset{+}{N}Me_2$ I^- O O

107 **108** **109** **61**

As mentioned above 7-hydroxyaporphines oliveroline (**110**) and ushinsunine (**111**) were synthesized from the corresponding bromoisoquinolines (**112** and **113**) by application of the photocyclization in acid solution (*342*). Similarly,

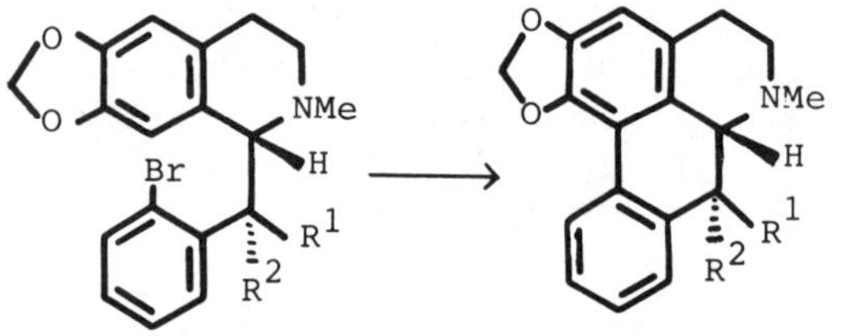

O O NMe H Br R^1 R^2 → O O NMe H R^1 R^2

112 R^1=OH, R^2=H
113 R^1=H, R^2=OH

110 R^1=OH, R^2=H
111 R^1=H R^2=OH

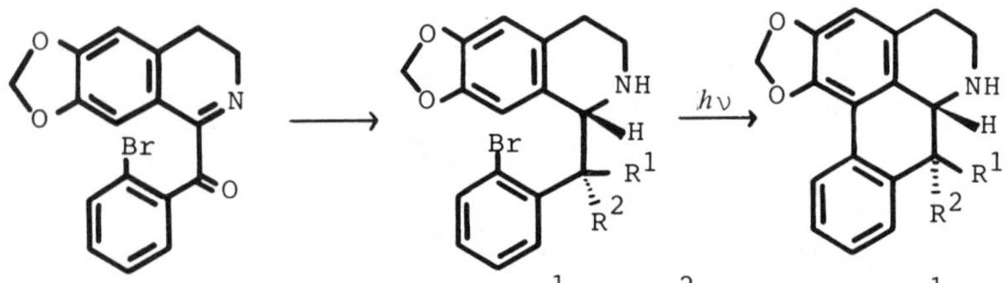

O O N Br O → O O NH H Br R^1 R^2 $\xrightarrow{h\nu}$ O O NH H R^1 R^2

116

117 R^1=OH, R^2=H
118 R^1=H, R^2=OH

114 R^1=OH, R^2=H
115 R^1=H,R^2=OH

norushinsunine (**114**) and noroliveroline (**115**) were prepared from the corresponding bromoisoquinolines (**117** and **118**), which were obtained by the reduction of the ketimine **116** (*342*).

The 4-hydroxyaporphine steporphine was synthesized in a similar manner (*348*). Interestingly, the photolysis of the bromoalcohol **119** led to the synthesis of the oxoaporphine liriodenine (**120**) (*342*).

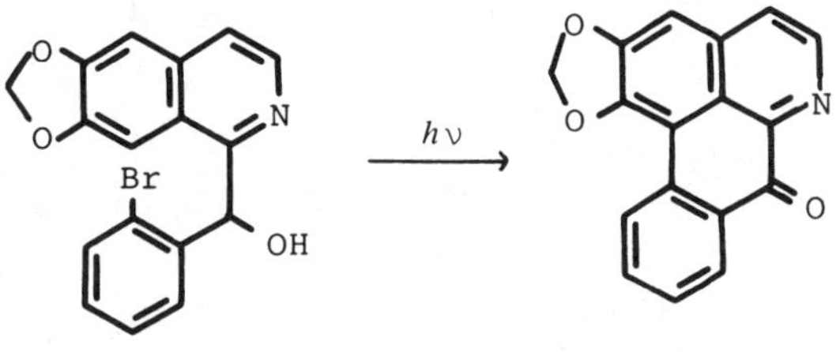

O O N Br OH $h\nu$ → O O N O

119 **120**

The biomimetic synthesis of boldine (**48**) via the proerythrinadienone, which had been postulated to be a precursor involved in the biosynthesis of aporphine alkaloids by Battersby, was achieved by Kupchan, employing a photolysis of the bromophenol **121.**

Thus the irradiation of **121** in ethanol in the presence of sodium hydroxide afforded the proerythrinadienone **122** in 34% yield, together with *N*-ethoxycarbonylnorboldine (**123**) in 5% yield (*350*). The former compound was further converted to the latter by irradiation in ethanol in the presence of sodium acetate in 44% yield. Lithium aluminum hydride reduction of **123** afforded boldine (**48**). Synthesis of the proerythrinadienone by photolysis of the corresponding bromophenolic isoquinoline had been reported by Kametani (*351*), and similar rearrangement of the proerythrinadienone to the aporphine was achieved by treatment with methyl fluorosulfonate (*352*).

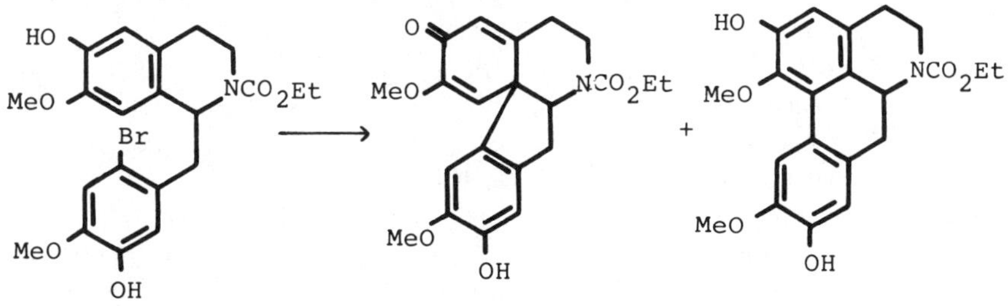

HO MeO NCO_2Et Br MeO OH → O MeO NCO_2Et MeO OH + HO MeO NCO_2Et MeO OH

121 **122** **123**

D. Phenolic and Nonphenolic Oxidation

Phenolic oxidative coupling has been shown to be involved in the biosynthesis of various types of isoquinoline alkaloids according to the hypothesis by Robinson (*353*) and Schöpf (*354*) and by the extensive tracer work by Barton and

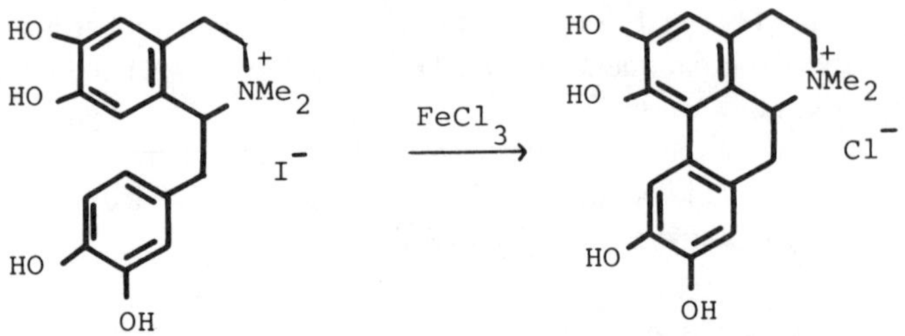

Battersby (*355–365*). Application of phenolic oxidative coupling for the synthesis of aporphine alkaloids as well as other isoquinoline alkaloids, as a biogenetic synthesis utilizing inorganic oxidants or some enzymatic system, has been widely developed. The first synthesis of the aporphine derivative **125** from laudanosoline methiodide (**124**) by the application of phenol oxidation on treatment with iron(III) chloride at 20°C for 20 hr in 60% yield has been demonstrated by Frank in 1962 (*366*). Similar syntheses of laurifoline (**126**) (*367*) and glaucine methopicrate (**127**) (*368*) have been achieved from the corresponding 1-benzylisoquinolines.

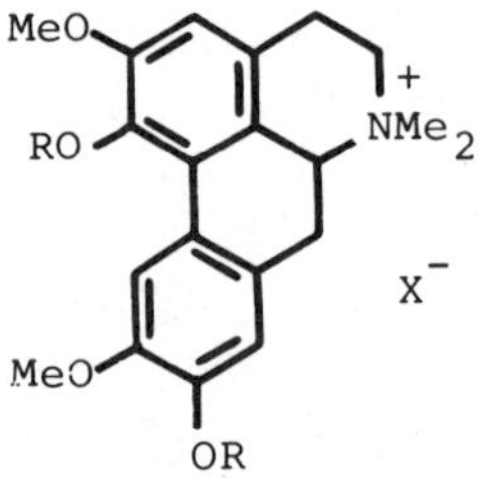

The one-electron-withdrawing oxidizing reagents used for the synthesis of aporphines have been thoroughly investigated, and iron(III) chloride, potassium fericyanide, manganese dioxide, and vanadium oxychloride are known to be effective reagents for this purpose, even though the yields are usually not satisfactory. The oxidation of (±)-reticuline (**46**) with various oxidants was found to produce (±)-isoboldine (**47**) together with (±)-pallidine by Kametani, and the results obtained are summarized in Table III (*369*).

When vanadium oxytrichloride was applied for the above conversion, isoboldine (**47**) was obtained in 55% yield (*370*). Furthermore, the reaction of **128** with thallium trifluoroacetate in methylene chloride at −78 to −20°C yielded the aporphine **129** in 7% yield, together with the morphinandienone **130** (*370*).

Recently, a number of new oxidants involving one-electron- or two-electron-withdrawing oxidants were further investigated to improve the yield, though two-electron-withdrawing oxidation for the above intramolecular coupling of 1-benzylisoquinolines is not involved in the biosynthetic route of aporphines.

Intramolecular oxidative coupling of monophenolic isoquinolines was also

TABLE III

CONVERSION OF RETICULINE (**46**) TO ISOBOLDINE (**47**), USING VARIOUS OXIDIZING AGENTS

Reagent (mol)	Base	Reaction conditions		Yield (%)
		Temperature (°C)	Time (hr)	
$K_3Fe(CN)_6$ (2.5)	5% $NaHCO_3$–$CHCl_3$	5–10	0.5	0.4
$K_3Fe(CN)_6$ (4.0)	5% $NaHCO_3$–$CHCl_3$	room temperature	1	5
$K_3Fe(CN)_6$ (4.0)	5% $NaHCO_3$–$CHCl_3$	5–10	0.25	2.6
$K_3Fe(CN)_6$ (2.5)	5% $NaHCO_3$–$CHCl_3$	5–10	1	1.1
Ag_2CO_3–Celite (2.5)	5% $NaHCO_3$–$CHCl_3$	room temperature	1.5	3
$VOCl_3$	$CHCl_3$	room temperature	2	trace

$Tl(OCOCF_3)_3$, CH_2Cl_2

128 → **129** + **130**

studied by Kupchan. Thus the monophenolic isoquinoline **131** on treatment with ceric sulfate in acidic solution furnished the quinonoid aporphine **132** in 25% yield, whereas on treatment with $MoOCl_4$ in trifluoroacetic acid–chloroform solution, the aporphine **132** was obtained in the highest yield. The oxidants attempted in the above conversion and the yields are summarized in Table IV (*371*).

TABLE IV

OXIDATION OF MONOPHENOLIC ISOQUINOLINE (**131**) WITH VARIOUS OXIDANTS

Oxidant	Medium	Temperature (°C)	Yield (%)
$Ce(SO_4)_2$	10% aq H_2SO_4	0	25
$Co(OH)_3$	10% aq H_2SO_4	25	15
MnO_2	CF_3CO_2H	0	30
CrO_3	aq H_2SO_4–AcOH	0	25
$Tl(OCOCF_3)_3$	CF_3CO_2H	25	12
Pb_3O_4	CF_3CO_2H	0	22
VOF_3	CF_3CO_2H	0	59
$MoOCl_4$	CF_3CO_2H-$CHCl_3$	25	62

In 1973, Kupchan reported the nonphenolic oxidative coupling of 1-benzylisoquinolines, employing a VOF_3–trifluoroacetic acid system as an oxidant. When laudanosine (**64**) was treated with VOF_3 in trifluoroacetic acid–methylene chlo-

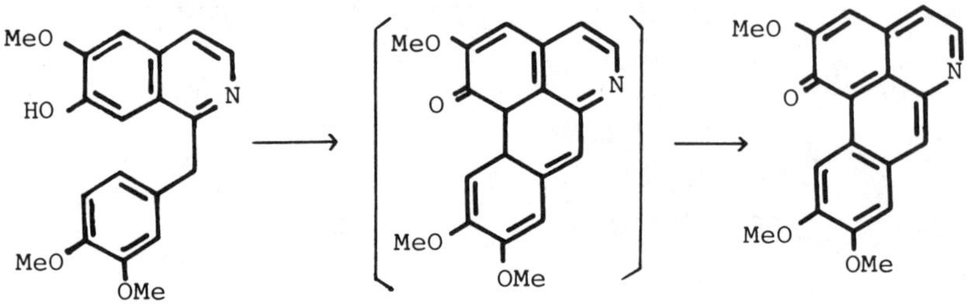

MeO HO N MeO OMe **131** → [MeO O N MeO OMe] → MeO O N MeO OMe **132**

ride–fluorosulfonic acid solution at −30°C, glaucine (**57**) was isolated via the morphinandienone **65** and subsequently the aryl-migrated proerythrinadienone **66** in 43% yield as described before (*8, 30*).

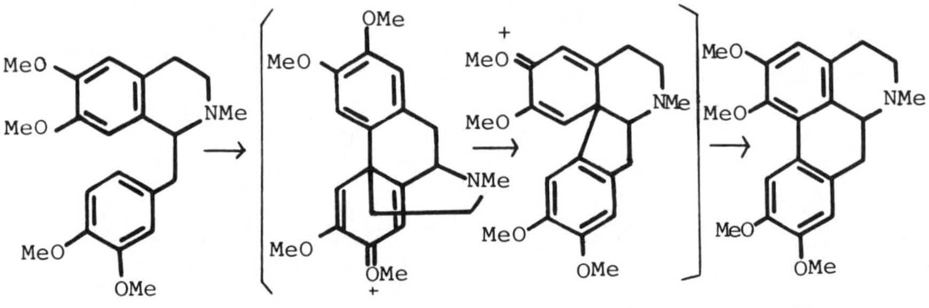

MeO MeO NMe MeO OMe **64** → [MeO MeO NMe OMe **65** → MeO MeO NMe MeO OMe **66**] → MeO MeO NMe MeO OMe **57**

Similarly, the treatment of *N*-trifluoracetylnorcodamine (**133**) with the VOF_3–TFA system gave *N*-trifluoroacetylnorthaliporphine (**134**) in 70% yield.

Moreover, the oxidation of the codamine–borane complex (**135** and **136**) with the above oxidant afforded thaliporphine (**137**) and the bracteoline derivative **105**, respectively, in 70–80% yields (*371*).

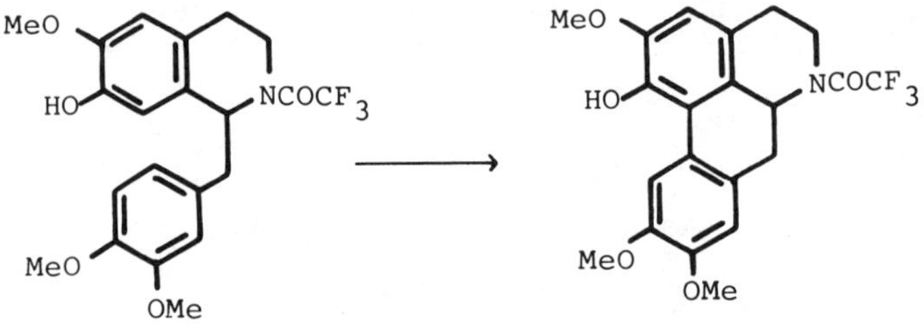

MeO HO $NCOCF_3$ MeO OMe **133** → MeO HO $NCOCF_3$ MeO OMe **134**

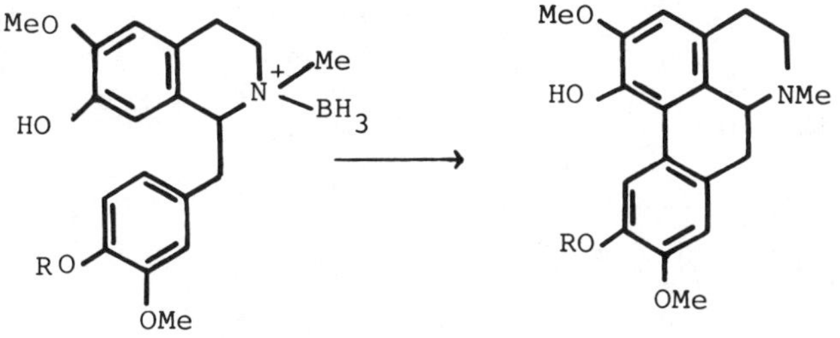
MeO
HO
Me
N
BH3
RO
OMe
MeO
HO
NMe
RO
OMe

135 R=Me
136 R=CH_2Ph

137 R=Me
105 R=H

Diphenyl selenoxide has also been used for the oxidation of the phenolic isoquinoline **138** to the aporphine **139** in high yield, whereas the oxidation of **138** with chloranil provided **139** in less than 10% yield (*372*).

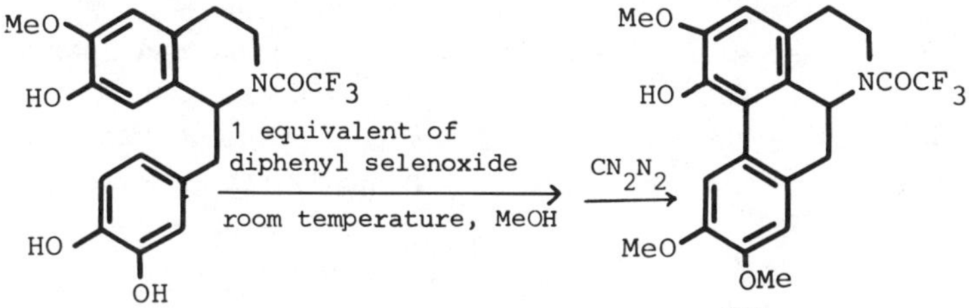
MeO
HO
NCOCF3
1 equivalent of
diphenyl selenoxide
room temperature, MeOH
HO
OH
CN2N2
MeO
HO
NCOCF3
MeO
OMe

138 **139**

Other effective oxidants are the thallium(III) species. Thus the tetrahydroisoquinoline **140** was treated with thallium(III) trifluoroacetate in methanol to give rise to ocoteine (**141**) in 46% yield. Whereas this reagent was used for the oxidation of the N-acylated isoquinoline **142**, the aporphine **143** and its dehydro derivative (**144**) were formed in 40 and 31% yields, respectively (*373*).

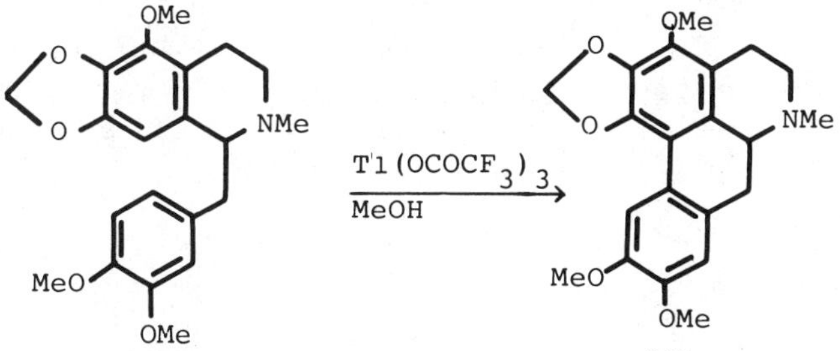
OMe
O
O
NMe
Tl (OCOCF3)3
MeOH
MeO
OMe
OMe
O
O
NMe
MeO
OMe

140 **141**

An alternative oxidant, tetraethylammonium diacyloxyiodate brings about phenolic oxidative coupling in good yield. For example, the reaction of norreticuline derivatives **145** and **146** with this reagent afforded N-acylated isoboldine derivatives **147** and **148** in good yield (*374*).

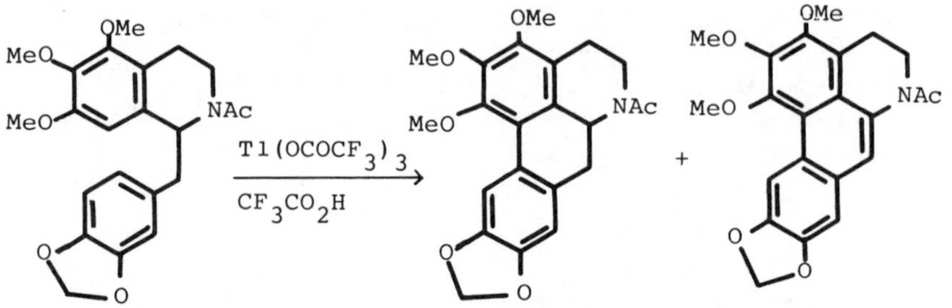

$Tl(OCOCF_3)_3$, CF_3CO_2H

142 **143** **144**

The transformation of tetrahydroisoquinolines to the aporphines, employing enzymatic oxidation is of considerable importance from the pharmacological and biosynthetic points of view. It has been known that tyrosinases, laccases, and peroxidases mainly conducted enzymatic phenol oxidation in plants. Tyrosinases contain copper ion, which remains in the monovalent state throughout the oxidation reaction, and the actual oxidizing species is activated molecular oxygen. On the other hand, the oxidation state changes during the reaction of laccases, which oxidize the phenol to the phenoxy radical by divalent copper. Peroxidases involve iron porphyrin and require hydrogen peroxide.

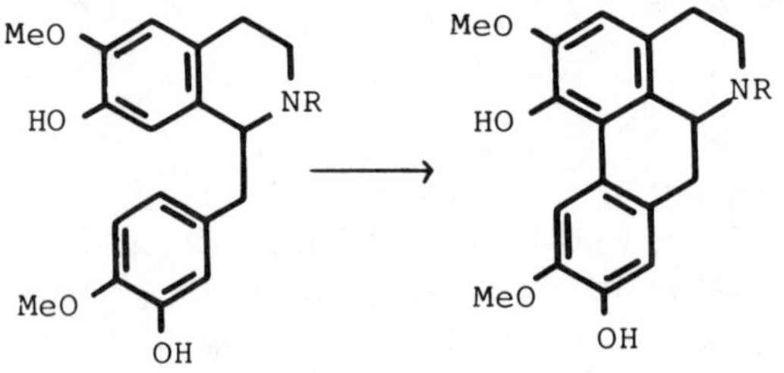

145 $R{=}CO_2Et$
146 R=CHO

147 $R{=}CO_2Et$
148 R=CHO

In 1967, Fromming announced the conversion of laudanosoline methobromide (**149**) to 1,2,9,10-tetrahydroxyaporphine methobromide (**150**) by oxidation with tyrosinase, laccase, and peroxidase in poor yield (*375*).

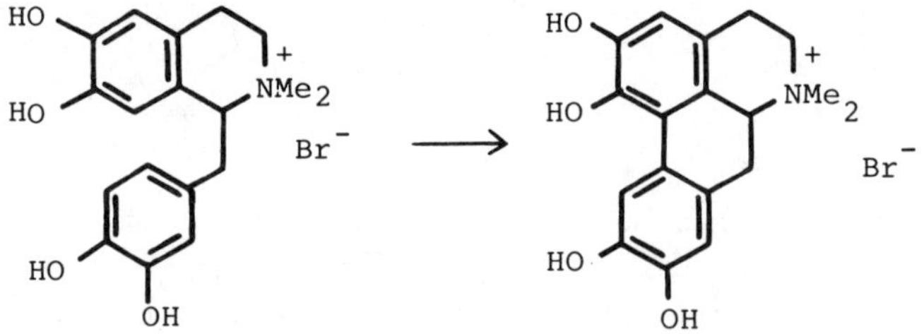

149 **150**

In contrast, the oxidation of (1*S*)-(+)-laudanosoline hydrobromide (**151**) and (1*R*)-(−)-laudanosoline methiodide (**153**), using purified enzyme, namely, horseradish peroxidase under controlled reaction conditions, afforded the quaternary dibenzopyrrocoline **152** in 81% yield and the quaternary aporphine **154** in 50% yield, respectively, with retention of configuration (*376*).

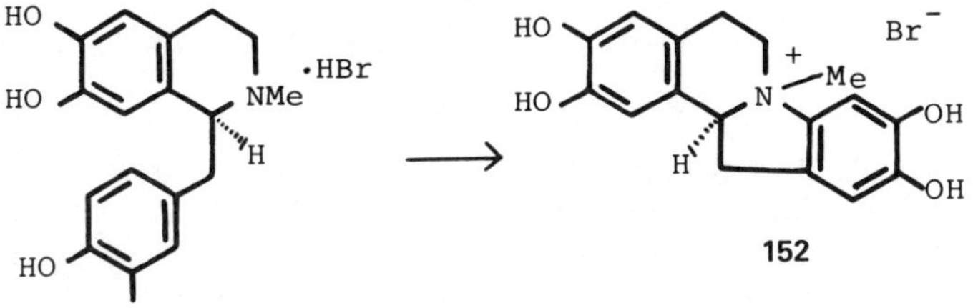

151 → **152**

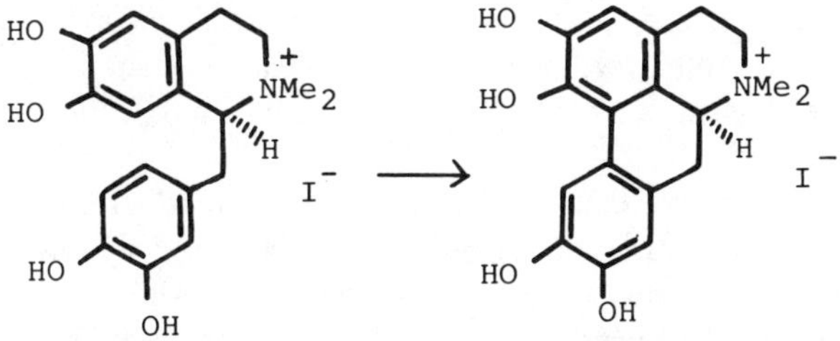

153 → **154**

Biotransformation of reticuline (**46**) to isoboldine (**47**) with mammalian enzymes was also reported (*377*), and the use of isolated enzymes in the formation of natural products including aporphines was reviewed (*378*).

Since tyrosinases and laccases are copper-containing enzymes, as already discussed, application of such an enzymatic model (copper–amine–oxygen system) *in vitro* provided an alternative biogenetic synthesis of aporphines.

Thus the treatment of the dark-green solution of cuprous chloride in pyridine under a current of oxygen with (+)-reticuline (**46**) perchlorate afforded (+)-corytuberine (**17**) and (+)-isoboldine (**47**) in 28 and 8% yields, respectively, together with the morphinandienone pallidine (**155**) (*379*).

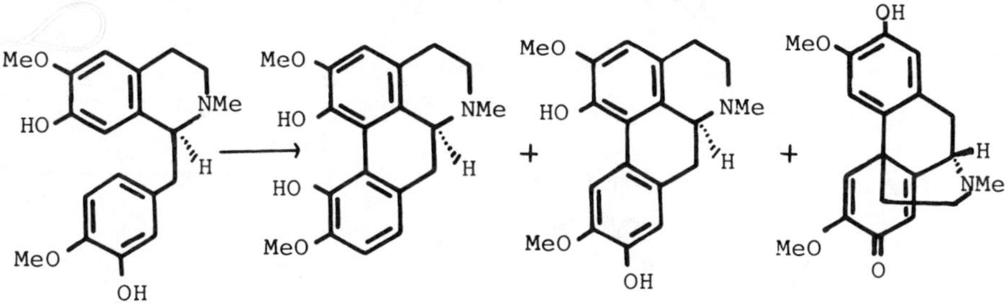

46 → **17** + **47** + **155**

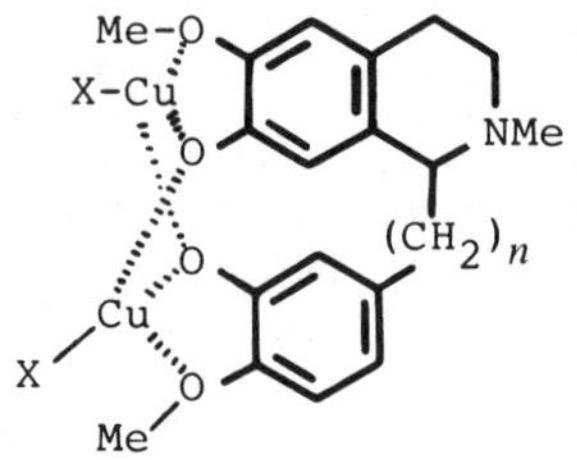

156

A mixture of cupric chloride, potassium superoxide, and pyridine also formed the dark-green solution even in the absence of oxygen, and this solution also converted (+)-reticuline to the above products in almost the same ratio as above. Moreover, the treatment of (+)-reticuline with a divalent copper complex [pryidine·CuCl(OMe)]$_2$ yielded corytuberine as a major product. Formation of corytuberine was rationalized by assuming that two associated copper ions (**156**) hold the two hydroxyl groups leading to the predominant formation of the ortho–ortho coupling products, which could not be formed by the oxidation with usual chemical reagents (*379*).

On the basis of the above considerations, Kametani reached the conclusion that the reaction mechanism of the enzymatic model using cuprous chloride–oxygen–pyridine involved divalent copper as the actual oxidizing species and that the oxidation could be regarded as a simulation of the laccases as shown in Fig. 1.

Furthermore, the redox reaction of isoquinoline *N*-oxides was investigated by Kametani. The reaction of reticuline *N*-oxide (**157**), easily prepared from reticuline by oxidation with *m*-chloroperbenzoic acid, with cuprous chloride in methanol under nitrogen yielded corytuberine (**17**) in 61% yield after treatment of the reaction mixture with sodium hydrosulfite. On the other hand, treatment of orientaline *N*-oxide (**158**) with cuprous chloride in methanol gave orientalinone (**53**) as a diastereoisomeric mixture. In this reaction, the *N*-oxides oxidized copper(I) chloride to give an active copper(II) species, which brings about the ortho–ortho phenolic oxidative coupling via the copper ions (**156**).

Thus this reaction could be regarded as an intramolecular redox cyclization (*380*).

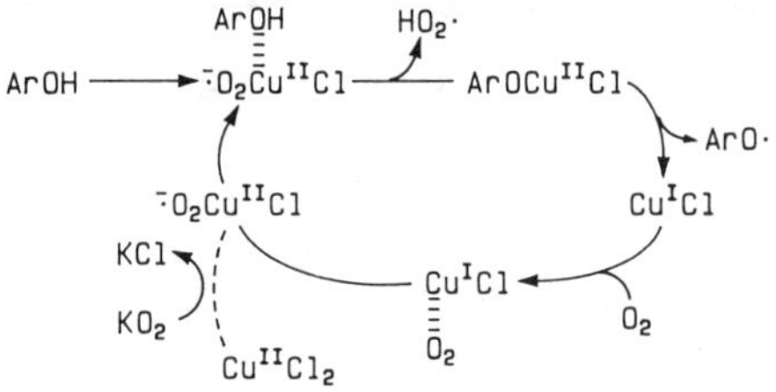

FIG. 1

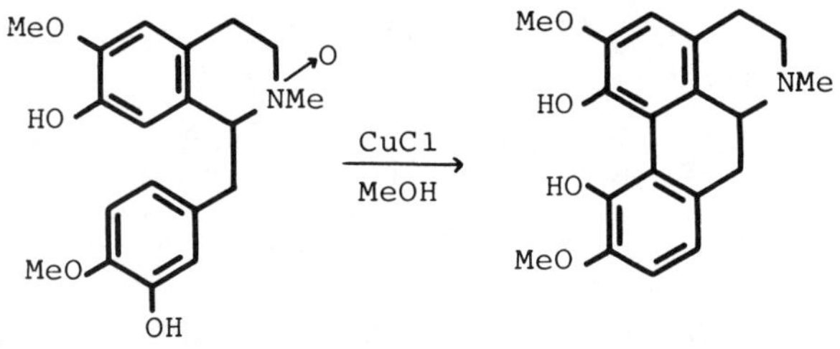

157 → 17 (CuCl, MeOH)

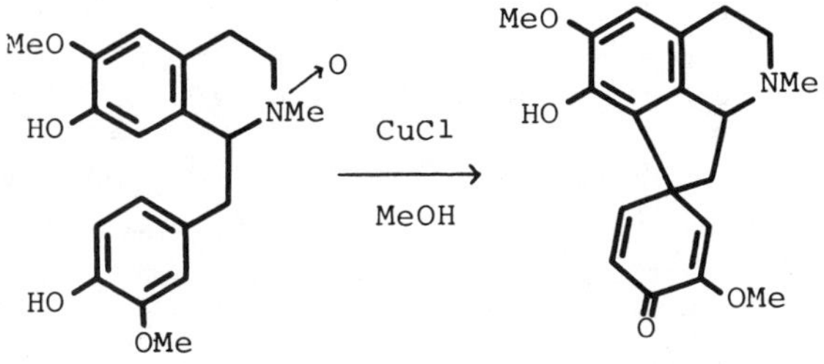

158 → 53 (CuCl, MeOH)

Electrooxidation provides an alternative superior synthetic route to aporphines. For example, electrolysis of laudanosine (**64**) in TFA afforded glaucine (**57**) in 17% yield (*381*). Alternatively, cathodic electrooxidation of the quaternary isoquinolinium salts **159** and **160** in acetonitrile in the presence of tetra-*N*-

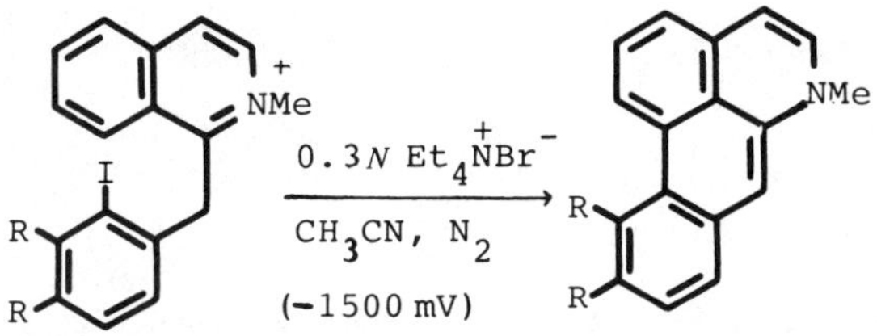

$0.3N\ Et_4\overset{+}{N}Br^-$, CH_3CN, N_2 (−1500 mV)

159 R=H
160 R=OMe

161 R=H
162 R=OMe

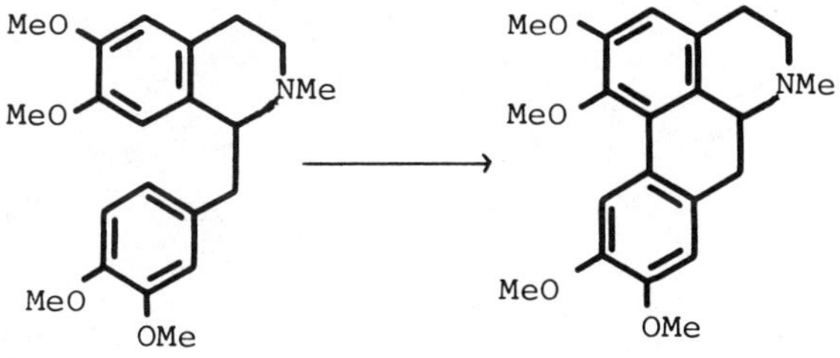

64 → **57**

ethylammonium bromide furnished dehydroaporphines **161** and **162** in 86 and 74% yields, respectively (*382*).

E. Benzyne-Mediated Syntheses

An alternative effective synthetic route to aporphine alkaloids involves a benzyne intermediate as a reactive species. Thus the phenolic bromoisoquinoline **104** was treated with a strong base (sodium or potassium amide) to afford domesticine (**106**) together with the morphinandienone amurine (**164**) via **163** (*383, 384*).

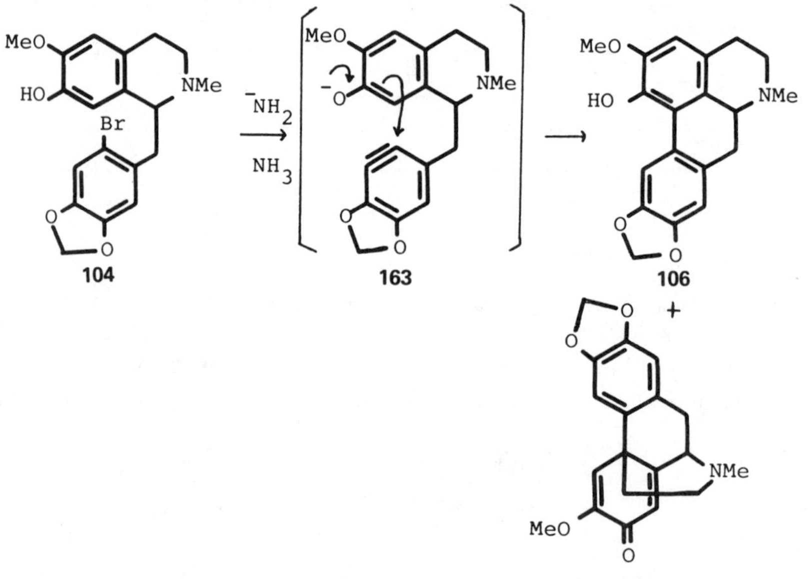

MeO HO NMe Br O O **104** $\xrightarrow[NH_3]{^-NH_2}$ MeO ^-O NMe O O **163** → MeO HO NMe O O **106** + O O NMe MeO O **164**

The existence of a benzyne as an intermediate has been proved by the cine-substitution reaction, using the bromoisoquinoline **165** as a starting material, which leads to the synthesis of thaliporphine (**137**), together with the dibenzopyrrocoline salt cryptaustoline iodide (**166**) (*385, 386*).

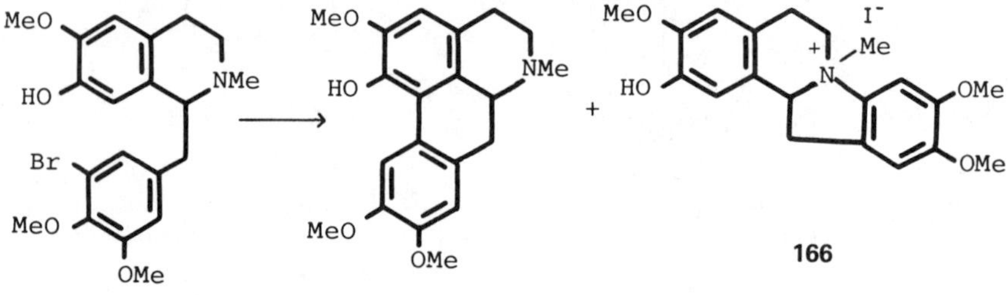

MeO HO NMe Br MeO OMe **165** → MeO HO NMe MeO OMe **137** + MeO HO I^- + N Me OMe OMe **166**

Alternative syntheses of thaliporphine, domesticine, thalicmidine, and *N*-methylcaaverine via benzyne intermediates were reported (*383, 387*).

This reaction was also applicable to the synthesis of dehydroaporphines. Treatment of the 3,4-dihydroisoquinolinium salt **167** with sodium methylsulfinylmethanide in dimethyl sulfoxide gave dehydrothaliporphine (**168**), whose reduction with sodium borohydride afforded thaliporphine (**137**) (*388*). Similarly, the phenolic bromoisoquinolinium salt **169** was converted to domesticine (**104**) by treatment with dimsyl sodium in dimethyl sulfoxide and subsequent reduction (*389*). Again, an intermediacy of a benzyne was shown by the conversion of the isoquinolinium salt **170** to the corresponding aporphine **171** (*389*).

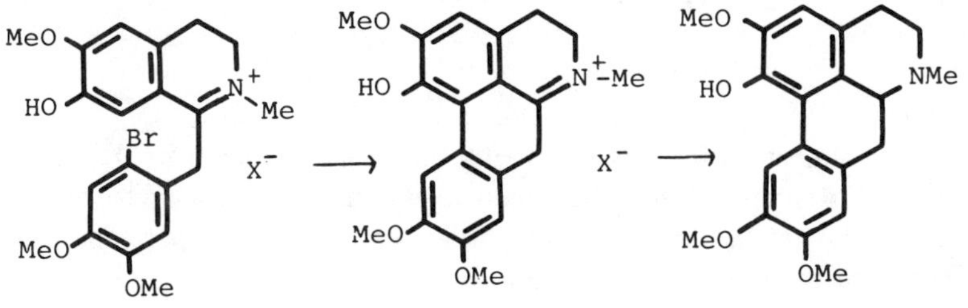

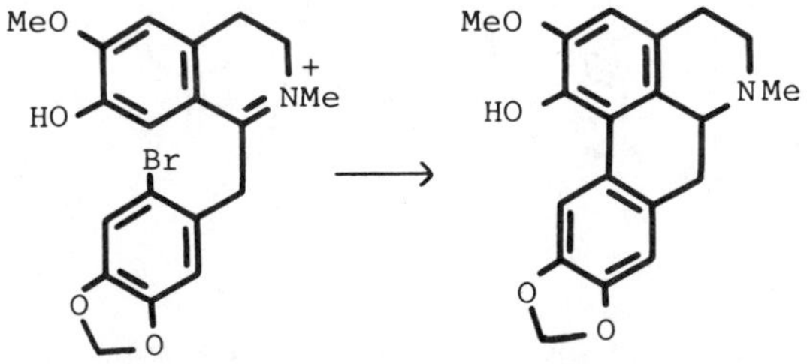

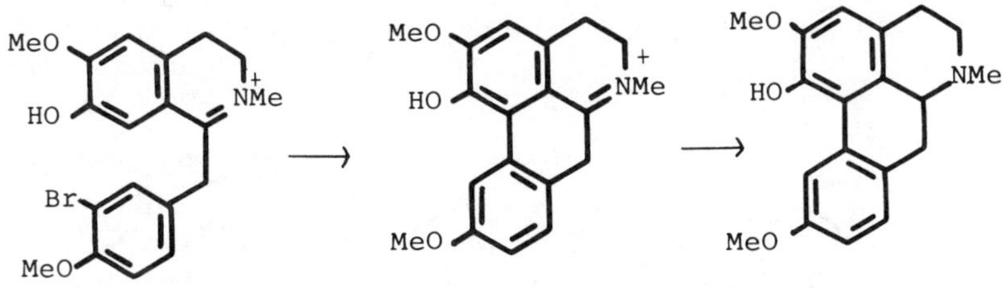

The reaction of an isoquinoline derivative, which does not bear the necessary phenolic function in ring A, with a strong base did not bring about the intramolecular cyclization. For instance, the treatment of 2′-bromolaudanosine (**172**) with the amide anion gave 2′-aminolaudanosine (**173**) (*390*).

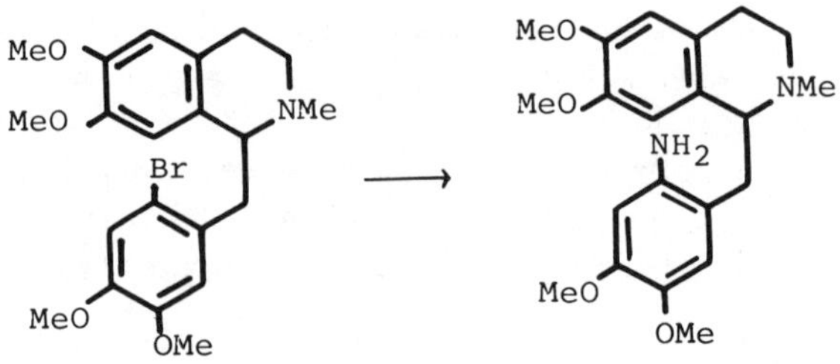

172 **173**

Thus treatment with potassium amide in liquid ammonia of the aminoisoquinoline **176,** prepared from the bromoisoquinoline (**174**) via the benzyne intermediate **175,** was further transformed to the corresponding aporphine laureline (**177**) by the Pschorr reaction (*391*).

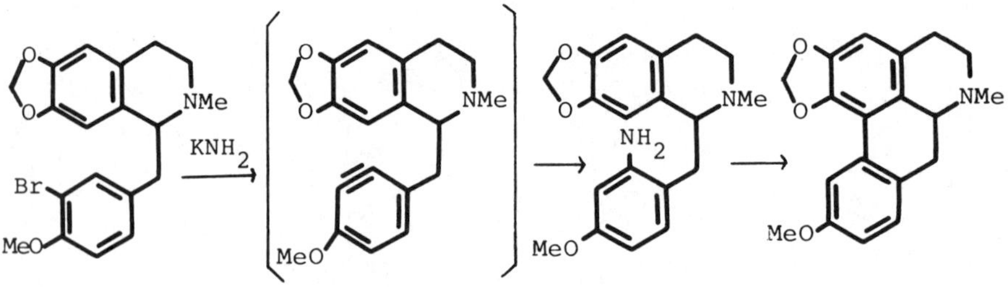

174 **175** **176** **177**

Thus thaliporphine (*387*), domesticine (*387*), and *N*-methylcaaverine (*387*) were synthesized via benzyne intermediates.

The intermolecular Diels–Alder-type cycloaddition between benzyne and appropriate dienes provided an alternative synthesis of aporphines. For example, *N*-acetyldehydronornuciferine (**179**) was synthesized by employing the cycloaddition of 2-acetyl-6,7-dimethoxy-1-methyleneisoquinoline (**178**) with benzyne (*392*).

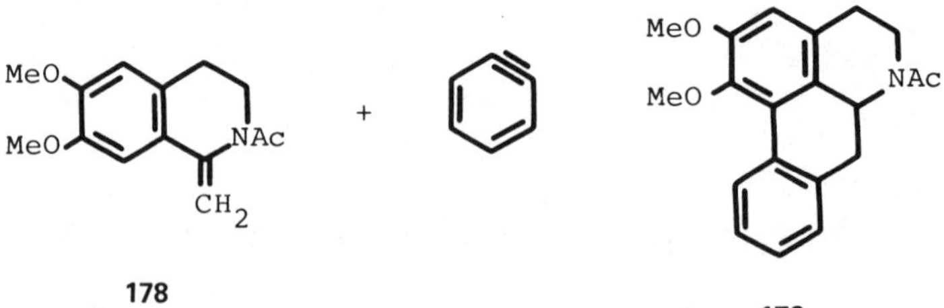

178 **179**

F. Other Reactions

1. Synthesis of Aporphines via Quinol Acetate (*393, 394*)

A transformation of phenols on treatment with lead tetraacetate (LTA) to quinol acetates has been well known. Application of this reaction to the synthesis of aporphine alkaloids was first reported by Umezawa and Hoshino in 1971. Thus the 7-hydroxyisoquinoline (±)-codamine (**180**) was treated with LTA to give the *p*-quinol acetate **181** whose treatment with acetic anhydride–sulfuric acid gave rise to (±)-acetylthaliporphine (**182**) and (±)-4-acetoxy-*O*-acetylthaliporphine (**183**) in 14 and 6% yields, respectively (*394*).

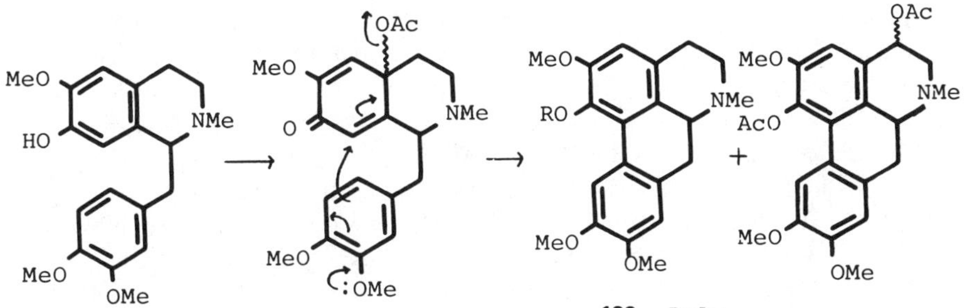

MeO HO NMe MeO OMe **180** → MeO O OAc NMe MeO :OMe **181** → MeO RO NMe MeO OMe **182** R=Ac **137** R=H + OAc MeO AcO NMe MeO OMe **183**

Hydrolysis of compound **182** afforded (±)-thaliporphine (**137**), which is also obtained from **183** by reduction with lithium aluminum hydride (*394*). Similar treatment of **184** with LTA, followed by treatment with acetic anhydride–sulfuric acid, afforded *O*-acetyldomesticine (**185**) and 4-acetoxy-*O*-acetyldomestine (**186**) as a mixture of 4β and 4α derivatives in 18 and 10.2% yields (4β:4α = 4.7:5.5), respectively. Compound **185** was again hydrolyzed to yield (±)-domesticine (**106**), which was further converted to (±)-nantenine (**187**) by treat-

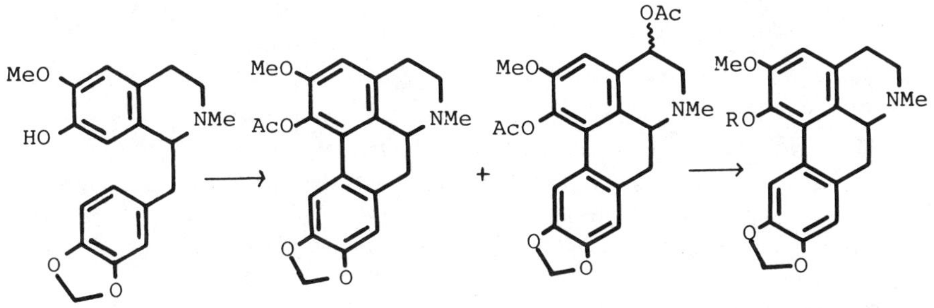

MeO HO NMe O O **184** → MeO AcO NMe O O **185** + OAc MeO AcO NMe O O **186** → MeO RO NMe O O **106** R=H **187** R=Me ment with diazomethane (*395*). Investigation of various types of Lewis and Brønsted acids for the conversion of *p*-quinol acetates to aporphine showed that trifluoroacetic acid (TFA) was the most effective reagent, and that this conversion did not require the hydrolysis of the phenolic acetates. The yields of the above conversion of the *p*-quinol acetates to the corresponding aporphines, using TFA, are summarized in Table V (*396, 397*).

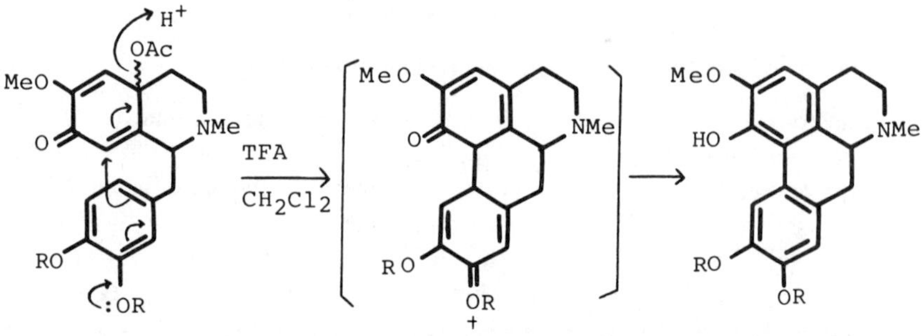

Interestingly, treatment of the isolated but unpurified *p*-quinol acetates with TFA instead of acetic anhydride–sulfuric acid, did not give the corresponding 4-hydroxyaporphine. However, the reaction of the 1-benzylisoquinoline (**180**) with LTA in methylene chloride in the presence of TFA brought about the formation of 4β-hydroxyaporphine, and this one-step conversion leads to the synthesis of 4β-hydroxythaliporphine (**188**), (*398*), which was also synthesized by the acetoxylation of thaliporphine with LTA in acetic acid.

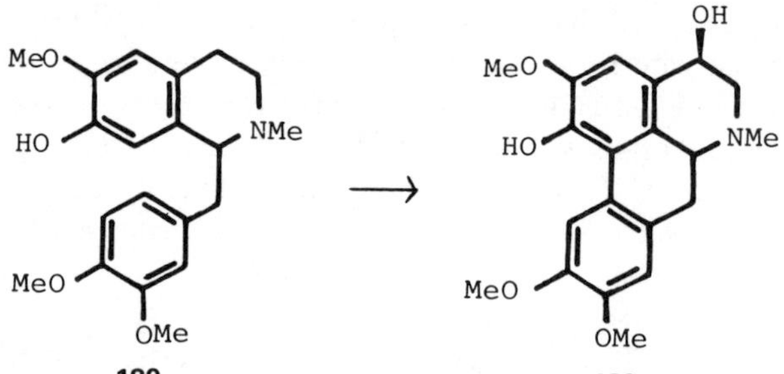

180 **188**

Thaliporphine (**137**) was further converted to a 4β-hydroxyaporphine cataline (**190**) by hydrolysis of the acetate and methylation with diazomethane (*399*).

Again, treatment of the *p*-quinol acetate (**191**), derived from 1-(3-methoxybenzyl)-1,2,3,4-tetrahydroisoquinoline, with TFA in methylene chloride furnished (±)-isothebaine (**55**) and the aporphine **192** in 22 and 53% yields, respectively (*400*).

The nonidentity of **192** prepared by this method, with natural lirinine (**193**) confirmed the new structure for the natural product. Bracteoline, isoboldine, *N*-methyllaurotetanine, and related aporphines were synthesized by this method (*397*).

TABLE V
YIELDS OF 1-HYDROXYAPORPHINES FROM *p*-QUINOLS BY TREATMENT WITH TRIFLUOROACETIC ACID IN METHYLENE CHLORIDE

Aporphines	Yields (%)
Thaliporphine	55.8
Domesticine	41.5
1-Hydroxy-2,9,10,11-tetramethoxyaporphine	48
9-Benzyloxy-1-hydroxy-2,10-dimethoxyaporphine	44
10-Benzyloxy-1-hydroxy-2,9-dimethoxyaporphine	48

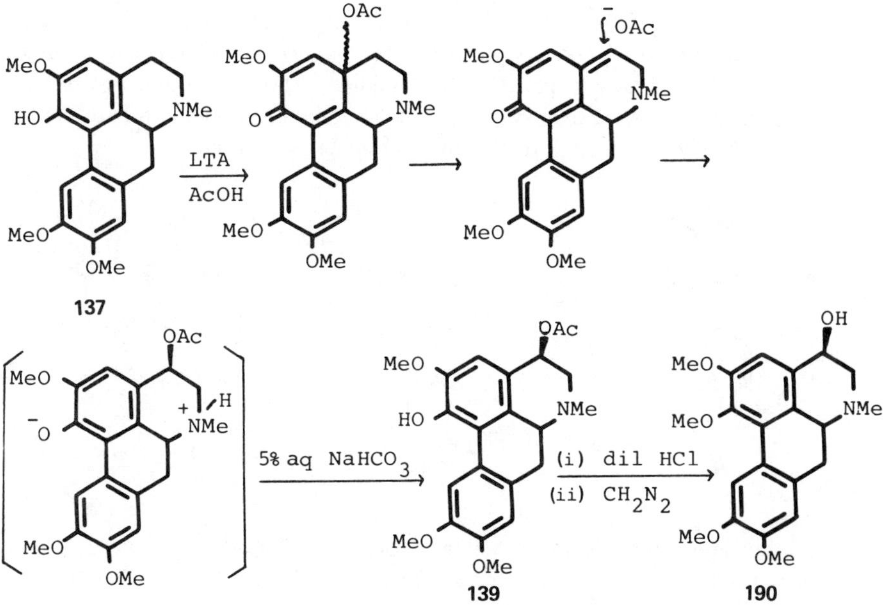

137 139 190

The formation of *p*-quinol acetate required the hydroxyl group at the 7-position of isoquinoline nucleus, whereas the same reaction for 6-hydroxyisoquinolines brought about the formation of *o*-quinol acetates. Thus the reaction of

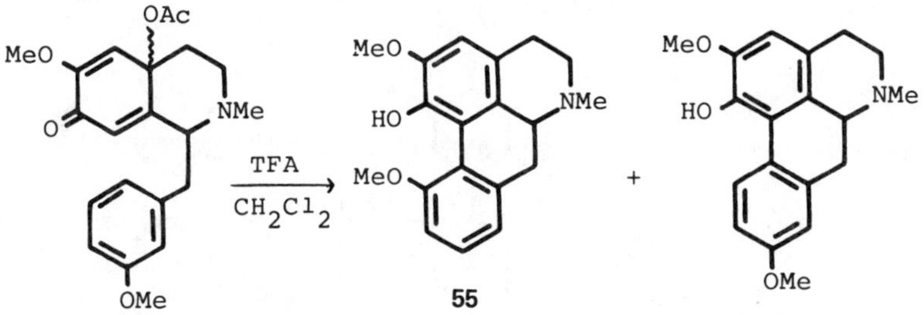

191 55 192

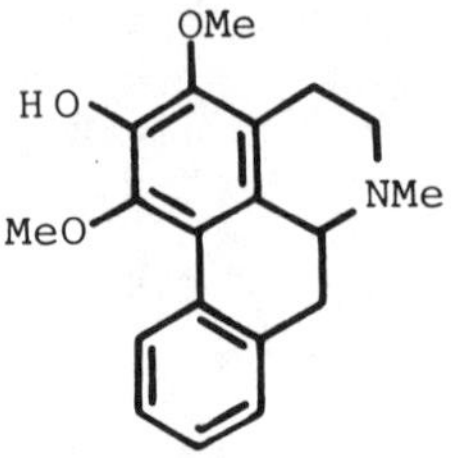

OMe
HO
MeO
NMe

193

194 with LTA in methylene chloride afforded the diastereomeric mixture of *o*-quinol acetates (**195**), whose treatment with acetic anhydride in concentrated sulfuric acid provided *O*-acetylpredicentrine. Hydrolysis of **196** led to predicentrine (**197**) (*401*). Isodomesticine, boldine, and 2,10-dihydroxy-1,9-dimethoxyaporphine were also synthesized from the corresponding 6-hydroxyisoquinoline by the application of this method (*402*).

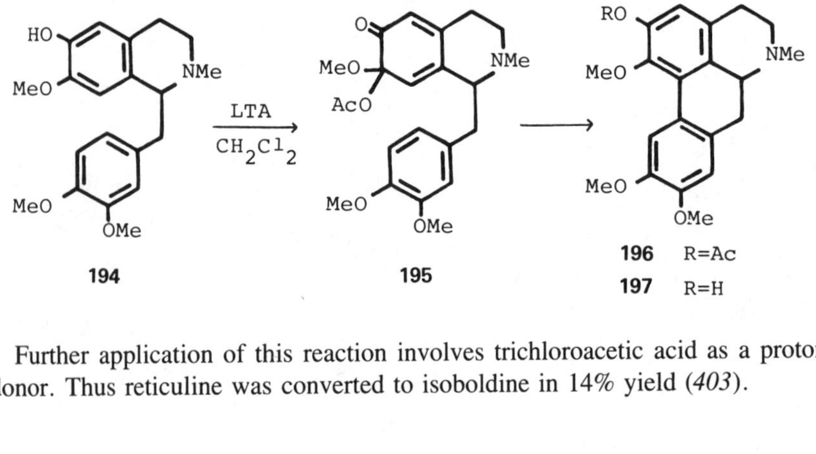

HO
MeO
NMe
MeO
OMe
194

LTA
CH_2Cl_2

O
MeO
AcO
NMe
MeO
OMe
195

RO
MeO
NMe
MeO
OMe
196 R=Ac
197 R=H

Further application of this reaction involves trichloroacetic acid as a proton donor. Thus reticuline was converted to isoboldine in 14% yield (*403*).

2. Alternative Syntheses of Aporphines

An alternative stepwise synthesis of the aporphine **198** was published, and the synthetic scheme is summarized as follows (*404*).

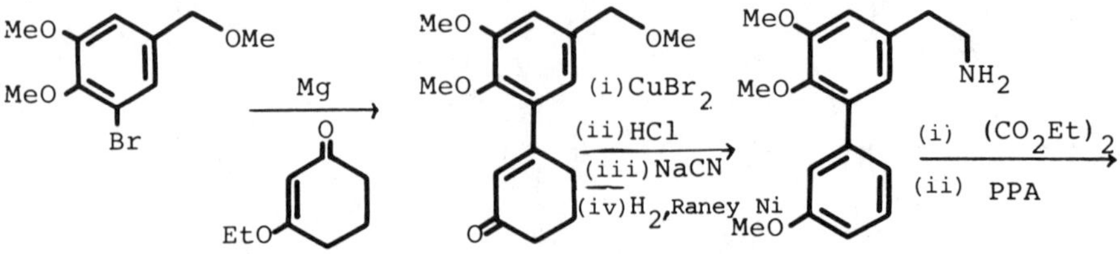

MeO
MeO
Br
OMe
Mg
O
EtO

MeO
MeO
OMe
O

(i) $CuBr_2$
(ii) HCl
(iii) NaCN
(iv) H_2, Raney Ni

MeO
MeO
MeO
NH_2

(i) $(CO_2Et)_2$
(ii) PPA

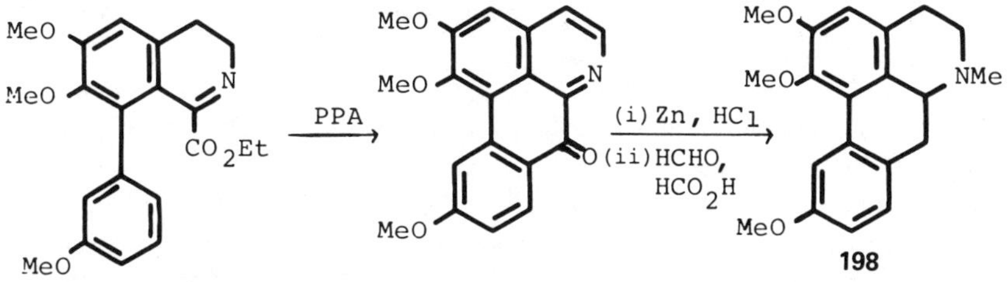

PPA; (i) Zn, HCl (ii) HCHO, HCO_2H

198

The 4,5-dioxoaporphine ponteverine (**199**) was synthesized from the ketocarboxylic acid **200** (*405*).

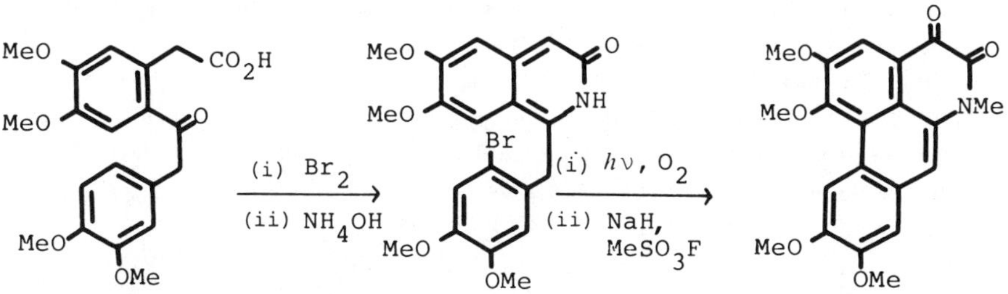

(i) Br_2 (ii) NH_4OH; (i) $h\nu$, O_2 (ii) NaH, $MeSO_3F$

200 **199**

An unusual aporphine formation was reported by Kametani, employing the acid-catalyzed Grewe cyclization of the partially reduced benzylisoquinoline **201** (*406*).

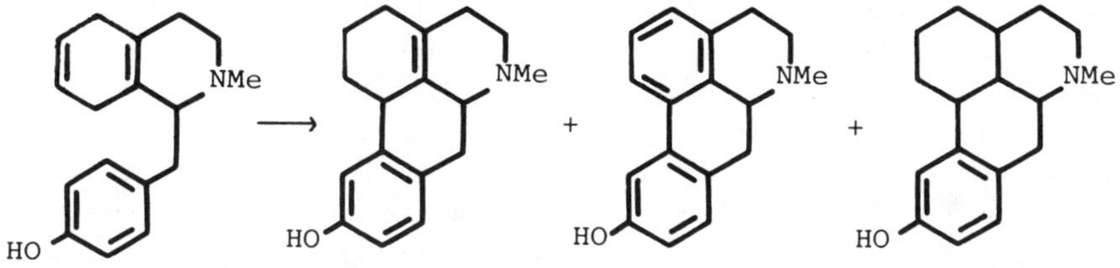

201

Phosphoric acid-catalyzed cyclization of **202** afforded the aporphine **203** (*407*).

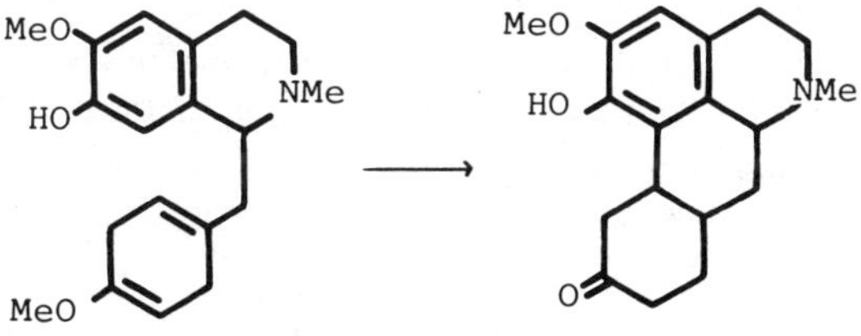

202 **203**

VI. Pharmacology

Interest in evaluating biological activities of aporphines stems from its structural relationship to apomorphine, which has been demonstrated to produce a dopamine-like renal vasodilation in dogs (*408*) and to have a hypertensive effect in cats (*409*). Apomorphine has also been shown to stimulate the dopaminergic system in the rat and mouse corpus striatum (*410, 411*). Based on its dopamine receptor-stimulating properties, apomorphine has been investigated for the treatment of Parkinson's disease and chronic manganism (*412*).

Pharmacology of aporphines and related compounds including apomorphine has already been reviewed (*413, 414*); therefore, the pharmacology of natural aporphines is only briefly mentioned in this review.

Oliveroline has been shown to display antiparkinsonian activity while oliveridine and oliverine cause a dose-dependent hypotension in normal rats followed by a secondary hypertension (*415*). *N*-Methylsparsiflorine methiodide also causes hypotension (*416*), while dehydroglaucine exhibits antimicrobial activity (*23, 24*). The bisisoquinoline alkaloids thalirevolutine and thalirevoline also possess hypotensive activity in normotensive rabbits. The latter shows antibiotic activity against *Mycobacterium smegmatis* at 100 μg/ml (*417*). Thalicarpine, thaliadine, and *O*-demethyladiantifoline have similar hypotensive activity, while thaliadanine, thalicarpine, thalmelatine, *O*-methylthalmelatine, and pennsylvanine were found to be active against *Mycobacterium smegmatis* (*259, 418*). *N*-Demethylthalphenine and bisnorthalphenine are effective against *M. smegmatis, Staphylococcus aureus,* and *Candida albicans* (*318*). Glaucine was said to exhibit antitussive activity (*419*), and its tetrahydroxy derivative shows antithrombic action (*420*). The oxoaporphines were evaluated for their antibacterial and antifungal activity against several microorganisms, and liriodenine showed good activity against *Trichophyton mentagraphytes* and *Syncephalestrum racemosum* compared to griseofulvin and candicidin. Its activity against a number of plant diseases was also tested and shown to be effective against barley net blotch, chocolate spot of broad beans, and rice blast. Liriodenine is relatively nontoxic in acute toxicity tests (*421*). Oxoglaucine is devoid of antifungal and antibacterial activity, while its methiodide shows antimicrobial activity (*421*). The interaction between horse-liver alcohol dehydrogenase (ADH) and aporphine alkaloids has been studied, and it has been demonstrated that the aporphine alkaloids inhibit ADH. The inhibitory effect depends on the position and type of substituents in the aporphine nucleus. The results indicate that the aporphine alkaloids bind to the active center of alcohol dehydrogenase with stoichiometry 1:1 (*422*).

VII. Reactions of Aporphines

A. Dealkylation

O-Dealkylation of aporphines could generally be carried out by employing a Lewis acid. For example, treatment of (+)-bulbocapnine methyl ether (**204**) with an excess of boron trichloride in methylene chloride containing a small amount of ethanol afforded the cleavage product of the methylenedioxy group (**205**) together with the monophenolic compound **206.** After the methylation of the

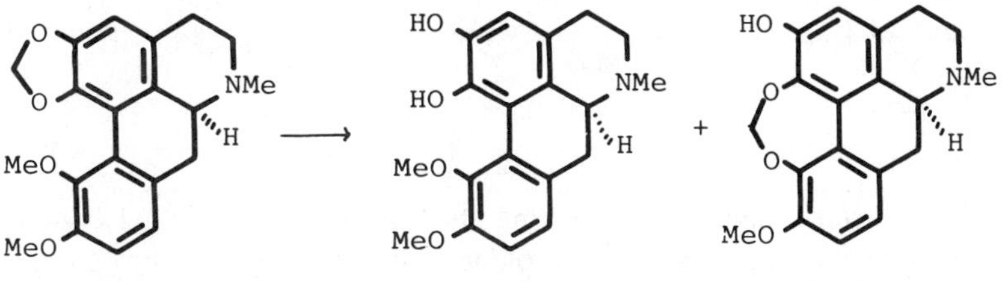

204 **205** **206**

latter compound with diazomethane, the resulting **207** was hydrolyzed with hydrochloric acid to give (+)-corytuberine (**17**).

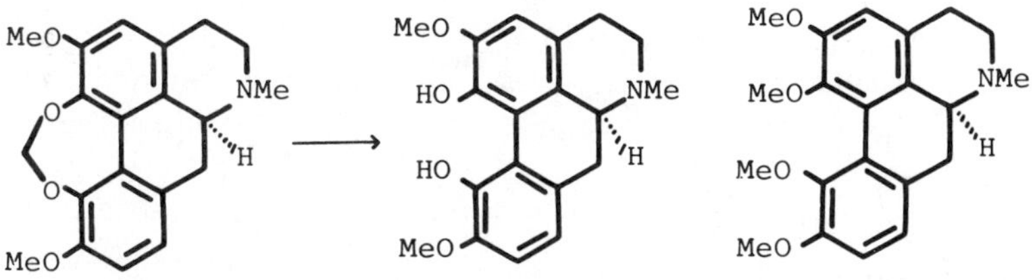

207 **17** **70**

The methylation of **205** with diazomethane yielded (+)-corydine methyl ether (**70**). Reaction of **204** with boron tribromide, however, gave the diphenolic compound **208,** which on similar treatment as above led to (+)-corytuberine (**17**). (*423*).

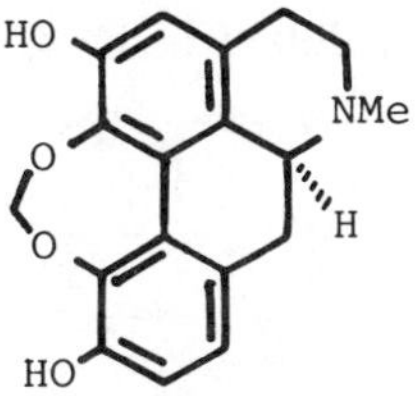

208

Interestingly, boron trichloride demethylates a methoxyl group adjacent to a phenolic group selectively to give rise to the corresponding catechol. Thus treatment of predicentrine (**197**) with boron trichloride afforded the catechol **209**, which was then converted to dicentrine (**58**) by methylenation with methylene chloride in DMSO containing NaOH (*424*).

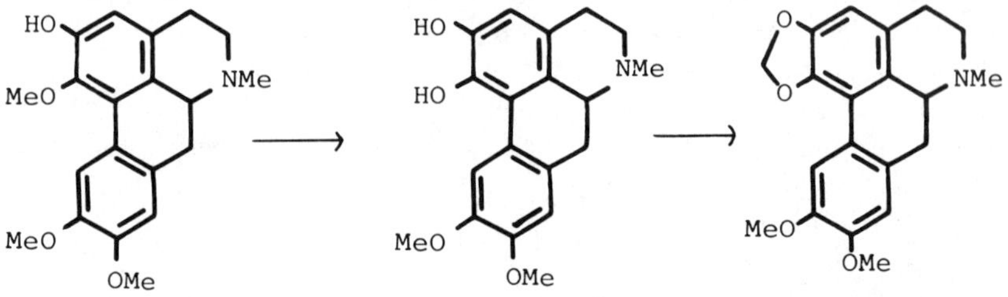

197 **209** **58**

Sodium benzeneselenoate in refluxing DMF resulted in demethylation at the hindered 1, 8, and 11 positions of aporphines, but the methylenedioxy group was not cleaved by this reagent (*425*). Similarly, sodium thioethoxide was employed in the O-dealkylation of 10,11-dimethoxyaporphine (*426*). O-Dealkylation of aporphines with mineral acids were also reported. Thus glaucine (**57**) was transformed to bracteoline (**105**) on treatment with hydrobromic acid in acetic acid (*427*).

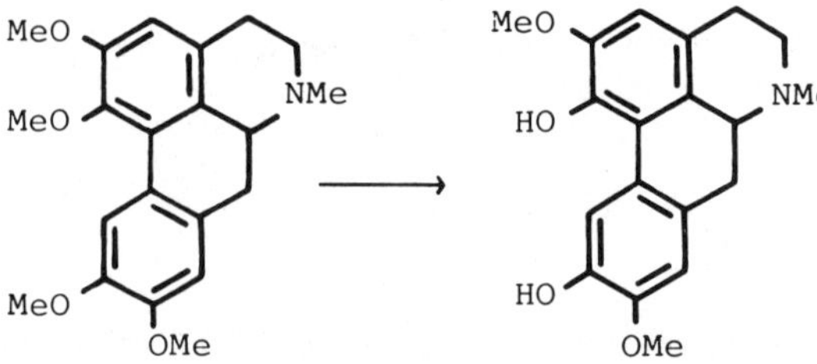

57 **105**

Sulfuric acid can also be utilized for the O-demethylation of dehydroaporphines and 7-hydroxylated aporphines (*428*). Another superior reagent for O-demethylation is trimethylsilyl iodide, which does not cleave a methylenedioxy

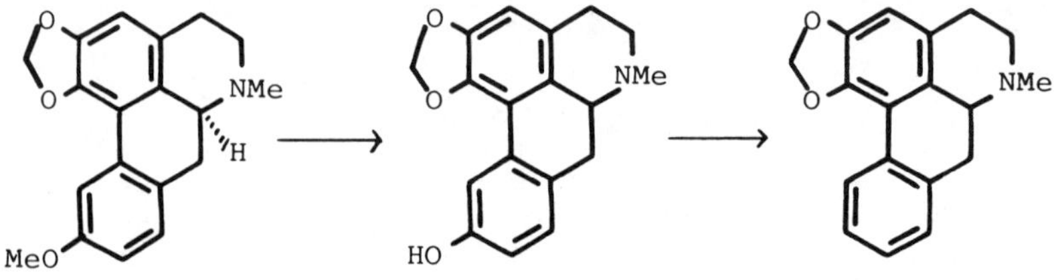

177 **210** **211**

group present in the same molecule. Hence, (+)-laureline (**177**), on treatment with trimethylsilyl iodide in heated dichlorobenzene in the presence of DABCO, was converted to (+)-mecambroline (**210**) which was further transformed to (+)-roemerine (**211**) by the application of Musliner–Gates deoxygenation (*429*).

Demethylation of the 1-benzylisoquinoline (**212**) with concentrated hydrobromic acid at 140°C gave 3,9,10-trihydroxynoraporphine (**214**), probably via the quinone methide intermediate **213** (*430*).

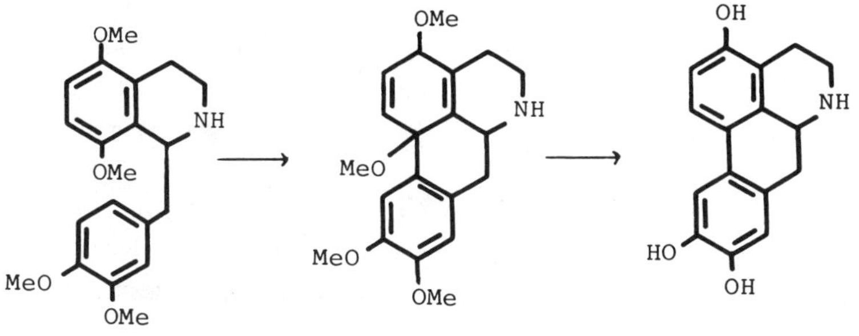

212 **213** **214**

An N-demethylation of aporphine was reported by treatment with methyl chloroformate, followed by cleavage of the urethane with hydrazine (*431*). Microbial N-demethylation and O-demethylation were also reported (*432*). Furthermore, dehydronuciferine (**88**) was N-demethylated by treatment with oxalyl chloride in ether and tetrahydrofuran in the presence of potassium carbonate to give the pentacyclic compound **215**, which was then converted to nornuciferine (**99**) by a few steps (*433*).

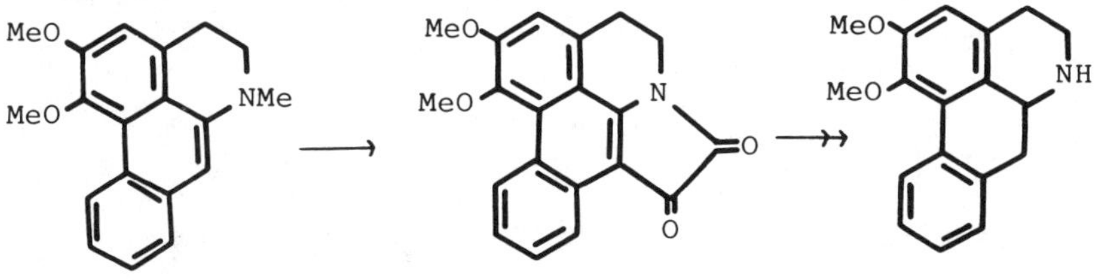

88 **215** **99**

B. Dehydrogenation

Nonphenolic aporphine alkaloids, such as nuciferine (**89**), dicentrine (**58**), ocopodine (**216**), and thalicarpine (**37**) could be converted to the corresponding dehydroaporphines by catalytic dehydrogenation, using 10% Pd–C in refluxing acetonitrile; however, this method is not applicable to noraporphines (*434*). Alternatively, the photooxidation of aporphines led to dehydroaporphines in moderate yield (*435*).

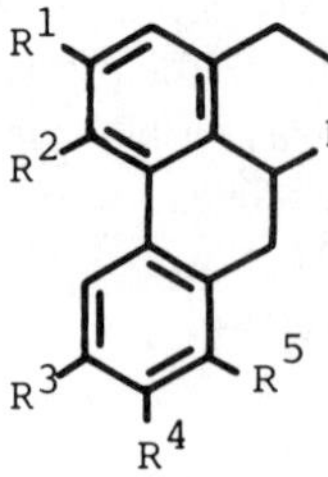

89 $R^1=R^2=OMe$, $R^3=R^4=R^5=H$

58 $R^1+R^2=OCH_2O$, $R^3=R^4=OMe$, $R^5=H$

216 $R^1+R^2=OCH_2O$; $R^3=R^4=R^5=OMe$

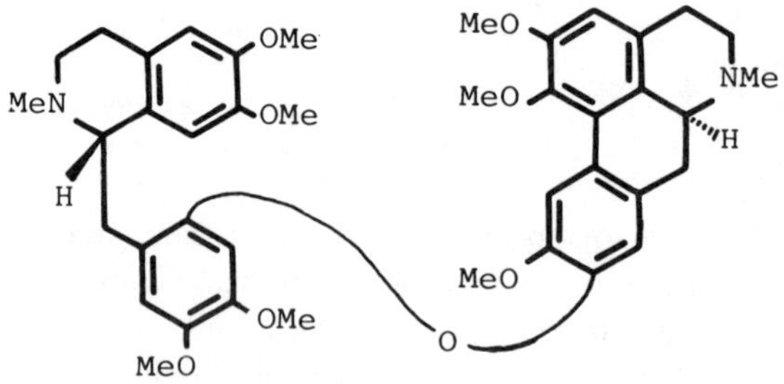

37

It is also worth noting that air oxidation of glaucine (**57**) gave rise to dehydroglaucine (**217**) together with the 7-oxoaporphine **218,** while photooxidation of **57** led to ponteverine (**199**), dihydroponteverine (**219**), glaucine *N*-oxide (**220**), and corunnine (**221**) (*436*).

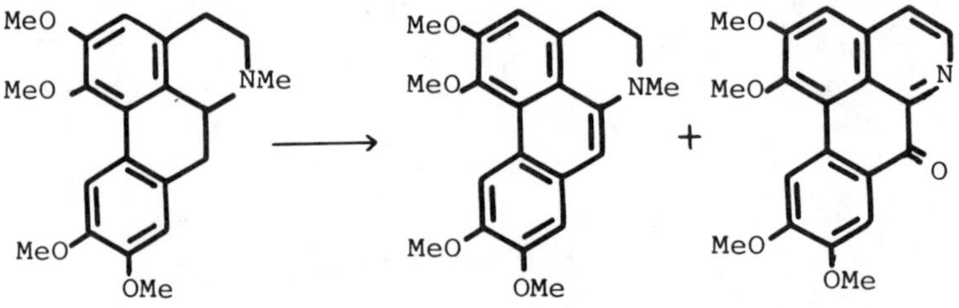

57 → **217** + **218**

Air oxidation of dehydroaporphines in the presence of alkali was also reported to furnish the corresponding oxoaporphines, 4,5-dioxoaporphines, and *N*-methylaristolactams, and this method was applied to the synthesis of 4,5-dioxodehydronantenine (*437*). The 7-oxoaporphine glaunidine (**14**) was derived from

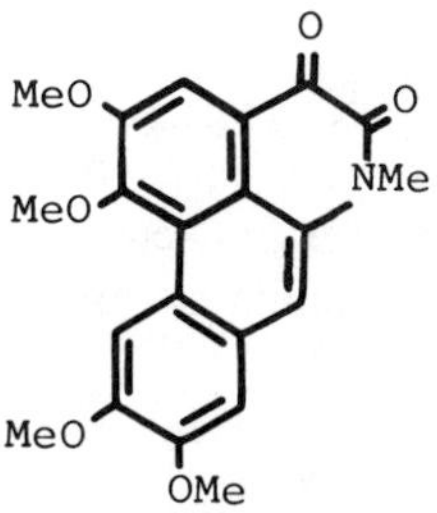

199 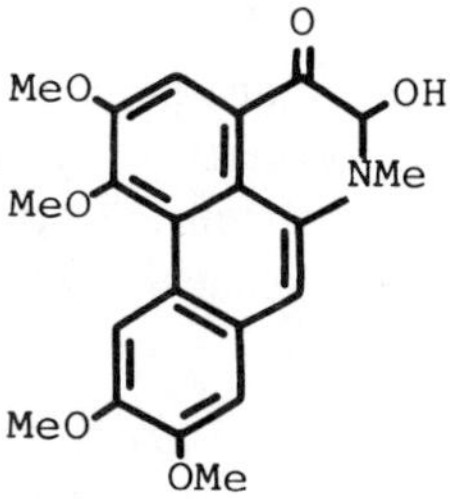 219 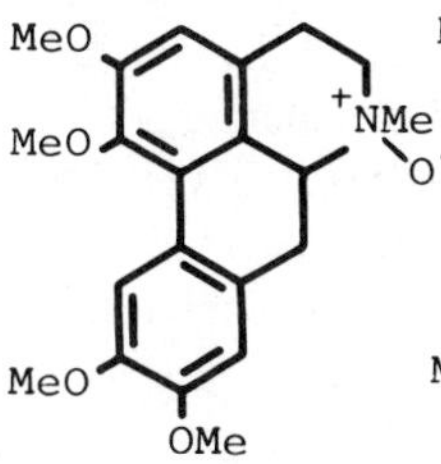 220 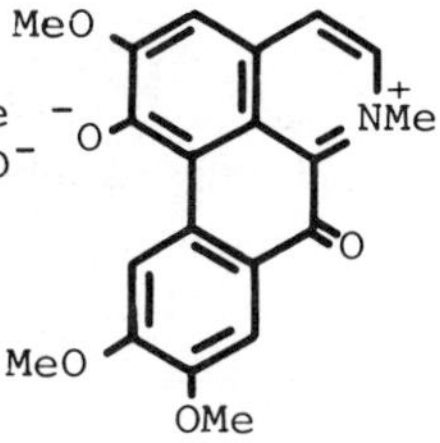 221

corydine (**11**) by oxidation with iodine and sodium acetate in dioxane in 30% yield (*438*).

11 14

Elimination of the *N*-oxide function of aporphines also provides dehydroaporphines, hence bulbocapnine *N*-oxide (**222**) was treated with acetic anhydride to give rise to dehydrobulbocapnine (**223**) whose reduction with zinc in acid solution yielded racemic bulbocapnine (**50**) (*439*).

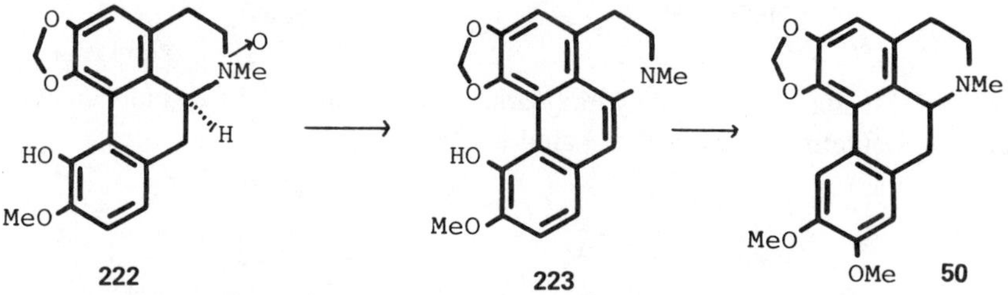

222 223 50

Microbial transformation of (+)-glaucine (**57**) with *Fusarium solani* provides dehydroglaucine (**217**) quantitatively (*440*).

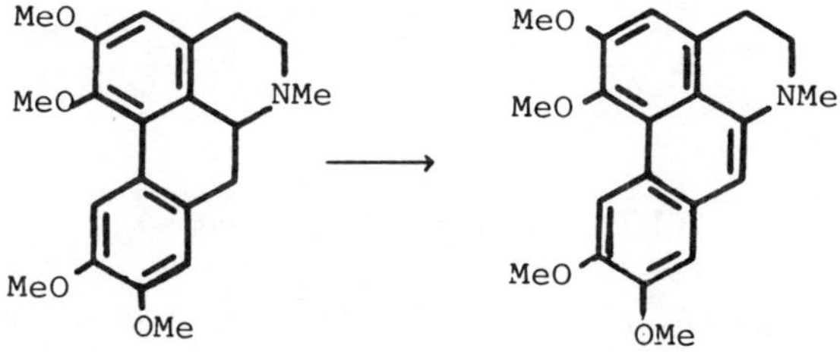

57 217

C. Dealkoxylation and Dehydroxylation

Hydrogenolysis of a phenyltetrazolyl ether of aporphine produces a dealkoxylated aporphine in moderate yield. Thus the bisphenyltetrazolyl ether of the diphenol **205**, obtained from (+)-bulbocapnine methyl ether by demethylenedioxylation with boron trichloride, was hydrogenated over palladium to give (±)-10,11-dimethoxyaporphine (**224**) (*441*).

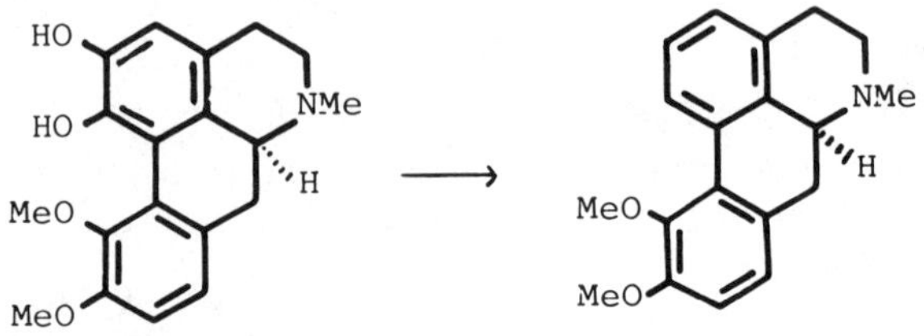

205 → **224**

Reduction of the diethyl phosphate ester of (+)-bulbocapnine (**225**) with alkali metal in liquid ammonia also brought about dehydroxylation to give the phenolic aporphine **226** (*442*).

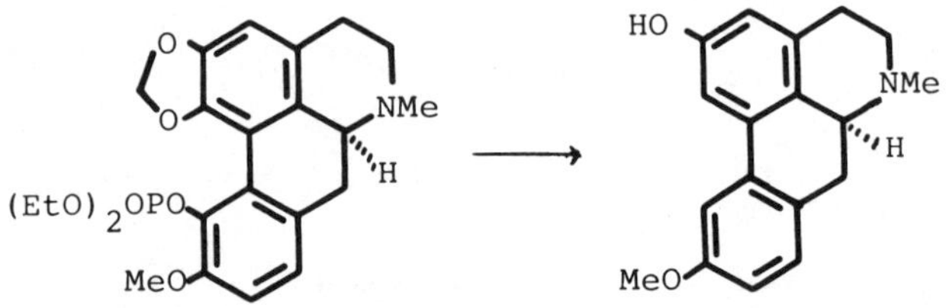

225 → **226**

When Musliner–Gates dehydroxylation was applied to (+)-bulbocapnine (**50**), (+)-laureline (**177**) was prepared (*442a*).

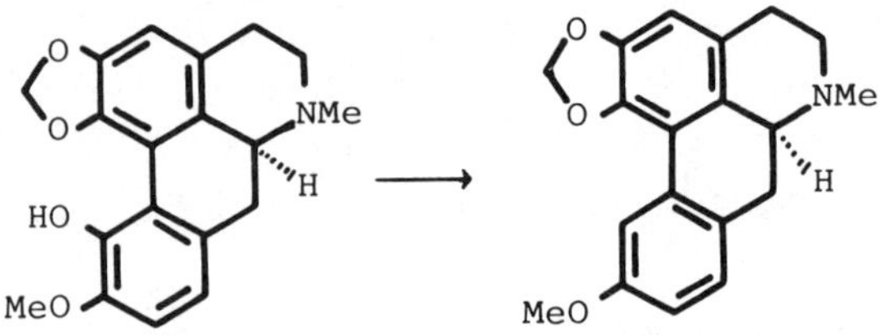

50 → **177**

VIII. Properties of Aporphines

Great development in the field of aporphines has been achieved by using carbon-13 nuclear magnetic resonance studies, and several reviews have appeared (*443–445*). Applications of these studies may assist in locating phenolic functions in newly isolated aporphines.

Determination of the enantiomeric purity of isoquinoline alkaloids by the use of chiral lanthanide nuclear magnetic resonance shift reagents were studied (*446*). Two characteristic signals were used for the determination of enantiomeric composition for glaucine: (a) the methoxyl signal resonance at 3.68 ppm and (b) the strongly deshielded C-11 aromatic proton at 8.11 ppm. Addition of $Eu(facam)_3$ to a solution of (±)-glaucine in $CDCl_3$ caused both of the singlet resonances to be split into two equal signals, with the *S* enantiomer shifting 0.05 ppm in each case.

It has been well known that UV spectroscopy is a valuable tool for the structure elucidation of aporphine alkaloids. For instance, 1,2,9,10-tetrasubstituted aporphines show absorption maxima in 95% ethanol near at 220, 282, and 305 nm. On the other hand, in the 1,2,10,11-series the maxima lie near 220, 270, and 305 nm. Moreover, the UV spectra of monophenolic aporphine were investigated (*447*), and the presence of a phenolic function at C-9 results in a bathochromic shift upon the addition of base, accompanied by a strong hyperchromic effect between 315 and 330 nm and a minimum between 269 and 274 nm. Similar phenomena were observed in the case of phenolic aporphines bearing a phenolic function at C-3.

IX. Addendum

Recently the complete listing of aporphinoid alkaloids has appeared (*448*).

References

1. M. Shamma, *Alkaloids (London)* **9,** 1 (1967); M. Shamma and R. L. Castenson, *ibid.* **14,** 225 (1973).
2. S.-T. Lu, S.-J. Wang, P.-H. Lai, C.-M. Lin, and L.-C. Lin, *J. Pharm. Soc. Jpn.* **92,** 910 (1972).
3. S. R. Johns, J. A. Lamberton, C. S. Li, and A. A. Sioumis, *Aust. J. Chem.* **23,** 363 (1970).
4. K. L. Stuart and C. Chambers, *Tetrahedron Lett.* 4135 (1967).
5. A. K. Kiang and K. Y. Sim, *J. Chem. Soc. C* 282 (1967).
6. S. R. Johns and J. A. Lamberton, *Aust. J. Chem.* **20,** 1277 (1967).
7. C. Casagrande and G. Ferrari, *Farmaco, Ed. Sci.* **25,** 442 (1970).
8. S. R. Johns, J. A. Lamberton, A. A. Sioumis, and H. J. Tweeddale, *Aust. J. Chem.* **22,** 1277 (1969).
9. K. Heydenreich and S. Pfeifer, *Pharmazie* **22,** 124 (1967).
10. S. R. Johns, J. A. Lamberton, and A. A. Sioumis, *Aust. J. Chem.* **20,** 1457 (1967).
11. S. R. Johns, J. A. Lamberton, and A. A. Sioumis, *Aust. J. Chem.* **19,** 2331 (1966).
12. K. S. Soh, F. N. Lahey, and R. Greenhalgh, *Tetrahedron Lett.* 5279 (1966).

13. M. P. Cava, Y. Watanabe, K. Bessho, M. J. Mitchell, A. I. da Rocha, B. Hwang, B. Douglas, and J. A. Weisbach, *Tetrahedron Lett.* 2437 (1968).
14. Z. F. Ismailov, M. V. Telezhenetskaya, and S. Yu. Yunusov, *Khim. Prir. Soedin.* **4,** 136 (1968).
15. P. E. Sonnet and M. Jacobson, *J. Pharm. Sci.* **60,** 1254 (1971).
16. S. Pfeifer and L. Kühn, *Pharmazie* **23,** 267 (1968).
17. M. P. Cava, M. Behforouz, and M. J. Mitchell, *Tetrahedron Lett.* 4647 (1972).
18. S. Pfeifer and L. Kühn, *Pharmazie* **23,** 199 (1968).
19. P. C. Patnaik and K. W. Gopinath, *Indian J. Chem.* **13,** 195 (1975).
20. R. Ziyaev, A. Abdusamatov, and S. Yu. Yunusov, *Khim. Prir. Soedin.* 760 (1973).
21. R. Ziyaev, A. Abdusamatov, and S. Yu. Yunusov, *Khim. Prir. Soedin.* 685 (1974).
22. I. R. C. Bick, N. W. Preston, and P. Potier, *Bull. Soc. Chim. Fr.* 4596 (1972).
23. C. D. Hufford and M. J. Funderburk, *J. Pharm. Sci.* **63,** 1338 (1974).
24. C. D. Hufford, M. J. Funderburk, J. M. Morgan, and L. W. Robertson, *J. Pharm. Sci.* **64,** 789 (1975).
25. M. P. Cava and A. Venkateswarlu, *Tetrahedron* **27,** 2639 (1971).
26. M. P. Cava, K. V. Rao, B. Douglas, and J. A. Weisbach, *J. Org. Chem.* **33,** 2443 (1968).
27. M. Sivakumaran and K. W. Gopinath, *Indian J. Chem., Sect. B* **14B,** 150 (1976).
28. C.-L. Chen, H.-M. Chang, E. B. Cowling, C.-Y. Huang Hsu, and R. P. Gates, *Phytochemistry* **15,** 1161 (1976).
29. R. Ahmad and M. P. Cava, *Heterocycles* **7,** 927 (1977).
30. B. T. Salimov, N. D. Abdullaev, M. S. Yunusov, and S. Yu. Yunusov, *Khim. Prir. Soedin.* 235 (1978).
31. S. U. Karimova, I. A. Israilov, M. S. Yunusov, and S. Yu. Yunusov, *Khim. Prir. Soedin.* 814 (1978); *Chem. Nat. Compd. (Engl. Transl.)* **14,** 699 (1978).
32. N. Borthakur and R. C. Rastogi, *Phytochemistry* **18,** 910 (1979).
33. S. Abduzhabbarova, S. Kh. Maekh, S. Yu. Yunusov, M. R. Yagudaev, and D. Kurbakov, *Khim. Prir. Soedin.* 472 (1978); *Chem. Nat. Compd. (Engl. Transl.)* **14,** 400 (1978).
34. I. A. Israilov, O. N. Denisenko, M. S. Yunusov, D. A. Muraveva, and S. Yu. Yunusov, *Khim. Prir. Soedin.* 474 (1978); *Chem. Nat. Compd. (Engl. Trans.)* **14,** 402 (1978).
35. V. Vecchietti, C. Casagrande, G. Ferrari, and G. Severinicca, *Farmaco, Ed. Sci.* **34,** 829 (1979).
36. K. Yakushijin, S. Sugiyama, Y. Mori, H. Murata, and H. Furukawa, *Phytochemistry* **19,** 161 (1980).
37. S. U. Karimova, I. A. Israilov, M. S. Yunusov, and S. Yu. Yunusov, *Khim. Prir. Soedin.* 224 (1980); *Chem. Nat. Compd. (Engl. Transl.)* **19,** 998 (1980).
38. R. C. Rastogi and N. Borthakur, *Phytochemistry* **19,** 998 (1980).
39. D. S. Bhakuni and R. S. Singh, *J. Nat. Prod.* **45,** 252 (1982).
40. K. Bernauer, *Helv. Chim. Acta* **50,** 1583 (1967).
41. R. Ziyaev, A. Abdusamatov, and S. Yu. Yunusov, *Khim. Prir. Soedin.* 505 (1973); *Chem. Nat. Compd. (Engl. Transl.)* 475 (1975).
42. K. Ito and S. Asai, *J. Pharm. Soc. Jpn.* **94,** 729 (1974).
43. M. Hamonniere, M. Leboeuf, and A. Cavé, *C. R. Hebd. Seances Acad. Sci., Ser. C* **278,** 921 (1974).
44. A. N. Tackie, D. Dnuma-Badu, P. A. Lartey, P. L. Schiff, Jr., J. E. Knapp, and D. J. Slatkin, *Lloydia* **37,** 6 (1974).
45. L. Cleaver, S. Nimgirawath, E. Ritchie, and W. C. Taylor, *Aust. J. Chem.* **29,** 3002 (1976).
46. I. A. Israliov, O. N. Denisenko, M. S. Yunusov, and S. Yu. Yunusov, *Khim. Prir. Soedin.* 799 (1976); *Chem. Abstr.* **86,** 171679 (1977).
47. C. D. Hufford, *Phytochemistry* **15,** 1169 (1976).

48. C. C. Hsu, R. H. Dobberstein, G. A. Cordell, and N. R. Farnsworth, *Lloydia* **40,** 505 (1977).
49. R. Hocquemiller, P. Cabalion, A. Bouquet, and A. Cavé, *C. R. Hebd. Seances Acad. Sci., Ser. C* **285,** 447 (1977).
50. J. Kunitomo, Y. Murakami, M. Oshikata, T. Shingu, S.-T. Lu, I.-S. Chen, and M. Akasu, *J. Pharm. Soc. Jpn.* **101,** 431 (1981).
51. J. Kunitomo, M. Oshikata, and Y. Murakami, *Chem. Pharm. Bull.* **29,** 2251 (1981).
52. J. Slavík, L. Slavíková, and L. Dolejš, *Collect. Czech. Chem. Commun.* **33,** 4066 (1968).
53. M. H. A. Zarga and M. Shamma, *J. Nat. Prod.* **45,** 471 (1982).
54. S. A. Atti, H. A. Ammar, C. H. Phoebe, Jr., P. L. Schiff, Jr., and D. J. Slatkin, *J. Nat. Prod.* **45,** 476 (1982).
55. J. Kunitomo, Y. Okamoto, E. Yuge, and Y. Nagai, *Tetrahedron Lett.* 3287 (1969).
56. A. Urzúa and B. K. Cassels, *Phytochemistry* **21,** 773 (1982).
57. I. Ribas, J. Sueiras, and L. Castedo, *Tetrahedron Lett.* 2033 (1972).
58. A. Shafiee, A. Ghanbarpour, I. Lalezari, and S. Lajevardi, *J. Nat. Prod.* **42,** 174 (1979).
59. A. Urzúa and B. K. Cassels, *Tetrahedron Lett.* 2649 (1978).
60. W. D. Smolnycki, J. L. Moniot, D. M. Hindenlang, G. A. Miana, and M. Shamma, *Tetrahedron Lett.* 4617 (1978).
61. I. A. Israilov, S. U. Karimova, M. S. Yunusov, and S. Yu. Yunusov, *Khim. Prir. Soedin.* 104 (1979).
62. F. Bévalot, M. Leboeuf, and A. Cavé, *Plant. Med. Phytother.* **11,** 315 (1977).
63. S. K. Talapatra, A. Patra, D. S. Bhar, and B. Talapatra, *Phytochemistry* **12,** 2305 (1973).
64. S. K. Talapatra, A. Patra, and B. Talapatra, *Tetrahedron* **31,** 1105 (1975).
65. F. Bévalot, M. Leboeuf, and A. Cavé, *C. R. Hebd. Seances Acad. Sci., Ser. C* **282,** 865 (1976).
66. M. Hamonnière, M. Leboeuf, and A. Cavé, *Phytochemistry* **16,** 1029 (1977).
67. F. Bévalot, M. Leboeuf, A. Bouquet, and A. Cavé, *Ann. Pharm. Fr.* **35,** 65 (1977).
68. A. Cavé, H. Guinaudeau, M. Leboeuf, A. Ramahatra, and J. Razafindrazaka, *Planta Med.* **33,** 243 (1978).
69. R. Hocquemiller, S. Rasamizafy, C. Moretti, H. Jacquemin, and A. Cavé, *Planta Med.* **41,** 48 (1981).
70. H. Guinaudeau, M. Shamma, B. Tantisewie, and K. Pharadai, *J. Nat. Prod.* **45,** 355 (1982).
71. J. Slavík, *Collect. Czech. Chem. Commun.* **33,** 323 (1968).
72. V. Preininger, J. Appelt, L. Slavíkova, and J. Slavík, *Collect. Czech. Chem. Commun.* **32,** 2682 (1967).
73. K. H. B. Duchevska, A. S. Orahovats, and N. M. Mollov, *Dokl. Akad. Nauk Uzb. SSR* **26,** 899 (1973); *Chem. Abstr.* **80,** 27410 (1974).
74. A. Abdusamatov, R. Ziyaev, and S. Yu. Yunusov, *Khim. Prir. Soedin.* 813 (1975); *Chem. Abstr.* **84,** 150806 (1976).
75. I. R. C. Bick and W. Sinchai, *Heterocycles* **9,** 903 (1978).
76. C.-L. Chen, H. Chang, and E. B. Cowling, *Phytochemistry* **15,** 547 (1976).
77. C.-L. Chen and H.-M. Chang, *Phytochemistry* **17,** 779 (1978).
78. H. Guinaudeau, A. Ramahatra, M. Leboeuf, and A. Cavé, *Plant. Med. Phytother.* **12,** 166 (1978); *Chem. Abstr.* **90,** 118062 (1979).
79. R. Hocquemiller, A. Cavé, and A. Raharisololalao, *J. Nat. Prod.* **44,** 551 (1981).
80. K. Ito and H. Furukawa, *Tetrahedron Lett.* 3023 (1970).
81. F. N. Lahey and K. F. Mak, *Aust. J. Chem.* **24,** 671 (1971).
82. F. Roblot, R. Hocquemiller, H. Jacquemin, and A. Cavé, *Plant Med. Phytother.* **12,** 259 (1978); *Chem. Abstr.* **91,** 2517 (1979).
83. H. G. Kiryakov, *Chem. Ind. (London)* 1807 (1968); H. G. Kiryakov and P. Panov, *Dokl. Bolg. Akad. Nauk* **22,** 1019 (1969); *Chem. Abstr.* **72,** 51776 (1970).

84. F. Baralle, N. Schvarzberg, M. Vernengo, and J. Comin, *Experientia* **28,** 875 (1972).
85. J. Kunitomo, M. Juichi, Y. Ando, Y. Yoshikawa, S. Nakamura, and T. Shingu, *J. Pharm. Soc. Jpn.* **95,** 445 (1975).
86. N. C. Franca, A. M. Giesbrecht, O. R. Gottlieb, A. F. Magalhães, and J. G. S. Maia, *Phytochemistry* **14,** 1671 (1975).
87. K. H. B. Duchevska, A. Orahovats, and N. M. Mollov, *Dokl. Bolg. Akad. Nauk* **26,** 899 (1973).
88. M. Kurbanov, Kh. Sh. Khusainova, M. Khodzhimstov, A. E. Vezen, K. Kh. Khaidarov, and V. K. Burichenko, *Dokl. Akad. Nauk Tadzh. SSR* **18,** 20 (1975); *Chem. Abstr.* **84,** 180440 (1976).
89. R. Ziyaev, M. S. Yunusov, and S. Yu. Yunusov, *Khim. Prir. Soedin.* 715 (1977); *Chem. Nat. Compd. (Engl. Transl.)* **14,** 602 (1978).
90. J. Kunitomo, Y. Murakami, M. Oshikata, T. Shingu, M. Akasu, S.-T. Lu, and I.-S. Chen, *Phytochemistry* **19,** 2735 (1980).
91. Z.-D. Min and S.-M. Zhong, *Yao Hsueh Hsueh Pao* **15,** 532 (1980).
92. J. Kunitomo, M. Oshikata, and M. Akasu, *J. Pharm. Soc. Jpn.* **101,**951 (1981).
93. M. Hasegawa, M. Sojo, A. Lira, and C. Marquez, *Acta Cient. Venez.* **23,** 165 (1972); *Chem. Abstr.* **79,** 42716 (1973).
94. J. W. Skiles and M. P. Cava, *J. Org. Chem.* **44,** 409 (1979).
95. Y. Watanabe, M. Matsui, M. Iibuchi, and S. Hiroe, *Phytochemistry* **14,** 2522 (1975).
96. R. Ziyaev, A. Abdusamatov, and S. Yu. Yunusov, *Khim. Prir. Soedin.* **11,** 528 (1975); *Chem. Abstr.* **84,** 44478 (1976).
97. A. Urzúa and B. K. Cassels, *Heterocycles* **4,** 1881 (1976).
98. C. C. Hsu, R. H. Dobberstein, G. A. Cordell, and N. R. Farnsworth, *Lloydia* **40,** 152 (1977).
99. P. D. Senter and C.-L. Chen, *Phytochemistry* **16,** 2015 (1977).
100. O. R. Gottlieb, A. F. Magalhães, E. G. Magalhães, J. G. S. Maia, and A. J. Marsaioli, *Phytochemistry* **17,** 837 (1978).
101. J. W. Skiles, J. M. Saá, and M. P. Cava, *Can. J. Chem.* **57,** 1642 (1979).
102. I. A. Israilov, S. U. Karimova, M. S. Yunusov, and S. Yu. Yunusov, *Khim. Prir. Soedin.* 415 (1979).
103. C. H. Phoebe, Jr., P. L. Schiff, Jr., J. E. Knapp, and D. J. Slatkin, *Heterocycles* **14,** 1977 (1980).
104. M. D. Menachery and M. P. Cava, *J. Nat. Prod.* **44,** 320 (1981).
105. L. Castedo, R. Suau, and A. Mouriño, *Tetrahedron Lett.* 501 (1976).
106. I. Ribas, J. Sueiras, and L. Castedo, *Tetrahedron Lett.* 3093 (1971).
107. J. Kunitomo, Y. Murakami, and M. Akasu, *J. Pharm. Soc. Jpn.* **100,** 337 (1980); J. Kunitomo and Y. Murakami, *Shoyakugaku Zasshi* **33,** 84 (1979).
108. D. Zhu, B. Wang, B. Huang, R. Xu, Y. Qui, and X. Chen, *Heterocycles* **17,** 345 (1982).
109. R. BrazFilho, S. J. Gabriel, C. M. R. Gomes, O. R. Gottlieb, M. das G. A. Bichara, and J. G. S. Maia, *Phytochemistry* **15,** 1187 (1976).
110. M. Shamma, J. L. Moniot, S. Y. Yao, and J. A. Stanko, *Chem. Commun.* 408 (1972).
111. M. Shamma and J. L. Moniot, *Heterocycles* **2,** 427 (1974).
112. J. Wu, J. L. Beal, W.-N. Wu, and R. W. Doskotch, *Lloydia* **40,** 292 (1977).
113. S. T. Akramov and S. Yu. Yunusov, *Khim. Prir. Soedin.* **4,** 199 (1968); *Chem. Abstr.* **69,** 87254 (1968).
114. M. Shamma and J. L. Moniot, *Heterocycles* **3,** 297 (1975).
115. R. W. Doskotch and J. E. Knapp, *Lloydia* **34,** 292 (1971).
116. K. P. Tiwari and M. Masood, *Phytochemistry* **17,** 1068 (1978).
117. J. Slavík and L. Slavíkova, *Collect. Czech. Chem. Commun.* **44,** 2261 (1979).
118. M. Nieto, A. Cavé, and M. Leboeuf, *Lloydia* **39,** 350 (1976).

119. V. G. Khozhdaev, S. Kh. Maekh, and S. Yu. Yunusov, *Khim. Prir. Soedin.* 631 (1972).
120. A. Karimov, M. V. Telezhenetskaya, K. L. Lutfullin, and S. Yu. Yunusov, *Khim. Prir. Soedin.* 419 (1978); *Chem. Nat. Compd. (Engl. Transl.)* **14,** 360 (1978).
121. V. Zabel, W. H. Watson, C. H. Phoebe, J. E. Knapp, P. L. Schiff, Jr., and D. J. Slatkin, *J. Nat. Prod.* **45,** 94 (1982).
122. R. Hoequemiller, C. Debitus, F. Roblot, and A. Cavé, *Tetrahedron Lett.* **23,** 4247 (1982).
123. R. Hocquemiller, S. Rasamizafy, and A. Cavé, *Tetrahedron* **38,** 911 (1982).
124. F. Roblot, R. Hocquemiller, and A. Cavé, *C. R. Hebd. Seances Acad. Sci., Ser. B* **293,** 373 (1981).
125. L. D. Yakhontova, V. I. Sheichenko, and O. N. Tolkachev, *Khim. Prir. Soedin.* 214 (1972); L. D. Yakhontova, O. N. Tolkachev, and D. A. Pakalin, *ibid.* 684 (1973).
126. L. Castedo, R. Suau, and A. Mouriño, *Heterocycles* **3,** 449 (1975).
127. I. Ribas, J. Saá, and L. Castedo, *Tetrahedron Lett.* 3617 (1973); S. M. Kupchan and P. F. O'Brein, *Chem. Commun.* 915 (1973).
128. L. Castedo, J. M. Saá, R. Suau, and C. Villaverde, *Heterocycles* **14,** 1131 (1980).
129. J. Kunitomo, M. Juichi, Y. Yoshikawa, and H. Chikamatsu, *Experientia* **29,** 518 (1973); *J. Pharm. Soc. Jpn.* **94,** 97 (1974).
130. L. Castedo, D. Dominguez, J. M. Saá, and R. Suau, *Tetrahedron Lett.* 4589 (1979).
131. H. Guinaudeau, M. Shamma, B. Tantisewie, and K. Pharadai, *Chem. Commun.* 1118 (1981).
132. H. Guinaudeau, M. Leboeuf, and A. Cavé, *Lloydia* **38,** 275 (1975); **42,** 325 (1979).
133. T. Kametani, "The Chemistry of the Isoquinoline Alkaloids." Elsevier, New York, 1969.
134. S. M. Kupchan and N. Yokoyama, *J. Am. Chem. Soc.* **85,** 1361 (1963).
135. H. Guinaudeau, M. Leboeuf, and A. Cavé, *J. Nat. Prod.* **42,** 133 (1979).
136. S. F. Hussain, L. Khan, and M. Shamma, *Heterocycles* **15,** 191 (1981); S. F. Hussain, L. Khan, K. K. Sadozai, and M. Shamma, *J. Nat. Prod.* **44,** 274 (1981).
137. S. F. Hussain and M. Shamma, *Tetrahedron Lett.* 21, 3315 (1980); S. F. Hussain, M. T. Siddiqui, and M. Shamma, *ibid.* 4573.
138. N. W. Wu, J. L. Beal, and R. W. Doskotch, *J. Nat. Prod.* **43,** 567 (1980); *Lloydia* **40,** 593 (1977).
139. H. Guinaudeau, A. J. Freyer, R. D. Minard, and M. Shamma, *Tetrahedron Lett.* **23,** 2523 (1982).
140. A. Jossang, M. Leboeuf, and A. Cavé, *Tetrahedron Lett.* **23,** 5147 (1982).
141. H. Guinaudeau, A. J. Freyer, R. D. Minard, M. Shamma, and K. H. C. Başer, *J. Org. Chem.* **47,** 5406 (1982).
142. H. Guinaudeau, M. Shamma, K. Hüsnü, and C. Başer, *J. Nat. Prod.* **45,** 505 (1982).
143. M. Leboeuf and A. Cavé, *Plant. Med. Phytother.* **14,** 143 (1980).
144. I. Khokhar, *Pak. J. Sci. Res.* **30,** 81 (1978).
145. C. Casagrande and G. Ferrari, *Farmaco, Ed. Sci.* **30,** 749 (1975).
146. J. Ellis, E. Gellert, and R. E. Summons, *Aust. J. Chem.* **25,** 2735 (1972).
147. E. Gellert and R. Rudzats, *Aust. J. Chem.* **25,** 2477 (1972).
148. M. Leboeuf, A. Cavé, J. Provost, and P. Forgacs, *Plant. Med. Phytother.* **13,** 262 (1979).
149. P. C. Patnaik and K. W. Gopinath, *Indian J. Chem.* **13,** 197 (1975).
150. H. Uprety, D. S. Bhakuni, and M. M. Dhar, *Phytochemistry* **11,** 3057 (1972).
151. M. Lavault, M. M. Debray, and J. Bruneton, *Bull. Mus. Natl. Hist. Nat., Sect. B* **2,** 387 (1980).
152. I. R. C. Bick, H. M. Leow, N. W. Preston, and J. J. Wright, *Aust. J. Chem.* **26,** 455 (1973).
153. P. Forgacs, G. Buffard, J. F. Desconclois, A. Jehanno, J. Provost, R. Tiberghien, and A. Touch, *Plant. Med. Phytother.* **15,** 80 (1981).
154. A. Urzúa, B. Cassels, E. Sanchez, and J. Comin, *An. Assoc. Quim. Argent.* **63,** 259 (1975).
155. L. Ren and F. Xie, *Yaoxue Xuebao* **16,** 672 (1981); *Chem. Abstr.* **96,** 48974 (1982).

156. A. Urzúa and B. K. Cassels, *Rev. Lationam. Quim.* **8,** 133 (1977).
157. M. Leboeuf, J. Streith, and A. Cavé, *Ann. Pharm. Fr.* **33,** 43 (1975).
158. C. R. Chen, J. L. Beal, R. W. Doskotch, L. A. Mitscher, and G. H. Svoboda, *Lloydia* **37,** 493 (1974).
159. J. Kunitomo, Y. Yoshikawa, S. Tanaka, Y. Imori, K. Isoi, Y. Masada, K. Hasimoto, and T. Inoue, *Phytochemistry* **12,** 699 (1973).
160. O. N. Denisenko, I. A. Israilov, D. A. Muravera, and M. S. Yunusov, *Chem. Nat. Compd. (Engl. Transl.)* **14,** 456 (1978).
161. I. A. Israilov, M. U. Ibragimova, M. S. Yunusov, and S. Yu. Yunusov, *Khim. Prir. Soedin.* 612 (1975).
162. A. Shafiee, I. Lalezari, and O. Rahimi, *Lloydia* **40,** 352 (1977).
163. I. Lalezari, A. Shafiee, and M. Mahjour, *J. Pharm. Sci.* **65,** 923 (1976).
164. N. N. Margvelashvili, O. N. Tolkachev, N. P. Prisyazhnyuk, and A. T. Kiryanova, *Khim. Prir. Soedin.* 592 (1978); *Chem. Nat. Prod. (Engl. Transl.)* **14,** 509 (1978).
165. Kh. G. Kiryakov, I. A. Israilov, and S. Yu. Yunusov, *Khim. Prir. Soedin.* 411 (1974); *Chem. Nat. Prod. (Engl. Transl.)* **10,** 418 (1975).
166. Kh. Kiryakov, E. Iskrenova, B. Kuzmanov, and L. Evstatieva, *Planta Med.* **41,** 298 (1981).
167. G. Verzar-Petri and P. T. Ming Hoang, *Sci. Pharm.* **46,** 169 (1978).
168. V. Preininger, R. S. Thakur, and F. Šantavý, *J. Pharm. Sci.* **65,** 294 (1976).
169. Kh. Kiryakov, E. Iskrenova, B. Kuzmanov, and L. Evstatieva, *Planta Med.* **43,** 51 (1981).
170. Kh. Sh. Khusainova and Yu. D. Sadykov, *Dokl. Akad. Nauk Tadzh. SSR* **24,** 489 (1981); *Khim. Prir. Soedin.* 670 (1981).
171. Kh. Kiryakov, E. Iskrenova, E. Daskalova, B. Kuzmanov, and L. Evstatieva, *Planta Med.* **44,** 168 (1982).
172. V. Vecchietti, C. Casagrande, and G. Ferrari, *Farmaco, Ed. Sci.* **32,** 767 (1977).
173. T. Kametani, M. Takemura, and M. Ihara, *Phytochemistry* **15,** 2017 (1976).
174. W. F. Xin and M. Lin, *Chung Ts'ao Yao* **12,** 1 (1981); *Chem. Abstr.* **95,** 20945 (1981).
175. J. Slavík, L. Slavíková, and L. Dolejš, *Collect. Czech. Chem. Commun.* **46,** 2587 (1981).
176. J. Slavík and L. Slavíková, *Collect. Czech. Chem. Commun.* **46,** 1534 (1981).
177. S.-T. Lu, T.-L. Su, and C.-Y. Duh, *T'ai-wan Yao Hsueh Tsa Chih* **31,** 23 (1979).
178. S. Mukhamedova, S. Kh. Maekh, and S. Yu. Yunusov, *Khim. Prir. Soedin.* 251 (1981).
179. D. M. Holloway and F. Scheinmann, *Phytochemistry* **12,** 1503 (1973).
180. J. A. Lamberton and V. N. Vashist, *Aust. J. Chem.* **25,** 2737 (1972).
181. S. F. Hussain, A. Amin, and M. Shamma, *J. Chem. Soc. Pak.* **2,** 157 (1980).
182. B. K. Chowdhury, M. L. Sethi, H. A. Lloyd, and G. J. Kapadia, *Phytochemistry* **15,** 1803 (1976).
183. R. T. Suess and F. R. Stermitz, *J. Nat. Prod.* **44,** 680 (1981).
184. L. Castedo, J. M. Saá, R. Suau, C. Villaverde, and P. Potier, *An. Quim., Ser. C* **14,** 49 (1978).
185. A. Shafiee, I. Lalezari, and M. Mahjour, *J. Pharm. Sci.* **66,** 593 (1977).
186. L. D. Yakhontova, O. N. Tolkachaev, and Yu. V. Baranova, *Khim. Prir. Soedin.* 686 (1973).
187. T. Gözler and S. Ünlüyol, *Doga, Seri C* **6,** 21 (1982).
188. Polish Patent 82,864; *Chem. Abstr.* **90,** 43798 (1979).
189. L. Bubeva-Ivanova, N. Donev, E. Mermerska, B. Avramova, P. Ioncheva, and S. Stefanov, *Postepy Dziedzinie Leku Rosl., Ref. Dosw. Wygloszone Symp.* 104 (1970); *Chem. Abstr.* **78,** 88550 (1973).
190. R. Marini-Bettòlo and M. L. Scarpati, *Phytochemistry* **18,** 520 (1979).
191. P. K. Mahanta, R. K. Mathur, and K. W. Gopinath, *Indian J. Chem.* **13,** 306 (1975).
192. H. Takayama, A. Tokita, M. Ito, S. Sakai, F. Kurosaki, and T. Okamoto, *J. Pharm. Soc. Jpn.* **102,** 245 (1982).
193. D. Y. Zhu, C.-Q. Song, Y.-L. Gao, and R.-S. Xu, *Hua Hsueh Hsueh Pao* **39,** 280 (1981); *Chem. Abstr.* **95,** 93849 (1981).

194. G. Sariyar and A. Oztekin, *Plant. Med. Phytother.* **15,** 160 (1981).
195. Z. Min, X. Liu, and W. Sun, *Yaoxue Xuebao* **16,** 557 (1981); *Chem. Abstr.* **97,** 3595 (1982).
196. M. Diaz, C. Schreiber, and H. Ripperger, *Rev. Cuband Farm.* **15,** 93 (1981).
197. H.-J. Chang, H.-H. Wang, and K.-E. Ma, *Yao Hsueh T'ung Pao* **16,** 52 (1981).
198. T. Gózler and S. Ünlüyol, *Doga, Seri C* **5,** 25 (1981).
199. D. A. Murav'eva, I. A. Israilov, and F. M. Melikov, *Farmatsiya (Moscow)* **30,** 25 (1981).
200. H. Ripperger, *Pharmazie* **34,** 435 (1979).
201. M. Shamma and A. S. Rothenberg, *Lloydia* **41,** 171 (1978).
202. M. Shamma and S. S. Salgar, *Phytochemistry* **12,** 1505 (1973).
203. V. N. Plugar, Ya. V. Rashkes, M. G. Zhamierashvii, V. A. Tel'nov, M. S. Yunusov, and S. Yu. Yunusov, *Khim. Prir. Soedin.* 80 (1982).
204. F. Věžník, E. Taborska, and J. Slavík, *Collect. Czech. Chem. Commun.* **46,** 926 (1981).
205. D. Dwuma-Badu, J. S. K. Ayim, C. A. Mingle, A. N. Tackie, D. J. Slatkin, J. E. Knapp, and P. L. Schiff, Jr., *Phytochemistry* **14,** 2520 (1975).
206. I. A. Benages, M. E. A. de Juarez, S. M. Albonico, A. Urzúa, and B. K. Cassels, *Phytochemistry* **13,** 2891 (1974).
207. S.-T. Lu and T.-L. Su, *J. Chin. Chem. Soc. (Taipei)* **20,** 75 (1973); *Chem. Abstr.* **79,** 123625 (1973).
208. I. Borup-Grochtmann and D. G. I. Kingston, *J. Nat. Prod.* **45,** 102 (1982).
209. A. Abdusamatov, R. Ziyaev, and S. Yu. Yunusov, *Khim. Prir. Soedin.* 112 (1974).
210. D. Dwuma-Badu, J. S. K. Ayim, A. N. Tackie, J. E. Knapp, D. J. Slatkin, and P. L. Schiff, Jr., *Phytochemistry* **14,** 2524 (1975).
211. R. Huls, *Bull. Soc. R. Sci. Liege* **41,** 686 (1972); *Chem. Abstr.* **79,** 15841 (1973).
212. O. R. Gottlieb, A. F. Magalhaes, E. G. Magalhaes, J. G. S. Maia, and A. J. Marsaioli, *Phytochemistry* **17,** 837 (1978).
213. G. B. Lockwood, *Phytochemistry* **20,** 1463 (1981).
214. A. Shafiee, I. Lalezari, F. Assadi, and F. Khalafi, *J. Pharm. Sci.* **66,** 1050 (1977).
215. J. D. Phillipson, A. Scutt, A. Baytop, M.Ozhatay, and G. Sariyar, *Planta Med.* **43,** 261 (1981).
216. A. A. L. Gunatilaka, S. Sotheeswaran, S. Sriyani, and S. Balasubramanian, *Planta Med.* **43,** 309 (1981).
217. W. N. Wu, J. L. Beal, and R. W. Doskotch, *J. Nat. Prod.* **43,** 372 (1980).
218. S. F. Hussain, R. D. Minard, A. J. Freyer, and M. Shamma, *J. Nat. Prod.* **44,** 169 (1981).
219. M. Alimova and I. A. Israilov, *Khim. Prir. Soedin.* 602 (1981).
220. I. R. C. Bick, T. Sevenet, W. Sinchai, B. Skelton, and A. H. White, *Aust. J. Chem.* **34,** 195 (1981).
221. B. T. Salimov, N. D. Abdullaev, M. S. Yunusov, and S. Yu. Yunusov, *Khim. Prir. Soedin.* 235 (1977).
222. D. H. R. Barton, A. A. L. Gunatilaka, R. M. Letcher, A. M. F. T. Lobo, and D. A. Widdowson, *J. Chem. Soc., Perkin Trans. 1* 874 (1973).
223. M. Leboeuf, J. Parello, and A. Cavé, *Plant. Med. Phytother.* **6,** 112 (1972); *Chem. Abstr.* **77,** 111494 (1972).
224. J. Bruneton, *C. R. Hebd. Seances Acad. Sci., Ser. C* **291,** 187 (1980).
225. S. El-Masry, M. G. El-Ghazooly, A. A. Omar, S. Khafagy, and J. D. Phillipson, *Planta Med.* **41,** 61 (1980).
226. J. D. Phillipson, O. O. Thomas, A. I. Gray, and G. Sariyar, *Planta Med.* **41,** 105 (1981).
227. V. A. Chelombitko, V. A. Mnatsakanyan, and L. V. Salnikova, *Khim. Prir. Soedin.* 270 (1978).
228. V. A. Mnatskanyan, M. A. Manuschakyan, and N. E. Mesropyan, *Chem. Nat. Compd. (Engl. Transl.)* 361 (1978).
229. A. Ikuta and H. Itokawa, *Phytochemistry* **15,** 577 (1976).

230. M. Leboeuf and A. Cavé, *Plant. Med. Phytother.* **6,** 87 (1972); *Chem. Abstr.* **77,** 98777 (1972).
231. N. K. Saxena and D. S. Bhakuni, *J. Indian Chem. Soc.* **56,** 1020 (1979).
232. F. Fish, A. I. Gray, P. G. Waterman, and F. Donachie, *Lloydia* **38,** 268 (1975).
233. S. Campos Flor, N. J. Doorenbos, G. H. Svoboda, J. E. Knapp, and P. L. Schiff, Jr., *J. Pharm. Sci.* **63,** 618 (1974).
234. S.-T. Lu and T.-L. Su, *J. Chin. Chem. Soc. (Taipei)* **20,** 87 (1973); *Chem. Abstr.* **79,** 123626 (1973).
235. S.-L. Liu, *J. Chin. Chem. Soc. (Taipei)* **24,** 209 (1977).
236. P. K. Bhaumik, B. Mukherjee, J. P. Juneau, N. S. Bhacca, and R. Mukherjee, *Phytochemistry* **18,** 1584 (1979).
237. M. Leboeuf and A. Cavé, *Phytochemistry* **11,** 2833 (1972).
238. B. Talapatra, P. Mukhopadhyay, and L. N. Dutta, *Phytochemistry* **14,** 589 (1975).
239. P. W. LeQuesne, J. E. Larrahondu, and R. F. Raffauf, *J. Nat. Prod.* **43,** 353 (1980).
240. L. A. Djakoure, D. Kone, and L. L. Douzoua, *Ann. Univ. Abidjan, Ser. C* **14,** 49 (1978).
241. J. Faust, A. Ripperger, D. Sandoval, and K. Schreiber, *Pharmazie* **36,** 713 (1981).
242. S. Y. C. Chiu, R. H. Dobberstein, H. H. S. Fong, and N. R. Farnsworth, *J. Nat. Prod.* **45,** 229 (1981).
243. D. Dwuma-Badu, J. S. K. Ayim, S. F. Withers, N. O. Agyemang, A. M. Ateya, M. M. Al-Azizi, J. E. Knapp, D. J. Slatkin, and P. L. Schiff, Jr., *J. Nat. Prod.* **43,** 123 (1980).
244. S. Dasgupta, A. B. Ray, S. K. Bhattacharya, and R. Bose, *J. Nat. Prod.* **42,** 399 (1979).
245. H. Ishii, T. Ishikawa, S.-T. Lu, and I.-S. Chen, *J. Pharm. Soc. Jpn.* **96,** 1458 (1976).
246. B. F. Bowden, K. Picker, E. Ritchie, and W. C. Taylor, *Aust. J. Chem.* **28,** 2681 (1975).
247. T. Kametani, H. Terasawa, M. Ihara, and J. Iriarte, *Phytochemistry* **14,** 1884 (1975).
248. K. V. Rao and T. L. Davis, *J. Nat. Prod.* **45,** 283 (1982).
249. J. Siwon, R. Verpoorte, G. F. A. van Essen, and A. B. Svendsen, *Planta Med.* **38,** 24 (1980).
250. J. A. Swinehart and F. R. Stermitz, *Phytochemistry* **19,** 1219 (1980).
251. G. Sariyar and J. D. Phillipson, *J. Nat. Prod.* **44,** 239 (1981).
252. M. A. Manushakyan, I. A. Israilov, V. A. Mnatsakanyan, M. S. Yunusov, and S. Yu. Yunusov, *Khim. Prir. Soedin.* 224 (1980).
253. D. Dwuma-Badu, J. S. K. Ayim, O. Rexford, A. M. Ateya, D. J. Slatkin, J. E. Knapp, and P. L. Schiff, Jr., *Phytochemistry* **19,** 1564 (1980).
254. R. Tschesche, C. Spilles, and G. Eckhardt, *Chem. Ber.* **107,** 1329 (1974).
255. T.-H. Yang, C.-M. Chen, C.-S. Lu, and C.-L. Liao, *J. Chin. Chem. Soc. (Taipei)* **19,** 143 (1972); *Chem. Abstr.* **77,** 161937 (1972).
256. J. R. Merchant and H. K. Desai, *Indian J. Chem.* **11,** 342 (1973).
257. S. V. Bhat, H. Dornauer, and N. J. Desurza, *J. Nat. Prod.* **43,** 588 (1980).
258. C. H. Chen, T. M. Chen, and C. Leu, *J. Pharm. Sci.* **69,** 1061 (1980).
259. W.-N. Wu, J. L. Beal, and R. W. Doskotch, *Lloydia* **40,** 508 (1977).
260. W.-N. Wu, J. L. Beal, R.-P. Leu, and R. W. Doskotch, *Lloydia* **40,** 284 (1977).
261. K. H. C. Baser, *Doga, Seri A* **5,** 163 (1981).
262. I. Ciulei and P. A. Ionescu, *Farmacia (Bucharest)* **21,** 17 (1973); *Chem. Abstr.* **79,** 123633 (1973).
263. C. W. Geiselman, S. A. Gharbo, J. L. Beal, and R. W. Doskotch, *Lloydia* **35,** 296 (1972).
264. S. A. Gharbo, J. L. Beal, R. W. Doskotch, and L. A. Mitscher, *Lloydia* **36,** 349 (1973).
265. W.-N. Wu, J. L. Beal, L. A. Mitscher, K. N. Salman, and P. Patil, *Lloydia* **39,** 204 (1976).
266. W.-N. Wu, J. L. Beal, R.-P. Leu, and R. W. Doskotch, *Lloydia* **40,** 281 (1977).
267. M. A. Elsohly, J. E. Knapp, P. L. Schiff, Jr., and D. J. Slatkin, *J. Pharm. Sci.* **65,** 132 (1976).
268. D. J. Slatkin, N. J. Doorenbos, J. E. Knapp, and P. L. Schiff, Jr., *J. Pharm. Sci.* **61,** 1825 (1972).

269. R. T. Boulware and F. R. Stermitz, *J. Nat. Prod.* **44,** 200 (1981).
270. P. Pachaly and C. Schneider, *Arch. Pharm. (Weinheim, Ger.)* **314,** 251 (1981).
271. L. Slavíková and J. Slavík, *Collect. Czech. Chem. Commun.* **45,** 761 (1980).
272. T. Furuya, A. Ikuta, and K. Syono, *Phytochemistry* **11,** 3041 (1972).
273. R. Verpoorte, J. Siwon, M. E. M. Tieken, and A. B. Svendsen, *J. Nat. Prod.* **44,** 221 (1981).
274. S. R. Hemingway, J. D. Phillipson, and R. Verpoorte, *J. Nat. Prod.* **44,** 67 (1981).
275. B. Podolesov and Z. Zdravkovski, *Acta Pharm. Jugosl.* **30,** 161 (1980).
276. R. H. Burnell, A. Chapelle, and P. H. Bird, *J. Nat. Prod.* **44,** 238 (1981).
277. J. Siwon, R. Verpoorte, T. van Beck, H. Meerburg, and A. B. Svendsen, *Phytochemistry* **20,** 323 (1981).
278. M. Matsui, M. Uchida, I. Usuki, Y. Saionji, H. Murata, and Y. Watanabe, *Phytochemistry* **18,** 1087 (1978).
279. R. S. Singh, P. Kumar, and D. S. Bhakuni, *J. Nat. Prod.* **44,** 664 (1981).
280. F. Fish and P. G. Waterman, *Phytochemistry* **11,** 3007 (1972).
281. A. Ikuta, K. Syono, and T. Furuya, *Phytochemistry* **13,** 2975 (1974).
282. V. Novák and J. Slavík, *Collect. Czech. Chem. Commun.* **39,** 3352 (1974).
282a. J. Slavík and L. Dolejs, *Collect. Czech. Chem. Commun.* **38,** 3514 (1973).
283. G. Klughardt and F. Zymalkowski, *Arch. Pharm. (Weinheim, Ger.)* **315,** 7 (1982).
284. D. Kostalova, B. Brazdovicova, and J. Tomko, *Chem. Zvesti* **35,** 279 (1981).
285. F. K. Duah, P. D. Owusu, J. E. Knapp, D. J. Slatkin, and P. L. Schiff, Jr., *Planta Med.* **42,** 275 (1981).
286. D. Kostalova, B. Brazdovicova, and H.-Y. Jin, *Farm. Obz.* **51,** 213 (1982); *Chem. Abstr.* **97,** 36105 (1982).
287. I. Khokhar, *Pak. J. Sci. Res.* **30,** 81 (1978).
288. R. Ziyaev, A. Abdusamatov, and S. Yu. Yunusov, *Khim. Prir. Soedin.* 108 (1974); *Chem. Nat. Compd. (Engl. Transl.)* **10,** 119 (1975).
289. M. A. Manushakyan and V. A. Mnatsakanyan, *Khim. Prir. Soedin.* 713 (1977); *Chem. Nat. Compd. (Engl. Transl.)* **13,** 599 (1977).
290. J. Slavík, K. Picka, L. Slavíkova, E. Täborská, and F. Věžnik, *Collect. Czech. Chem. Commun.* **45,** 914 (1980).
291. P. G. Gorovoi, A. A. Ibragimov, S. Kh. Maekh, and S. Yu. Yunusov, *Khim. Prir. Soedin.* 533 (1975).
292. J. Kunitomo, M. Juichi, Y. Yoshioka, and H. Chikamatsu, *J. Pharm. Soc. Jpn.* **94,** 97 (1974).
293. V. G. Khozhdaev, S. Kh. Maekh, and S. Yu. Yunusov, *Khim. Prir. Soedin.* 631 (1972); *Chem. Abstr.* **78,** 108183 (1973).
294. C. Casagrande, L. Canonica, and G. Severini-Ricca, *J. Chem. Soc., Perkin Trans. I* 1659 (1975).
295. K.-J. Cheng, K.-C. Wang, and Y.-H. Wang, *Yao Hsueh T'ung Pao* **16,** 49 (1981); *Chem. Abstr.* **95,** 138460 (1981).
296. W. N. Wu, W. T. Liao, Z. F. Mahmoud, J. L. Beal, and R. W. Doskotch, *J. Nat. Prod.* **43,** 472 (1980).
297. W. N. Wu, J. L. Beal, G. W. Clark, and L. A. Mitscher, *Lloydia* **39,** 65 (1976).
298. X. A. Dominguez, O. R. Franco, C. G. Cano, S. Garcia, and R. S. Tamez, *Rev. Latinoam. Quim.* **12,** 61 (1981).
299. F. Baralle, N. Schvarzberg, M. J. Vernengo, G. Y. Moltrasio, and D. Giacopello, *Phytochemistry* **12,** 948 (1973).
300. E. Brochmann-Hanssen, C.-C. Fu, and L. Y. Misconi, *J. Pharm. Sci.* **60,** 1880 (1971).
301. D. S. Bhakuni, S. Tewari, and R. S. Kapil, *J. Chem. Soc., Perkin Trans. 1* 706 (1977).
302. G. Blaschke, G. Waldheim, V. von Schantz, and P. Peura, *Arch. Pharm. (Weinheim, Ger.)* **307,** 122 (1974).

303. O. Prakash, D. S. Bhakuni, and R. S. Kapil, *J. Chem. Soc., Perkin Trans. 1* 622 (1978).
304. E. Brochmann-Hanssen, C.-H. Chen, H.-C. Chiang, and K. McMurtrey, *Chem. Commun.* 1269 (1972).
305. D. S. Bhakuni, S. Jain, and R. S. Singh, *Tetrahedron* **36,** 2525 (1980).
306. A. R. Battersby, R. T. Brown, J. H. Clements, and G. G. Iverach, *Chem. Commun.* 230 (1965); A. R. Battersby and R. T. Brown, *ibid.* 170 (1966): A. R. Battersby, T. J. Brockson, and R. Ramage, *ibid.* 464 (1979).
307. A. R. Battersby, J. L. McHugh, J. Staunton, and M. Todd, *Chem. Commun.* 985 (1971).
308. S. M. Kupchan, V. Kameswaran, J. T. Lynn, D. K. Williams, and A. J. Liepa, *J. Am. Chem. Soc.* **97,** 5622 (1975).
309. N. Murugesan and M. Shamma, *Tetrahedron Lett.* 4521 (1979).
310. S. M. Kupchan, V. Kameswaran, and J. W. A. Findley, *J. Org. Chem.* **37,** 405 (1972); **38,** 475 (1973).
311. M. P. Cava, I. Noguchi, and K. T. Buck, *J. Org. Chem.* **38,** 2394 (1973).
312. P. Kerekes, K. Délenk-Heydenreich, and S. Pfeifer, *Chem. Ber.* **105,** 609 (1972).
313. J. R. Merchant and H. K. Desai, *Indian J. Chem.* **11,** 342 (1973).
314. I. Ribas, J. Saá, and L. Castedo, *Tetrahedron Lett.* 3617 (1973).
315. J. L. Neumeyer, M. McCarthy, S. P. Battista, F. J. Rosenberg, and D. G. Teiger, *J. Med. Chem.* **16,** 1228 (1973).
316. T. Kametani, S. Shibuya, R. Charubala, M. S. Premila, and B. R. Pai, *Heterocycles* **3,** 439 (1975).
317. V. Sharma and R. S. Kapil, *Indian J. Chem., Sect. B* **20B,** 70 (1981).
318. T. R. Suess and F. R. Stermitz, *J. Nat. Prod.* **44,** 688 (1981).
319. C. R. Ghoshal and S. K. Shah, *Chem. Ind. (London)* 889 (1972).
320. S. Chackalamannil and D. R. Dalton, *Tetrahedron Lett.* **21,** 2029 (1980).
321. C. D. Hufford, A. S. Sharma, and B. O. Oguntimein, *J. Pharm. Sci.* **69,** 1180 (1980).
322. T. Kametani and K. Fukumoto, *Heterocycles* **8,** 465 (1977).
323. T. Kametani and K. Fukumoto, *Acc. Chem. Res.* **5,** 212 (1972).
324. T. Kametani, K. Fukumoto, and K. Shishido, *Chem. Ind. (London)* 1566 (1970); T. Kametani, M. Koizumi, K. Shishido, and K. Fukumoto, *J. Chem. Soc. C* 1923 (1970).
325. T. Kametani, T. Sugahara, and K. Fukumoto, *Tetrahedron* **27,** 5367 (1971).
326. T. Kametani, H. Sugi, S. Shibuya, and K. Fukumoto, *Chem. Pharm. Bull.* **19,** 1513 (1971).
327. T. Kametani, R. Charubala, M. Ihara, K. Koizumi, K. Takahashi, and K. Fukumoto, *Chem. Commun.* 289 (1971); *J. Chem. Soc. C* 3315 (1971).
328. M. P. Cava, S. C. Havlicek, A. Lindert, and R. J. Spangler, *Tetrahedron Lett.* 2937 (1966); M. P. Cava, M. J. Mitchell, S. C. Havlicek, A. Lindert, and R. J. Spangler, *J. Org. Chem.* **35,** 175 (1970).
329. N. C. Yang, G. R. Lenz, and A. Shani, *Tetrahedron Lett.* 2941 (1966).
330. G. R. Lenz, *Synthesis* 489 (1978).
331. M. P. Cava, P. Stern, and K. Wakisaka, *Tetrahedron* **29,** 2245 (1973).
332. M. P. Cava and S. S. Libsch, *J. Org. Chem.* **39,** 577 (1974).
333. G. Y. Moltrasio, R. M. Sotelo, and D. Giacopello, *J. Chem. Soc., Perkin Trans. 1* 349 (1973).
334. S. M. Kupchan and R. M. Kanojia, *Tetrahedron Lett.* 5353 (1966).
335. S. M. Kupchan and F. P. O'Brien, *Chem. Commun.* 915 (1973).
336. R. J. Spangler and D. C. Boop, *Tetrahedron Lett.* 4851 (1971).
337. T. Kametani, S. Shibuya, H. Sugi, O. Kusama, and K. Fukumoto, *J. Chem. Soc. C* 2446 (1971).
338. T. Kametani, H. Sugi, S. Shibuya, and K. Fukumoto, *Tetrahedron* **27,** 5375 (1971).
339. T. Kametani, K. Fukumoto, S. Shibuya, H. Nemoto, T. Nakano, T. Sugahara, T. Takahashi, Y. Aizawa, and M. Toriyama, *J. Chem. Soc., Perkin Trans. 1* 1435 (1972).

340. B. R. Pai, H. Suguna, S. Natarajan, and G. Manikumar, *Heterocycles* **6,** 1993 (1977).
341. Y. P. Gupta, V. S. Yadav, and T. Mohammad, *Indian J. Chem., Sect. B* **22,** 429 (1983).
342. S. V. Kessar, Y. P. Gupta, V. S. Yadav, M. Narula, and T. Mohammad, *Tetrahedron Lett.* **21,** 3307 (1980).
343. M. S. Premila and B. R. Pai, *Indian J. Chem.* **13,** 13 (1975).
344. T. Kametani, R. Nitadori, H. Terasawa, K. Takahashi, M. Ihara, and K. Fukumoto, *Tetrahedron* **33,** 1069 (1977).
345. B. R. Pai, S. Natarajan, H. Suguna, and G. Manikumar, *Indian J. Chem., Sect. B* **15,** 1042 (1977).
346. H. Suguna and B. R. Pai, *Indian J. Chem., Sect. B* **15,** 416 (1977).
347. M. S. Premila and B. R. Pai, *Indian J. Chem., Sect. B* **14,** 134 (1976).
348. J. Kunitomo, M. Oshikata, K. Nakayama, K. Suwa, and Y. Murakami, *Chem. Pharm. Bull.* **30,** 4283 (1982).
349. M. Shamma and D.-Y. Hwang, *Heterocyles* **1,** 31 (1973); *Tetrahedron* **30,** 2279 (1974).
350. S. M. Kupchan, C. Kim, and K. Miyano, *Chem. Commun.* 91 (1976); *J. Org. Chem.* **41,** 3210 (1976).
351. T. Kametani, T. Honda, M. Ihara, and K. Fukumoto, *Chem. Ind. (London)* 119 (1972); T. Kametani, K. Takahashi, T. Honda, M. Ihara, and K. Fukumoto, *Chem. Pharm. Bull.* **20,** 1793 (1972).
352. T. Kametani, K. Takahashi, K. Ogasawara, and K. Fukumoto, *Chem. Pharm. Bull.* **21,** 766 (1973).
353. R. Robinson, "The Stractural Relationships of Natural Products." Oxford Univ. Press (Clarendon), London and New York, 1955.
354. C. Schöpf and K. Thierfelder, *Justus Liebigs Ann. Chem.* **497,** 22 (1932).
355. D. H. R. Barton and T. Cohen, *Festschr. Prof. Dr. Arthur Stoll Siebzigsten Geburtstag,* 1957 117 (1957).
356. D. H. R. Barton, *Proc. Chem. Soc., London* 293 (1963).
357. B. Frank, G. Blaschke, and G. Schlingloff, *Angew. Chem.* **75,** 957 (1963).
358. H. Musso, *Angew. Chem.* **75,** 965 (1963).
359. A. I. Scott, *Q. Rev., Chem. Soc.* **19,** 1 (1965).
360. A. R. Battersby, *Q. Rev., Chem. Soc.* **15,** 259 (1981).
361. T. Kametani and K. Fukumoto, "Phenolic Oxidation," p. 121. Gihodo, Tokyo, 1970.
362. T. Kametani and K. Fukumoto, *J. Heterocycl. Chem.* **8,** 341 (1971).
363. T. Kametani and K. Fukumoto, *Synthesis* 657 (1972).
364. T. Kametani, K. Fukumoto, and F. Satoh, *Bioorg. Chem.* **3,** 430 (1974).
365. T. Kametani and M. Ihara, *Heterocycles* **13,** 497 (1979).
366. B. Frank and G. Schlingloff, *Justus Liebigs Ann. Chem.* **659,** 132 (1962).
367. S. M. Albonico, A. M. Kuck, and V. Deulofeu, *Chem. Ind. (London)* 1580 (1964); *Justus Liebigs Ann. Chem.* **685,** 200 (1965).
368. T. Kametani and I. Noguchi, *J. Chem. Soc. C* 1440 (1967).
369. T. Kametani, A. Kozuka, and K. Fukumoto, *J. Chem. Soc. C* 1021 (1971).
370. M. A. Schwartz, *Synth. Commun.* **3,** 33 (1973).
371. S. M. Kupchan, O. P. Dhingra, and C.-K. Kim, *J. Org. Chem.* **41,** 4049 (1976).
372. J. P. Marino and A. Schwartz, *Tetrahedron Lett.* 3253 (1979).
373. E. C. Taylor, J. C. Andrade, G. J. H. Rall, and A. McKillop, *J. Am. Chem. Soc.* **102,** 6513 (1980).
374. C. Szántay, G. Blasko, M. Barczai-Beke, P. Pechy, and G. Dornyei, *Tetrahedron Lett.* **21,** 3509 (1980).
375. K.-H. Frömming, *Arch. Pharm. (Weinheim, Ger.)* **300,** 977 (1967).
376. A. Brossi, A. Ramel, J. O'Brien, and S. Teitel, *Chem. Pharm. Bull.* **21,** 1839 (1973).

377. T. Kametani, Y. Ohta, M. Takemura, M. Ihara, and K. Fukumoto, *Bioorg. Chem.* **6,** 249 (1977).
378. K. L. Stuart, *Heterocycles* **5,** 701 (1976).
379. T. Kametani, M. Ihara, M. Takemura, Y. Satoh, H. Terasawa, Y. Ohta, K. Fukumoto, and K. Takahashi, *J. Am. Chem. Soc.* **99,** 3805 (1977).
380. T. Kametani and M. Ihara, *Heterocycles* **12,** 893 (1979); *J. Chem. Soc., Perkin Trans. 1* 629 (1980).
381. S. M. Kupchan and C.-K. Kim, *J. Am. Chem. Soc.* **97,** 5623 (1975).
382. R. Gotllieb and J. L. Neumeyer, *J. Am. Chem. Soc.* **98,** 7108 (1976).
383. S. V. Kessar, S. Batra, and S. S. Gandhi, *Indian J. Chem.* **8,** 468 (1970).
384. T. Kametani, S. Shibuya, K. Kigasawa, M. Hiiragi, and O. Kusama, *J. Chem. Soc. C* 2712 (1971).
385. T. Kametani, K. Fukumoto, and T. Nakano, *J. Heterocycl. Chem.* **9,** 1363 (1972); *Tetrahedron* **28,** 4667 (1972).
386. T. Kametani, A. Ujiie, T. Takahashi, T. Nakano, T. Suzuki, and K. Fukumoto, *Chem. Pharm. Bull.* **21,** 766 (1973).
387. S. V. Kessar, S. Batra, U. K. Nadir, and S. S. Ghandi, *Indian J. Chem.* **13,** 1109 (1975).
388. T. Kametani, S. Shibuya, and S. Kano, *J. Chem. Soc., Perkin Trans. 1* 1212 (1973).
389. S. Kano, Y. Takahagi, E. Komiyama, T. Yokomatsu, and S. Shibuya, *Heterocycles* **4,** 1013 (1976).
390. I. Ahmad and M. S. Gibson, *Can. J. Chem.* **53,** 3660 (1975).
391. M. S. Gibson and J. M. Walthew, *Chem. Ind. (London)* 185 (1965); M. S. Gibson, G. W. Prenton, and J. W. Walthew, *J. Chem. Soc. C* 2234 (1970).
392. L. Castedo, E. Guitian, J. Saa, and R. Suau, *Tetrahedron Lett.* **23,** 457 (1982).
393. B. Umezawa and O. Hoshino, *Heterocycles* **3,** 1005 (1975); *J. Synth. Org. Chem., Jpn.* **36,** 858 (1978).
394. O. Hoshino, T. Toshioka, and B. Umezawa, *Chem. Commun.* 1533 (1971); *Chem. Pharm. Bull.* **21,** 1302 (1974).
395. O. Hoshino, H. Hara, N. Serizawa, and B. Umezawa, *Chem. Pharm. Bull.* **23,** 2048 (1975).
396. H. Hara, O. Hoshino, and B. Umezawa, *Heterocycles* **3,** 123 (1975).
397. H. Hara, O. Hoshino, and B. Umezawa, *Chem. Pharm. Bull.* **24,** 262, 1921 (1976).
398. O. Hoshino, H. Hara, M. Ogawa, and B. Umezawa, *Chem. Pharm. Bull.* **23,** 2578 (1975).
399. O. Hoshino, H. Hara, M. Ogawa, and B. Umezawa, *Chem. Commun.* 306 (1975).
400. H. Hara, O. Hoshino, T. Ishige, and B. Umezawa, *Chem. Pharm. Bull.* **29,** 1083 (1981).
401. O. Hoshino, M. Ohtani, and B. Umezawa, *Chem. Pharm. Bull.* **26,** 3920 (1978).
402. O. Hoshino, M. Ohtani, and B. Umezawa, *Chem. Pharm. Bull.* **27,** 3101 (1979).
403. C. Szántay, M. Barczai-Beke, P. Pechy, G. Blaskó, and G. Dörnyei, *J. Org. Chem.* **47,** 594 (1982).
404. M. Gerecke and A. Brossi, *Helv. Chim. Acta* **62,** 1549 (1979).
405. L. Castedo, R. Estévez, J. M. Saá, and R. Suau, *Tetrahedron Lett.* 2179 (1978).
406. T. Kametani, T. Uryu, and K. Fukumoto, *J. Chem. Soc., Perkin Trans. 1* 383 (1977).
407. W. V. Curran, *J. Heterocycl. Chem.* **10,** 307 (1973).
408. L. I. Goldberg, P. F. Sonneville, and J. L. McNay, *J. Pharmacol. Exp. Ther.* **163,** 188 (1968).
409. A. Barnett and J. W. Fiove, *Eur. J. Pharmacol.* **14,** 206 (1971).
410. A. M. Ernst and P. G. Smelik, *Experientia* **22,** 837 (1966); N. E. Andén, A. Rubenson, K. Fuxe, and T. Hökfelt, *J. Pharm. Pharmacol.* **19,** 627 (1967).
411. V. J. Lotti, *Life Sci.* **10,** 781 (1971).
412. G. C. Cotzias, P. S. Papavasiliou, C. Fehling, B. Kaufman, and E. Mena, *N. Engl. J. Med.* **282,** 31 (1970); J. Braham, I. Savova-Pinho, and I. Goldhammer, *Br. Med. J.* **3,** 768 (1970); S. Daley, G. C. Cotzias, A. Steck, and P. S. Papavasiliou, *Fed. Proc., Fed. Am. Soc. Exp.*

Biol. **30,** 216 (1971); P. Castaigne, D. Laplane, and G. Dordain, *Res. Commun. Chem. Pathol. Pharmacol.* **2,** 154 (1971).
413. F. C. Colpaert, W. F. M. van Bever, and J. E. Leysen, *Int. Rev. Neurobiol.* **19,** 225 (1976).
414. J. L. Neumeyer, S. J. Law, and J. S. Lamont, *in* "Apomorphine and Other Dopaminomimetics" (G. L. Gessa and G. U. Corsini, eds.), Vol. I, p. 209. Raven Press, New York, 1981.
415. A. Quevauviller and M. Hamonniere, *C. R. Hebd. Seances Acad. Sci., Ser. D* **284,** 93 (1977).
416. M. P. Dubey, R. C. Srimal, and B. N. Dhawan, *Indian J. Pharmacol.* **7,** 73 (1969).
417. W.-N. Wu, J. L. Beal, and R. W. Doskotch, *Tetrahedron* **33,** 2919 (1977).
418. W.-T. Liao, J. L. Beal, W.-N. Wu, and R. W. Doskotch, *Lloydia* **41,** 271 (1978).
419. Ger. Offen. 2,717,062; *Chem. Abstr.* **90,** 43806 (1979).
420. Ger. Offen. 2,717,001; *Chem. Abstr.* **90,** 43807 (1979).
421. C. D. Hufford, A. S. Sharma, and B. O. Oguntimein, *J. Pharm. Sci.* **69,** 1180 (1980).
422. D. Walterová, J. Kovár, V. Preininger, and V. Šimánek, *Collect. Czech. Chem. Commun.* **47,** 296 (1982).
423. M. Gerecke, R. Borer, and A. Brossi, *Helv. Chim. Acta* **59,** 2551 (1976).
424. J. P. O'Brien and S. Teitel, *Heterocycles* **11,** 347 (1978).
425. R. Ahmad, J. M. Saá, and M. P. Cava, *J. Org. Chem.* **42,** 128 (1977).
426. J. C. Kim, *Taehan Hwahakhoe Chi.* **24,** 266 (1980).
427. N. Mollov and S. Philipov, *Chem. Ber.* **112,** 3737 (1979).
428. L. Castedo, A. Rodoriguez de Lera, J. M. Saá, R. Suau, and C. Villaverde, *Heterocycles* **14,** 1135 (1980).
429. J. Minamikawa and A. Brossi, *Can. J. Chem.* **57,** 1720 (1979).
430. F. C. Copp, A. R. Elphick, and K. W. Franzmann, *Chem. Commun.* 507 (1979).
431. J. C. Kim, *Org. Prep. Proced. Int.* **9,** 1 (1977); *Chem. Abstr.* **87,** 23574 (1977).
432. P. J. Davis, D. Wiese, and J. P. Rosazza, *J. Chem. Soc., Perkin Trans. 1* 1 (1977).
433. R. Ahmad, *Islamabad J. Sci.* **4,** 36 (1977).
434. M. P. Cava, D. L. Edie, and J. M. Saá, *J. Org. Chem.* **40,** 3601 (1975).
435. L. Castedo, T. Iglesias, A. Puga, J. M. Saá, and R. Suau, *Heterocycles* **15,** 915 (1981).
436. V. Chervenkova, N. Mollov, and S. Paszyc, *Phytochemistry* **20,** 2285 (1981).
437. J. Kunitomo, Y. Murakami, and M. Akasu, *J. Pharm. Soc. Jpn.* **100,** 337 (1980).
438. L. Castedo, D. Dominguez, J. M. Saá, and R. Suau, *Tetrahedron Lett.* 4589 (1979).
439. M. Gerecke, R. Borer, and A. Brossi, *Helv. Chim. Acta* **62,** 1543 (1979).
440. P. J. Davis and J. P. Rosazza, *Bioorg. Chem.* **10,** 971 (1981).
441. S. Teitel and J. P. O'Brien, *Heterocycles* **5,** 85 (1976).
442. K. C. Rice and A. Brossi, *Syn. Commun.* **8,** 391 (1978).
442a. A. Brossi, M. F. Rahman, K. C. Rice, M. Gerecke, R. Borer, J. P. O'Brien, and S. Teitel, *Heterocycles* **7,** 277 (1977).
443. E. Wenkert, B. L. Buckwalter, I. R. Burfitt, M. I. Gasic, H. E. Gottlieb, E. W. Hagaman, F. M. Schell, and P. M. Wovkulich, *Top. Carbon-13 NMR Spectrosc.* **2,** 105 (1976).
444. M. Shamma and D. M. Hindenlang, "Carbon-13 NMR Shift Assignments of Amines and Alkaloids." Plenum, New York, 1979.
445. L. M. Jackman, J. C. Trewella, J. L. Moniot, M. Shamma, R. L. Stephens, E. Wenkert, M. Levoeuf, and A. Cavé, *J. Nat. Prod.* **42,** 437 (1979).
446. N. A. Shaath and T. O. Soine, *J. Org. Chem.* **40,** 1987 (1975).
447. M. Shamma and S. Y. Yao, *J. Org. Chem.* **36,** 3253 (1971).
448. H. Guinaudeau, *J. Nat. Prod.* **46,** 761 (1983).

—CHAPTER 5—

PHTHALIDEISOQUINOLINE ALKALOIDS AND RELATED COMPOUNDS

D. B. MACLEAN

Department of Chemistry
McMaster University
Hamilton, Ontario, Canada

I. Introduction

The term phthalideisoquinoline alkaloid has been historically associated with compounds of the structure represented in Table I. In recent years, however, a relatively large number of compounds have been isolated that are formally derived from the classical structures by ring-cleavage reactions. These compounds, known as *seco*-phthalideisoquinoline alkaloids, are also included in this review. There are several structural variants within this group, examples of which may be

THE ALKALOIDS, VOL. XXIV

ISBN 0-12-469524-8

found in Table V. It is now believed that the imides of the seco group may be artefacts of isolation but they are included here for the sake of completeness.

In the period since the last review by Sǎntavý (*1*) many significant advances have been made in the chemistry of these alkaloids. The advances in synthesis have been particularly noteworthy and have come about to some extent as a result of the greater accessibility of substituted phthalides and phthalide derivatives. Transformation of the berberine system into the phthalideisoquinoline system, because of the biogenetic connection, has also been given much attention as has the transformation of the phthalideisoquinolines into other ring systems present in the isoquinoline alkaloids and into other alkaloids of the series. The examination of new plant species or the reexamination of previously examined species has led to the isolation of several new alkaloids as well as the detection of compounds of established structure. Tables listing the alkaloids in alphabetical order and giving their structures and their physical properties will be found in the text. All of the known alkaloids are included and as a result there is some duplication of references with those appearing in earlier reviews. For the most part, however, this work is a sequel to the section on the phthalideisoquinolines found in Santavý's chapter on the Papaveraceae alkaloids and covers the literature until mid-1983.

This group of alkaloids is reviewed regularly in the series on alkaloids (*2*) in the specialist reports of the Chemical Society and a chapter is devoted to them in a recent book by Shamma and Moniot (*3*). A compendium of the alkaloids extant in mid-1981 including a listing of their physical properties became available in 1982 (*4*).

II. The Phthalideisoquinoline Alkaloids and Their Occurrence

An alphabetical listing of the alkaloids along with a key to their structure and their relative configuration may be found in Table I. The sources of the alkaloids are listed in Table II in which the literature has been covered from 1975 to mid-1983. References prior to 1975 are given only in those instances where a reference to the isolation of the alkaloid has not appeared since that date. Information on the stereochemical form, (+), (−), or (±), in which the alkaloid was isolated may also be found in Table II (*5–60*).

Four new alkaloids of this series have been discovered since the previous review by Sǎntavý (*1*). Hydrastidine (**10**) and isohydrastidine (**11**), isolated from *H. canadensis* L., are isomeric phenolic alkaloids that yield (−)-β-hydrastine on methylation with diazomethane. This reaction defines their relative and absolute configuration as well as their structure (*51*). The positions of the OH and OCH_3 groups in the phthalide moiety were deduced from their IR spectra. In hydrasti-

TABLE I

THE STRUCTURES OF THE ALKALOIDS AND THEIR RELATIVE CONFIGURATIONS[a]

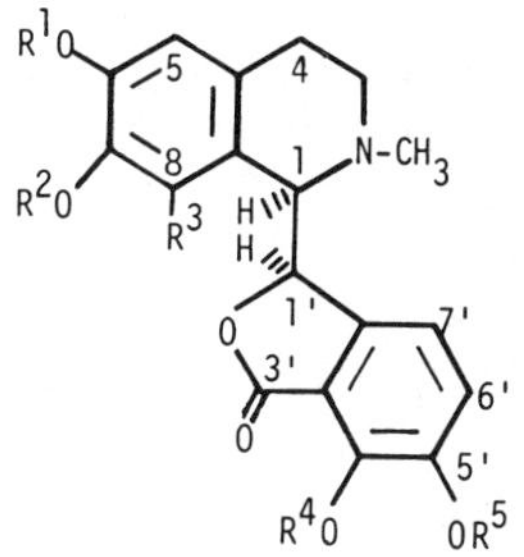

Alkaloid	R^1	R^2	R^3	R^4	R^4	Relative configuration
1 Adlumidine[b]	CH_2		H	CH_2		threo
2 Adlumine	CH_3	CH_3	H	CH_2		threo
3 Bicuculline	CH_2		H	CH_2		erythro
4 Capnoidine[b]	CH_2		H	CH_2		threo
5 Cordrastine I	CH_3	CH_3	H	CH_3	CH_3	threo
6 Cordrastine II	CH_3	CH_3	H	CH_3	CH_3	erythro
7 Corledine	H	CH_3	H	CH_2		threo
8 Corlumidine	CH_3	H	H	CH_2		erythro
9 Corlumine	CH_3	CH_3	H	CH_2		erythro
10 Hydrastidine	CH_2		H	H	CH_3	erythro
11 Isohydrastidine	CH_2		H	CH_3	H	erythro
12 α-Hydrastine	CH_2		H	CH_3	CH_3	threo
13 β-Hydrastine	CH_2		H	CH_3	CH_3	erythro
14 α-Narcotine	CH_2		OCH_3	CH_3	CH_3	erythro
15 β-Narcotine	CH_2		OCH_3	CH_3	CH_3	threo
16 Narcotoline	CH_2		OH	CH_3	CH_3	erythro
17 Severtzine[c]	CH_3	H	H	CH_2		threo

[a] For purpose of reference the configuration shown at the chiral centers 1 and 1′ is (*R*,*R*), threo.

[b] (−)-Capnoidine and (+)-adlumidine are enantiomeric but each has been given a trivial name.

[c] Also spelled severcine and severzine.

dine the OH group was internally hydrogen bonded ($\nu_{max}^{CHCl_3} = 3500\ cm^{-1}$) while in isohydrastidine it was not ($\nu_{max}^{CHCl_3} = 3620\ cm^{-1}$). There were also differences in the chemical shift of the methoxyl group in both the ^1H- and ^{13}C-NMR spectra that supported the structural assignments based on IR examination.

In the same year, Israilov *et al.* (*46*) reported the isolation of corftaline, an isomer of the hydrastidines, from *C. pseudoadunca*. Methylation yielded (+)-β-hydrastine. The ^1H-NMR spectrum of corftaline is very similar to that of iso-hydrastine, as would be expected of an enantiomer. However their reported melting points differ, which is not expected of enantiomers, and their specific

TABLE II
OCCURRENCE OF THE ALKALOIDS

Adlumidines: *Corydalis gigantea* Troutv. and Mey. (+) (*5*); *C. ochotensis* Turcz. (+) (*6,7*); *C. ochotensis* var. *raddeana* (+) (*8,9*); *C. paniculigera* (*10*); *C. remota* (+) (*5*); *C. rosea* (±) (*5*); *Fumaria indica* Pugsley (+) (*11*); *F. parviflora* Lam. (+) (*12*); *F. parviflora* (*13*); *F. vaillantii* (*14*) *C. vaginans* (+) and (±) (*5*).

Adlumines: *C. gigantea* Troutv. and Mey. (*5*), (−) (*15*); *C.* lineariodes (+) (*16*); *C. paniculigera* (*10*); *C. sempervirens* (L.) Pers. (*C. glauca* Pursh.) (−) (*17*); *C. sempervirens* (*18*); *C. vaginans* Royle (±) (*5*); *F. kralikii* Jord. (−) (*19*); *F. parviflora* (*13*); *F. schrammii* (−) (*20,21*); *F. vaillantii* (*14*)

N-Methyladlumine: *F. parviflora* (*13*); *F. vaillantii* (*14*)

Bicucullines: *C. bulbosa* (*22*); *D. decumbens* (*23*); *C. gigantea* Troutv. and Mey (+) (*15*); *C. govaniana* Wall. (+) (*24*); *C. lutea* (L.) D.C. (+) (*25*); *C. mucronifera* (*16*); *C. ochotensis* var *raddeana* (*9*); *C. paniculigera* (*10*); *C. remota* (*5*); *C. repens* (*26*); *C. sempervirens* (L.) Pers. (*C. glauca* Pursh) (*17*); *C. suaveolens* Hance (*27*); *C. taliensis* (*28*); *C. vaginans* Royle (*5*); *F. judaica* (*29*); *F. muralis* (*30*); *F. parviflora* Lam (+) (*12,31*); *F. parviflora* (*13*); *F. schleicherii* Soy–Will (±) (*32,33*); *F. vaillantii* Loisl. (−) (*34*); *F. vaillantii* (*14*)

(−)-Capnoidine: *C. bulbosa* (*22*); *C. cava* (L.) Sch. and K. (*C. tuberosa* D.C.) (*35,36*); *C. gigantea* Trautv. and Mey. (*15*); *C. gortschakovii* Schrenk (*37*); *C. sempervirens* (L.) Pers. (*C. glauca* Pursh.) (*17*); *F. muralis* (*30*)

Cordrastines: Cordrastine was reported only once from a natural source, *C. aurea* Willd., and its configuration was not assigned (*38*). It has been synthesized several times and separated into *threo*-cordrastine-I and *erythro*-condrastine II forms (*39–44*); (−)-cordrastine I and (−)-cordrastine II have been prepared synthetically (*45*).

(+)-Corftaline: *C. pseudoadunca* (*46*)

(−)-Corledine: *C. ledebouriana* K. et K. (*47*); *C. ledebouriana* (*48*).

(+)-Corlumidine: *C. decumbens* (*23*); *C. lineariodes* (*16*).

Corlumines: *C. severtozovii* Rgl. (+) (*49,50*); *F. parviflora* Lam. (−) (*12*).

(−)-Hydrastidine: *Hydrastis canadensis* L. (*51*).

(−)-Isohydrastidine: *H. canadensis* L. (*51*).

α-Hydrastines: *F. parviflora* (+) (*13*); *F. schleicherii* Soy–Will (+) (*32*), (±) (*33*); *F.* vaillantii (+) (*14*).

β-Hydrastines: *C. pseudoadunca* (+) (*52*); *C. stricta* (+) (*16*); *F. schleicherii* Soy–Will (±) (*33*); *H. canadensis* L. (*51*).

α-Narcotines and β-Narcotines: *C. cava* (L.) Sch. & K. (*C. tuberosa* D.C.) (*35*); *Papaver cylindricum* (*53*); *P. persicum* Lindl. (*54*); *P. rhoeas* Linn. (*55*); *P. somniferum* Linn (*56*); *P. somniferum* (*57*); Poppy straw (*58*).

(−)-Narcotoline: *P. somniferum* (*59,60*); Poppy straw (*58*)

(−)-Severtzine: *C. severtzovii* Rgl. (*50*); *C. severtzovii* (*48*)

rotations differ not only in sign as expected but in magnitude, and more so than might be expected because of a difference in solvent. The two alkaloids should be examined under identical conditions in order to resolve the discrepancies between their physical properties. It would appear however that corftaline is (+)-isohydrastidine and this name is recommended for the alkaloid until its structure is proved otherwise.

(−)-Corlumine has been isolated from *F. parviflora* (*12*). It is the first time that this enantiomer has been reported.

An alkaloid designated decumbenine, isomeric with the bicucullines, has been isolated from *C. decumbens* (*23*). It is stated that one methylenedioxy group is situated at C-6 and C-7 in the isoquinoline moiety, which is normal, and the second at C-6′ and C-7′ in the phthalide moiety, which is unprecedented and unlikely on biogenetic grounds. Acceptance of this structural formula must await verification. Unfortunately, the original paper was not available to the reviewer at the time of writing.

N-Methyladlumine has been reported in two species, *F. parviflora* (*13*) and *F. vaillantii* (*14*).

III. Physical Properties of Phthalideisoquinoline Alkaloids

In Table III (*61–81*) the absolute configuration, the melting points, and specific rotations of the alkaloids are collated, and references are given to sources providing ^{1}H- and ^{13}C-NMR data. In the case of melting points and specific rotations only a single value is recorded. Readers are referred to the more complete listing of Blaskó *et al.* (*4*), which also provides UV, IR, CD, and ORD data.

The relative configuration of the alkaloids (Table I) was first derived by ^{1}H-NMR spectroscopy (*82*) and subsequent studies have proved the reliability of this method (*1, 83*). In the cases examined to date, ^{13}C-NMR spectra are also diagnostic of configuration provided both diastereomeric forms are available for examination (*65, 69*). Once the relative configuration has been established, there appears to be a correlation between rotation and absolute configuration as intimated by Šantavý (*1*). Analysis of the data of Table III shows that in both the threo and erythro series the enantiomer of negative rotation has the *R* configuration at C-1. The absolute configurations of the alkaloids have been established by ORD and CD studies and corroborated by X-ray and chemical correlation as previously discussed (*1*). CD studies have now been reported on severtzine and corledine and on several other alkaloids of the series (*84*).

The conformations of the alkaloids in solution in the erythro and the threo series have been discussed by Elango *et al.* (*85*). Their studies, using a 200-MHz instrument and nuclear Overhauser enhancement studies, have apparently clarified earlier conflicting interpretations of the data (*70, 86*). It is now believed that the favored conformation of the erythro compounds may be represented by structure A and for the threo compounds by structure B in Fig. 1 (*85*).

TABLE III
PHYSICAL PROPERTIES OF THE ALKALOIDS

	Molecular formula	Configuration (1,1′)	Melting point (°C)	Specific rotation, $[\alpha]_D^T$ [T°C, $([\alpha]_D^T)°$ (concn, solvent)]	^{1}H NMR	^{13}C NMR
(+)-Adlumidine[a]	$C_{20}H_{17}NO_6$	S,S	239–240(*7*)	20, +116.2(*c* 2, $CHCl_3$)(*61*)	(*4,12*)	
(±)-Adlumidine	—	—	205(*61*)	—	—	—
(+)-Adlumine	$C_{21}H_{21}NO_6$	S,S	180(*62*)	—, +42($CHCl_3$)(*63,64*)	(*4*)	(*4,65*)
(−)-Adlumine	—	R,R	180(*66*)	20, −42(*c* 1.8, $CHCl_3$)(*67*)	—	—
(±)-Adlumine	—	—	190(*68*)	—	—	—
(+)-Bicuculline	$C_{20}H_{17}NO_6$	S,R	193–194(*45*)	20, +132.7(*c* 0.49,$CHCl_3$)(*9*)	(*4*)	—
(−)-Bicuculline	—	R,S	193–194(*45*)	33, −128(*c* 0.27,$CHCl_3$)(*45*)	(*4*)	(*69*)
(±)-Bicuculline	—	—	217–220(*70*)	—	—	—
(−)-Capnoidine[a]	$C_{20}H_{17}NO_6$	R,R	239(*62*)	20, −116(*c* 0.52, $CHCl_3$)(*67*)	(*67*)	(*4,69,71*)
(−)-Cordrastine I[b]	$C_{22}H_{25}NO_6$	R,R	196(*38*), 189–90(*45*)	—, −99(*c* 1,$CHCl_3$)(*45*)	(*4,45*)	—
(±)-Cordrastine I[b]	—	—	156–157(*39*)	—	(*39,41*)	—
(−)-Cordrastine II[b]	—	R,S	90(*45*)	—, −10(*c* 1.0,$CHCl_3$)(*45*)	—	—
(±)-Cordrastine II[b]	—	—	117–118(*39*)	—	(*4,39,41*)	—
(+)Corftaline[c]	$C_{20}N_{19}NO_6$	S,R	173–174(*46*)	—, +33(*c* 0.32,MeOH)(*46*)	(*46*)	—
(−)-Corledine	$C_{20}H_{19}NO_6$	R,R	210–212(*47*)	—, −100(*c* 0.2,MeOH)(*47*)	(*4,47*)	—
(+)-Corlumidine	$C_{20}H_{19}NO_6$	S,R	236(*62*)	23, +80(*c* 0.4,$CHCl_3$)(*64*)	—	—
(+)-Corlumine	$C_{21}H_{21}NO_4$	S,R	162(*72*)	25, +77(*c* 1.0, $CHCl_3$)(*72*)	(*4,72*)	(*4,65,69*)

(−)-Corlumine	—	R,S	oil	—	(*4,12*)	—
(±)-Corlumine[b]	—	—	178–181(*70*)	—	—	—
(−)-Hydrastidine	$C_{20}H_{19}NO_6$	R,S	172–174(*51*)	20, −60.3(*c* 0.7,$CHCl_3$)(*51*)	(*4,51*)	(*4,51*)
(−)-Isohydrastidine[c]	$C_{20}H_{19}NO_6$	R,S	163–166(*51*)	20, −77.7(*c* 0.6,$CHCl_3$)(*51*)	(*4,51*)	(*4,51*)
(+)-α-Hydrastine	$C_{21}H_{21}NO_6$	S,S	159–161(*12*)	—, +127.7(*c* 1.1,$CHCl_3$)(*73*)	(*4,74*)	(*4,65,69,71*)
(−)-α-Hydrastine	—	R,R	162–163.5(*75*)	—, −163(*c* 1.0,$CHCl_3$)(*76*)	—	—
(±)-α-Hydrastine	—	—	150–154(*70*)	—	—	—
(+)-β-Hydrastine	$C_{21}H_{21}NO_6$	S,R	131–132(*77*)	18, +63($CHCl_3$)(*77*)	—	—
(−)-β-Hydrastine	—	R,S	141–143(*78*)	25, −59.9(*c* 1.0,$CHCl_3$)(*78*)	(*4,74,78*)	(*4,65,69*)
(±)-β-Hydrastine	—	—	136–140(*70*)	—	—	—
(+)-α-Narcotine[b]	$C_{22}H_{23}NO_7$	S,R	179(*79*)	—, +199.9(79)	—	—
(−)-α-Narcotine	—	R,S	176(*64*)	—, −200($CHCl_3$)(*63*)	(*4*)	—
(±)-α-Narcotine[d]	—	—	230–233(*64*)	—	(*4,80*)	—
(−)-β-Narcotine	$C_{22}H_{23}NO_7$	R,R	181–182(*80*)	20, −87.5(*c* 1,$CHCl_3$)(*80*)	(*4,80*)	
(±)-β-Narcotine[d]	—	—	184–186(*80*)	—	—	—
(−)-Narcotoline	$C_{21}H_{21}NO_7$	R,S	202(*81*)	—, −189($CHCl_3$)(*64*)	—	—
(−)-Severtzine	$C_{20}H_{19}NO_6$	R,R	94–95(*50*)	—, −52(*c* 0.91,$CHCl_3$)(*50*)	(*4,50*)	—

[a] (+)-Adlumidine and (−)-Capnoidine are enantiomers.
[b] Synthetic compounds; not found in nature.
[c] (+)-Corftaline and (−)-Isohydrastidine are enantiomers.
[d] Racemic narcotines are also called gnoscopines.

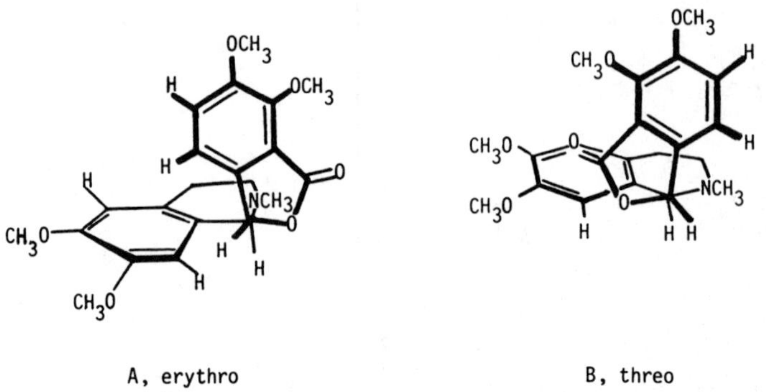

A, erythro B, threo

FIG. 1. Preferred conformation of *erthro*- and *threo*-phthalideisoquinolines.

IV. Synthesis of the Alkaloids

A. From 3,4-Dihydroisoquinolinium Salts

The direct condensation of a 2-methyl-3,4-dihydroisoquinolinium salt with a phthalide derivative is, in terms of synthetic strategy, the most direct route to this class of alkaloids. This approach was used by Perkin and Robinson (*79*) in the first successful synthesis of a phthalideisoquinoline alkaloid but was not developed into a practicable method until a few years ago. Both halophthalides and phthalide anions are now known to be effective phthalide derivatives for use in the condensation. These approaches to the alkaloids have gone hand in hand with improvements in the synthesis of the phthalides themselves (*87–90*). It is now a relatively simple matter to prepare a substituted phthalide whereas only a few years ago this was a formidable task.

The reaction with halophthlides was investigated in this laboratory, a preliminary report appearing in symposium papers in 1979 (*91*) and a full account of the work in 1981 (*70*). In this study it was shown that 3-halophthalides condense readily with 3,4-dihydroisoquinolinium salts in the presence of Zn metal or Zn–Cu couple as outlined in Scheme 1. Both erythro and threo forms of the alkaloids are formed. (In all syntheses discussed, racemates are obtained.) The reaction was used successfully in the synthesis of the diastereomers adlumidine (**1**) and bicuculline (**3**), adlumine (**2**) and corlumine (**9**), α- and β-hydrastine, **12** and **13,** and the isocordrastines (**18** and **19**). Shono *et al.* (*44*) have used the same route under different experimental conditions to prepare the cordrastines (**5** and **6**), hydrastines, narcotines (**14** and **15**), and isocordrastines. It is worthwhile noting that Shono *et al.* found a preference for formation of the erythro isomer, whereas our studies, although not consistent, showed a preference for threo isomers.

Shono *et al.* have also condensed halophthalides with dihydroisoquinolinium salts, using an electrochemical method (*43*). In this way they have prepared the a(i) X=Cl or Br; Zn or Zn-Cu in DMF (70)

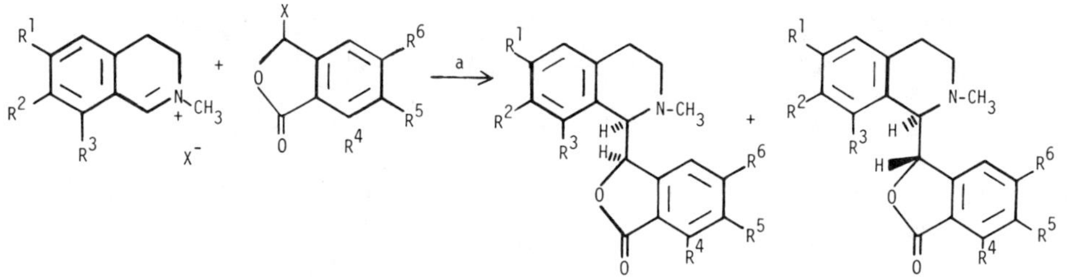

a(ii) X=Br; Zn in CH_3CN (44)

a(iii) X=Br; electrochemical reduction (43)

a(iv) X=Li; THF, -40° (92)

COMPOUNDS*	R^1	R^2	R^3	R^4	R^5	R^6
1 and **3**	OCH_2O		H	OCH_2O		H
2 and **9**	OCH_3	OCH_3	H	OCH_2O		H
5 and **6**	OCH_3	OCH_3	H	OCH_3	OCH_3	H
12 and **13**	OCH_2O		H	OCH_3	OCH_3	H
15 and **14**	OCH_2O		OCH_3	OCH_3	OCH_3	H
18 and **19**	OCH_3	OCH_3	H	H	OCH_3	OCH_3

SCHEME 1. *The numbers in the first column represent threo isomers, in the second, erythro isomers.

cordrastines, hydrastines, and narcotines. The mechanism of the electrochemical and Zn-promoted coupling reactions has not been clarified.

The coupling of phthalide anions with dihydroisoquinolinium salts has now been realized (*92*). The overall yields are not high but there is compensation in the simplicity of the reaction itself. This and the previous reactions are all outlined in Scheme 1.

B. FROM ISOQUINOLINES

Several methods have been developed in which isoquinolines or isoquinoline derivatives are coupled with a phthalide or phthalide derivative to construct the framework of the phthalideisoquinoline system.

Prager *et al.* (*93*) have condensed phthalides and isoquinolinium salts with sodium methoxide in refluxing methanol. The condensation product was then reduced to a mixture of the threo and erythro isomers, as shown in Scheme 2. A variety of substituted and unsubstituted compounds were prepared in their study but none of the natural bases. The erythro isomer was predominant in the mixtures. This method might be expected to be generally applicable, however, to the synthesis of the alkaloids.

Kerekes *et al.* (*42, 94*) have demonstrated that anions derived from Reissert compounds (**20**) react with phthalaldehydic acid esters (**21**) to yield 1-benzyliso-

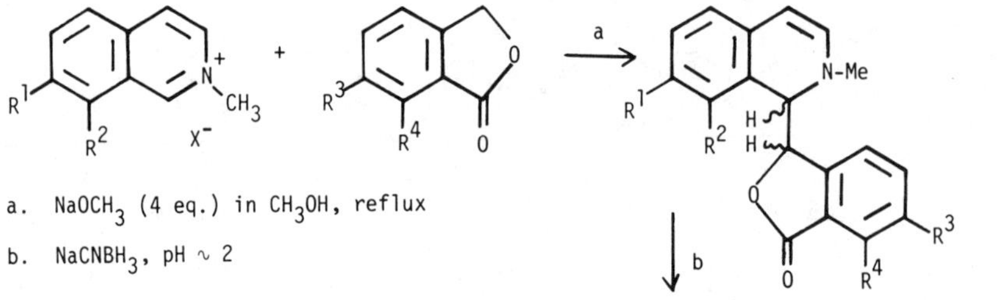

a. $NaOCH_3$ (4 eq.) in CH_3OH, reflux

b. $NaCNBH_3$, pH ~ 2

threo- and erythro-phthalidetetrahydroisoquinolines

SCHEME 2

quinolines (**22**) carrying a benzoyloxy group at the α-position. These compounds have been converted by conventional steps to phthalideisoquinoline alkaloids, as shown in Scheme 3. In this way α- and β-hydrastine and cordrastine I and II were prepared along with a number of compounds with substitution patterns not found in nature.

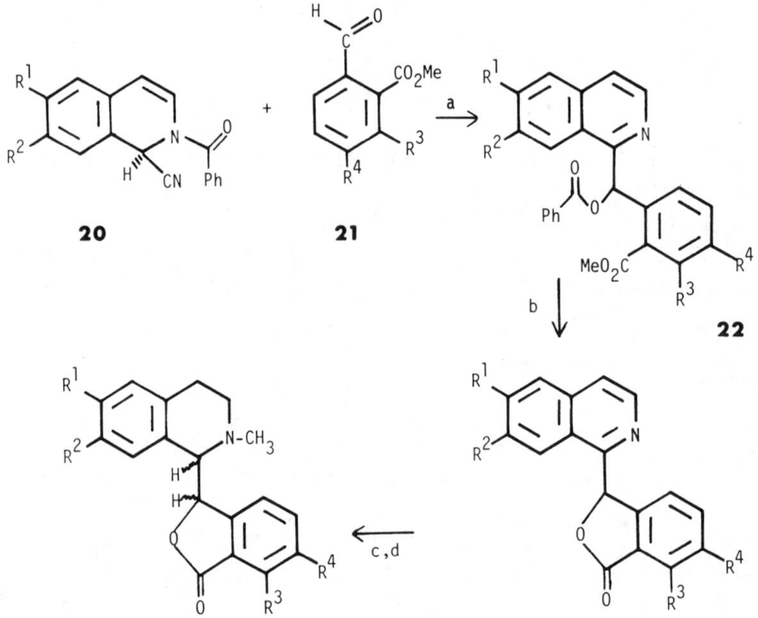

20 **21** **22** **23**

5,6; $R^1=R^2=R^3=R^4=OCH_3$

12,13; $R^1+R^2=OCH_2O$, $R_3=R_4=OCH_3$

a, NaH, DMF, -20°C, 3h; b, (i) OH^-, H_2O, (ii) H_3O^+; c, H_2/cat.; d, $CH_2=O$, HCO_2H

SCHEME 3

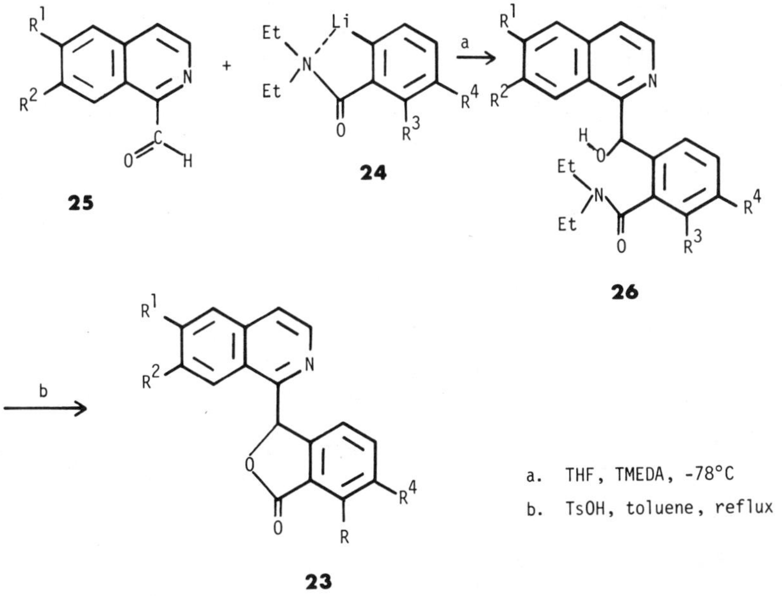

a. THF, TMEDA, -78°C

b. TsOH, toluene, reflux

SCHEME 4

Hung *et al.* (*95*) have used the same approach to prepare aromatic analogs of the alkaloids **23** and have investigated the condensation of Reissert compounds with phthalaldehydic acids, using phase-transfer conditions.

Snieckus *et al.* (*90, 96*) have developed an interesting approach to compounds of structure **23.** They found that the lithiated carboxamide **24** reacted readily with the 1-formylisoquinoline (**25**), yielding compound **26.** The latter on hydrolysis provided the lactone **23** ($R^1 = R^2 = R^3 = R^4 = OCH_3$), a compound that had been already converted to a mixture of cordrastine I and II so that their synthesis constituted a formal synthesis of these alkaloids. The reaction was also applied to compounds containing two methylenedioxy groups in place of the four methoxy groups (Scheme 4).

C. FROM *N*-PHENYLETHYLPHTHALIDE-3-CARBOXAMIDES

In the synthetic approach to the alkaloids developed by Haworth and Pinder (*97*), phenylethylamines (**27,** R = CH_3) were condensed with phthalide-3-carboxylic acid derivatives (**28**, X = Cl) in the preparation of the amides (**29,** R = CH_3) required for cyclization to the dehydrophthalides (**30,** R = CH_3) (see Scheme 5). The latter were then reduced to the alkaloids. Several new routes have been described recently for the preparation of the intermediate amides. Falck and Manna (*40*) have applied the Passerini reaction in an innovative way to accomplish this end. They have shown that piperonyl bromide will react with the

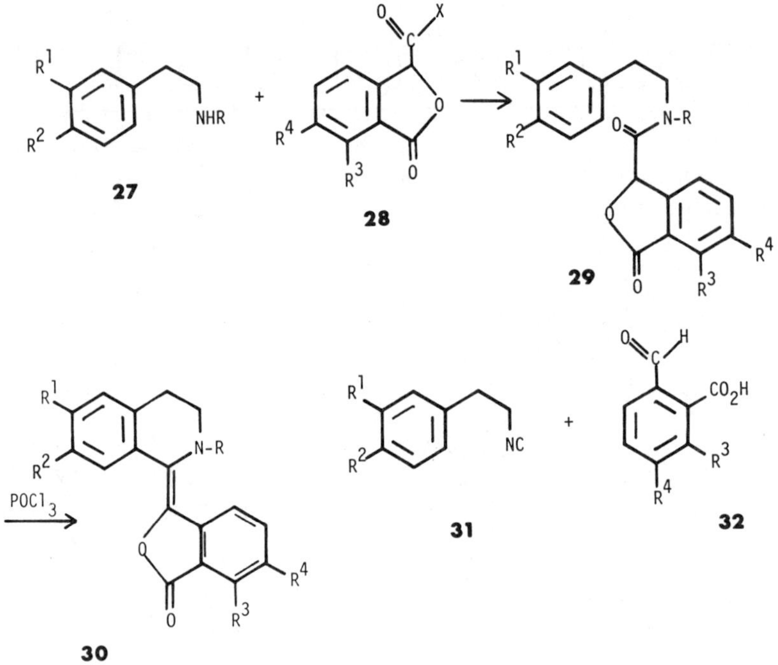

SCHEME 5

lithium derivative of methyl isocyanide to generate the isonitrile **31** ($R^1 + R^2 = OCH_2O$). The latter on treatment with phthalaldehydic acids **32** ($R^3 = R^4 = OCH_3$) in an intramolecular Passerini reaction yielded the amide **29** (R = H). This compound under the usual Haworth–Pinder conditions may be converted to the hydrastines, N-methylation being the final step. Snieckus and co-workers (*41, 96*) have shown that α-bromohomophthalic anhydrides react with phenylethylamines (**27**, $R = CH_3$) to provide the amide (**29**, $R = CH_3$, $R^1 = R^2 = R^3 = R^4 = OCH_3$) which may be converted to the cordrastines.

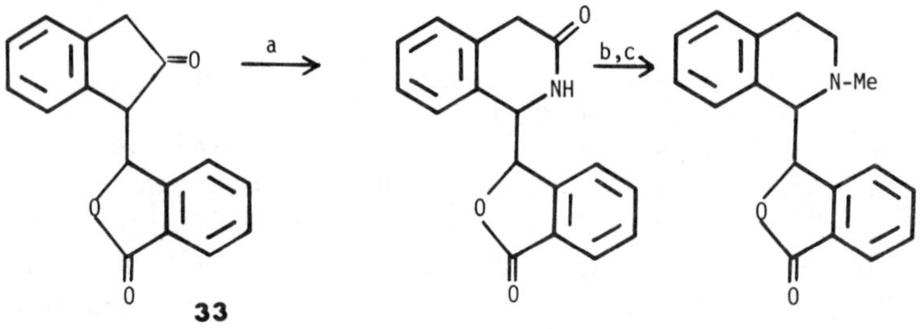

a, HN_3-H_3PO_4-P_2O_5; b, Xs FSO_3Me, K_2CO_3; c, $NaBH_4$

SCHEME 6

D. Miscellaneous Methods

Hutchison *et al.* (*98*) have condensed 2-indanones with phthalaldehydic acid **32** to produce the expected condensation product **33.** This in turn was converted by the steps indicated in Scheme 6 to a phthalidetetrahydroisoquinoline to which the authors assign the erythro configuration. It is stated that indanones substituted in the aromatic ring with alkoxy groups failed to condense in the first step so that this approach leads only to analogs of the alkaloids.

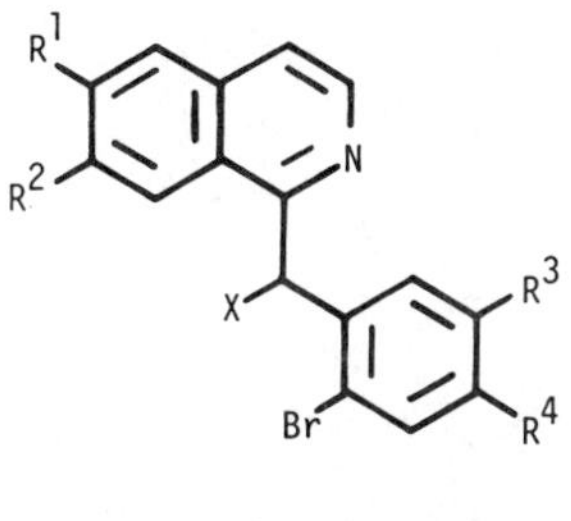

34 X=OH, $R^1=R^2=R^3=R^4=H$

35 X=H, $R^1=R^2=R^3=R^4=OCH_3$

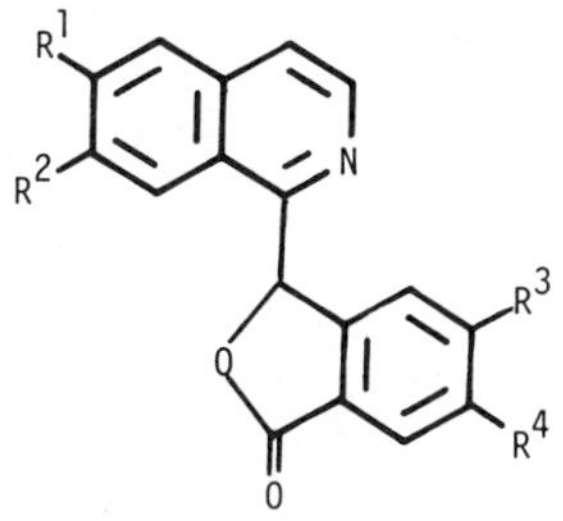

36 $R^1=R^2=R^3=R^4=OCH_3$

Scheme 7

Hung *et al.* (*87*) have prepared phthalideisoquinoline systems from compounds such as **34** by metalation and carboxylation in hexane solution (see Scheme 7). The relatively low yields obtained in the reactions are ascribed to competitive deprotonation of the benzylic position. In an alternative approach they have treated 2′-bromopapaverine (**35**) with *n*-BuLi at −78°C and converted it to the acid with CO_2. Reduction to the hydroxymethyl compound and oxidation by the method of Shamma and St. Georgiev (*86*) gave the expected phthalide **36.** These methods, although novel in approach are not as direct nor as versatile as methods discussed previously.

V. Transformations among Isoquinoline Alkaloids Involving Phthalideisoquinolines

A. Conversion of Protoberberines to Phthalideisoquinolines

The transformation of the protoberberine system into the phthalideisoquinoline system has been a topic of lively interest over the past decade. Three groups, those of Shamma, Kondo, and Hanaoka, have made significant contributions to these studies.

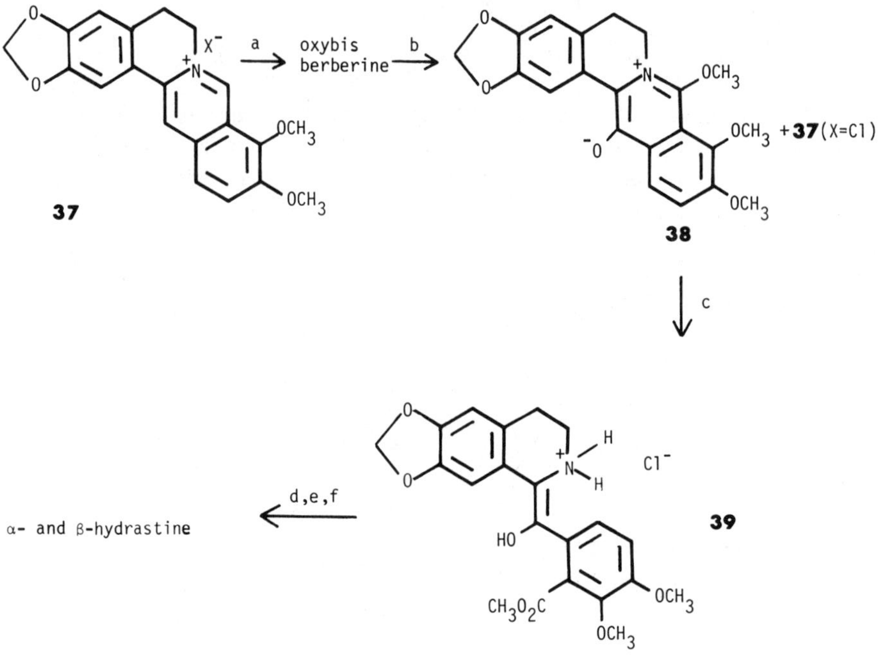

X⁻ a → oxybis berberine b →

OCH_3 OCH_3

37

^-O OCH_3 OCH_3 OCH_3 + **37**(X=Cl)

38

↓ c

H H Cl^-

α- and β-hydrastine ← d,e,f

HO CH_3O_2C OCH_3 OCH_3

39

a, $K_3Fe(CN)_6$, H_2O; basify, extract; b, MeOH, HCl; c, wet ether, HCl;
d, CH_3I, CH_3CN; e, $NaBH_4$; f, HX, H_2O

SCHEME 8

In 1976 Moniot and Shamma (*99*) reported that berberine (**37**) on oxidation with ferricyanide yielded a dimeric product of oxidation that was named oxybisberberine (Scheme 8). The latter, on treatment with methanolic HCl, yielded a mixture of berberine chloride and 8-methoxyberberinephenolbetaine (**38**). Compound **38** was transformed in wet ether into the methyl ester **39,** which was converted by N-methylation, reduction, and cyclization to a mixture of α- and β-hydrastines, respectively. A more complete account of this work has also appeared (*100*).

A second method (*101*) was developed to effect the same conversion and yielded β-hydrastine with only traces of the α isomer (Scheme 9). Oxoberberine (**40**) on photolysis yielded a lactol (**42**), which was postulated to form through the intermediacy of a peroxy lactone (**41**) and a keto acid, a tautomer of **42.** Three routes were developed to convert the lactol to β-hydrastine: $NaBH_4$ reduction followed by acidification and N-methylation, N-methylation followed by $NaBH_4$ reduction, and O- and N-methylation followed by borohydride reduction.

It was also shown (*102, 103*) that 8,13-dioxo-14-hydroxyberberine (**43**), de-a, hν, O_2, C_6H_6, 1h; b, Py.HCl in Py-H_2O; c, 25% H_2SO_4; d, 25% H_2SO_4 heat, 70°, 1 h, neutralize, extract.

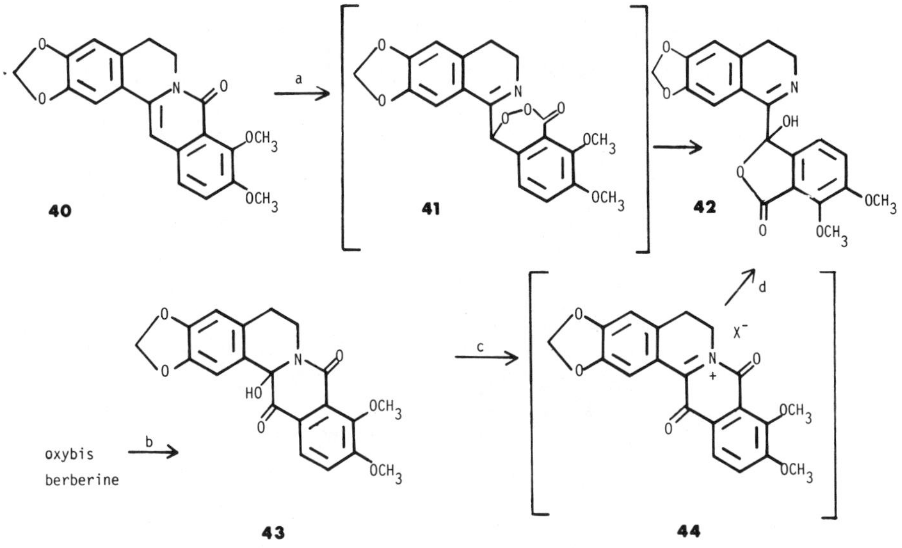

SCHEME 9

rived from oxybisberberine by cleavage with pyridine·HCl in aqueous pyridine, was converted by 25% H_2SO_4 to the iminium salt **44.** Heating of the acidic solution, followed by neutralization, gave the same lactol (**42**) described above. The reader is referred to further studies on 8,13-dioxoberberines, which have also led to the hydrastines (*104*).

Kondo and co-workers in a similar time period studied the oxidation of berberines and in 1977 effected their conversion to phthalideisoquinolines. In one route (*105–107*), outlined in Scheme 10, norcoralyne (**45**) was reduced to its dihydro derivative (**46**) and oxidized with *m*-chloroperbenzoic acid to the betaine (**47**). The latter, on photochemical oxidation with oxygen, in the presence of Rose Bengal, followed by borohydride reduction, yielded the lactone (**48**). In another approach, coralyne (**49**) was reduced to its dihydro derivative (**50**) and subjected to photochemical oxidation, yielding 6-acetylpapaveraldine (**52**). Autoxidation of **50** in hot ethanol gave the phenolbetaine (**51**), which was also oxidized readily to **52.** The latter was converted with alkaline hypobromite to papaveraldine-6-carboxylic acid, which gave the lactone (**48**) on borohydride reduction and acidification.

It was found (*106–108*) that 7,8-dihydroberberine is readily oxidized photochemically in the presence of Rose Bengal to 13-hydroxyberberinephenolbetaine (**53**). By further oxidation of **53** at low concentration in MeOH, it was a, Zn/AcOH, HCl; b, MCPBA, Et_3N, -20°; c, (i) hν/O_2/Rose Bengal, 0°C, 10 min; (ii), $NaBH_4$; d, EtOH/O_2/reflux, 2h; e, MCPBA or hν/O_2/Rose Bengal; f, hν/O_2/Rose Bengal; g, NaOBr, H_3O^+, $NaBH_4$.

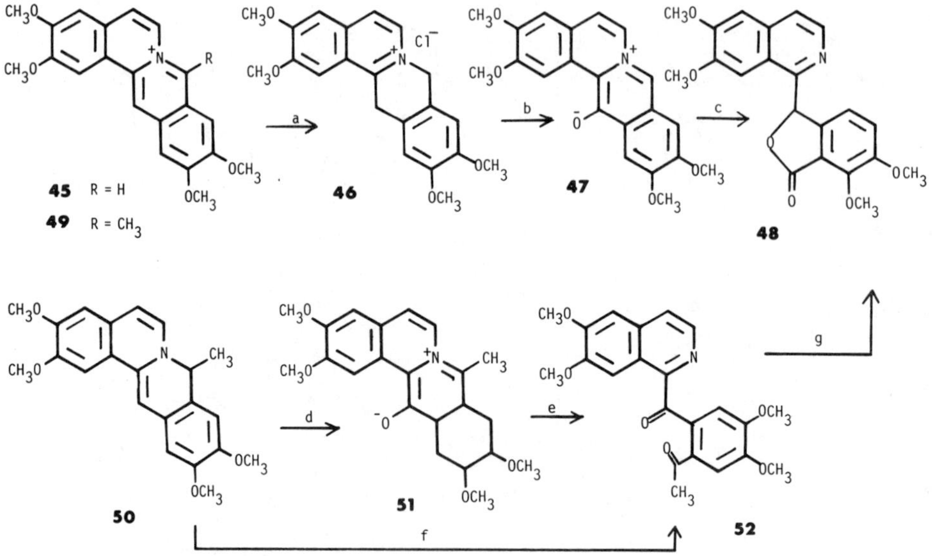

SCHEME 10

possible to isolate the 8,14-epidioxy compound **54.** A structure of this type is considered to be an intermediate in the oxidation of the phenolbetaines of Scheme 10 and a mechanism is suggested for its formation. Treatment of **54** with pyridine·HCl in pyridine yielded three products: noroxohydrastinine (9%) and compounds **55** (42%) and **43** (40%) (Scheme 11). The latter are interconvertible, and **43** has been assigned the same structure as the compound reported by Shamma, although Kondo *et al.* (*107*) state the two have different physical properties; their reactions are, however, similar. Similarly, **55** is a tautomer of Shamma's lactol (**42**). By routes analogous to those employed by Shamma, compound **55** was converted in good yield to β-hydrastine.

Hanaoka *et al.* (*109, 110*) have developed a simple method for the preparation of 8-methoxyberberinephenolbetaine. Irradiation of berberine chloride in methanol containing NaOMe, Rose Bengal, and oxygen afforded an intermediate that yielded **38** on recrystallization from methanol, the overall yield being ~60%.

In a nonoxidative method, Hanaoka *et al.* (*111*) have converted (±)-*O*-acetylophiocarpine (**56**) and (±)-*O*-acetylepiophiocarpine (**57**) to (±)-α- and β-hydrastines, respectively. The N—C-8 bond was cleaved with ethyl chloroformate, and in a series of reactions the resulting CH_2Cl and NCO_2Et groups were

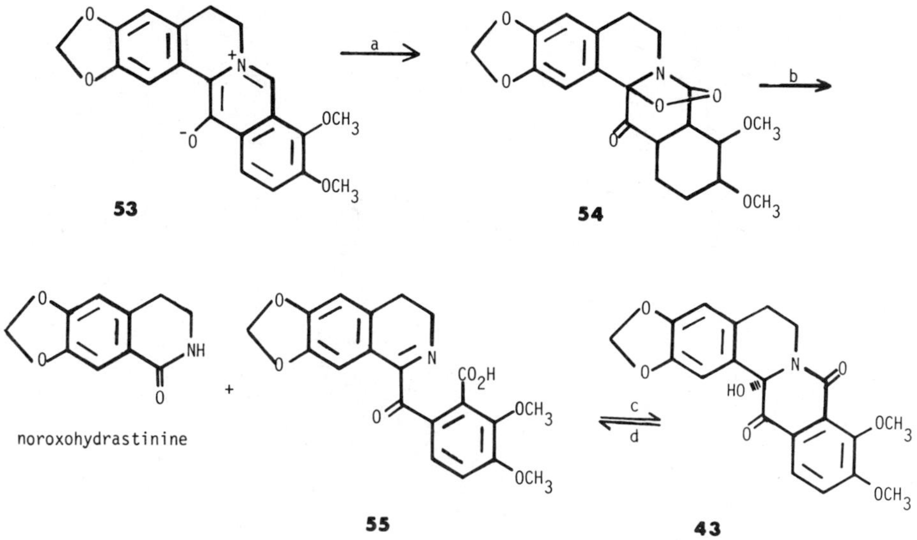

53 **54**

noroxohydrastinine + **55** **43**

a, hν/O_2/Rose Bengal, 0°C, MeOH; b, Py.HCl, Py; c, Ac_2O; d, 25% H_2SO_4

SCHEME 11

transformed to the appropriate functionality without affecting the configuration at C-13 and C-14 in **56** and **57**.

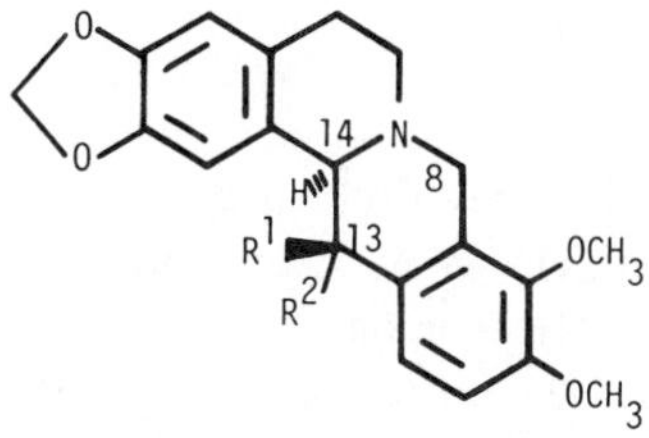

56 R^1 = OAc, R^2 = H
57 R^1 = H, R^2 = OAc

B. CONVERSION OF PHTHALIDEISOQUINOLINES TO OTHER ISOQUINOLINE ALKALOIDS

Nalliah *et al.* have demonstrated that dehydrophthalideisoquinolines (**30**), intermediates in the Haworth synthesis, may be converted to spirobenzylisoquinoline alkaloids of the corydaine–sibiricine type (*1, 112*). In a subsequent paper (*68*) they have reported that the phthalide alkaloids themselves may be converted via a Polonovski reaction to dehydrophthalides (**30**) and thereby serve as starting materials for the synthesis.

Nalliah and MacLean (*113*) have achieved the conversion of dehydrobicuculline (**58**) to the *seco*-phthalideisoquinoline bicucullinine (**59**) and to hypera, $LiAlH_4$/THF/0°C; b, $Hg(OAc)_2$/AcOH/H_2O; c, CH_3OSO_2F/CH_2Cl_2

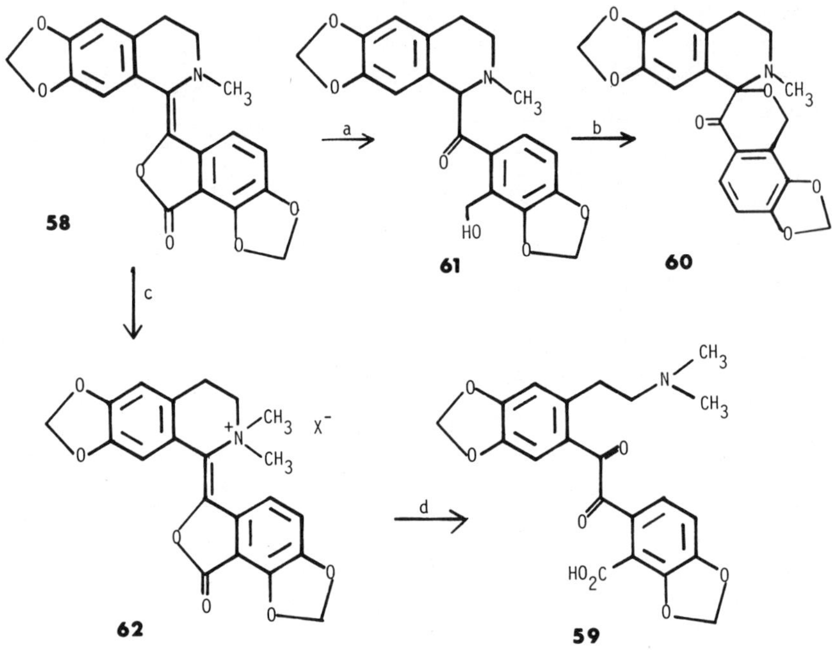

d, MeOH/Triton B, KI, HCl

SCHEME 12

corinine (**60**) as shown in Scheme 12. Dehydrobicuculline was reduced with $LiAlH_4$ to the keto alcohol **61,** which on air oxidation or oxidation with $Hg(OAc)_2$ yielded hypercorinine. N-Methylation of **58** yielded **62,** which on treatment with triton B in methanol in the presence of KI underwent ring opening, air oxidation, and reduction. Bicucullinine was isolated as its hydrochloride.

C. Conversions within the Phthalideisoquinoline Series

Kerekes *et al.* (*78*) have converted (−)-α-narcotoline (**16**) to (−)-β-hydrastine by removal of the OH group at C-8 of narcotoline. They employed the procedure developed by Teitel and O'Brien (*114*) in which the phenol was treated with 5-chloro-1-phenyl-1*H*-tetrazole to generate the corresponding ether. Hydrogenolysis of the ether over Pt yielded (−)-β-hydrastine. They reported that there was no isomerization at the chiral centers in the reaction.

Rao and Kapicak (*115*) have reported that (−)-β-hydrastine on treatment with

N-bromosuccinimide yielded 4-bromo-6,7-methylenedioxy-2-methylisoquinolinium bromide, the phthalide moiety being expelled in the process.

Kerekes and Gaál (*116*) have studied the reaction of cyanogen bromide with narcotines under solvolytic conditions (THF–H_2O or $CHCl_3$–EtOH). In both cases ring opening to the cyanamide occurred with attachment of OH or OC_2H_5 at C-1. The surprising feature of the reaction was that (−)-α-narcotine (1*R*,1′*S*) and (+)-β-narcotine (1*S*,1′*S*) gave the same compound, namely, the *seco*-(1*S*, 1′*S*)-cyanamide. Hydrolysis of the cyanamide in aqueous HCl was accompanied by recyclization of the seco compound, yielding mainly (+)-β-narcotine (15 parts) with a small amount (1 part) of (−)-α-narcotine. The enantiomeric compounds on similar treatment gave, as expected, (−)-β-narcotine with a small amount of (+)-α-narcotine. The authors provide an explanation for these results based on conformational and mechanistic arguments. The practical consequence of these studies lies in the provision of a method to convert the erythro to the threo isomer.

VI. *seco*-Phthalideisoquinoline Alkaloids

A. The Alkaloids and Their Occurrence

An alphabetical listing of the alkaloids and the plant species in which they have been found is provided in Table IV (*117–132*). The seco compounds are less widely distributed than the parent alkaloids and have been found only in species of the *Fumariaceae* and *Papaveraceae*.

A key to the structure of the alkaloids and several synthetic analogs is given in Table V. The alkaloids are organized into four groups for ease of tabulation. The first and second groups comprise enol lactones and ene lactams in the *Z* and *E* configurations, respectively. The third group is made up of mono and diketo carboxylic acids, and the final group of two is characterized by the presence of a saturated lactone ring.

It has been pointed out by others (*117, 125, 128*) that all of these compounds may be formally derived from quaternary *N*-methylphthalideisoquinolinium salts through Hofmann ring-opening reactions of the nitrogen-containing ring. This reaction would lead to the formation of the dimethylaminoethyl group, which is present in all of the compounds except nornarceine. Blaskó *et al.* (*125*) have pointed out that nornarceine may be an artefact of isolation resulting from N-demethylation, presumably of narceine, in the acidic media used in extraction. However, its formation directly from narcotine cannot be ruled out.

Enol lactones would be expected to be the initial products of the base-catalyzed ring-fission reaction. They could be transformed into the keto acids of the third group by opening of the lactone ring of the enol lactones and into the diketo

TABLE IV

OCCURRENCE OF THE *seco*-PHTHALIDEISOQUINOLINES

Adlumiceine (**63**): *Corydalis sempervirens* (L.) Pers. (*C. glauca* Pursh) (*17,117*); *Fumaria schrammii* (Ascherson) Velen (*20,21,118*).
Adlumiceine enol lactone (**64**): *F. schrammii* (Ascherson) Velen (*118*).
Adlumidiceine (**65**): *C. cava* Sch. & K (*C. tuberosa* D.C.) (*35*); C. lutea (L.) D.C. (*25,119*); *C. sempervirens* (L.) Pers (*17,117*); *F. kralikii* (*120,121*); F. parviflora (*13,120,121*); *F. schrammii* (*20,21,118*); *F. vaillantii* (*14*); *Papaver rhoeas* (L.) (*117*).
Adlumidiceine enol lactone (**66**): *C. sempervirens* (L.) Pers. (*117*); *F. schramii* (*118*).
Aobamidine (**67**): *C. lutea* (L.) D.C. (*25*); *C. ochotensis* var. *raddeana* (*8,9*).
Bicucullinidine (**68**): *F. schrammii* (*20,21*).
Bicucullinine (narceimine) (**59**): *C. ochroleuca* Koch (*122*); *F. indica* Pugsley (*11*); *F. schrammii* (*20,21*).
Fumaramidine (**69**): *F. parviflora* Lam. (*31*)
Fumaramine (**70**): *C. ochroleuca* Koch (*122*); *F. parviflora* Lam. (*31,73,123*); *F. vaillantii* Loisl. (*123,124*).
Fumaridine (**71**): *F. parviflora* Lam. (*31,73,123*); *F. schleicherii* Soy–Will. (*123*); *F. vaillantii* Loisl. (*123,124*).
E-Fumaridine (**72**): Synthetic (*125*).
Fumschleicherine (**73**): *F. schleicherii* Soy–Will (*126,127*); *F. schrammii* (*21*).
N-Methylhydrasteine (**74**): *C. lutea* (L.) D.C. (*128*); *C. solida* (L.) Swart (*C. densiflora* (*125*); *F. officinalis* (*119*); *F. parviflora* Lam. (*128*); *F. vaillantii* Loisl. (*128*).
N-Methylhydrastine (**75**): *C. lutea* (L.) D.C. (*128*); *F. officinalis* (*119*); *F. parviflora* Lam. (*31,128*); *F. vaillantii* (*128*).
E-N-Methylhydrastine (**76**): Synthetic (*125*).
N-Methyloxohydrasteine (**77**): *F. microcarpa* Boiss. (*125*); *F. officinalis* (*119*).
Narceine (**78**): *P. somniferum* L. (*55,63*).
[a]Narceine enol lactone (**79**): synthetic (*129*).
Narceine imide (**80**): P. *somniferum* L. (*130*).
E-Narceine imide (**81**): synthetic (*131*).
Narlumidine (**82**): *F. indica* Pugsley (*11*).
Nornarceine (**83**): *P. somniferum* L. (*81*); Synthetic (*132*).

[a] Original reference unavailable to author at time of writing, Also called aponarceine.

acids by air oxidation of the monoketo acids. The lactams of groups 1, 2, and 4 may well be artefacts of isolation resulting from reaction of the enol lactones or keto acids with ammonia, a reagent frequently used in isolation processes (*125*, *133*). Narlumidine is the only alkaloid that does not fit conveniently into this pattern.

The presence of *N*-methylphthalideisoquinolinium salts in several plant species is well documented (*133*) (see Section II), and it is not improbable that all alkaloids of this group may be artefacts of isolation resulting from ring opening of quaternary salts under the alkaline conditions often used in the extraction process. Despite these misgivings about their authenticity as constituents of plants the seco compounds nevertheless deserve treatment in this review, if not as a separate group of alkaloids, then surely as a significant part of the chemistry of the phthalideisoquinoline group.

Since the previous review of Săntavý (*1*) five new alkaloids have been isolated, three diastereomers of alkaloids of established structure have been synthesized, and narceine enol lactone has been prepared. Adlumiceine enol lactone (**64**) has been identified in the extracts of *F. schrammii* (*118*) and has been assigned the *Z* configuration. Bicucullinidine (**68**) has also been found in *F. schrammii*. Its structure was elucidated by spectroscopic methods (*20, 21*) and by comparison of its properties with those of the structurally related bicucullinine and *N*-methyloxohydrasteine (**77**). Fumaramidine, found in the extract of *F. parviflora* (*31*), was assigned its structure on the basis of spectroscopic examination and by comparison of its properties with those of the isomeric fumaridine (**71**) of established structure (*134*). Fumschleicherine (**73**), found in *F. schleicheiri* (*126, 127*) and in *F. schrammii* (*21*), is readily dehydrated to fumaramine (**70**) (*133*) on treatment with trifluoroacetic acid. The remaining structural problem, the assignment of the OH group to the 1′- or the 1-position, was based on analysis of its ^{1}H- and ^{13}C-NMR spectra (*127*). Narlumidine (**82**), has been isolated from *F. indica* (*11*). It has been assigned a structural formula (Table V) that does not fit the pattern expected of *seco*-phthalideisoquinolines. Unfortunately, the authors have not provided a complete listing of the ^{1}H-NMR signals of narlumidine nor have they provided supporting ^{13}C-NMR evidence that would substantiate their structural assignment. This structural formula of narlumidine should be treated therefore with some scepticism until a more detailed examination has been carried out. *E*-Fumaridine (**72**) (*125*), *E*-*N*-methylhydrastine (**76**) (*125*) and *E*-narceine imide (**81**) (*131*) are not natural compounds but have been prepared synthetically as discussed in a following section. *Z*-Narceine enol lactone (**79**) as its hydrochloride has been reported (*129*).

B. Physical Properties

The alkaloids, their melting points, and references (where available) to their ^{1}H- and ^{13}C-NMR spectra may be found in Table VI. Information on their IR, UV, and mass spectra is recorded in the review of Blaskó *et al.* (*4*).

The stereochemistry of the enol lactones has been correlated with that of an analog of aobamidine in which one *N*-methyl group has been replaced by a *p*-bromophenoxycarbonyl group. The structure of this compound was established by X-ray analysis (*135*).

Blaskó *et al.* (*125*) and Proska and Voticky (*131*) have found a correlation between the stereochemistry of the enol lactones and ene lactams and their UV and ^{1}H-NMR spectra. The absorption bands in the UV spectrum of the *E* isomer of *N*-methylhydrastine appear at shorter wavelength than those of the *Z*-isomer (**75**) (*125*), which is not unexpected because the *E* isomer is the more hindered of the two and unable to assume a planar conformation. It was noted, however, that the spectra change rapidly in methanol solution so that one must take care in recording the spectra of these compounds. Caution should therefore be exercised

TABLE V

seco-PHTHALIDEISOQUINOLINES

	X	R^1	R^2	R^3	R^4	R^5
64 Adlumiceine enol lactone	O	CH_3	CH_3	H	CH_2	
67 Aobamidine	O	CH_2		H	CH_2	
75 *N*-Methylhydrastine	O	CH_2		H	CH_3	CH_3
79 Narceine enol lactone	O	CH_2		OCH_3	CH_3	CH_3
69 Fumaramidine	NH	CH_3	CH_3	H	CH_2	
70 Fumaramine	NH	CH_2		H	CH_2	
71 Fumaridine	NH	CH_2		H	CH_3	CH_3
80 Narceineimide	NH	CH_2		OCH_3	CH_3	CH_3
66 Adlumidiceine enol lactone	O	CH_2		H	CH_2	
76 *E*-*N*-Methylhydrastine	O	CH_2		H	CH_3	CH_3
72 *E*-Fumaridine	NH	CH_2		H	CH_3	CH_3
81 *E*-Narceineimide	NH	CH_2		OCH_3	CH_3	CH_3

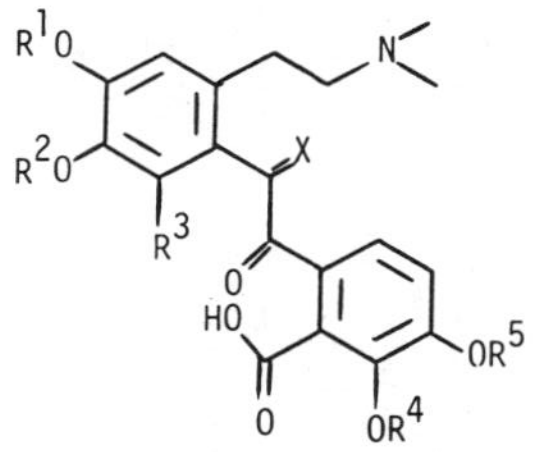

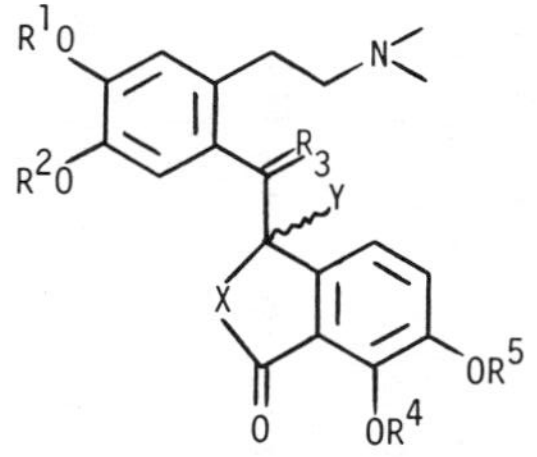

Compound	X	R^1	R^2	R^3	R^4	R^5
63 Adlumiceine	H,H	CH_3	CH_3	H	CH_2	
65 Adlumidiceine	H,H	CH_2		H	CH_2	
68 Bicucullinidine	O	CH_3	CH_3	H	CH_2	
59 Bicucullinine	O	CH_2		H	CH_2	
74 *N*-Methylhydrasteine	H,H	CH_2		H	CH_3	CH_3
77 *N*-Methyloxohyrasteine	O	CH_2		H	CH_3	CH_3
78 Narceine	H,H	CH_2		OCH_3	CH_3	CH_3
83 Nornarceine[a]	H,H	CH_2		OCH_3	CH_3	CH_3
82 Narlumidine (Y = H)	O	CH_2		O	CH_2	
73 Fumschleicherine (Y = OH)	NH	CH_2		H,H	CH_2	

[a] Lacks one *N*-methyl group relative to narceine.

TABLE VI
PHYSICAL PROPERTIES OF *seco*-PHTHALIDEISOQUINOLINES

Compound	Molecular formula	Melting point (°C)	^{1}H NMR	^{13}C NMR
63 Adlumiceine	$C_{22}H_{25}NO_7$	226–228 (*118*)	(*4,117*)	—
64 Adlumiceine enol lactone	$C_{22}H_{23}NO_6$	170–175 (*118*)	—	—
65 Adlumidiceine	$C_{21}H_{21}NO_7$	244–246 (*25,117*)	(*4,117*)	—
66 Adlmidiceine enol lactone	$C_{21}H_{19}NO_6$	200–203 (*117*)	(*4,117*)	—
67 Aobamidine	$C_{21}H_{19}NO_6$	195–197 (*8*)	(*4,8*)	—
68 Bicucullinidine	$C_{22}H_{23}NO_8$	265–266 (*20*)	(*4,20*)	(*20*)
59 Bicucullinine	$C_{21}H_{19}NO_8$	268 (*122*)	(*4,20,122*)	(*4,20,122*)
69 Fumaramidine	$C_{22}H_{24}N_2O_5$		(*4,31*)	—
70 Fumaramine	$C_{21}H_{20}N_2O_5$	227 (*122*)	(*4,31*)	(*127*)
71 Fumaridine	$C_{22}H_{24}N_2O_5$	188–189 (*123*)	(*4,31,125*)	—
72 *E*-Fumaridine	$C_{22}H_{24}N_2O_5$	193–194 (*125*)	(*4,125*)	—
73 Fumschleicherine	$C_{21}H_{22}N_2O_6$	224–226 (*127*)	(*4,127*)	(*4,127*)
74 *N*-Methylhydrasteine	$C_{22}H_{25}NO_7$	223 (C_2H_5OH) (*119*)	(*4,119,125*)	—
		150 (H_2O) (*119*)		—
75 *N*-Methylhydrastine	$C_{22}H_{23}NO_6$	156 (*119*)	(*4,119,125*)	—
76 *E*-*N*-Methylhydrastine	$C_{22}H_{23}NO_6$		(*4,125*)	—
77 *N*-Methyloxohydrasteine	$C_{22}H_{23}NO_8$	168 (CH_3OH) (*119*)	(*4,119,125*)	—
		203 (H_2O) (*119*)		—
		234–235 (*125*)		
78 Narceine	$C_{23}H_{27}NO_8$	145 (81)	(*4,125*)	—
		155–157 ($3H_2O$) (*63*)		
79 Narceine enol lactone	$C_{23}H_{25}NO_7$			—
80 Narceine imide	$C_{23}H_{26}N_2O_6$	150–152 (*129*)	(*4,129*)	—
81 *E*-Narceine imide	$C_{23}H_{26}N_2O_6$	142–144 (*130*)	—	—
82 Narlumidine	$C_{21}H_{19}NO_7$	248–250 (*11*)	(*4,11*)	—
83 Nornarceine	$C_{22}H_{25}NO_8$	229 (*63,81*)	(*4,132*)	—
		223–225 (*132*)		

in recording spectra of these compounds. A similar correlation was observed for the isomeric ene lactams, *E*- and *Z*-furmaridine (*125*) and *E*- and *Z*-narceine imide (*130*). The ^{1}H-NMR spectra of the ene lactams and lactones were also informative (*125*). In the case of the isomeric *N*-methylhydrastines, the signals at H-8 and H-7′ were diagnostic of the stereochemistry. They appeared, respectively, at δ7.71 and 7.49 in the *Z* isomer and at δ6.85 and 6.90 in the *E* isomer. In the case of the isomeric fumaridines (**71** and **72**) only the H-7′ proton was diagnostic of structure appearing at δ7.48 in the *Z* isomer and δ6.80 in the *E* isomer. It should be noted that some proton assignments for the ene lactams and enol lactones quoted above differ from those found in the earlier review of Blaskó *et al.* (*4*). These correlation studies have placed on a firm foundation the stereochemical assignments given for all of the ene lactams and for the three enol lactones that Blaskó *et al.* considered (*125*). The assignment of the *Z* configuration to adlumiceine enol lactone (*118*) must be considered tentative since its diastereomer is unknown and no ^{1}H-HMR data are available.

C. Synthesis of seco Compounds

Blaskó *et al.* (*125*) have investigated the stereochemistry of the well-established conversion of *N*-methylphthalidetetrahydroisoquinolinium salts to the *seco*-phthalide enol lactones. Under the basic reaction conditions that they employed in their study they found that the reaction was highly stereospecific. For example, α-hydrastine (the threo isomer) yielded mainly *E*-*N*-methylhydrastine, while β-hydrastine (the erythro isomer) gave mainly *Z*-*N*-methylhydrastine. The reactions proceed via a syn rather than the more common anti elimination, presumably because of the less crowded transition state for syn elimination in these systems. These findings are in agreement with earlier studies that showed that adlumidine methiodide (**1**·MeI) (threo) (*9*) and the enantiomeric caponidine methiodide (**4**·MeI) (*117*) are both converted to adlumidiceine enol lactone (**66**) with the *E* configuration and that bicuculline (erythro) and β-hydrastine (erythro) are converted in a related reaction to *Z*-enol lactones (*135*).

The enol lactones isomerize on photolysis and in the case of **75** and **76** yield an equilibrium mixture of *Z* and *E* isomers in a 3:2 ratio. A similar equilibration of the ene lactams **71** and **72** has been observed by the same group (*125*).

The conversion of *N*-methylhydrasteine to *N*-methyloxohydrasteine has been effected by air oxidation (*125*), and in Section V,B the conversion of dehydrobicuculline to bicucullinine was discussed. An analog of fumschleicherine (**73**) was prepared by treating **74** with dilute aqueous ammonia at room temperature for 3 hr. The saturated hydroxy lactam so formed was readily transformed to **74** by acid treatment just as **73** was converted to **70.** The authors conclude from these and other studies that the hydroxy lactams and probably also the ene lactams are artefacts of isolation (*125*). It has also been shown that narceine enol lactone hydrochloride **79** reacts with amines and presumably also with ammonia to yield analogs of fumschleircherine and that these compounds are readily dehydrated to the corresponding ene lactams (*129*).

D. Reactions of seco Compounds

Several reactions of seco compounds derived from narcotine and narcotoline have been reported. 8-Benzylnarcotoline has been converted to **85** by the action of acetic acid (*136*) (Scheme 13). The product is an analog of nornarceine (**83**), which may form from narcotine in an analogous manner during the acid workup. *N*-Benzyl-8-ethylnarcotoline, on treatment with methanolic KOH, followed by acidification, gave **86.** Hydrogenolysis of **86** followed by air oxidation gave the benzazepinone **87** (*137*).

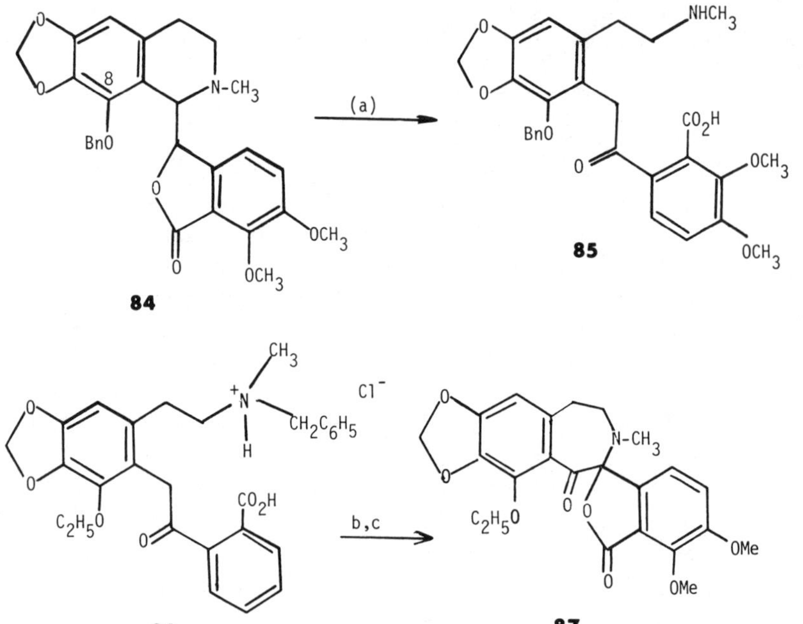

84 → (a) → 85

86 → b,c → 87

a, AcOH, 100°;

b, H_2/Pd; c, OH^-, air oxid^n.

SCHEME 13

α- or 1-Bromonarceinimide has been prepared and its properties and reactions have been examined (*138*).

The preparation and chemistry of narceineimide *N*-oxide (**88**) has been investigated (*130, 139, 140*). On treatment with acetic anhydride, **88** is converted to a 1:1 mixture of *N*-acetylnornarceine and, depending on the reaction conditions,

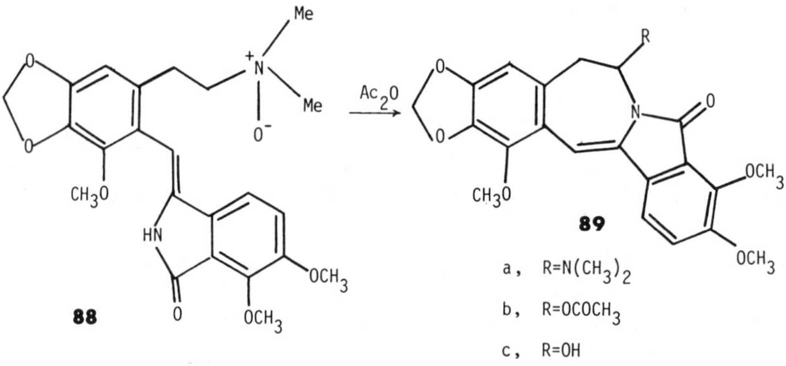

88 → Ac_2O → 89

a, $R=N(CH_3)_2$

b, $R=OCOCH_3$

c, $R=OH$

SCHEME 14

various substituted 7,8-dihydro-5*H*-isoindolo[1,2b][3]benzapein-5-ones (**89**) (see Scheme 14). Dehydration of **89c** or Hofmann elimination of **89a** yielded the expected unsaturated compound.

VII. Biosynthesis

It has been known for some 20 years that tyrosine and methionine are the basic building units in the biosynthesis of the isoquinoline alkaloids (*141, 142*) and that the protoberberberine alkaloid scoulerine (**90**), itself derived from reticuline (**91**) via tyrosine and methionine, is a precursor of α-narcotine in *P. somniferum* (*143*). Only 14-(*S*)-scoulerine is incorporated and H-14(C) retains its stereochemical integrity. Experiments with scoulerine stereospecifically labeled at C-13 have shown that the 13-*pro-S* hydrogen of scoulerine is removed in its

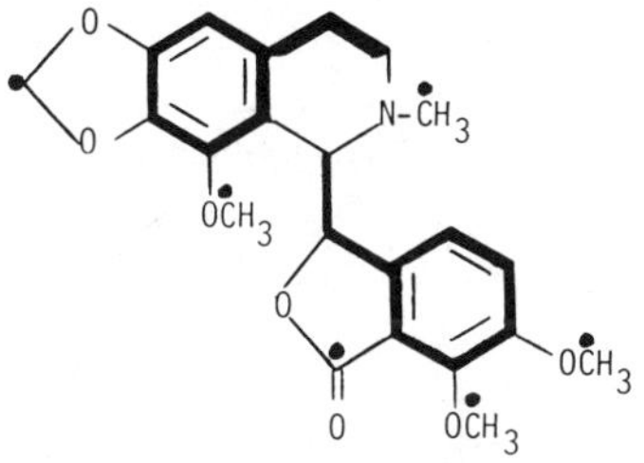

Tyrosine-derived C-atoms are represented by heavy lines. Methionine-derived C-atoms by •.

narcotine

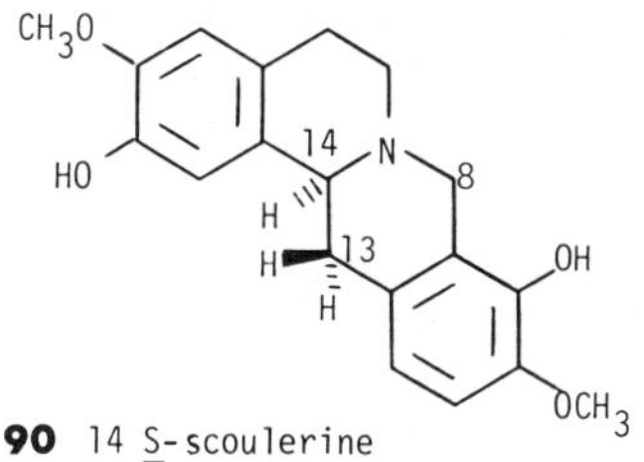

90 14 <u>S</u>-scoulerine

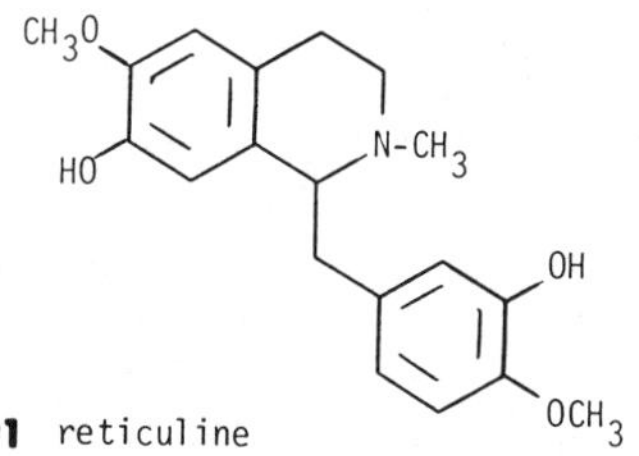

91 reticuline

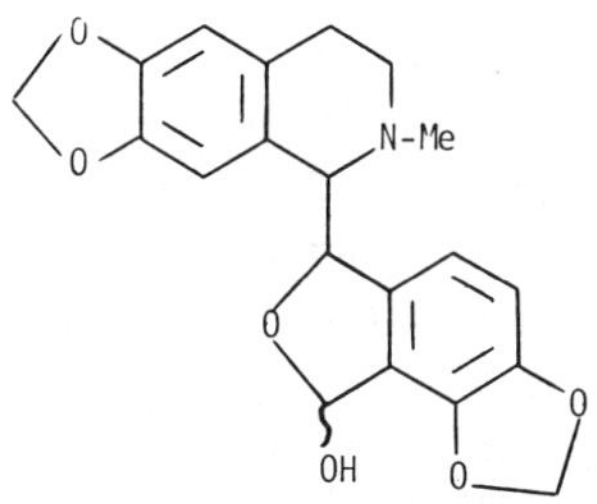

92 (+)-egenine

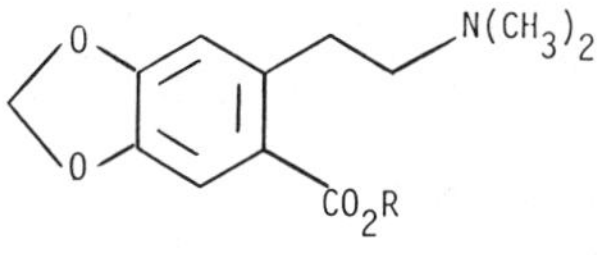

93a R = C_2H_5, parviflorine

93b R = H

conversion to narcotine in *P. somniferum*. The intermediates between scoulerine and narcotine have not yet been identified although it has been suggested that ophiocarpine [**56** (R^1 = OH, R^2 = H)] may be on the pathway (*144*). In this connection, it is interesting to note that the alkaloid (+)-egenine (**92**), containing a hemiacetal group, has been isolated by Gözler *et al.* (*145*). They have observed that compounds of this functionality may be the immediate precursors of the phthalideisoquinolines in the plant.

Hussain and Shamma (*146*) have discussed the catabolism of the phthalideisoquinolines. They have isolated the alkaloid fumariflorine (**93a**) from *F. parviflora* Lam. and consider it, or the amino acid **93b,** of which it is a derivative, to be the end product of catabolism of the isoquinoline portion of the molecule. The formation of **93b** and 3,4-methylenedioxyphthalic acid by oxidation of the seco alkaloid bicucullinine is readily envisaged (see Section VI).

VIII. Pharmacology

In the early 1970s Curtis *et al.* (*147–149*) reported that (+)-bicuculline is an antagonist of the inhibitory neurotransmitter γ-aminobutyric acid (GABA). These studies were extended to various analogs of bicuculline of which the most effective proved to be bicuculline methochloride (*150*). In the same publication it was pointed out that (+)-corlumine, which has the same absolute configuration as bicuculline, possessed similar activity. Bicuculline methochloride has the advantage of water solubility over the free base, and like the free base, only the methochloride of (+)-bicuculline is an effective GABA antagonist (*151*); the methiodide is reported to behave similarly (*152*). Bicuculline has now become widely used in neurological research and references to it in this regard abound in the current literature.

Kardos *et al.* (*153*) have recently examined some 45 phthalideisoquinolines and *seco*-phthalideisoquinolines as GABA antagonists in rat brain synaptic membranes (some seco compounds surprisingly showed evidence of binding affinity). It was found that there was a relationship between structure and physiological activity. In particular, the authors observed that erythro isomers were more active than were threo and that compounds with the 1*S* configuration were better binding agents than those with 1*R* in the phthalide systems. The conformation of the molecules in solution was considered to be an important factor affecting binding affinity. The N—CO distance, which is a function of the solution conformation, was postulated to have a major influence on binding to the receptor site. The results also suggested that the conformation of the receptor site was the same for GABA agonists and phthalideisoquinolines.

Hydrastine and its salts have been reported in two patents (*154*) to be effective in ophthalmology for dilation of pupils and in topical anesthesia, among other things. The antitussive properties of narcotine, its salts, and several analogs have been examined and compared with other agents (*155*).

Acknowledgments

I wish to express my thanks to Gayle Griffin for her patience and good nature in typing this manuscript and preparing the diagrams. My thanks are also extended to Drs. I. M. Piper and R. Marsden, postdoctoral fellows, who pointed out references that I missed or correlations that I failed to observe and to my students, R. Gerard and Jahangir, who proofread the manuscript.

References

1. F. Šantavý, *in* "The Alkaloids" (R. H. F. Manske and R. G. A. Rodrigo, eds.), Vol. 17, p. 386. Academic Press, New York, 1979.
2. Specialist Periodical Reports, "The Alkaloids," Vols. I–XIII. Chemical Society, London, 1971–1983.
3. M. Shamma and J. L. Moniot, "Isoquinoline Alkaloids Research 1972–1977," p. 307. Plenum, New York, 1978.
4. G. Blaskó, D. J. Gula, and M. Shamma, *J. Nat. Prod.* **45,** 105 (1982).
5. N. N. Margvelashvili, O. N. Tolkachev, N. P. Prisyazhnyuk and A. T. Kir'yanova, *Khim. Prir. Soedin.* 592 (1978); *Chem. Abstr.* **90,** 69088k (1979).
6. S.-T. Lu, T.-L. Su, T. Kametani, and M. Ihara, *Heterocycles* **3,** 301 (1975).
7. S.-T. Lu, T.-L. Su, T. Kametani, A. Ujiie, M. Ihara, and K. Fukumoto, *J. Chem. Soc., Perkin Trans. 1* 63 (1976).
8. T. Kametani, M. Takemura, M. Ihara, and K. Fukumoto, *Heterocycles* **4,** 723 (1976).
9. T. Kametani, M. Takemura, M. Ihara, and K. Fukumoto, *J. Chem. Soc., Perkin Trans. 1* 390 (1977).
10. M. Alimova, I. A. Israilov, M. S. Yunusov, N. D. Abdullaev, and S. Yu. Yunusov, *Khim. Prir. Soedin.* 727 (1982): *Chem. Abstr.* **98,** 122821n (1983).
11. K. K. Seth, V. B. Pandey, A. B. Ray, B. Dasgupta, and S. A. H. Shah, *Chem. Ind. (London)* 744 (1979).
12. G. Blaskó, S. F. Hussain, and M. Shamma, *J. Nat. Prod.* **44,** 475 (1981).
13. M. Alimova, I. A. Israilov, M. S. Yunusov, and S. Yu. Yunusov, *Khim. Prir. Soedin.* 642 (1982); *Chem. Abstr.* **98,** 104282h (1983).
14. M. Alimova, and I. A. Israilov, *Khim. Prir. Soedin* 602 (1981); *Chem. Abstr.* **96,** 48966d (1982).
15. N. N. Margvelashvili, N. P. Prisyazhnyuk, L. D. Kislov, and O. N. Tolkachev, *Khim. Prir. Soedin.* 832 (1976): *Chem. Abstr.* **86,** 117592m (1977).
16. C.-C. Fang, M. Lin, C.-M. Weng, C.-T. Chu, and H. Liu, *Yao Hsueh T'ung Pao* **16,** 49 (1981); *Chem. Abstr.* **95,** 138459z (1981).
17. V. Preininger, J. Veselý, O. Gašić, V. Šimánek, and L. Dolejš, *Collect. Czech. Chem. Commun.* **40,** 699 (1975).
18. A. M. Rabinovich, N. P. Tarasova, T. N. Kruzhalina, N. N. Margverashvili, B. A. Krivut, O. N. Tolkachev, and N. A. Fedyunina, *Khim.-Pharm. Zh.* **14,** 83 (1980); *Chem. Abstr.* **93,** 201014b (1980).
19. Kh. Kiryakov and P. Panov, *Dokl. Bolg. Akad. Nauk* **29,** 677 (1976); *Chem. Abstr.* **85,** 119633d (1976).

20. H. G. Kiryakov, Z. H. Mardirossian, D. B. MacLean, and J. P. Ruder, *Phytochemistry* **20,** 1721 (1981).
21. Kh. Kiryakov, Z. Mardirossian, and P. Panov, *Dokl. Bolg. Akad. Nauk* **34,** 43 (1981); *Chem. Abstr.* **95,** 58063a (1981).
22. Kh. Kiryakov, E. Iskrenova, B. Kuzmanov, and L. Evstatieva, *Planta Med.* **43,** 51 (1981); *Chem. Abstr.* **96,** 31623s (1982).
23. T.-Y. Chu, S.-C. Sung, Y.-L. Kao, J.-S. Hsu P.-H. Tai, L. Chen, and S.-S. Teng, *Chung Ts'ao Yao* **11,** 341 (1980); *Chem. Abstr.* **94,** 109170x (1981).
24. K. Mehra, H. S. Garg, D. S. Bhakuni, and N. M. Khanna, *Indian J. Chem., Sect. B* **14B,** 844 (1976); *Chem. Abstr.* **87,** 68513w (1977).
25. V. Preininger, J. Novák, V. Simánek, and F. Šantavý, *Planta Med.* **33,** 396 (1978); *Chem. Abstr.* **89,** 176318m (1978).
26. Q. Fang, M. Lin, J. Zhou, and X. Liu, *Yaoxue Tongbao* **17,** 3 (1982); *Chem. Abstr.* **97,** 20713r (1982).
27. W. F. Xin and M. Lin, *Chung Ts'ao Yao* **12,** 1 (1981); *Chem. Abstr.* **95,** 209450c (1981).
28. S. Luo and S. Wu, *Yaoxue Xuebao* **17,** 699 (1982); *Chem. Abstr.* **98,** 31431u (1983).
29. A. H. A. Abou-Donia, S. El-Masry, M. R. I. Saleh, and J. D. Phillipson, *Planta Med.* **40,** 295 (1980); *Chem. Abstr.* **94,** 27371z (1981).
30. A. Loukis and R. D. Waigh, *J. Pharm. Pharmacol.* **33,** Suppl., 16P (1981).
31. S. F. Hussain, R. D. Minard, A. J. Freyer, and M. Shamma, *J. Nat. Prod.* **44,** 169 (1981).
32. S. S. Markosyan, T. A. Tsulikyan, and V. A. Mnatsakanyan, *Arm. Khim. Zh.* **29,** 1053 (1976); *Chem. Abstr.* **86,** 136375j (1977).
33. S. S. Markosyan, *Tezisy Dokl.—Molddezhnaya Konf. Org. Sint. Bioorg. Khim.* 59 (1976); *Chem. Abstr.* **88,** 133268g (1978).
34. M. Shamma, P. Chinnasamy, S. F. Hussain, and F. Khan, *Phytochemistry* **15,** 1802 (1976).
35. V. Preininger, R. S. Thakur, and F. Šantavý, *J. Pharm. Sci.* **65,** 294 (1976).
36. J. Slavík and L. Slavíková, *Collect. Czech. Chem. Commun.* **44,** 2261 (1979).
37. I. A. Israilov, T. Irgashev, M. S. Yunusov, and S. Yu. Yunusov, *Khim. Prir. Soedin.* 834 (1977); *Chem. Abstr.* **88,** 117772m (1978).
38. R. H. F. Manske, *Can. J. Res., Sect. B* **16,** 81 (1938).
39. V. Smula, N. E. Cundasawmy, H. L. Holland, and D. B. MacLean, *Can. J. Chem.* **51,** 3287 (1973).
40. J. R. Falck and S. Manna, *Tetrahedron Lett.* 619 (1981).
41. S. O. de Silva, I. Ahmad, and V. Snieckus, *Can. J. Chem.* **57,** 1598 (1979).
42. P. Kerekes, R. Gaál, R. Bognár, T. Törö, and B. Costisella, *Acta Chim. Acad. Sci. Hung.* **105,** 283 (1980).
43. T. Shono, Y. Usui, and H. Hamaguchi, *Tetrahedron Lett.* 1351 (1980).
44. T. Shono, H. Hamaguchi, M. Sasaki, S. Fujita, and K. Nagami, *J. Org. Chem.* **48,** 1621 (1983).
45. S. Teitel, J. O'Brien, and A. Brossi, *J. Org. Chem.* **37,** 1879 (1972).
46. I. A. Israilov, T. Irgashev, M. S. Yunusov, and S. Yu. Yunusov, *Khim. Prir. Soedin.* 851 (1980); *Chem. Abstr.* **94,** 188630y (1981).
47. I. A. Israilov, M. S. Yunusov, N. D. Abdullaev, and S. Yu. Yunusov, *Khim. Prir. Soedin.* 536 (1975); *Chem. Abstr.* **84,** 44481w (1976).
48. I. A. Israilov, T. Irgashev, M. S. Yunusov, and S. Yu. Yunusov, *Tezisy Dokl.—Sove-Sheh. Ind. Simp. Khim. Prir. Soedin., 5th, 1978* 33 (1978); *Chem. Abstr.* **93,** 146291n (1980).
49. I. A. Israilov, M. U. Ibragimova, M. S. Yunusov, and S. Yu. Yunusov, *Khim. Prir. Soedin.* 612 (1975); *Chem. Abstr.* **84,** 86727m (1976).
50. I. A. Israilov, M. S. Yunusov, and S. Yu. Yunusov, *Khim. Prir. Soedin.* 811 (1975); *Chem. Abstr.* **84,** 180442m (1976).

51. I. Messana, R. LaBua, and C. Galeffi, *Gazz. Chim. Ital.* **110,** 539 (1980).
52. A. Z. Sadikov, D. A. Rakhimov, T. Sadikov, E. K. Dobronravova, and T. T. Shakirov, *Khim. Prir. Soedin.* 815 (1978); *Chem. Abstr.* **91,** 16667n (1979).
53. G. Sariyar, *Planta Med.* **46,** 175 (1982); *Chem. Abstr.* **98,** 31427x (1983).
54. J. D. Phillipson, A. I. Gray, A. A. R. Askari, and A. A. Khalil, *J. Nat. Prod.* **44,** 296 (1981).
55. P. Khanna and G. L. Sharma, *Indian J. Exp. Biol.* **15,** 951 (1977); *Chem. Abstr.* **88,** 3119y (1978).
56. P. Khanna and R. Khanna, *Indian J. Exp. Biol.* **14,** 628 (1976); *Chem. Abstr.* **85,** 189238c (1976).
57. Y. M. El Kheir, *Planta Med.* **27,** 275 (1975); *Chem. Abstr.* **83,** 75407k (1975).
58. P. Gorecki and M. Drozdzynska, *Herba Pol.* **21,** 263 (1975); *Chem. Abstr.* **84,** 132637g (1976).
59. B. Proksa, J. Cerny, and M. Stefék, Czech Patent CS 194,878 (1978); *Chem. Abstr.* **98,** 2895t (1983).
60. B. Proksa, J. Cerny, and J. Putek, *Pharmazie* **34,** 194 (1979); *Chem. Abstr.* **91,** 16713z (1979).
61. R. H. F. Manske, *J. Am. Chem. Soc.* **72,** 3207 (1950).
62. K. Bláha, J. Hrbek, Jr., J. Kovár, L. Pijewska, and F. Šantavý, *Collect. Czech. Chem. Commun.* **29,** 2328 (1964).
63. J. Holubek and O. Strouf, "Spectral Data and Physical Constants of the Alkaloids," Vol. I, Heyden, London, 1965; Vol. II, 1973.
64. J. Stanek and R. H. F. Manske, *in* "The Alkaloids" (R. H. F. Manske, ed.), Vol. 4, p. 167. Academic Press, New York, 1954.
65. D. W. Hughes, H. L. Holland, and D. B. MacLean, *Can. J. Chem.* **54,** 2252 (1976).
66. R. H. F. Manske, *Can. J. Res., Sect. B* **17,** 51 (1939).
67. N. N. Margvelashvili, A. T. Kir'yanova, and O. N. Tolkachev, *Khim. Prir. Soedin.* 127 (1972); *Chem. Abstr.* **77,** 58825d (1972).
68. B. C. Nalliah, D. B. MacLean, H. L. Holland, and R. Rodrigo, *Can. J. Chem.* **57,** 1545 (1979).
69. D. W. Hughes and D. B. MacLean, *in* "The Alkaloids" (R. H. F. Manske and R. G. A. Rodrigo, eds.), Vol. 18, p. 217. Academic Press, New York, 1981.
70. C. E. Slemon, L. C. Hellwig, J.-P. Ruder, E. W. Hoskins, and D. B. MacLean, *Can. J. Chem.* **59,** 3055 (1981).
71. M. Shamma and D. M. Hindenlang, "^{13}C-NMR Shift Assignments of Amines and Alkaloids." Plenum, New York, 1979.
72. O. E. Edwards and K. L. Handa, *Can. J. Chem.* **39,** 1801 (1961).
73. I. A. Israilov, M. S. Yunusov, and S. Yu. Yunusov, *Khim. Prir. Soedin.* 194 (1968); *Chem. Abstr.* **69,** 57449g (1968).
74. K. L. Seitanidi, M. R. Yagudaev, I. A. Israilov, and M. S. Yunusov, *Khim. Prir. Soedin.* 465 (1978); *Chem. Abstr.* **90,** 6585z (1979).
75. M. Ohta, H. Tani, and S. Morozumi, *Chem. Pharm. Bull.* **12,** 1072 (1964).
76. M. A. Marshall, F. L. Pyman, and R. Robinson, *J. Chem. Soc.* 1315 (1934).
77. M. U. Ibragimova, M. S. Yunusov, and S. Yu. Yunusov, *Khim. Prir. Soedin.* 438 (1970); *Chem. Abstr.* **73,** 127738j (1970).
78. P. Kerekes, Gy. Gaál, and R. Bognár, *Acta Chim. Acad. Sci. Hung.* **103,** 339 (1980).
79. W. H. Perkin, Jr., and R. Robinson, *J. Chem. Soc.* **99,** 775 (1911).
80. T. Kametani, H. Inoue, T. Honda, T. Sugahara, and K. Fukumoto, *J. Chem. Soc., Perkin Trans. 1* 374 (1977).
81. F. Săntavý, *in* "The Alkaloids" (R. H. F. Manske, ed.), Vol. 12, p. 333. Academic Press, New York, 1970.

82. S. Safe and R. Y. Moir, *Can. J. Chem.* **42,** 160 (1964).
83. M. Shamma, "The Isoquinoline Alkaloids," p. 359. Academic Press, New York, 1972.
84. G. P. Moiseeva, I. A. Israilov, M. S. Yunusov, and S. Yu. Yunusov, *Khim. Prir. Soedin.* 103 (1978); *Chem. Abstr.* **89,** 75411f (1978).
85. V. Elango, A. J. Freyer, G. Blaskó, and M. Shamma, *J. Nat. Prod.* **45,** 517 (1982).
86. M. Shamma and V. St. Georgiev, *Tetrahedron* **32,** 211 (1976).
87. T. V. Hung, B. A. Mooney, R. H. Prager, and J. M. Tippett, *Aust. J. Chem.* **34,** 383 (1981).
88. F. E. Ziegler and K. W. Fowler, *J. Org. Chem.* **41,** 1564 (1976).
89. H. P. Plauman, B. A. Keay, and R. Rodrigo, *Tetrahedron Lett.* 4921 (1979).
90. V. Snieckus, *Heterocycles* **14,** 1649 (1980).
91. D. B. MacLean, *Symp. Pap.—IUPAC Int. Symp. Chem. Nat. Prod., 11th, 1978* **4**(2), 204 (1978).
92. R. Marsden and D. B. MacLean, *Can. J. Chem.* **62,** 306 (1984).
93. R. H. Prager, J. M. Tippett, and A. D. Ward, *Aust. J. Chem.* **34,** 1085 (1981).
94. P. Kerekes, G. Horváth, Gy. Gaál, and R. Bognár, *Acta Chim. Acad. Sci. Hung.* **97,** 353 (1978).
95. T. V. Hung, B. A. Mooney, R. H. Prager, and A. D. Ward, *Aust. J. Chem.* **34,** 151 (1981).
96. S. O. de Silva, I. Ahmad, and V. Snieckus, *Tetrahedron Lett.* 5107 (1978).
97. R. D. Haworth and A. R. Pinder, *J. Chem. Soc.* 1776 (1950).
98. G. I. Hutchison, P. A. Marshall, R. H. Prager, J. M. Tippett, and A. D. Ward, *Aust. J. Chem.* **33,** 2699 (1980).
99. J. L. Moniot and M. Shamma, *J. Am. Chem. Soc.* **98,** 6714 (1976).
100. J. L. Moniot and M. Shamma, *J. Org. Chem.* **44,** 4337 (1979).
101. M. Shamma, D. M. Hindenlang, T.-T. Wu, and J. L. Moniot, *Tetrahedron Lett.* 4285 (1977).
102. M. Shamma, J. L. Moniot, and D. M. Hindenlang, *Tetrahedron Lett.* 4273 (1977).
103. J. L. Moniot, D. M. Hindenlang, and M. Shamma, *J. Org. Chem.* **44,** 4343 (1979).
104. J. L. Moniot, D. M. Hindenlang, and M. Shamma. *J. Org. Chem.* **44,** 4347 (1979).
105. J. Imai and Y. Kondo, *Heterocycles* **5,** 153 (1976); **6,** 959 (1977).
106. Y. Kondo, J. Imai, and H. Inoue, *J. Chem. Soc., Perkin Trans. 1* 911 (1980).
107. Y. Kondo, J. Imai, and S. Nozoe, *J. Chem. Soc., Perkin Trans. 1* 919 (1980).
108. Y. Kondo, H. Inoue, and J. Imai, *Heterocycles* **6,** 953 (1977).
109. M. Hanaoka, C. Mukai, and Y. Arata, *Heterocycles* **6,** 895 (1977).
110. M. Hanaoka, C. Mukai, and Y. Arata, *Chem. Pharm. Bull.* **31,** 947 (1983).
111. M. Hanaoka, K. Nagami, and T. Imanishi, *Chem. Pharm. Bull.* **27,** 1947 (1979).
112. B. C. Nalliah, D. B. MacLean, R. G. A. Rodrigo, and R. H. F. Manske, *Can. J. Chem.* **55,** 922 (1977).
113. B. C. Nalliah and D. B. MacLean, *Can. J. Chem.* **56,** 1378 (1978).
114. S. Teitel and J. P. O'Brien, *J. Org. Chem.* **41,** 1657 (1976).
115. K. V. Rao and L. S. Kapicak, *J. Heterocycl. Chem.* **13,** 1073 (1976).
116. P. Kerekes and Gy. Gaál, *Acta Chim. Acad. Sci. Hung.* **103,** 343 (1980).
117. V. Preininger, V. Šimánek, O. Gašić, F. Šantavý, and L. Dolejš, *Phytochemistry* **12,** 2513 (1973).
118. M. Popova, A. Boeva, L. Dolejš, V. Preininger, V. Šimánek, and F. Šantavý, *Planta Med.* **40,** 156 (1980).
119. P. Forgacs, J. Provost, R. Tiberghien, J.-F. Desconclois, G. Buffard, and M. Pesson, *C. R. Hebd. Seances Acad. Sci., Ser. D* **276,** 105 (1973).
120. M. Popova, L. Dolejš, V. Šimánek, and V. Preininger, *Int. Conf. Chem. Biotechnol. Biol. Act. Nat. Prod. [Proc.], 1st, 1981* **3,** 95 (1981); *Chem. Abstr.* **97,** 52537c (1982).

121. M. E. Popova, V. Simanek, L. Dolejs, B. Smysl, and V. Preininger, *Planta Med.* **45,** 120 (1982).
122. R. G. A. Rodrigo, R. H. F. Manske, H. L. Holland, and D. B. MacLean, *Can. J. Chem.* **54,** 471 (1975).
123. I. A. Israilov, M. S. Yunusov, and S. Yu. Yunusov, *Khim. Prir. Soedin.* 588 (1970); *Chem. Abstr.* **74,** 42528m (1971).
124. M. U. Ibragimova, I. A. Israilov, M. S. Yunusov, and S. Yu. Yunusov, *Khim. Prir. Soedin.* 476 (1974); *Chem. Abstr.* **82,** 28586n (1975).
125. G. Blaskó, V. Elango, B. Sener, A. J. Freyer, and M. Shamma, *J. Org. Chem.* **47,** 880 (1982).
126. Kh. Kiryakov, Z. Mardirossian, and P. Panov, *Dokl. Bolg. Akad. Nauk* **33,** 1377 (1980); *Chem. Abstr.* **94,** 188644f (1981).
127. H. G. Kiryakov, Z. H. Mardirossian, D. W. Hughes, and D. B. MacLean, *Phytochemistry* **19,** 2507 (1980).
128. P. Forgacs, J. Provost, J.-F. Desconclois, A. Jehanno, and M. Pesson, *C. R. Hebd. Seances Acad. Sci., Ser. D* **279,** 855 (1974).
129. Z. Koblicova, J. Kreckova, and J. Trojanek, *Cesk. Farm.* **30,** 177 (1981); *Chem. Abstr.* **96,** 85806x (1982); Czech Patent 165,700 (1976); *Chem. Abstr.* **86,** 171702k (1977); Czech Patent 165,701 (1976); *Chem. Abstr.* **86,** 171703m (1977).
130. J. Hodková, Z. Veselý, Z. Koblicová, J. Holubek, and J. Trojánek, *Lloydia* **35,** 61 (1972).
131. B. Proksa and Z. Voticky, *Collect. Czech. Chem. Commun.* **45,** 2125 (1980).
132. W. Klötzer, S. Teitel, and A. Brossi, *Monatsh. Chem.* **103,** 1210 (1972).
133. P. Forgacs, G. Bufford, A. Jehanno, J. Provost, R. Tiberghien, and A. Touche, *Plant. Med. Phytother.* **16,** 99 (1982).
134. M. Shamma and J. L. Moniot, *Chem. Commun.* 89 (1975).
135. W. Klötzer, S. Teitel, and A. Brossi, *Helv. Chim. Acta* **55,** 2228 (1972).
136. P. Gorecki and M. Drozdzynska, *Herba Pol.* **22,** 228 (1976); *Chem. Abstr.* **87,** 201834v (1977).
137. P. Gorecki and M. Drozdzynska, *Herba Pol.* **22,** 233 (1976); *Chem. Abstr.* **87,** 201835w (1977).
138. B. Proksa, M. Bobal, and S. Kovac, *Chem. Zvesti* **36,** 559 (1982); *Chem. Abstr.* **98,** 16898 (1983).
139. B. Proksa, Z. Voticky, and M. Stefek, *Chem. Zvesti* **34,** 248 (1980); *Chem. Abstr.* **93,** 220567a (1980).
140. B. Proksa, J. Fuska, and Z. Voticky, *Pharmazie* **37,** 350 (1982).
141. I. D. Spenser, *Compr. Biochem.* **20,** 231 (1968).
142. E. McDonald, *in* "The Chemistry of Heterocyclic Compounds" (G. Grethe, ed.), Vol. 38, Part 1, p. 275. Wiley (Interscience), New York, 1981.
143. A. R. Battersby, M. Hirst, D. J. McCaldin, R. Southgate, and J. Staunton, *J. Chem. Soc. C* 2163 (1968).
144. A. R. Battersby, J. Staunton, H. R. Wiltshire, B. J. Bircher, and C. Fuganti, *J. Chem. Soc., Perkin Trans. 1* 1162 (1975).
145. B. Gözler, T. Gözler, and M. Shamma, *Tetrahedron* **39,** 577 (1983).
146. S. F. Hussain and M. Shamma, *Tetrahedron Lett.* 1693 (1980).
147. D. R. Curtis, A. W. Duggan, D. Felix, and G. A. R. Johnston, *Nature (London)* **226,** 1222 (1970).
148. D. R. Curtis, A. W. Duggan, D. Felix, and G. A. R. Johnston, *Nature (London)* **228,** 676 (1970).
149. D. R. Curtis, A. W. Duggan, D. Felix, and D. A. R. Johnston, *Brain Res.* **32,** 69 (1971).

150. G. A. R. Johnston, P. M. Beart, D. R. Curtis, C. J. A. Game, R. M. McCulloch, and R. M. Maclachlan, *Nature (London), New Biol.* **240,** 219 (1972).
151. J. F. Collins and R. G. Hill, *Nature (London)* **249,** 845 (1974).
152. S. F. Pong and L. T. Graham, *Brain Res.* **42,** 486 (1972).
153. J. Kardos, G. Blaskó, P. Kerekes, I. Kovács, and M. Simonyi, *Biochem. Pharmacol.* (in press).
154. N. B. Medow and J. Greco, U.S. Patent 3,903,282 (1975); *Chem. Abstr.* **83,** 197831q (1975); U.S. Patent 3,943,251 (1976); *Chem. Abstr.* **84,** 169672r (1976).
155. A. Put, J. Wojcicki, S. Stanosz, and P. Gorecki, *Herba Pol.* **20,** 285 (1974); *Chem. Abstr.* **83,** 22520k (1975).

—CHAPTER 6—

THE STUDY OF ALKALOID STRUCTURES BY SPECTRAL METHODS

R. J. HIGHET

Laboratory of Chemistry
National Heart, Lung, and Blood Institute
Bethesda, Maryland

AND

JAMES W. WHEELER

Department of Chemistry
Howard University
Washington, D.C.

I. Introduction

Alkaloid chemists and organic chemists in general support structural investigations of compounds by study of their physical properties, melting points, and optical rotatory powers early providing essential data. With the advent of mass spectroscopy and nuclear magnetic resonance these two methods have come to

THE ALKALOIDS, VOL. XXIV

ISBN 0-12-469524-8

dominate alkaloid studies. Isolation procedures are followed by mass spectra and, when sufficient material is available, proton and C-13 NMR spectra. Often relatively simple applications of these methods suffice to demonstrate the requisite structure. When, however, crystalline material becomes available, and the problem appears to be of sufficient complexity—a somewhat subjective evaluation—the chemist may decide the most efficient use of time and labor requires the use of X-ray crystal studies, as recently reviewed in this treatise (*1*). All too often, paucity of material or unfortunate amorphic character may preclude this definitive method. The chemist must then exploit the growing powers of spectral methods.

We describe below the capabilities of these methods, both in simple application and the sophisticated methods currently evolving.

II. Mass Spectrometry

A. Mass Spectrometric Methods

1. Electron Impact Studies

The study of alkaloids by electron impact mass spectroscopy dates back to the classic studies of Biemann, who used the "shift" technique in the early 1960s to establish the structure of alkaloids variously substituted on aromatic rings (2). Because of the stability of aromatic systems to electron impact, the fragmentation of alkaloids such as ibogamine (**1**, $R^1 = R^2 = H$) largely involves cleavage of the alicyclic and heterocyclic systems. Alkaloids differing only in aromatic substitution show peaks at high mass shifted by the mass difference of the aromatic substituents. In the spectra (Fig. 1) of ibogamine and ibogaline, peaks corresponding to the loss of 15, 29, 85, and 124 appear in both. At lower mass,

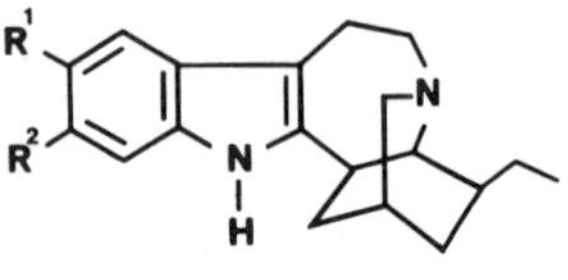

1 Ibogamine $R^1 = R^2 = H$
Ibogaline $R^1 = R^2 = OCH_3$

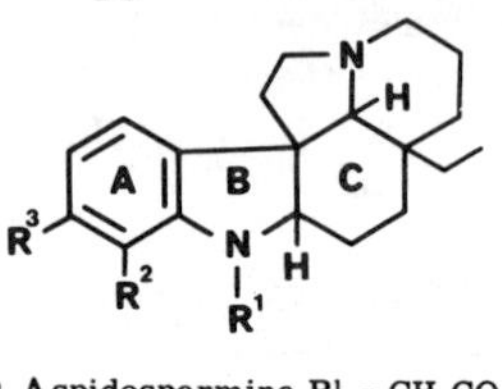

2 Aspidospermine $R^1 = CH_3CO$
$R^2 = CH_3O$, $R^3 = H$

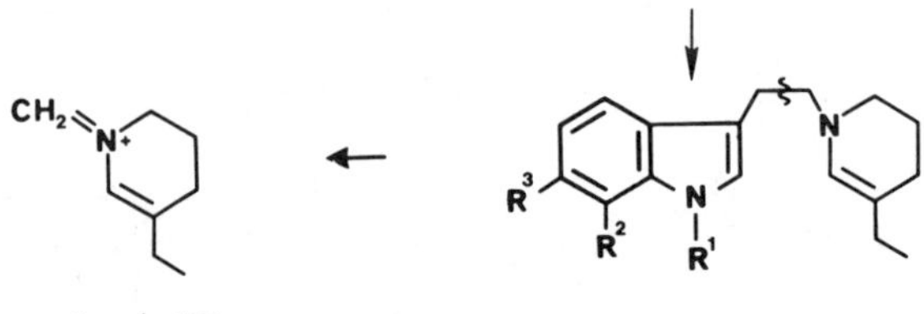

3 *m/z* 124

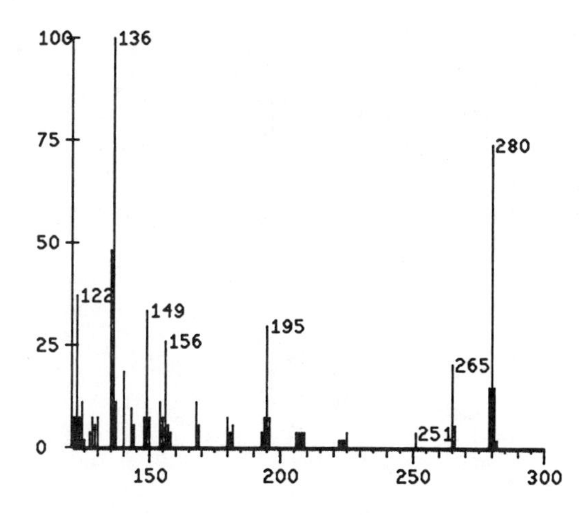

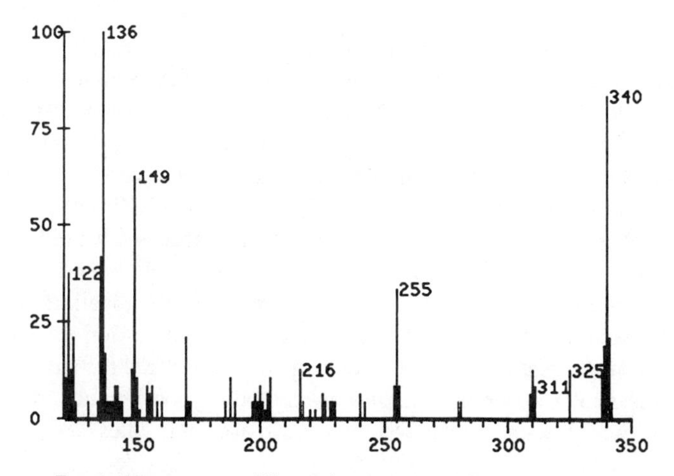

FIG. 1. Mass spectra of ibogamine (upper) and ibogaline (lower).

peaks arising from fragments without the aromatic system occur at the same position in each spectrum (m/z 122, 136, and 149). It is clear that the two alkaloids differ only in substitution on the aromatic ring.

The method is also suitable for alkaloids related to aspidospermine (**2**, R^1 = CH_3CO, R^2 = CH_3O, R^3 = H), which differ at R^1, R^2, and R^3. All gave a retro-Diels–Alder of ring C (M − 28), with further cleavage as shown to give a prominent ion at m/z 124 (**3**).

Biemann's shift method could also be used when the alkaloids differed in the alicyclic part of the molecule, but this depended strongly on whether the other groups present interfered with the fragmentation mode. Removal of that func-

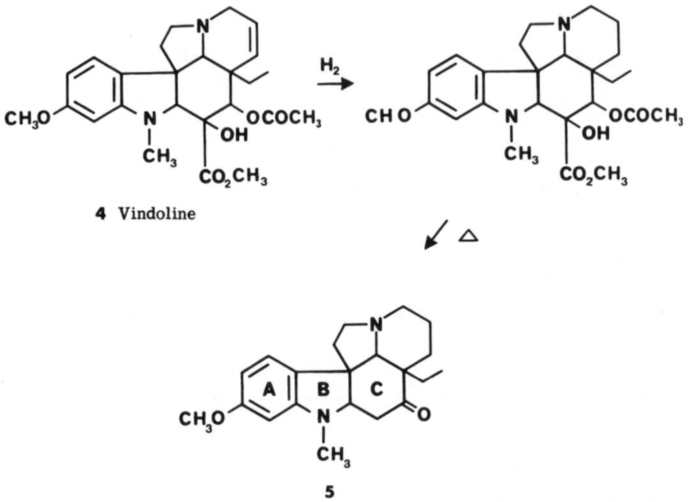

4 Vindoline

5

tional group or conversion of it to a group not interfering with the fragmentation process was necessary. The conversion of vindoline (**4**) to **5**, in which ketene (M − 42) is lost from the C ring and gives the *m/z* 124 fragment **3**, provides an example of this procedure. The studies established the power of mass spectrometry to elucidate the structures of alkaloids available only in very small quantities.

The advent of gas chromatography–mass spectroscopy (GC–MS) has allowed the study of the more volatile alkaloids such as tobacco alkaloids, the frog alkaloids, and the piperidine alkaloids in insects (see later), but electron-impact spectra have been used primarily for molecular weight determination of alkaloids that can be volatilized. The emphasis on newer methods of ion production has overshadowed the usefulness of electron-impact spectra (EI) for structure determination of smaller alkaloids. The newer methods of ion production have very limited utility for low-molecular-weight (<300) alkaloids unless they are quaternary salts (see below).

2. Chemical Ionization Methods

Molecules bearing readily eliminated substituents, such as primary alcohol groups, often fail to show molecular ions in electron-impact mass spectra. This difficulty can often be eliminated by chemical ionization (CI), in which the mass spectrometer source is filled with a gas at approximately 1 torr; as a consequence, the sample is ionized at much lower energies through proton transfer from ions of the gas (*3*). Fragmentation is less pronounced both because of the lower energy and because of the proton source of the charge; often only protonated molecular ions are observed.

Ammonia has been successfully used as a carrier gas for chemical-ionization mass spectroscopy in the study of an extensive number of alkaloids from neotropical poison frogs (*4*). The protonated molecular ion of each alkaloid in the mixture could be obtained. In some cases two (or more) alkaloids were eluted as a single peak in the gas chromatogram but could be distinguished by their different quasi-molecular ions. Comparison of the spectra obtained with deuterated ammonia establishes the number of readily exchangeable protons (NH, OH) (*5*). This sequence was followed by CI GC–MS, using methane as the carrier gas, giving fragmentation characteristic of various classes of these alkaloids (see later section).

More recently this group has used a combination of nitrogen and nitric oxide to obtain "pseudoelectron impact spectra." This combination provides the low energy advantage of CI spectra by using the low recombination energy of NO^+ (8.5–9.0 eV). It retains the fragmentation patterns initiated by radical cation species rather than the neutral molecule eliminations characteristic of MH^+ CI spectra. These analyses were followed by electron impact GC–MS for comparison and high resolution GC–MS to provide molecular formulas. Derivatization (perhydrogenation, acetylation, trimethylsilylation) was also useful for the combined GC–MS and characterization of these relatively low-molecular-weight (219–323) alkaloids.

Ammonia chemical ionization mass spectrometry, using a direct insertion probe, has been found applicable to plant extracts containing pyrrolizidine, quinolizidine, diterpenoid, bishordeninyl terpene, and indole alkaloids (*6*). These authors found no adduct formation and a lack of significant fragmentation. The molecular weights of the compounds ranged up to 510. Although fragmentation may be minimal, the reported peaks for the pyrrolizidine alkaloids at m/z 120 for senecionine (**6**) and at 122 for supinine (**7**) correspond to loss of the side chain(s) with retention of the ring system. Senecionine apparently suffers allylic cleavage and loss of the other ester function as the acid, forming a double bond. Supinine (the structure given in the paper is actually lycopsamine) suffers only allylic cleavage inasmuch as the other ring contains no oxygenated function.

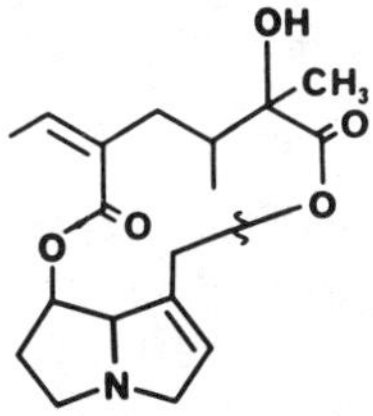

OH CH$_3$ O O O O N

6 Seneclonine

O O OH OH N

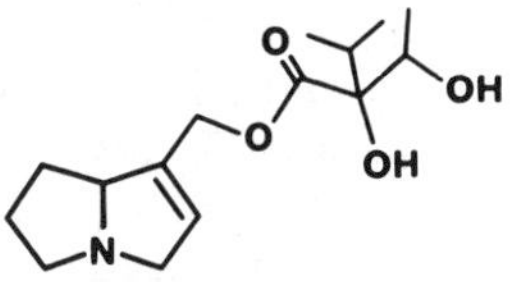

7 Supinine

3. The Study of Nonvolatile Materials

Although the primary methods of mass spectroscopy have been electron impact and more recently chemical ionization (*7*) other techniques have been developed since the late 1970s, which show substantial promise. These include field desorption (*8*), laser desorption (*9*), plasma desorption (*10*), organic SIMS (secondary ion mass spectrometry) (*11*), electrospray (*12*), and flash vaporization of dilute solution droplets (*13*). All of these methods are designed to address the continuing need for high mass, high sensitivity mass spectrometry, which is applicable to thermally labile molecules of low volatility. Many of these methods require specialized instruments, which have become commercially available only recently, limiting their application to alkaloids as well as other classes of molecules.

Field desorption mass spectroscopy has been used to correct the assigned structure (**8**) of agropine (**9**) obtained from crown gall tumor (*14*). Electron-impact spectra showed an apparent molecular ion at *m/z* 275, suggesting the molecular formula $C_{11}H_{17}NO_7$. On the basis of this it was formulated as the bicyclic derivative of glutamic acid and a hexitol sugar (**8**) (*15*). However, the field desorption mass spectrum showed at peak at 292, demonstrating that ammonia had been eliminated in the EI spectrum either by an electronic or thermal process. The structural formula was revised to **9**, which was proved by synthesis. It is ironic that the original workers used field desorption mass spectroscopy for derivatives of agropine but used electron impact for agropine itself.

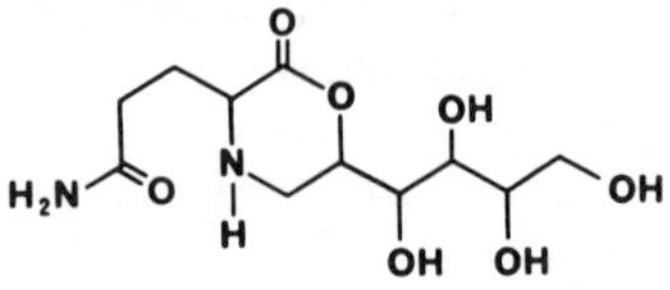

O OH H₂N O N H OH OH OH

9 **Agropine**

O O OH N OH O OH OH

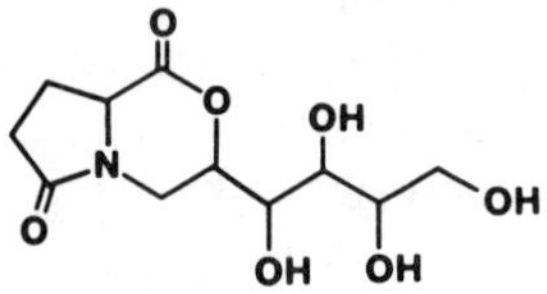

8

Laser desorption with detection by mass spectrometry–mass spectrometry (MS–MS) has been used for cactus alkaloids, providing spectra from the two quaternary alkaloids *O*-methylcandicine (**10**) and coryphanthine (**11**) (*16*). These authors also found that candicine volatilizes intact, using SIMS, and suggest that this is a useful technique for simple quaternary alkaloids. Other authors (*17*) have suggested the use of SIMS with sulfolane containing lithium iodide, sodium iodide, silver fluoroborate, or thallium fluoroborate as a solution to sputter ions. They indicate that this method has advantages over ions produced by fast-atom bombardment (FAB) from glycerol solutions (*18*). Addition of trifluoromethane-

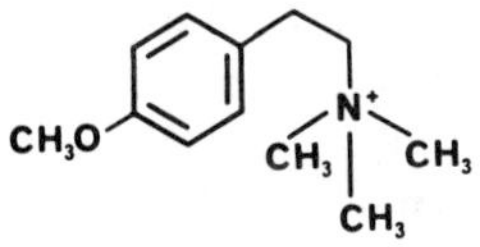

10 *O*-Methylcandicine

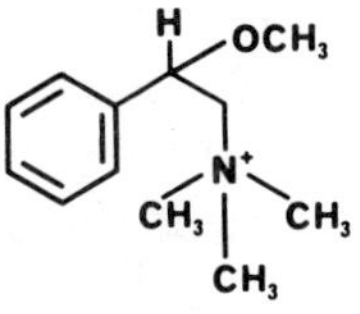

11 Coryphanthine

sulfonic acid to sulfolane enhanced the protonated molecular ion more than 10-fold for yohimbine. However, the same effect was observed with *p*-toluenesulfonic acid in glycerol. No other alkaloids were mentioned by name, nor were any other data presented for alkaloids. They indicate that accurate masses can be obtained for compounds in the mass range 400–2000.

Mass-analyzed ion kinetic energy spectrometry (MIKES) has been used to screen cactus species for alkaloids to map the distribution of alkaloids (*19*). An ethanolic extract containing only nonphenolic compounds results in improved sensitivity over crude plant material, as might be anticipated. These authors indicate that the *N*-methyl derivative of dimethoxytetrahydroisoquinoline (**12**) can be distinguished from the 1-methyl derivative (**13**) by MIKES. Electron-

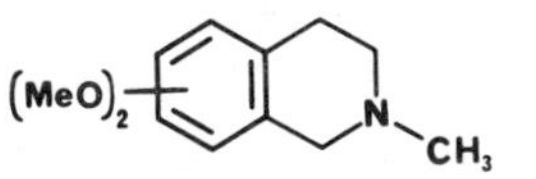

12

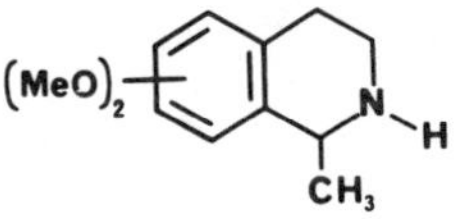

13

impact spectra should also clearly differentiate them. The related compounds without the methoxy groups show a striking difference in their EI spectra. The *N*-methyl compound exhibits a base peak at *m/z* 147 (M^+) with the loss of methyl only 10% of the base peak. The 1-methyl compound, on the other hand, exhibits its base peak at *m/z* 132, with the molecular ion at 20% of the base peak (*20*). It is difficult to see how the MIKES spectra improve on this distinction. These authors indicate that an isomerization occurs during the workup of the *N*-CH_3 to the 1-CH_3 compound, only the *N*-methyl being observed in the dried plant material. However, no independent evidence is presented for this unusual isomerization. Other cacti have been examined by the MIKES technique (*21*).

Negative-ion mass spectroscopy has also been used to obtain molecular ions of naturally occuring macrocyclic nitrogen-containing compounds such as nortrewiasine (**14**) from *Trewia nudiflora* (*22*). Its EI mass spectrum exhibits no molecular ion, losing H_2O and HCNO as well as the ester side chain. Its negative ion chemical ionization mass spectrum shows a molecular ion (M^-) as well as the losses mentioned above.

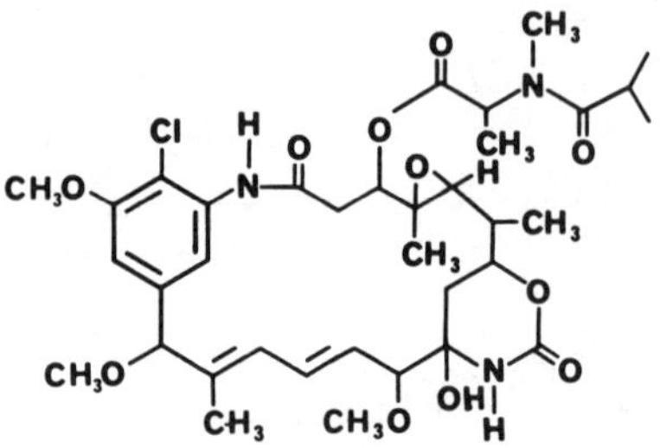

14 Nortrewiasine

B. Dendrobatid Alkaloids

Mass spectroscopy has been an important tool in the structural elucidation of alkaloids from the neotropical poison frogs. First characterization of the alkaloids was achieved by the observations that a half-dozen structurally similar groups could be recognized by similar ions and fragmentation patterns. Although the ''apparent'' molecular ion for both batrachotoxin and homobatrachotoxin was observed at *m/z* 399 and had the formula $C_{24}H_{33}NO_4$, this ion was actually a fragment ion corresponding to the loss of a pyrrole-3-carboxylic acid from the 20-position of the alkaloid **15.**

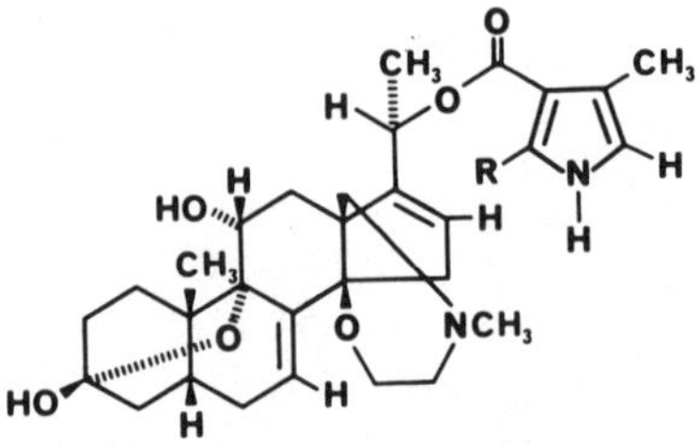

15 R = CH_3 Batrachotoxin;
R = CH_3CH_2 Homobatrachotoxin

The structure of the related batrachotoxin A (same structure with an alcohol instead of the 20-pyrrole-3-carboxylate ester) had been proved by X-ray analysis of its *p*-bromobenzoate ester (*23, 24*). It was now possible to go back to the mass spectra of these two alkaloids and find fragment ions indicative of the pyrrole side chain. It was even possible to detect the true molecular ion of batrachotoxin at *m/z* 538, even though it amounted to only 2% of the apparent molecular ion at 399. The pyrrole side chain was responsible for ions at *m/z* 139 ($C_7H_9NO_2$), the acid itself, 95 (C_6H_9N) and 94 (C_6H_8N) in batrachotoxin and ions at *m/z* 153 ($C_8H_{11}NO_2$), 138 ($C_7H_8NO_2$), 109 ($C_7H_{11}N$), and 94 (C_6H_8N) in homobatrachotoxin (*4*). This was rationalized after the structure was known by other means, however, and is a good example of the limitations of mass spectroscopy of complex unknown molecules.

More recently, the skin of the frog *Phyllobates terriblis* has been shown to contain these three alkaloids plus a hydroxy analog of homobatrachotoxin having the formula $C_{32}H_{44}N_2O_7$ and a hydroxy analog of batrachotoxin with the formula $C_{31}H_{42}N_2O_7$ (*25*). The mass spectra were more useful here, inasmuch as the same loss of the pyrrole side chain occurred, giving an ion at m/z 415 ($C_{24}H_{33}NO_5$), which indicated that the additional oxygen was on the steroid ring, and not on the C-20 side chain. In addition to these steroidal alkaloids, noranabasamine ($C_{15}H_{17}N_3$) (**16**) calycanthine (**17**), and chimonanthine (**18**) were isolated from the skins.

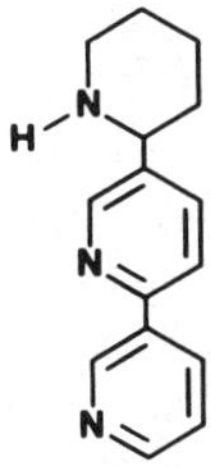

16 Noranabasamine

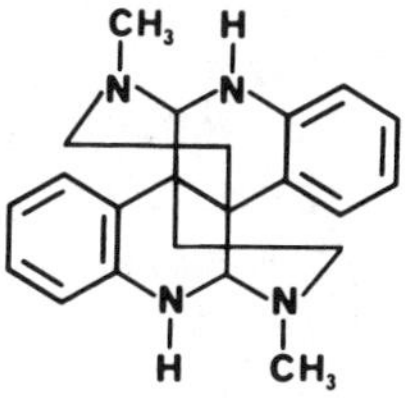

17 Calycanthine

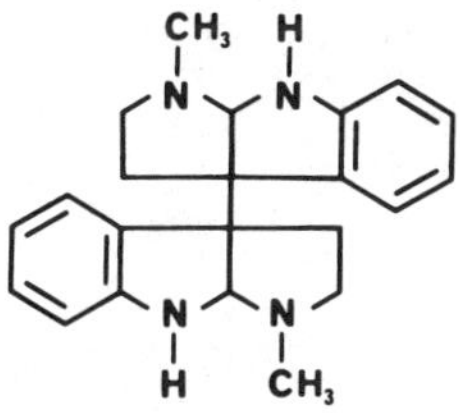

18 Chimonanthine

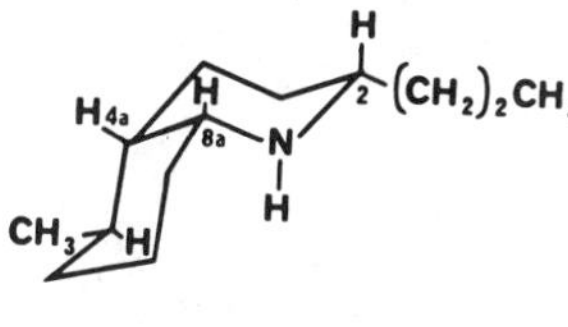

19 Pumiliotoxin C

Noranabasamine exhibits two prominent ions in addition to the molecular ion, these being at m/z 157 ($C_{10}H_9N_2$), loss of the piperidine ring with hydrogen transfer, and m/z 84 ($C_5H_{10}N$), cleavage at the piperidine ring with retention of charge on it.

Other alkaloids isolated from neotropical frog skins are not nearly as complex as the batrachotoxins. The simplest of these is pumiliotoxin C (**19**), a *cis*-decahydroquinoline. The major ion in its mass spectrum occurs at m/z 152, arising from loss of the propyl side chain as well as a weak molecular ion and other fragment ions (*26, 27*). Other related alkaloids have been assigned tentative structures based on a base peak at 152, indicating the same ring system with a side chain different from propyl (*28*). Another series tentatively identified exhibits a base

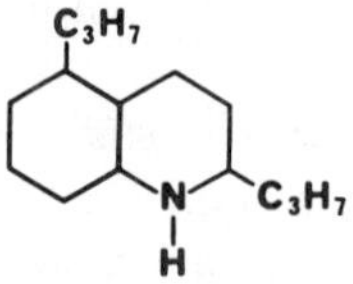

20

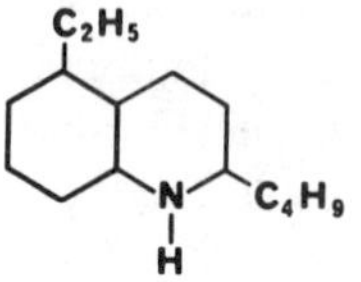

21

peak at m/z 138, indicating a decahydroquinoline system without the methyl group and with different side chains at C-2. A third series shows a base peak at m/z 180, **20** being an example, and a fourth series has a base peak at m/z 166, **21** being an example of this type of decahydroquinoline. All exhibit cleavage α to the nitrogen atom, as expected, with loss of the alkyl group.

A third type of alkaloid has been isolated from *Dendrobates histrionicus* frogs from Colombia. These are the histrionicotoxins, unusual spiropiperidine alkaloids containing acetylene or allenic groups in the side chains. Histrionicotoxin (2*R*,6*R*,7*S*,8*S*)-7-(1-*cis*-buten-3-ynyl)-2-(*cis*-2-penten-4-ynyl)-1-azaspiro-5,5 undecan-8-ol (**22**) illustrates this type. Although much of the characterization of this class of alkaloid has been accomplished by proton and carbon-13 NMR, they all show a major fragment of m/z 96 ($C_6H_{10}N$), which in many cases is the base peak of the spectrum. This ion cannot arise from simple α cleavage, but could correspond to **23** with cleavage of three bonds, two of which are allylic.

The gephyrotoxins represent a fourth type of frog alkaloid (*29*). Here again the base peak arises from α cleavage with loss of CH_2CH_2OH. The only other major loss is the highly unsaturated side chain C_5H_5 from the molecular ion **24.** Structure determination was again based on X-ray analysis.

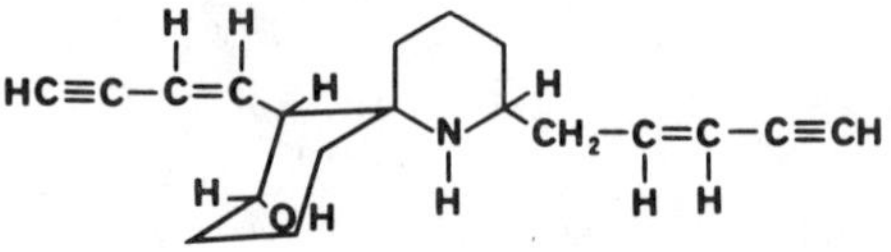

22 Histrionicotoxin

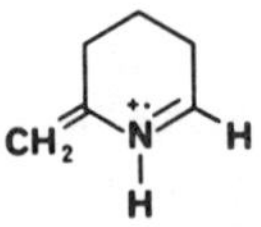

23 m/z 96

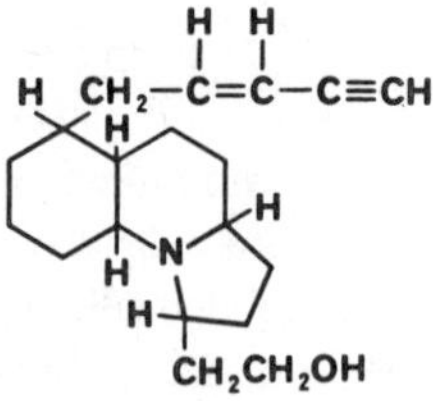

24 Gephyrotoxin

Although gephyrotoxin itself is a tricyclic perhydrobenzoindolizidine, other gephyrotoxins are bicyclic indolizidines. They also exhibit α cleavage with loss of the appropriate alkyl group (propyl and butyl groups in **25**).

Pumiliotoxin A and B are representatives of some 23 frog alkaloids related by mass spectral similarities. This fifth type of alkaloid has a prominent ion at *m/z* 70 (C_4H_8N), accompanied by an ion at either *m/z* 166 ($C_{10}H_{16}NO$) or 182 ($C_{10}H_{16}NO_2$). Eight alkaloids besides pumiliotoxins A and B have the $C_{10}H_{16}NO$ ion, while 11 others showed the $C_{10}H_{16}NO_2$ ion (*28*). X-ray analysis of the related pumiliotoxin 251D (**26**) led to this ring system for pumiliotoxins A and B (*30*).

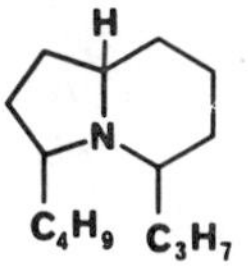

25

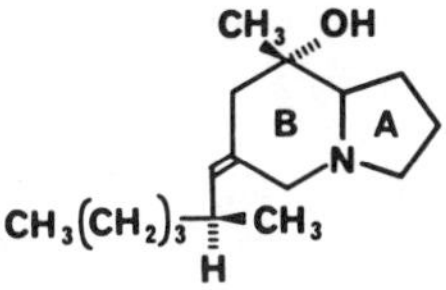

26 251D

The base peak at *m/z* 70 (C_4H_8N) is formed from ring A by cleavage of two bonds and hydrogen transfer. A large peak (20%) at *m/z* 194 constitutes loss of the butyl side chain by allylic cleavage. The ions at 166 (or 182) also arise from the indolizidine ring. The 166 ion ($C_{10}H_{16}NO$) was originally thought to arise from loss of the butyl side chain followed by loss of ethylene from ring A (*30*). However, synthesis of a norpumiliotoxin not containing the allylic methyl group showed the same ion at *m/z* 166 (*5*). It would then appear that the ion at 166 does not involve rupture of ring A and most probably comes from rearrangement of the double bond to ion **27,** followed by a normal allylic cleavage to the stabilized ion **28.** This rearrangement would keep the ring system intact and be similar to the cleavages in the other frog alkaloids. The tertiary alcohol present in these pumiliotoxins plays little part in the fragmentation process, the molecular ion showing the loss of hydroxyl (4%) at 234 and the loss of water (5%) at 233. That the double bond provides the trigger for the loss of the butyl and/or hexyl side chains is shown from the mass spectrum of the reduced compound. The principal ions here are at *m/z* 70 (C_4H_8N) as before, 84 ($C_5H_{10}N$) and 110 ($C_7H_{12}N$), all of which apparently involve cleavage of ring B.

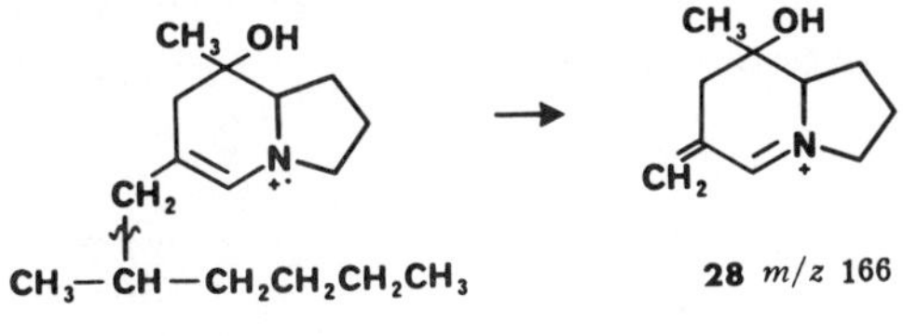

27 Rearranged molecular ion of 251D

28 *m/z* 166

The "allo" series of the pumiliotoxin A class (**27**), which contain an additional hydroxy group at C-7 between the double bond and the tertiary hydroxy group of **26,** exhibit a peak similar to the 166 ion at *m/z* 182 (*31*). Since the dihydro derivatives of this series have large ions at *m/z* 70 and 84, substitution of the additional hydroxy group in positions 1, 2, and 3 was eliminated from consideration (*m/z* 70). The peak at *m/z* 84 is attributed to cleavage in the original series of the B ring between carbons 5 and 6 as well as 8 and 8a. Therefore the additional hydroxy group cannot be at position 5, leaving only the 7 position. Later NMR studies confirmed that the additional hydroxy group is at C-7, but also showed that the naturally occuring alkaloids constitute epimeric mixtures at C-7 as well as at C-15 [e.g., 323B′ and 323B″ (**29**)].

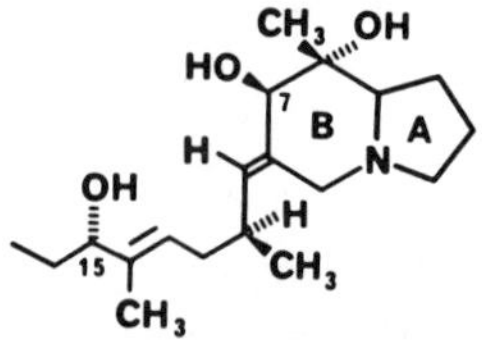

29

C. Tobacco Alkaloids

The mass spectrum of nicotine (**30**) has been studied in detail, using deuterium substituted in different positions on the pyridine and the pyrrolidine rings. Nicotine itself shows a molecular ion at *m/z* 162, M − 1, M − 29, and a base peak at *m/z* 84 (M − 78). The base peak has been assigned structure **31** from cleavage α to the pyrrolidine nitrogen and loss of the pyridine ring (*32, 33*). This assignment has been verified from the deuterated nicotine, the base peak moving to *m/z* 86 with two deuterium atoms at either C-4 or C-5. Appropriate shifts are also observed when deuterium is placed either on the *N*-methyl group or the C-2 position. No shift is seen in the base peak when deuterium is placed in the α position of the pyridine ring. The nature of the M − 1, M − 29 (C_2H_5), and M − 43 (C_3H_7) peaks is much less obvious. The results of ethyl loss with deuterated molecules will illustrate this point. When two deuterium atoms are incorporated into the 5 position of the pyrrolidine ring, 20% of the peak at *m/z* 133 moves to 134, while 80% moves to *m/z* 135, indicating that ethyl loss partially involves

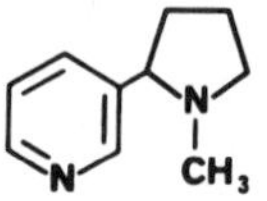

30 Nicotine

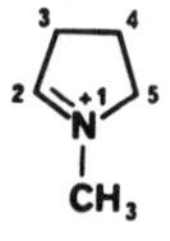

31 *m/z* 84

C-5. When two deuterium atoms are incorporated at C-4 of the same ring, ethyl loss removes both deuterium atoms. When the *N*-methyl group contains three deuterium atoms, none are lost in the ethyl loss. Most surprisingly, similar deuterium scrambling is seen when ethyl is lost from nicotine containing a single deuterium in the α position of the pyridine ring adjacent to the pyrrolidine ring. It appears that the loss of ethyl is complex. It is equally unclear from where the propyl is lost and the deuterium studies indicate that the M − 1 ion is not just loss of hydrogen from C-2 of the pyrrolidine ring.

Nicotine has recently been reported from an insect source as a defensive material in *Sclerobunus robustus* (a daddy-long-legs) (*34*).

Anabasine (**32**) and anabaseine (**33**) exhibit the same loss of ethyl as well as a methyl loss (*32, 35*). Only **32** has been studied by deuterium incorporation, but these studies also indicate that the fragmentations are more complex than originally thought.

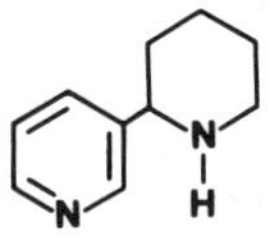

32 Anabasine

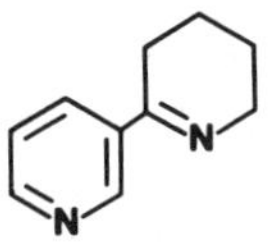

33 Anabaseine

D. Alkaloids from Insects

1. Simple Amines

Because of their volatility, simple amines are well suited to study by GC–MS. Their fragmentations are dominated by cleavage α to the nitrogen atom, but cleavage characteristic of branched alkyl groups can also provide structural information.

A few simple amines have been reported from insects. The ponerine ants *Mesoponera castanea* and *M. castaneicolor* contain 2,5-dimethyl-3-isopentylpyrazine in extracts of heads. In addition to this compound, however, they contain a series of aliphatic amines and amides in extracts of combined gasters and thoraxes. *N*-Isoamylnonylamine is the major constituent accompanied by *N*-isoamylnonenylamine, *N,N*-diisoamylnonylamine, *N*-acetylnonylamine, *N*-formylisoamylnonylamine, *N*-isovaleroylnonylamine, and other secondary and tertiary amines (*36*). *N*-Isoamylnonylamine shows a molecular ion at *m/z* 213, intense fragments at 156 (M − 57) (100) and 100 (M − 113) (80), with an ion at *m/z* 44 $(CH_2{=}NHCH_3)^+$ resulting from cleavage and rearrangement with transfer of a hydrogen. The last ion is typical of aliphatic secondary amines. Although these losses are consistent with two straight-chain alkyl groups, synthesis of the isoamyl and 1-methylbutyl isomers indicated that the three compounds could be distinguished from the intensities of the fragment ions. The isoamyl isomer showed a greater loss of butyl and a slightly greater loss of methyl than the straight-chain C_5 compound. The 1-methylbutyl isomer exhibits three modes of α cleavage, losing, as expected, methyl, propyl, and octyl.

Although the 2-methybutyl isomer was not synthesized, the retention times of the isoamyl isomer and mass spectrum were identical to that natural product. There is little doubt about this identification because the spectrum of the 2-methylbutyl compound would be expected to be more similar to that of the isoamyl than to that of the 1-methylbutyl. Similar α cleavages are observed for the other amines. The *N*-formylisoamylnonylamine and *N*-acetylisoamylnonylamine were suspected to be amides from their late retention times although the mass spectrum of *N*-ethylisoamylnonylamine was quite similar to that of *N*-formylisoamylnonylamine. The loss of methyl from the *N*-ethyl derivative by α cleavage gives an ion at *m/z* 226 that is no more intense than the loss of methyl from the side chains of its *N*-formyl analog. The ion at *m/z* 72 (44 + 28) attributable to double rearrangement is intense in both derivatives. The *N*-acetyl compound lost 42 mass units (ketene) from all of its even-electron α-cleavage ions as well as the double rearrangement ion at *m/z* 86 (44 + 42). Another possibility, *N,N*-diisovaleroylisoamylamine, exhibited a base peak at *m/z* 57 ($C_4H_9^+$) rather than the observed 156 (loss of ketene and butyl) as well as no ketene loss. *N*-Isovaleroylnonylamine showed a molecular ion at 227, a base peak at 57 ($C_4H_9^+$), and many fragment ions. The α-cleavage loss of octyl is a small ion (~40%) while the isovaleroyl fragment is surprisingly large (85%), being the second largest peak in the spectrum. Loss of isopropyl (185, 35%) and isobutyl (170, 30%) are also large.

A tertiary amine, *N,N*-dimethyl-β-phenylethylamine (**34**) has been identified (*34*) in the defensive secretion of the harvestman (*Sclerobunus robustus*). Aside from the base peak at *m/z 58* [$(CH_3)_2NCH_2^+$] few other fragment ions are present. Small ions are present at 65 (2), 77 (2), 91 (2) and 105 (0.5), suggesting either a tertiary or secondary amine. Although there were clusters of ions at 116, 117, 118 and 132, 133 and 146, 147, 148, and 149, the molecular ion at 149 was not clearly defined. Proof of structure was accomplished by comparison of retention time and mass spectrum with an authentic compound.

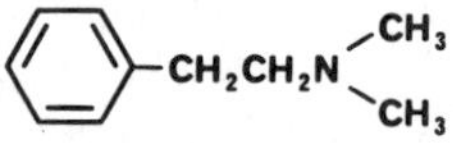

$C_6H_5CH_2CH_2N(CH_3)_2$

34

2. Pyrroles

Only one pyrrole has been isolated from insect sources, but it has major importance in the leaf-cutting ants *Atta texana* (*37, 38*), *A. cephalotes* (*39*), *A. sexdeus,* and *Acromyrmex octospinosus* (*40, 41*). This trail pheromone, methyl 4-methylpyrrole-2-carboxylate (**35**), exhibits a molecular ion at *m/z* 139 (93),

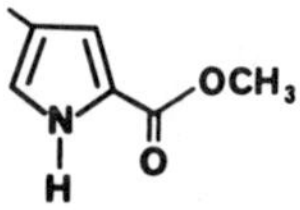

35

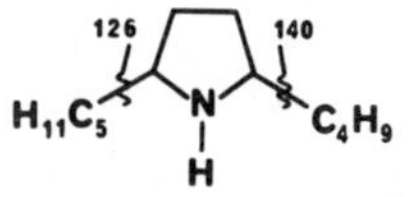

126 140 $H_{11}C_5$ C_4H_9

36

with losses of OCH_3 at 108 (100), CH_3 and H_2O at 106 (24), CH_3CO_2 at 80 (73) and CH_3CO_2H at 79 (97), and other large ions at 53 (81) and 52 (70) (*38*).

Partially reduced pyrroles (pyrrolines) and fully reduced pyrroles (pyrrolidines) have been indentified in several species of fire ants (*Solenopsis*) (*42*), the European thief ant (*Solenopsis fugax*) (*43*), as well as two species of *Monomorium* ants (*44, 45*). The pyrrolidines exhibit weak molecular ions of odd mass and two intense peaks resulting from α cleavage of the side chains to the pyrrolidine ring (e.g., **36**). The pyrrolines exhibit more complex mass spectra, having more intense molecular ions and two (or more) principal cleavage ions. Thus 5-ethyl-2-pentyl-1-pyrroline (**37**) and 2-ethyl-5-pentyl-1-pyrroline (**38**) can be distinguished by the following fragmentation: for **37,** the base peak appears at *m/z* 82, with an odd-electron ion at *m/z* 111. With the double bond directed toward the five-carbon side chain, a McLafferty rearrangement of the γ hydrogen leads to the ion at *m/z* 111 (**39**). Subsequent α cleavage of the ethyl side chain provides an ion at *m/z* 82 (6) (**40**) and a metastable ion at an apparent mass of 60.6 confirms this fragmentation. Isomer **38,** on the other hand, has intense peaks at 110, 97, 96, and 82. The authors suggest that the 96 peak results from allylic cleavage of the C-5 side chain, giving **41** and the 97 peak from allylic cleavage with hydrogen transfer to nitrogen, forming an allylic radical ion (**42**).

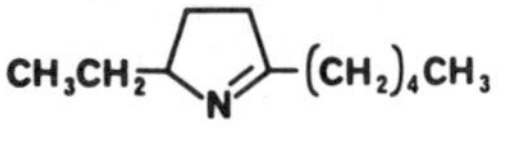

CH_3CH_2 $(CH_2)_4CH_3$

37

CH_3CH_2 $(CH_2)_4CH_3$

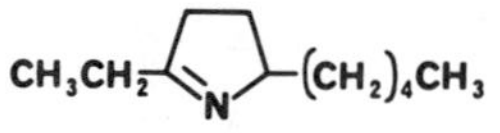

38

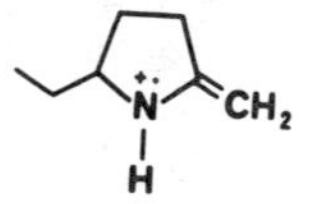

39 *m/z* 111

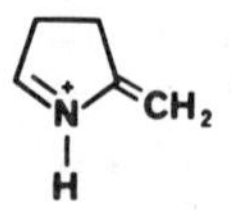

40 *m/z* 82

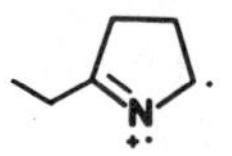

41 *m/z* 96

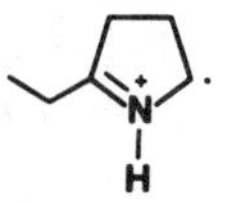

42 *m/z* 97

This ion then loses a methyl group, producing the ion at *m/z* 82 as before. Although the ion at *m/z* 110 is not rationalized, it appears to correspond to **43,** in accord with β rather than α cleavage. The *N*-methylpyrrolidines **44** and **45** exhibit the same type of cleavages, with **44** having α-cleavage ions at 166 and 210, while 45 has α-cleavage ions at 166 and 208 (*44*).

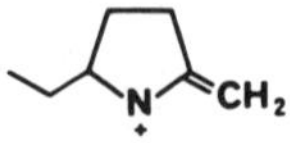

43 m/z 110

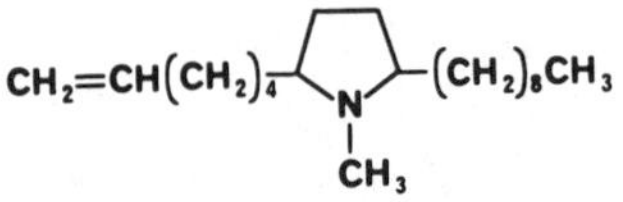

$CH_2{=}CH(CH_2)_4$–[N–CH_3 pyrrolidine]–$(CH_2)_8CH_3$

44

$CH_2{=}CH(CH_2)_4$–[N–CH_3 pyrrolidine]–$(CH_2)_7CH{=}CH_2$

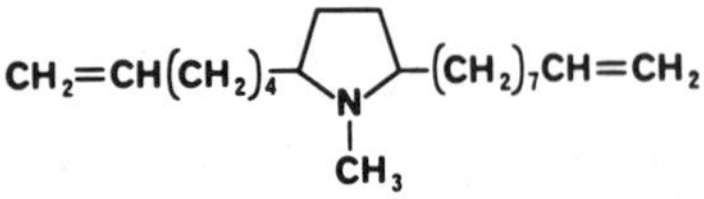

45

3. Pyridines

Two substituted pyridines have been isolated from insect sources. These are the monoterpene actinidine (**46**) (*46*) and the tobacco-like alkaloid anabaseine (**33**) (*35*). The base peak of actinidine at m/z 132 is formed by loss of methyl from the molecular ion at 147 (50). Additional peaks are seen at 146 (26), 131 (20), 130 (18), 117 (42), 105 (5), 103 (17), 91 (15), 89 (6), 77 (33), 65 (18), 55 (17), 53 (22), and 51 (22). The spectrum reported earlier for actinidine (*20, 47*) is incorrect; the sample used to obtain the former spectrum (*47*) when rerun produces a spectrum that agrees with that of reference *35* (*48*). Anabaseine exhibits a molecular ion at m/z 160 (85) and loses C_4H_8 to form a base peak at 104, corresponding to the nitrile. It also shows an appreciable loss of four hydrogens, m/z 156 (81%), presumably forming 2,3′-bipyridyl in the mass spectrometer. Peaks at m/z 145 (53) and 131 (97), corresponding to the loss of methyl and ethyl, require rearrangement, but this process is known for tobacco alkaloids of a similar type (*32*).

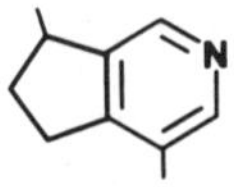

46 Actinidine

Partially reduced pyridines (piperideines) and fully reduced pyridines (piperidines) are found in several species of fire ants (*Solenopsis*). The structures of the piperideines have been determined by reduction to the corresponding piperidines and from the mass spectra of the piperideines themselves (*49*). The isomeric 2-methyl-6-undecyl-1-piperideine (**47**) and 2-methyl-6-undecyl-6-piperideine (**48**) exhibit very different mass spectra. McLafferty rearrangement of **47** results in the odd-electron ion **49** (m/z 97), whereas the rearrangement of **48** gives the odd-electron ion **50** (m/z 111), allowing them to be differentiated on the basis of their mass spectra. Such an inference must be used with caution, however, since monosubstituted piperideines such as **51** also give an ion at m/z 97 by rearrangement. Although 111 is the base peak in the fragmentation of **48, 52** (m/z 110) is

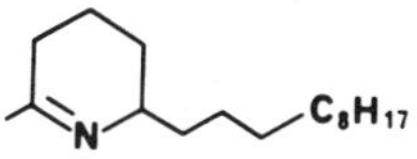

47 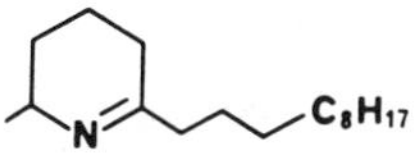 **48** 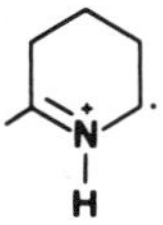 **49** *m/z* 97

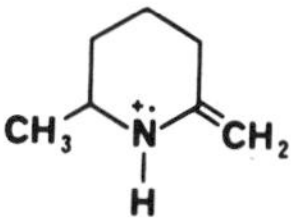

50 *m/z* 111 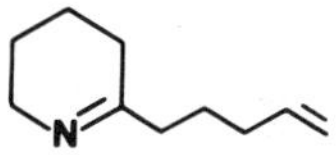 **51** 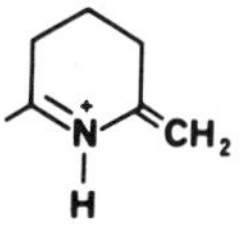 **52** *m/z* 110

the base peak in the fragmentation of **47,** corresponding to the loss of the side chain after transfer of two hydrogens.

2-Alkyl-6-methylpiperidines exhibit a base peak at *m/z 98* (**53**) arising from α cleavage of the alkyl group as well as M − CH_3 and M − 1 ions (*50*). This ion at *m/z* 98 dominates the spectrum. Mass spectra of the cis and trans isomers are indistinguishable, but they can be differentiated by gas chromatographic retention times, using polar phases such as Carbowax 20M or SP-1000. This method has shown that side-chain double bonds in 2-methyl-6-alkenylpiperidines are always cis and that trans 2,6-disubstitution in the piperidine ring predominated in *S. invicta* (*51*). In other *Solenopsis* species, however, the cis isomer predominates (*52*). The molecular ion of these dialkylpiperidines is weak unless the alkyl side chain is unsaturated.

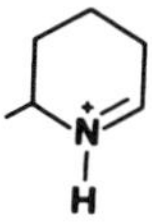

53 *m/z* 98

An *N*-ethylpiperidine, "stenusine," has been isolated from a staphylinid beetle, *Stennus comma*. This compound, *N*-ethyl-3-(2-methylbutyl)piperidine (**54**) loses a C_5 alkene, giving an ion at *m/z* 113 (**55**), occurring by direct rearrangement of the molecular ion as indicated by metastable ions (*53*). Synthesis confirmed this postulated structure (*49*).

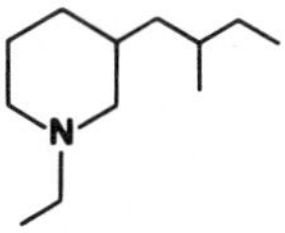

54 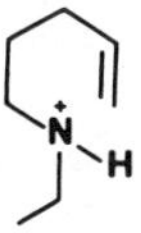 **55** *m/z* 113

4. Pyrrolizidines and Indolizidines

Cleavage α to the nitrogen on both sides of a heterocyclic system often allows the identification of the ring system. Most of the pyrrolizidines described here exemplify this useful property.

(5*Z*,8*E*)-3-Heptyl-5-methylpyrrolizidine (**56**) has been isolated from the poison gland of the thief ant *Solenopsis sp.* var. *tennesseensis* (*54*). This is the only example of a 3,5-dialkylpyrrolizidine from a natural source. It exhibits a molecular ion at 223 and significant ions at 208 (8) (loss of a methyl group), 124 (100) (loss of the C_7H_{15} side chain), and 110 (9) (loss of the methyl and heptyl side chains with a hydrogen transfer), and no other significant fragments.

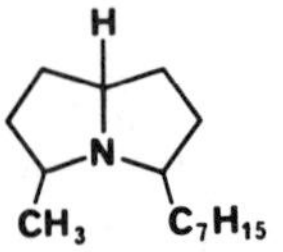

H

N

CH_3 C_7H_{15}

56

Pyrrolizidine alkaloids have also been found in adult male butterflies and moths, which use them as sex pheromones. The adult males feed on plants that contain compounds related to these nitrogen heterocycles, lycopsamine (**57**) (*55*) being a possible precursor to danaidone (**58**) produced by *Lycorea ceres* (*56, 57*) and *Danaus chrysippus*. Two other pyrrolizidine alkaloids, **59** and **60** (*58–61*) have been found in other moths and butterflies.

The mass spectrum of **58** was originally interpreted as a substituted pyridine before enough sample was available so that NMR data could be obtained. It shows a molecular ion at *m*/*z* 135 (100) with additional ions at 107, 79, 52, 27, 39, 65, 92, and 120 (reported in decreasing intensity). In this molecule, loss of methyl is a minor fragmentation, the loss of carbon monoxide and ethylene

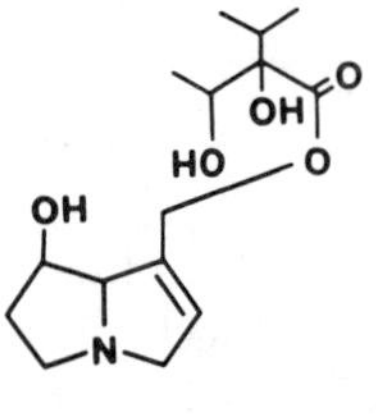

OH OH O HO O N

57 Lycopsamine

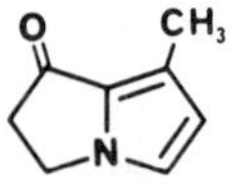

O CH_3 N

58

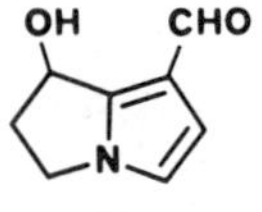

OH CHO N

59

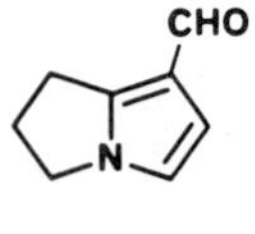

CHO N

60

representing the major fragments, as they do in the unsubstituted molecule lacking the methyl group. Details of the mass spectra of **59** and **60** have not been published.

Another pyrrolizidine alkaloid has been found in the millipede *Polyzonium rosalbum* and is called nitropolyzonamine (**61**) (*62*). It exhibits a molecular ion at *m/z* 238 (2) with loss of the nitro group at *m/z* 192 (8). The major losses at 122 (32), 108 (35), and 82 (100) (*62*) appear to represent loss of C_5H_{10} from the spiro ring system after the loss of NO_2, an additional CH_2 group, and the remainder of the ring leaving C_5H_8N, but exact masses have not been measured. Only the molecular ion has been reported for the related pyrroline polyzonamine (**62**) (*63*).

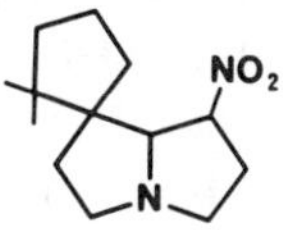

61 Nitropolyzonamine

62 Polyzonamine

Indolizidines were originally found in Pharaoh's ant (*Monomorium pharaonis*) **63** (and later **64**) having been reported from that ant (*64, 65*). More recently two indolizidines (**65** and **66**) have been reported in *Solenopsis* species (*66*). Although no detailed spectra are reported in any of the Ritter papers for **63** and no spectrum of **64,** the major losses for **63** are M − 1, M − CH_3 and M − 57 (*64*).

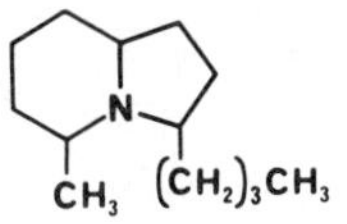

63

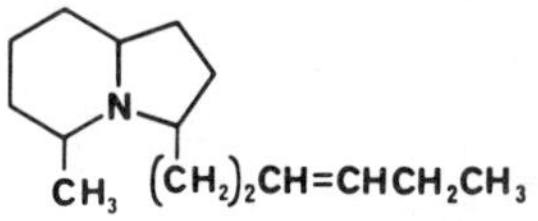

64

The cleavage patterns of **63, 65,** and **66** are all very similar, the base peaks being formed by cleavage of the longer alkyl chain to an ion of *m/z* 138. Although no spectra have been published for **64,** one would expect similar fragmentation. Aside from loss of the methyl group in each compound, there is little other fragmentation, the ring system being kept intact.

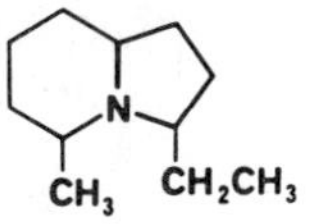

65

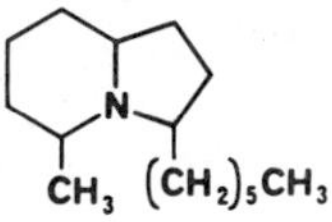

66

5. Coccinellines

Ladybugs (Coccinellidae) have been a rich source of tricyclic alkaloids called coccinellines. These are all closely related and are shown below.

One of these, myrrhine, has been reported in the boll weevil, but the identification is based wholely upon GC–MS data (*67*). Adaline, a homotropane alkaloid, has been found in the ladybug *Adalia bipunctata* (*68*). A soldier beetle (*Chauliognarthus pulchellus*) secretes coccinellines from paired glands on the prothorax and abdomen, whereas the ladybugs contain these compounds as blood-borne constituents throughout the body, which can be secreted by reflex bleeding.

Coccinelline (**67**) exhibits a molecular ion at *m/z* 209 ($C_{13}H_{23}NO$) as does convergine (**68**). Both lose oxygen from the molecular ion. Although it is stated that the mass spectra of coccinelline, convergine, and adaline are virtually identical (*69*), their structures are based on single crystal X-rays of their hydrochlorides, and mass spectral data are scanty. The mass spectrum of myrrhine (the trans-trans-trans isomer **69**) has been published. It exhibits *m/z* 193 (46), 192 (100), 178 (27), 164 (33), 151 (54), 150 (52), 137 (41), 136 (22), and 123 (33). Propyleine (**70**) exhibits *m/z* 191 (100), 176, 162, 149, and 148 (intensities not given) (*70*). Loss of methyl is similar to those of the tobacco alkaloids discussed earlier and loss of 42 would appear to be C_3H_6 from one of the six-membered ring systems present. The position of the double bond in propyleine is based on only end absorption in the ultraviolet spectrum analogous to that of anhydrolycoccernuin (**71**).

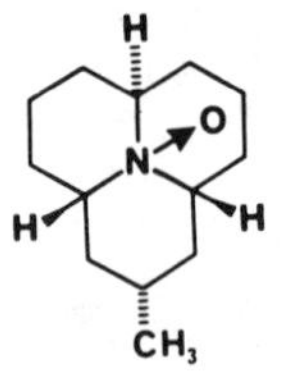

67 Coccinelline; (*N*-oxide)

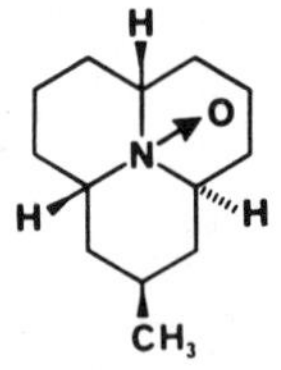

68 Convergine (*N*-oxide)

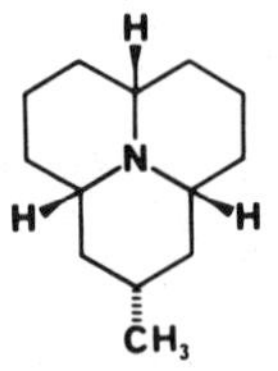

69 Myrrhine

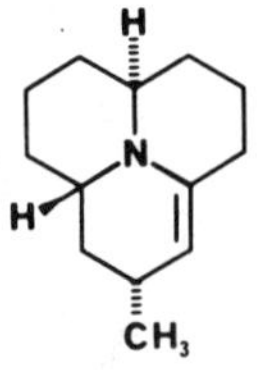

70 Propyleine

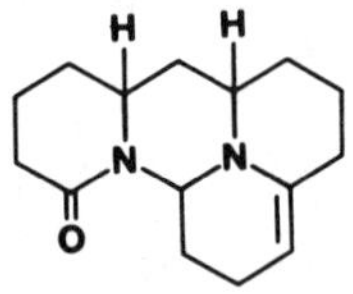

71 Anhydrolycoccernuin

6. Pyrazines

Alkyl substituted pyrazines occur extensively in Hymenoptera, particularly in the ant subfamilies Ponerinae, Dolichoderinae, and Myrmicinae. Recently such pyrazines also have been identified in several genera of the wasp families Eumenidae and Sphecidae (*71*). Most of these pyrazines are mandibular gland products having two methyl groups and a third alkyl group on the pyrazine ring. The base peak of these pyrazines has been useful in establishing the number of alkyl groups attached to the ring. Pyrazines that have methyl groups in the 2 and 5 (or 2 and 6) positions and contain a straight-chain alkyl group of three or more carbons exhibit a McLafferty rearrangement peak at *m/z* 122 (**72**).

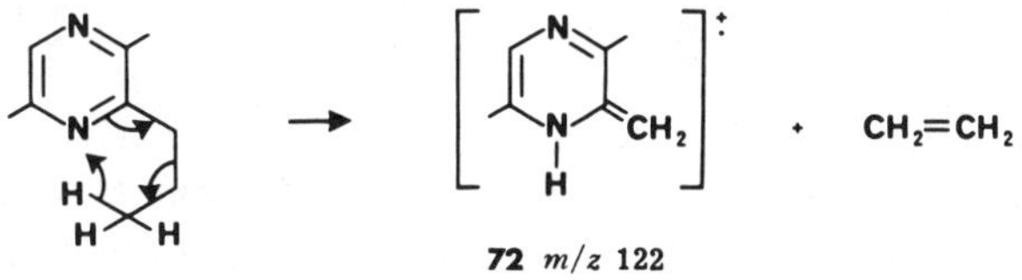

72 *m/z* 122

Although the 2,3-dimethyl-5-*n*-pentylpyrazine has been reported to exhibit a base peak at *m/z* 108 (*72*), this appears to be in error as other 2,3-dimethyl-5-alkylpyrazines show the normal base peak at *m/z* 122. Substitution at the α position (e.g., 2,3-dimethyl-5-*sec*-butyl) moves the rearrangement peak to a higher mass (136). However, when the side chain contains a double bond, as in **73,** the normal McLafferty rearrangement does not occur. Metastable ions indicated that the base peak at *m/z* 133 was formed by the loss of 29 (C_2H_5), followed by loss of 28 from the unsaturated side chain (*71*). This is a proximity effect and not characteristic of alkenyl chains conjugated to the ring as isomers of **73,** in which the propyl and butenyl groups on opposite sides of the ring show the normal rearrangement ion at *m/z* 162. The reduced compound with a butyl side chain also exhibits the normal rearrangement. It was also noted that the nature of the double bond had an effect on the base peak (e.g., *trans*-**73** = *m/z* 133, whereas *cis*-**73** = *m/z* 147). Other isomeric pyrazines that were expected to exhibit a base peak at *m/z* 133 by allylic cleavage actually showed a base peak at *m/z* 147 (e.g., **74**), possibly indicating that rearrangement takes place here before cleavage at the allylic position.

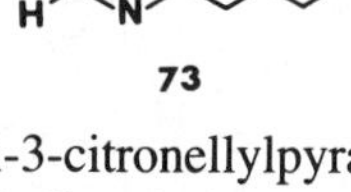

73 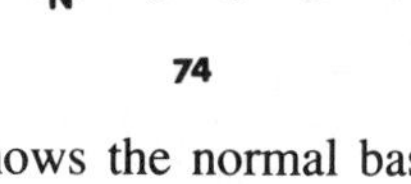 **74**

2,5-Dimethyl-3-citronellylpyrazine (**75**) shows the normal base peak at m/z 122 (*73*) but 2,5-dimethyl-3-styrylpyrazine (**76**) has a base peak at *m/z* 210 (the molecular ion) with a large peak at *m/z* 133 ($C_8H_9N_2$) corresponding to vinyl cleavage (*74*). The most common pyrazines from insects have methyl groups at the 2 and 5 positions, but 2,6 (*72*) and 2,3 (*75*) have also been reported. It is extremely difficult to determine the substitution pattern on the pyrazine ring from the mass spectrum alone since the 2,5 and 2,6 isomers are almost identical in

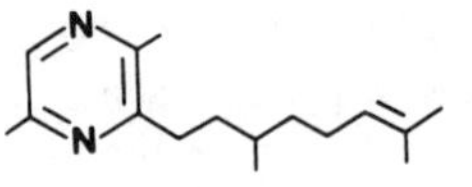

75

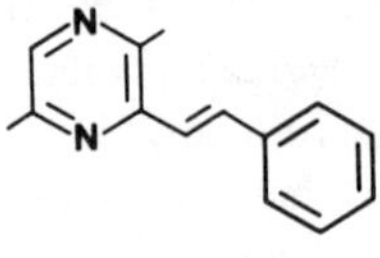

76

their fragmentation patterns. Therefore a comparison of gas-chromatographic retention times with authentic samples must be made. The 2,5 isomer is eluted before the 2,6 on polar columns, and this has been used for identification (*71*).

Two other methods have been used to determine substitution patterns on the pyrazine ring. Reduction of **73** with platinum and hydrogen gave, in addition to 5-methyl-3-propyl-2-butylpyrazine, the piperazine **77**. This fragmented in the manner shown below, allowing placement of the three alkyl groups on the ring,

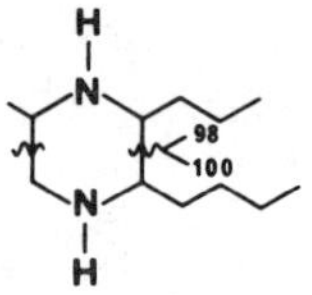

98

100

77

such that the top and bottom halves of the ring contained an equal number of carbons. This fragmentation corresponds to half of the ring with transfer of a hydrogen. Synthesis of both possible isomers having equal numbers of carbons confirmed the identification. A second method of distinguishing substitution on the pyrazine ring involves the quaternization of one nitrogen with methyl iodide, followed by reduction of the ring with sodium borohydride to *N*-methylpiperazines. The methylpiperazine from 2,5-dimethyl-3-isopentylpyrazine exhibits a base peak at *m*/*z* 72, while the methylpiperazine from 2,6-dimethyl-3-*n*-pentylpyrazine has its base peak at *m*/*z* 128 (*72*). Similar treatment of other isomers showed that the difference is due to ring substitution rather than differences in the side chain, even though fragmentation of the unmethylated piperazine **77** results in two halves of equal molecular weights. 2,6-Dimethyl-3-alkylpyrazines are alkylated preferentially on the nitrogen not bearing the methyl groups, and the base peak is derived from that half of the molecule (see **78**), while the corresponding 2,5-dimethyl isomer is alkylated on the nitrogen not bearing the large alkyl group, and the base peak is derived from that half (see **79**). It would appear

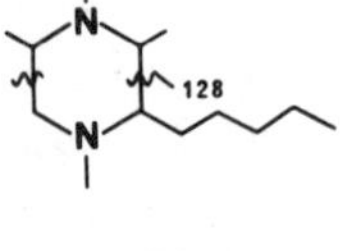

128

78

72

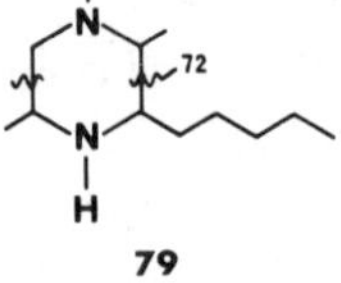

79

that this steric effect might be useful for the determination of structure in future unknown pyrazines having 2,5 or 2,6 substitution patterns.

Pyrazines are not restricted to insects; two of them have been found in the scent-gland secretion of the Canadian beaver. These are the trimethyl and the tetramethyl compounds (*76*). A recent review of pyrazines (*77*) indicates how extensively pyrazines occur as food flavors and contains mass spectra of representative types.

7. Quinolizidines

Castoramine (**80**) has been isolated from the Canadian beaver (*78, 79*) but was cited as an unproven structure as recently as 1981 (*80*). It and 14 other nitrogen-containing compounds have been indentified in the scent gland of the beaver. Seven of these are closely related to castoramine (*76*), being sesquiterpenoid quinolizidines. The other compounds are pyrazines, an isoquinoline, three tetrahydroquinoxalines, and an octahydrophenazine.

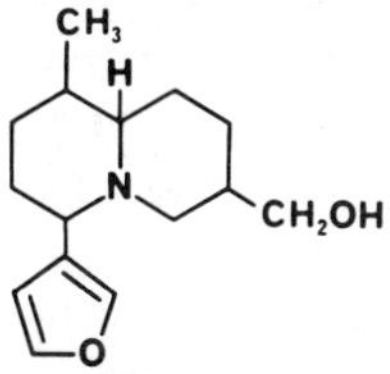

CH_3 H N CH_2OH O

80 Castoramine

The mass spectrum of castoramine (and many of the others) show a base peak at *m/z* 94 attributed to **81** and a peak at 114 (60) attributed to **82** with additional large ions at 136 (45) and 249 (30). These are closely related to the nuphar alkaloids isolated from the water lilies *Nuphar japonicum* and *N. lutem*. These plants contain dimeric sulfur-containing alkaloids of the same carbon system such as that in thiobinupharidine (**83**).

O

81 *m/z* 94

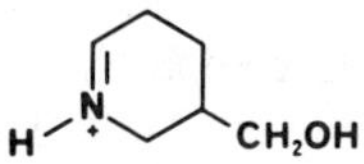

H N CH_2OH

82 *m/z* 112

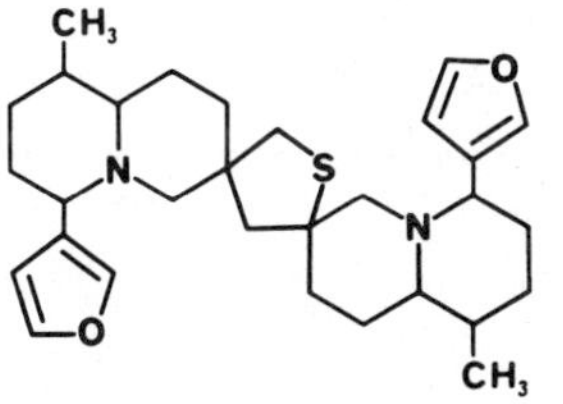

CH_3 N S O N O CH_3

83 Thiobinupharidine

III. Nuclear Magnetic Resonance Studies

A. Proton NMR

When the fundamental relationships between structures and NMR parameters had been established, the study of alkaloid structures was among the first applications of NMR spectroscopy. The development of NMR spectroscopy has been characterized by remarkably rapid evolution, such that no spectrometer that has remained unmodified for a few years has ever represented the state of the art. Alkaloid chemists have eagerly exploited each development as it was reduced to practice. Currently, the development of multiple-pulse methods is proceeding at such a rate that it is inevitable that an review be outmoded when it reaches print.

Early NMR spectrometers allowed the examination of the proton spectrum of small quantities, perhaps .03 mmol, of alkaloids, readily revealing the presence of methoxyl or methylenedioxy groups, the presence and substitution patterns of aromatic protons, and, in appropriate molecular environments, substantial information about alicyclic systems. Knowledge of the relation of coupling constants to the stereochemistry of vicinal protons evolved from the study of the theoretical origins of the couplings and those observed in sugars (*81, 82*). The Karplus curve quickly became the spectroscopist's most potent tool for studying solution conformations.

In addition to the application of general knowledge of the relationship of chemical shifts to structural characteristics, methods of particular value to alkaloid characterization were devised. It was found possible to distinguish methoxyl peaks from *N*-methyl peaks by comparison of spectra of neutral and acidic solutions (*83*). The relation of aromatic protons to methoxyl on an aromatic ring was shown by comparison of the spectra in chloroform and benzene solutions methoxyls with an ortho proton being shifted 0.3 ppm upfield in the second solution (*84*). Comparison of the chemical shifts of aromatic protons with those of the corresponding anion or acetate was shown to establish the relation of those protons with the phenol, to allow the differentiation of substituents in a polyoxygenated aromatic system (*85*). Close study with such spectrometers allowed investigators to demonstrate effects across space, such as that in the amaryllis alkaloid powelline (**84**), in which reduction of the A-ring double bond produced a 0.2-ppm upfield shift of the aromatic proton (*86*). In a similar study of the

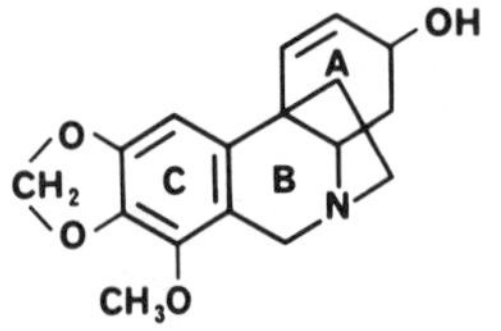

84 Powelline

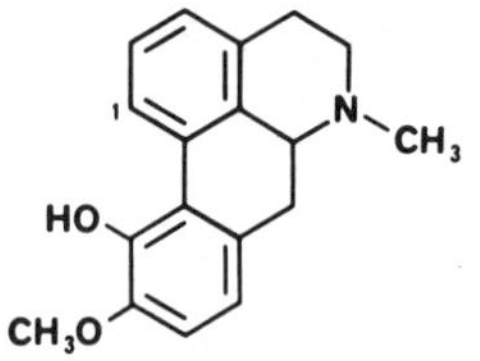

85

aporphine alkaloid **85,** formation of the anion caused a downfield shift of the C-1 proton of 0.8 ppm (*87*).

These studies on simple spectrometers established the relationships between structure and NMR parameters, which form the basis of all current structural work (*88–90*). The introduction of spectrometers with internal field lock abilities made double-resonance experiments facile, allowing simplification of complicated coupling patterns and the demonstration of coupling involving discretely visible groups with protons whose signal is obscured by those of other protons. Modern spectrometers incorporate two further improvements to facilitate the determination of NMR parameters: they employ the high field of a superconducting magnet and they allow the use of pulse techniques under computer control.

The upfield region of alicyclic systems gives rise to overlapping signals of closely coupled protons; the chemical shifts and coupling patterns of such systems are extracted only with great difficulty or not at all. Modern spectrometers operate at 4 to 10 times the field strength of earlier instruments; overlapping signals are much better separated, and the effects of close coupling are much reduced [cf. Fig. 2, spectra of undulatine (**86**)]. The use of higher fields also carries the additional advantage of higher sensitivity and possibly lower operating costs.* Such an instrument can produce useful spectra down to approximately millimolar concentrations.† When somewhat larger quantities are available, the interpretation of the structure will be aided by carbon-13 spectra; discussion of the process in this review will therefore be deferred.

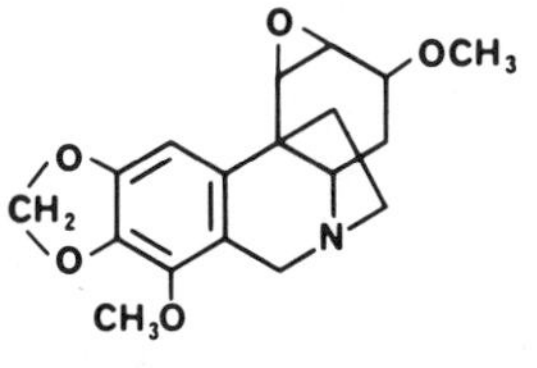

O OCH$_3$ O CH$_2$ O N CH$_3$O

86 Undulatine

However, it is worth noting that knowledge of the values of the proton chemical shifts to be anticipated within well studied groups of compounds has devel-

* The question of exactly what improvement in sensitivity should be obtained at high field can give rise to a vigorous discussion among NMR physicists, but a working rule for practicing spectrometrists is that the increase should be at least linear with field. Modern superconducting magnets require a charge of perhaps 25 liters of liquid helium, costing U.S. $80–100, every 1–3 months. Electromagnets producing fields of 1.5–2.5 Tesla (proton at 60–100 MHz) typically require 3.5–8 KVA, for an operating cost of $0.05/KWH of $120–270/month, with attendant cooling costs. The higher costs of helium elsewhere in the world may well invert this relationship. Unfortunately, the capital expenditure remains approximately $1/KHz.

† Common deuterated solvents are approximately 30 m*M* in proton concentration, although solvents of approximately 1 m*M* proton can be obtained at higher cost. Noncrystalline samples much below a milligram are quite generally contaminated with impurities from chromatographic media or congeners to the point of obscuring useful signals.

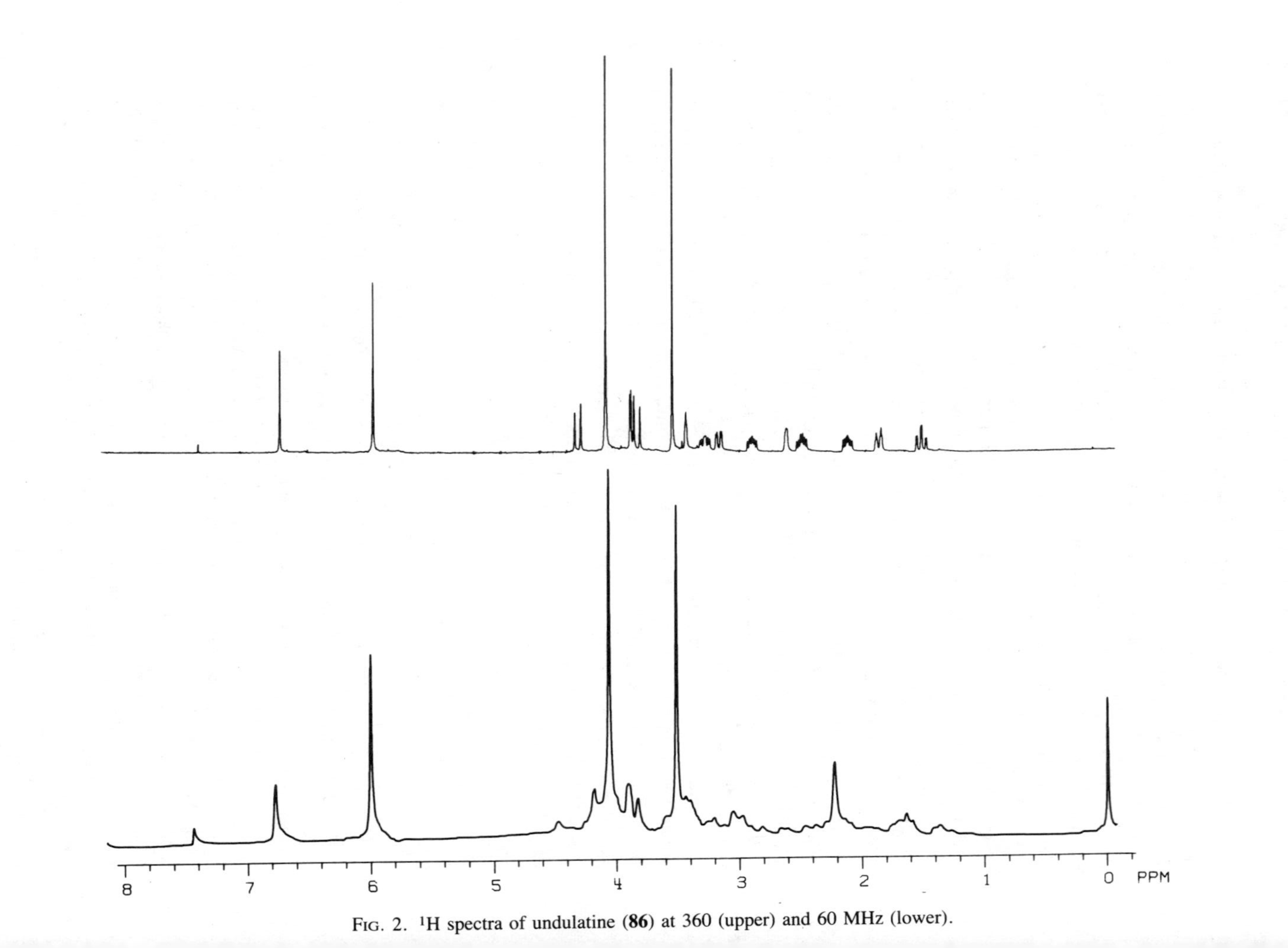

FIG. 2. ^{1}H spectra of undulatine (**86**) at 360 (upper) and 60 MHz (lower).

oped to a remarkable precision. Thus the alkaloid fumarofine was earlier assigned the spirobenzylisoquinoline structural formula **87,** the observed proton resonances (CH_3O's, 3.87 and 3.94; aromatic protons, 6.63, 7.15, 7.16, and 7.28; *N*-CH_3, 2.55 ppm) being well within the anticipated ranges (*91*). However, *O*-methylfumarofine had methoxyl resonances at 3.87 and 3.94 ppm; extensive study of such spirobenzylisoquinolines has shown that methoxyls at C-2 and C-3 are invariably separated by 0.2 ppm, that H-1 absorbs ~6.25 ppm, and that the *N*-CH_3 falls within the range of 2.25–2.40 ppm. Fumarofine was therefore assigned an indenobenzazepine structure (**88**) (*92*).

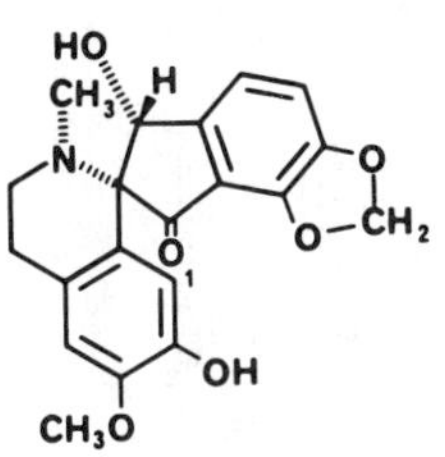

HO CH₃ H N O O O—CH₂ 1 OH CH₃O

87

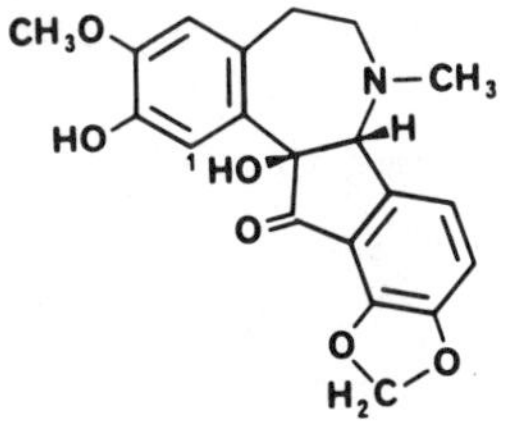

CH₃O HO N—CH₃ H 1 HO O O O H₂C

88 Fumarofine

B. Carbon-13 NMR Spectra

Carbon-13 spectra became useful for structural studies of alkaloids with the advent of Fourier transform spectrometers, which allowed the collecting and averaging of thousands of free induction decays. Such spectra are collected while the proton frequency is irradiated with a noise-modulated signal, effecting complete decoupling of protons from the carbon-13 nuclei and producing a spectrum of single peaks for each carbon. With such devices, it became possible to obtain a chemical shift of each carbon of the same 0.03-mol sample, which earlier had provided a proton spectrum from a single scan on a CW proton spectrometer.

In a complementary experiment, the free induction decays from pulses at carbon-13 frequencies are obtained, while a coherent (single-frequency) irradiation is maintained at a frequency a few hundred hertz from the frequency of tetramethylsilane. As a result of the effective field at each proton frequency, the couplings are reduced according to the formula:

$$J_r = (2\pi J_{CH}/\gamma_{H_2})\Delta_{f2} \qquad (1)$$

The irradiating field γ_{H_2} is chosen so that the reduced couplings J_r are approximately 30% of their normal values, J_{CH}. As a consequence, multibond couplings with J_{CH} 15 to 0 Hz, are effectively eliminated, leaving multiplets resulting from the one-bond couplings, ${}^1J_{CH}$ of 120–250 Hz; methyl groups appear as quartets, methylenes as triplets, methines as doublets, and quaternary carbons as singlets. The amaryllis alkaloid undulatine (**86**) (*93*), produces noise-decoupled and off-resonance decoupled spectra as shown in Fig. 3. The broaden-

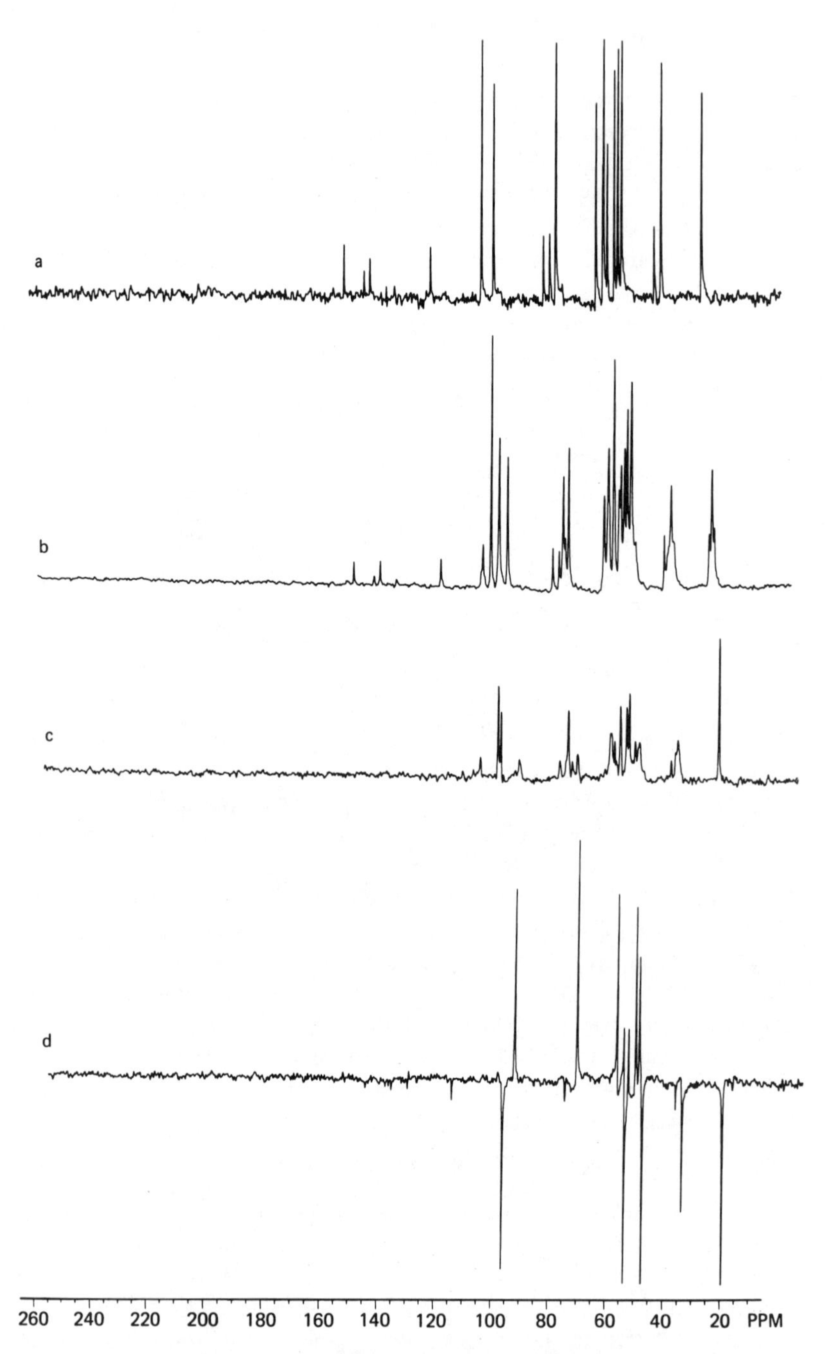

FIG. 3. Carbon-13 spectra of undulatine (**86**) at 15 MHz. (a) Noise-decoupled spectrum. (b) Off-resonance decoupled spectrum. (c) Single-frequency decoupled spectrum with irradiation at δ1.5 ppm at ^{1}H frequencies. (d) APT spectrum.

ing of the peaks by the residual couplings results in substantial loss in sensitivity. Commonly, the smaller peaks of quartets and triplets are lost in the noise. It is still generally possible to distinguish triplets and singlets from quartets and doublets by observing whether a peak remains in the off-resonance spectrum at a frequency corresponding to that observed in noise-decoupled spectra. The same distinction without loss of sensitivity can be made more readily by a two-pulse experiment described below.

A small bonus from an off-resonance experiment allows the recognition of methylene groups with protons of substantially different chemical shifts, such as those of the methylene groups α to the nitrogen atom of quinolizidines, in which the axial protons are shifted upfield by antiperiplanar electron pairs of the nitrogen atom. Because the Δ_{f2}s of the above equation are different, such protons effect different reduced couplings, producing a pattern in the carbon spectrum of doubled doublets.

By identifying the degree of substitution of each carbon, off-resonance spectra allow a count of (nonexchanging) protons. Because carbons attached to oxygen are generally readily recognized, these two carbon spectra and a low-resolution mass spectrum will generally provide the molecular formula of a substance.

It is often helpful in structure determinations to identify which carbon is coupled to a specific proton. This can be learned by a single-proton decoupling experiment, in which the frequency corresponding to the proton is irradiated by a coherent radio frequency. Only the carbon atom coupled to the proton will appear as a singlet. Such an experiment on undulatine is shown in Fig. 3. Differentiation of carbons attached to protons with frequencies approximately 30 Hz different can commonly be made. A two-dimensional method, chemical shift correlation mapping, described below, promises to provide superior resolution.

Because the chemical shift range of carbon-13 is substantially greater than proton—3500 versus 600 Hz at 1.5 Tesla—routine spectra such as those of Fig. 3 can be obtained as satisfactorily at low as at high field. The necessity of collecting free induction decays for a much longer time is often balanced by the familiar observation that the available instrument time is roughly inversely proportional to the field strength for shared instruments. In the spectra of compounds of modest molecular weight, the problem of overlapping peaks is seldom troublesome and can generally be resolved by the apparent intensities of the peaks or by repeating the spectrum in a different solvent. However, as the number of carbon atoms of similar character increases or as the amount of material decreases, the greater dispersion and sensitivity of high-field instruments become more important.

The spectra of Fig. 3 display an effect quite commonly observed of carbon-13 spectra, the differing height of peaks representing a single carbon atom, a result of relaxation effects. Following a pulse, carbon nuclei relax to their equilibrium states in a first-order process dependent largely on dipolar interactions with nearby protons. As stated below, the effects of these fields diminish by the sixth power of the distance between the nuclei. As a consequence, carbon atoms bearing no attached protons relax much more slowly. The relaxation time T_1, the inverse of the relaxation rate, of proton-bearing carbons in alkaloids commonly ranges from 0.3 to 1.5 sec. Carbon atoms more than one bond separated from a proton may well have relaxation times 10 times as great. Since routine spectra such as those of Fig. 3 are conveniently taken with pulse-repetition times of approximately 1 sec, quaternary carbon atoms often appear much smaller than methines and methylenes. Consequently, carbon counts based on such spectra must be made by comparison of carbons of similar type, as indicated by their off-resonance spectra and chemical shift.

Carbon atoms may also show a difference in the nuclear Overhauser effect of the proton decoupling if relaxation effects other than proton dipolar mechanisms are important. Such is not generally the case for alkaloids.

C. Coupling Constants

1. Geminal Proton Couplings ($^2J_{HH}$)

The relationship of geminal proton couplings to the structural environment of the methylene group is well understood (*94*) and an extensive tabulation of observed values is available (*95*), but these couplings are seldom quoted in support of the structures of alkaloids. Values of $^2J_{HH}$ encountered in alkaloids range from small positive values to -18 Hz. The signs are seldom determined, but by analogy to those which have been determined in similar structures, those of sp^3 methylene groups are assumed to be negative. In discussion, therefore, increments are treated as algebraic values; an increment diminishing the values of the coupling constant, producing a more negative value, is actually observed as a greater separation of the limbs of the AB, or AX system.

Geminal coupling increases with the *s* character of the bonds, those of olefins being small positive values. Within saturated and unsubstituted alkyl systems, they differ very little from that of methane (-12.5 Hz); α substitution by a negative group increases the coupling a few hertz. In cyclic systems, smaller rings, with increased *s* character in the C—H bonds, have more positive values. The effect can serve to identify a cyclopropyl system, but the chemical shifts at high field provide a more obvious characterization. Cyclobutyl systems reflect the effect but are probably more securely identified by the increased one-bond couplings $^1J_{CH}$.

The molecular orbital treatment of ^2J emphasizes the supply or withdrawal of electrons to orbitals with respect to the plane normal to the H—H axis. Withdrawal of electrons from orbitals symmetric to this plane increases ^2J; withdrawal of electrons from antisymmetric orbitals decreases the coupling. As a consequence, ^2Js of methylene groups bearing an atom with free electrons reflect the stereochemistry of the atom. The oxygen of tetrahydropyran has little effect on the coupling of the α-methylene group ($^2J = -12$ Hz) because the staggered pairs have little interaction with the antisymmetric orbitals; however, in the more nearly planar tetrahydrofuran, the C—H bonds eclipse the oxygen lone pairs, and the supply of electrons to the antisymmetric pairs increases the coupling to −8 Hz (*95*).

A detailed study of geminal couplings in 1,3-oxazine systems has shown that they can reveal the conformation of the ring (*96*). In the tricyclic perhydropyridobenzoxazine system, the ^2J of the trans,trans isomer **89,** with the nitrogen lone pair antiperiplanar to the C—H bond of the methylene, is increased to −7.5 Hz, constrasting with that of the cis,trans isomer **90,** −10.8 Hz, in which the nitrogen lone pair is staggered with the methylene group. The study supports the assigned stereochemistry of rosibiline (**91**) (*97*) in which the observed geminal coupling of the oxazine system, −9.0 Hz, reflects the staggered conformation of the nitrogen lone pair (*98*). Substitution of a methylene group by a freely rotating aromatic system lowers ^2J by 1.6 Hz. However, if the molecular geometry holds the aromatic system with the π bonds parallel to the H—H axis, the effect is much greater. The −17-Hz coupling of undulatine (**86**) results from two aspects of the cage structure: the nitrogen lone pair is held in a conformation staggered with the C—H bonds of the methylene group, and the π bonds of the aromatic system are parallel to the H—H axis of the group.

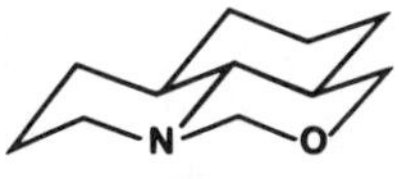

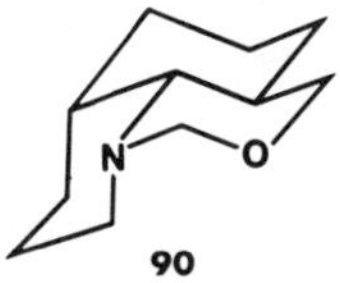

The effect of an adjacent π system can serve to identify a methylene group adjacent to a carbonyl group. The −17-Hz coupling of Coleon E (**92**) supports the presence of a methylene group α to the carbonyl group (*99*).

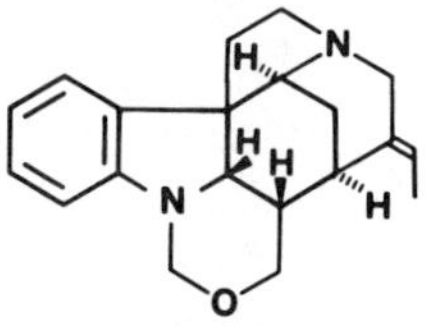

91 Rosibiline

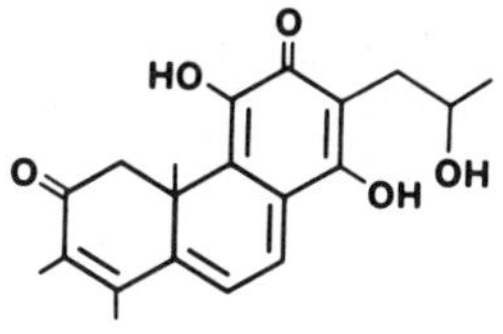

92 Coleon E

2. Vicinal Couplings ($^3J_{HH}$)

The relation of vicinal coupling constants to the dihedral angle of the three-bond system is one of the most useful aspects of NMR spectrometry. The relationship derived by Karplus fits apprxoimately the equation (*81*)

$$^3J = A + B\cos(\theta) + C\cos(2\theta) \tag{2}$$

The curve plotted in Fig. 4 corresponds to the values $A = 7$, $B = -1$ and $C = 5$ Hz (*100*). However, these values are altered by the precise nature of the group. Substitution by negative groups reduces them to an extent dependent on the stereochemistry of the substitution; a negative substituent in an antiperiplanar conformation to the bonds of the coupling path is particularly effective. If one of the carbon atoms is sp^2, the shape of the curve is altered as shown in Fig. 4 (*101*). Such fluctuations render it, in general, impossible to convert observed couplings to dihedral angles. Within extensive groups of closely related mate-

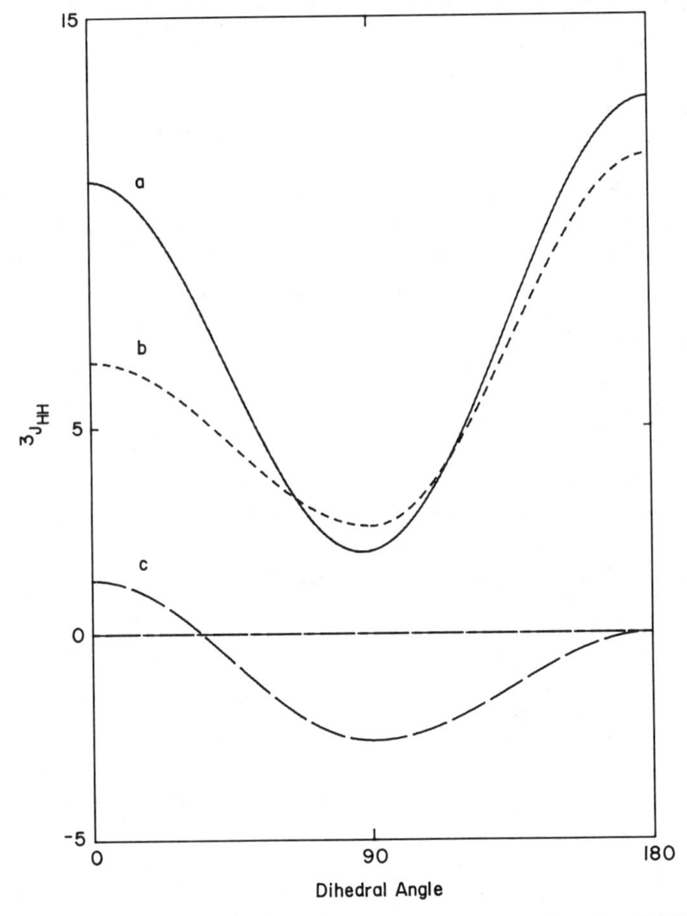

FIG. 4. Karplus curves for (a) R_2CHCHR_2, (b) $=CHCHRCHR_2$, and (c) $HRC=CRCHR_2$ from values quoted in references *100* and *101*.

rials, such as carbohydrates and peptides, reliable calibrations have been established, but no such system is applicable to a group so varied as alkaloids.

However, even with all these variations, it remains the case that protons in an antiperiplanar conformation and opposed protons have larger coupling constants than those in staggered conformations. Thus the stereochemistry of substituents in cyclohexyl systems can be reliably assigned, but within cyclopentyl systems these relationships are ambiguous. In cyclobutyl and cyclopropyl systems, cis protons give rise to larger couplings than trans. More extensive descriptions of couplings in complex systems will be found in such references as *88*.

3. Four-Bond Couplings ($^4J_{HH}$)

Allylic couplings are dependent on the conformation of the sp^3 carbon relative to the plane of the double bond (*101*). The corresponding curve in Fig. 4 retains the shape of the Karplus relation but is displaced downward; as a result, observed couplings are larger when the sp^3 C—H bond is normal to the plane. As a consequence, coupling of approximately 2 Hz is seen when the C—H bond of a methine or methylene group is normal to the plane of the ring. The effect is seen in the spectrum of brunsvigine (**93**), in which a 2-Hz coupling is seen between H-1 and H-4a (*102*).

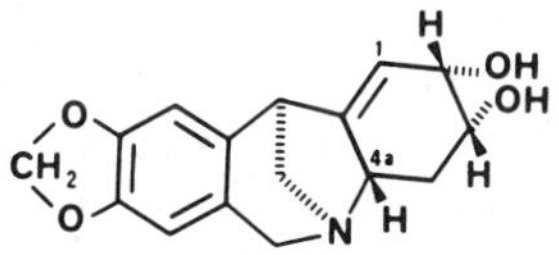

93 Brunsvigine

Coupling may be observed over four or more α bonds when all of the bonds lie in the same plane in an all-trans form—the "W" form. Such a coupling is observed in the spectrum of undulatine (**86**) between H-2 and H-4eq (above).

4. Carbon–Proton Couplings

Inasmuch as the relationship of carbon-13–proton couplings to organic structure is well understood (*103, 104*) these parameters should provide useful structural information not otherwise readily available. However, prior to the development of the two-dimensional Fourier transform method described in Section III,F,1, obtaining these couplings from molecules of any complexity was often nearly impossible. Coupled carbon-13 spectra display intricately overlapping peaks, the limbs subdivided and broadened by smaller couplings. The earlier work, which established the relationships quoted below, was performed on molecules chosen to avoid these difficulties and present readily understood coupled spectra. With the background of this work, the 2-D methods promise valuable results.

5. One-Bond Couplings ($^1J_{CH}$)

One-bond carbon–proton couplings vary from 100 to about 250 Hz, but most commonly are between 125 and 170 Hz. The variations are dominated by the hybridization of the carbon atom, conforming fairly well to the relation

$$^1J_{CH} = 500\ \rho_{CH} \tag{3}$$

in which ρ_{CH} is the fractional *s* character of the carbon orbital (*105, 106*). Ring strain increases the coupling constant (cyclopentane, 128; cyclobutane, 136; cyclopropane, 160 Hz); negative substitution also increases it (methane, 125; methanol, 140 Hz). The effects of conformational differences and lone pairs are small but well documented.

Many of these observations are redundant in simple structural studies. The effects of hybridization and negative substitution on proton and carbon chemical shifts serve to identify these systems. Although cyclopropyl systems are readily identified by high-field shifts, cyclobutyl rings are less readily detected. An Australian group were able to make use of this characteristic in their revision of the structural formula of grantaline (**94**) (*107*). The single high-field methine signal (48.6 ppm) could be assigned unambiguously to C-14; the $^1J_{CH}$ observed, 143 Hz, is suitable for a cyclobutyl system but well below that to be anticipated of an epoxide, ~170 Hz, and excluded the part structure **95.** The relatively low-field shifts of the oxygenated carbons, 82.3, 80.4, and 77.5 ppm, favored **94** over the epoxide in **96.** In the spectra of undulatine, above, the couplings observed of the C-1 and C-2 carbons serve to identify the epoxide carbons. It is clear that the possibility of studying $^1J_{CH}$ in detail in complex systems will extend the utility of such observations.

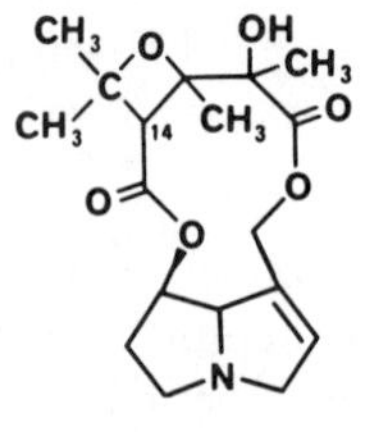

94 Grantaline

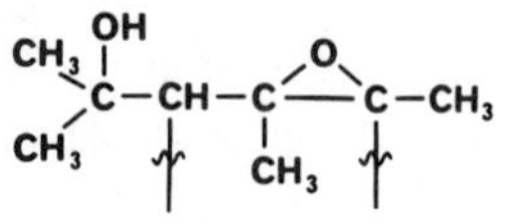

95

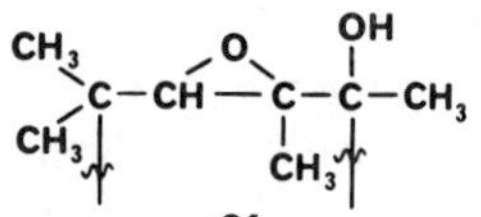

96

6. Geminal Couplings ($^2J_{CH}$)

Two-bond couplings range from about +60 to −10, the large values being those through acetylene bonds. Small negative values are most common. Knowledge of the structural dependence of these values is evolving rapidly, but little use has yet been made in the study of alkaloid structures.

7. Vicinal Couplings ($^3J_{CH}$)

Three-bond (CCCH) couplings follow the Karplus relation in much the same manner as HCCH couplings (*108*). Knowledge of the effect of substituents is still evolving, much complicated by the additional possibilities of substitution and conformation effects. The study of alkaloid structure and stereochemistry will clearly benefit greatly from these studies and the use of two-dimensional methods to provide $^3J_{CH}$s.

Trisubstituted double bonds constitute one such system that has been studied thoroughly enough to be useful (*109*). The trans vicinal coupling across a double bond is invariably larger than the cis, and generally the largest $^3J_{CH}$ present. The complexity of coupled spectra has largely prevented the study of the stereochemistry of double bonds by this method, but two-dimensional methods should make it a standard approach to the problem. A one-dimensional pulse sequence suitable for older spectrometers has been described, but suffers from drastic loss in sensitivity (*110*).

Three-bond couplings in aromatic systems obey the same generalization, trans couplings being larger than cis. In the study of the polybrominated bisindoles from the alga *Rivularia firma,* the couplings observed of the dimer **97** established the identity of several of the carbon atoms: C-3a, trans $^3J_{CH}$, 5.8 Hz; C-3, cis $^3J_{CH}$, 2.7; C-2, $^2J_{CH}$, 2.4 Hz; and C-7a, $^2J_{CH}$, 3.0 Hz (*111*). All of these couplings were eliminated by exchanging the NH with D_2O.

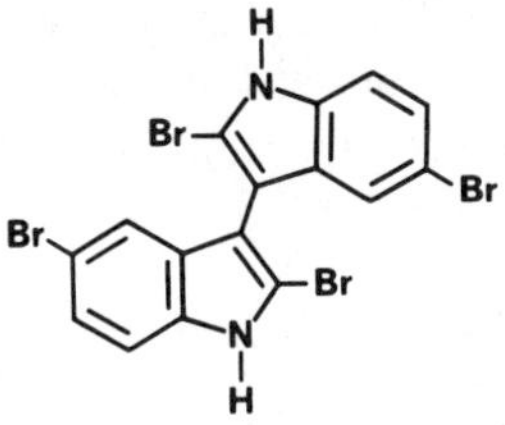

97

D. The Interpretation of NMR Spectra

Compilations of the ranges of chemical shifts anticipated for variously substituted carbon atoms and the protons they bear are widely available (*88, 89, 103, 112*). Reviews specifically limited to alkaloids provide more detailed information (*113–115*). Substitution by an electronegative element moves the chemical shifts of both nuclei somewhat downfield. Unsaturation, or aromatic character, produces chemical shifts still further downfield. Instances, of course, occur in which the chemical shifts deviate from the tabulated ranges, and can mislead even careful workers.

Recently in an investigation of a new class of indolosesquiterpene alkaloids from *Polyathia suaveolens,* structure **98** was assigned to polyavolensin largely

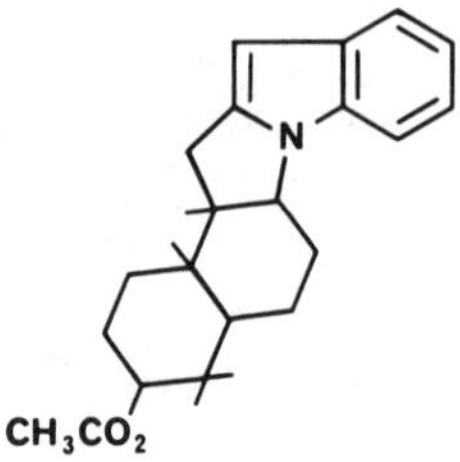

98

from spectral evidence (*116*). The presence of the indole system was inferred from a peak in the mass spectrum at *m*/*z* 130, the ultraviolet spectrum, and four closely coupled protons at 6.94–7.63 ppm. A single proton at 6.18 ppm showed that the β proton of the indole system was present. The structure of the sesquiterpene moiety was inferred from the resemblance of the NMR spectrum to that of polyathenol (**99**) from *P. oliversi* (*117*). The absence of exchangeable protons showed the nitrogen to be tertiary. Its site of incorporation into the sesquiterpenoid moiety was inferred from the presence in the carbon-13 spectrum of a peak at 65.1 ppm, which appeared as a doublet in the off-resonance decoupled spectrum. The chemical shift is well out of the range anticipated of methines attached only to carbon and quite suitable for a methine bearing a nitrogen atom. However, the deviation of the structure from the suggested biogenesis (*117*) prompted a reinvestigation by a crystal structure determination (*118*). All aspects of the structure were confirmed, except for the attachment of the methyl in the sesquiterpene portion (**100**); the misleading chemical shift evidently arises from the unusual number of β substituents surrounding the methine.

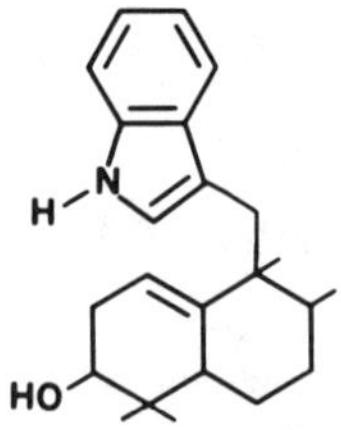

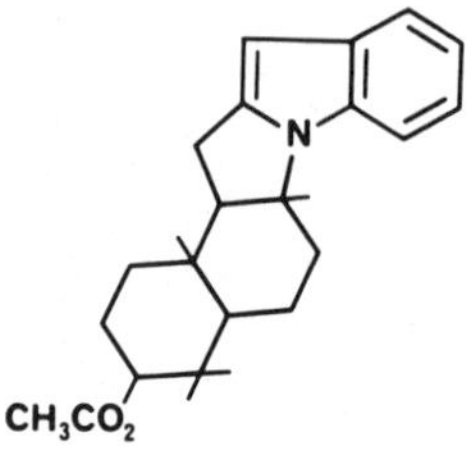

100 Polyavolensin

99 Polyathenol

In early studies it had been observed that substitution in acyclic or simple cyclic systems produced readily correlated effects on carbon atoms as remote as three bonds away. Carbon atoms β to the substitution are shifted 6–8 ppm downfield, while carbons three bonds away and in a gauche conformation are shifted somewhat less upfield. The observation carried the happy implication that reasonably accurate chemical-shift predictions might be made for postulated structures, and that the chemical shifts observed of an unknown material might lead directly to unique structures. Further study, however, revealed complications rather than solutions. It was observed, for instance, that substituents four bonds away held in close opposition, as the methyls of 1,2-dimethylnaphthalene, produced a downfield shift of ~4 ppm (*119*). The effects of electronegative substituents were found to depend on conformation (*120*). Spectroscopists correlating the shifts observed of closely related compounds often encounter substantial differences—perhaps 1 ppm—in the chemical shifts of carbon atoms in seemingly identical environments. There remains the possibility that extensive collections of structure and spectra can be correlated by computer. However, attempts to predict chemical shifts with the aid of data bases in computers have not yet produced predictions of useful accuracy, with the exception of narrowly restricted structural types as exemplified below (*121*).

However, the observed carbon-13 chemical shifts of an unknown material serve admirably to confirm the presence of a fragment of an alkaloid identical to a similar fragment of a known material. Differences in the carbon-13 spectrum from that of the known material, together with corresponding proton differences, may now allow complete structural deductions. This approach is exemplified in recent studies of the toxins from the skins of dendrobatid frogs.

Mass spectral studies of threse skin extracts showed that some 2 dozen of the alkaloids could be collected into the "Pumiliotoxin A" group, characterized by major peaks at *m/z* 70 and 166 or 182 (*28*). A crystal structure determination showed one member of the group, alkaloid 251D, to be **26** (*30*), the mass spectral fragmentations evidently resulting from cleavage of the allyl bond (*m/z* 166), and that of the pyrrolidine ring (*m/z* 70). Comparison of the carbon-13 spectrum with that reported for 2-methylhexane allowed the separation of shifts characteristic of the ring system from those of the side chain, as shown in Fig. 5 (*30*). Other members of the group now yielded structures from spectral observations. Pumiliotoxin A itself evidently possesses the same ring nucleus, since signals appear in both the carbon-13 and proton-NMR spectrum corresponding to those assigned to the ring system of 251D. The signals remaining unassigned could thereupon be readily interpreted as the side chain assigned in **101.** The presence of a double bond is shown by carbon-13 signals at 124.8 and 134.2 ppm; the appearance of its proton at 5.39 ppm as a triplet shows it to be attached to a methylene group. The substituted olefinic carbon bears the methyl group,

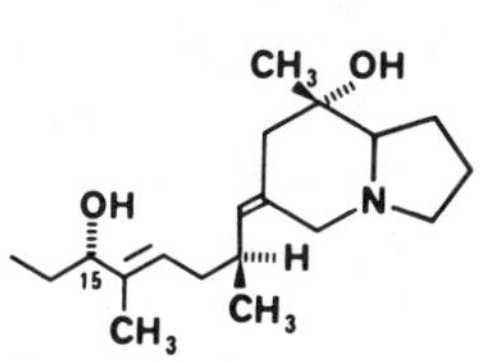

101 Pumiliotoxin A

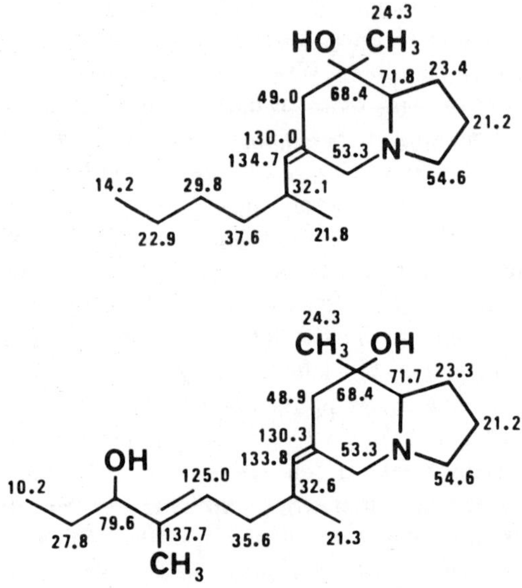

FIG. 5. Chemical shift assignments at alkaloid 251D (upper) and pumiliotoxin A (lower).

giving rise to a proton singlet at 1.55 ppm, and the carbinol systems are represented by a carbon signal at 79.4 ppm and a proton triplet at 3.88 ppm.

Alkaloids occurring in dendrobatid frogs show a major peak at *m/z* 182; they must bear a hydroxyl group at C-7 because the ion arising from the pyrrolidine ring (*m/z* 70) is unchanged; no peaks corresponding to hydroxymethylene groups appear, and the substances are stable in acid (*31*). That labeled 323B′ shows carbon-13 peaks corresponding to those arising from the side chain of pumiliotoxin A, while those remaining are suitable for the indolizidine ring system altered by the substitution of a hydroxyl group trans to that of **26** and **101.** C-7 is shifted 32 ppm downfield; carbons β to the substitution, C-6 and -8, are shifted approximately 2 ppm downfield; carbon atoms three bonds removed and in a gauche configuration, C-5, -8a, and -9, are shifted slightly upfield. Surprisingly, C-10 is shifted 3 ppm downfield. The coincidence of the side-chain carbon and proton absorptions with those observed of pumiliotoxin A and the understanding of the effects within the indolizidine ring system allowed the formation of 323B′ as **29.**

Thus structural studies of alkaloids commonly employ NMR in a context where much of the structure is known, and clues as to the unanticipated perturbations in structure are sought from those of the spectra. When the type of an alkaloid is known, reference to comprehensive complilations (*122, 123*) of carbon chemical shifts will generally allow many of the observed shifts to be assigned, making it possible to identify large portions of the molecule. When a very extensive library of closely related structures exists, very small variations can be interpreted with surprising effect.

A study of the C-19 diterpenoid alkaloids provided an early example of the ability of carbon-13 NMR spectra to aid in structural investigations. A compilation of the chemical shifts of seven alkaloids closely related to lycoctonine allowed the assignment of shifts to carbon atoms and the examination of an unknown alkaloid A from *Delphinium bicolor* (*124*). Comparison of the carbon-13 spectra with those of known compounds allowed the assignment of the structure **102.** However, when the number of available spectra for comparison had increased to 50, it became necessary to interchange the substituents at C-6 and C-8, because it could now be established that secondary and tertiary acetoxyl groups differ by 1 ppm within the group (*125*); alkaloid A showed acetoxyl absorption at 170.9 and 21.5 ppm, and must bear a secondary acetoxyl at C-6, rather than a tertiary acetoxyl at C-8, which should absorb at 169.9 and 22.2 ppm. Alkaloids within this group bearing an acetoxyl at C-8 have absorption at 85 ppm, absent in the spectrum of alkaloid A, for which an absorption at 79.9 ppm is assigned to C-8. Such close distinctions as these are impossible outside of a group of closely related materials.

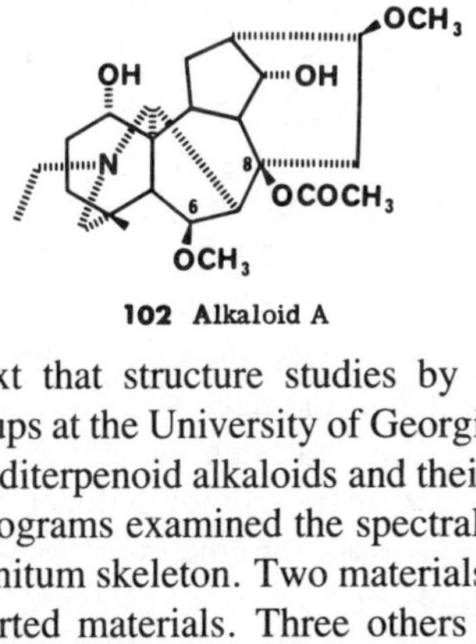

102 Alkaloid A

It is in such a context that structure studies by spectra can be aided by computer programs. Groups at the University of Georgia and Stanford University utilized a data base of 98 diterpenoid alkaloids and their derivatives to examine 6 alkaloids (*126*). Three programs examined the spectral data and produced structures restricted to the aconitum skeleton. Two materials were shown to be identical with previously reported materials. Three others led to pairs of structural alternatives that were differentiated by synthesis from known alkaloids. Two alternative structures for the sixth could be differentiated by comparison with the spectra of closely related structures. The achievement of such a program probably does not exceed that expected of an expert in a specific area, but a novice would surely find it of great value.

E. Multipulse Methods

The last few years have seen a spectacular development of nuclear-spin physics, and the proliferation of multipulse NMR experiments (*127*). Many of the newer methods are substantial aids in structural investigations. We shall first describe a useful one-dimensional multipulse method in some detail and then several two-dimensional methods.

1. The Attached Proton Test

In Section III,B above, off-resonance decoupled spectra were seen to characterize the signals from methyl groups, methylenes, methines, and quaternary carbons. The technique suffers from loss of sensitivity and uncertainty of the multiplicities observed as the result of the residual couplings from the larger three-bond couplings and as a result of the effects of closely coupled protons. The effect is particularly apparent in the spectra of the dendrobatid alkaloid pumiliotoxin C (**19**) (Fig. 6). The only methine giving rise to a clear doublet is that of C-8a, an observation that suggests the origin of the problem. All other methines bear a proton coupled to a vicinal proton by a large diaxial coupling, which proton is itself closely coupled to others of vicinal methylene groups. A repetition of the spectrum at high field shows satisfactory doublets for C-2, -4a, and -5, because increasing the field has increased the chemical-shift differences of the involved protons and reduced the close coupling.

A two-pulse method that allows differentiation of methines and methyls from methylenes and quaternary carbons without loss of resolution or substantial reduction in sensitivity has been suggested by several groups (*128–133*).* If the development of the isochronats of the signal arising from a methylene carbon is followed in a frame rotating at the frequency of the carbon, the smaller limbs of the triplet will be seen to diverge from the central peak at a rate of $2\pi J$ radians/sec (Fig. 7). After a period of $1/J$ sec, these limbs will coincide with the central peak. The limbs of the doublet arising from a methine carbon diverge from the central frequency at half the rate, πJ radians/sec; after $1/J$ sec, the two limbs coincide at π radians, 180° out of phase with the signal from the methylene carbon. If the decoupler power at proton frequencies is now turned on and decoupling begun with acquisition, the free induction decay collected will, after Fourier transformation, show the two peaks as singlets, 180° out of phase. It is convenient to pulse the carbon frequency with a 180° pulse at the initiation of decoupling and delay acquisition for another period of $1/J$ sec to allow an echo, which eliminates phase distortions. The resulting spectrum, shown in Fig. 8, shows the signals from methine (and methyl) carbons erect, while those from methylene (and quaternary) carbons are inverted. Distinction of the two groups is clear. This method is suitable for most older spectrometers, which were equipped

* This procedure is variously termed an INEPT, SEFT, or *J*-modulated experiment. The original INEPT sequence, which can be used for the same purpose, involved a substantially different pulse sequence (*134*). The title used here, which seems best suited to its purpose, is that of Patt and Shoolery (*129*).

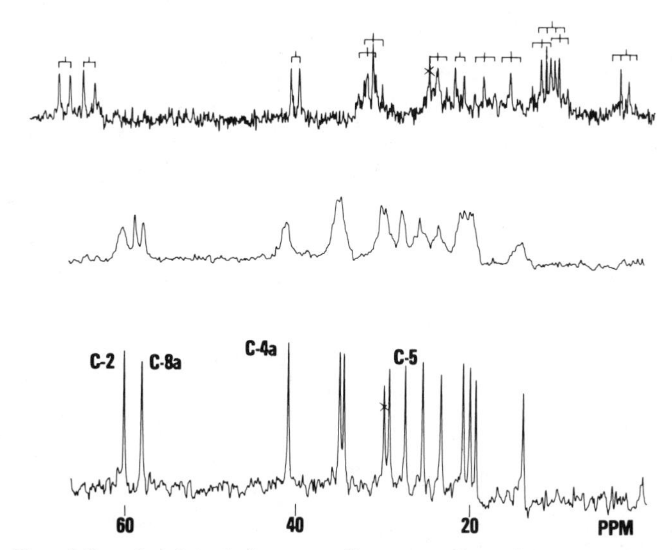

FIG. 6. Normal (lower) and single-frequency off-resonance decoupled spectra of pumiliotoxin C at 15 (middle) and 68 (upper) MHz.

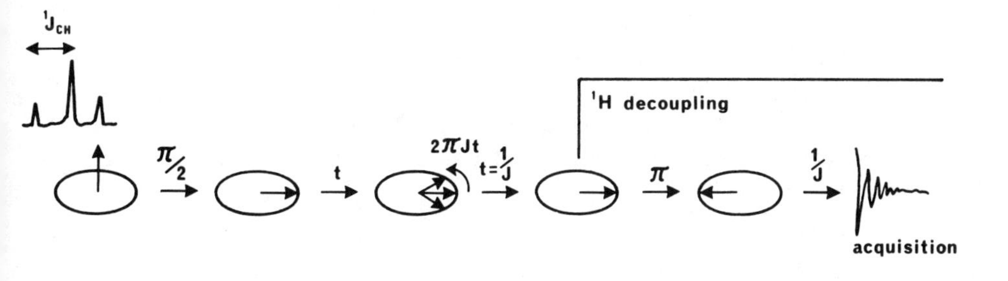

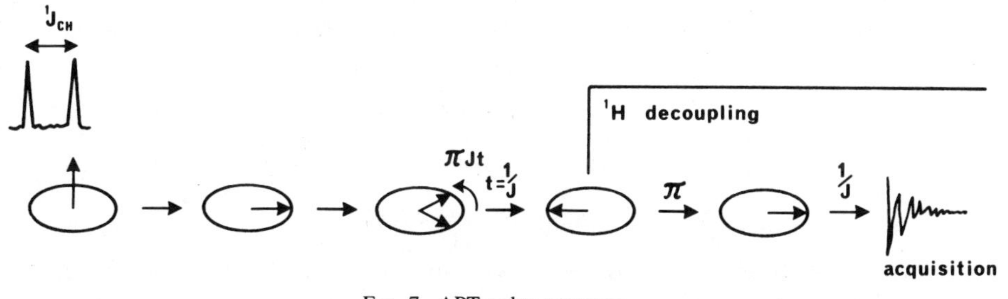

FIG. 7. APT pulse sequence.

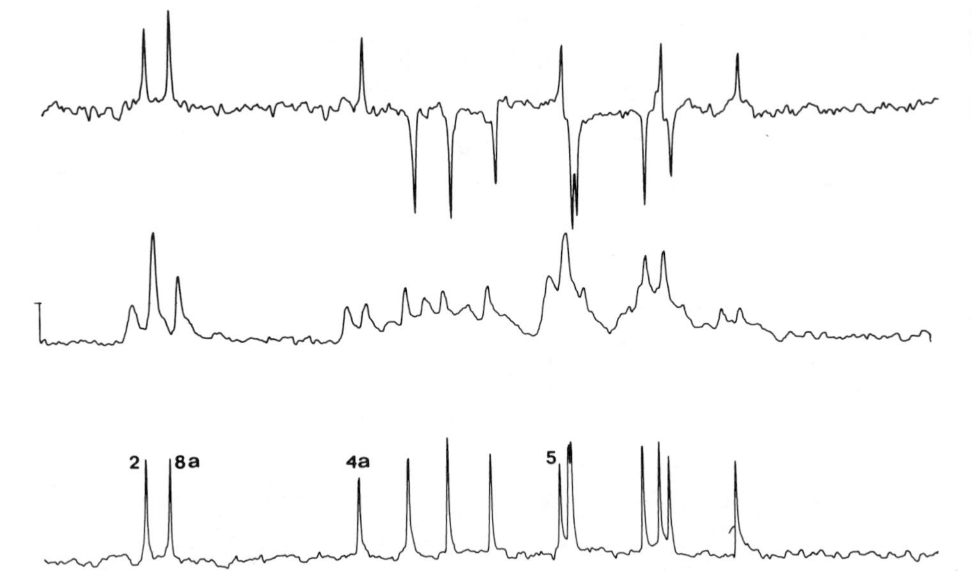

FIG. 8. A comparison of the normal carbon-13 spectrum of Pumiliotoxin C **19** (lower), the off-resonance decoupled spectrum (middle), and the APT spectrum (upper) at 15 MHz.

to perform two-pulse sequences (for spin-lattice relaxation measurement), and decoupling gated off except during collection (for collection of spectra without NOE). In practice, the choice of $1/J$ is commonly based on the one-bond coupling of aliphatic carbons, 125–140 Hz, using a 7-msec delay. The choice is not critical, however, unless carbons with unusual couplings, such as those in acetylenes, must be studied.

2. Determination of $^1J_{CH}$

The interval during which the nuclear spins were allowed to develop above, before decoupling is initiated, was chosen to result in the isochronats at π and 2π. However, if the development had been interrupted at a time t, the signal observed would be the projection of the spin vectors upon the y' axis, i.e., for the methine,

$$I_{CH} = I_o \cos(\pi t^1 J_{CH}) \quad (4)$$

and for the methylene,

$$I_{CH_2} = I_o[1 + \cos(2\pi t^1 J_{CH})] \quad (5)$$

Measurement of the intensities observed after several different development times should therefore allow the determination of the value of $^1J_{CH}$. The method

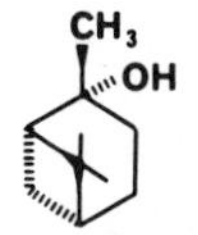

CH₃
OH

103 *cis*-Pinan-2-ol

has been used to differentiate the methylene group of the four-membered ring of *cis*-pinan-2-ol (**103**) ($^1J_{CH}$ = 136 Hz) from those in the six-membered ring ($^1J_{CH}$ = 128 Hz) (*135*).

F. Two-Dimensional Spectra (*136*)

1. Heteronuclear Couplings from 2-D Spectra

The experiment above incorporates the characteristics of pulse sequences used in two-dimensional Fourier transform experiments. The essential features are a preparation pulse followed by a development period t_1 before acquisition during time t_2. The pulse sequence of Fig. 9 provides a two-dimensional equivalent (*137*). The development period is incremented through a range of values suitable for the experiment. The result is an array of free induction decays. A Fourier transform of each of these (along t_2) will provide an array of spectra differing in phase. Fourier transform along the t_1 direction provides the coupling patterns for each carbon.* Such coupling patterns from the carbon-13 spectrum of undulatine are shown in Fig. 10.

2. Determination of J_{HH}

An analogous experiment allows the determination of homonuclear coupling constants (*138*). Again, the nucleus to be observed is subjected to a $\pi/2$ pulse and allowed to evolve coupled for a time $t_1/2$, before a π pulse; this pulse changes not only the orientation of the vector but also the spin state of the coupled nucleus. The vectors continue to evolve during the second $t_1/2$ period, while the broadening resulting from the inhomogeneity of the magnetic field is canceled. As a consequence, the coupled multiplets display resolution limited, in principle, only by the inherent spin–spin relaxation times. Because the protons cannot be decoupled from each other, the coupled multiplets occur along an axis tilted at 45° to the f_2 axis; the array must be tilted by this angle to provide multiplets normal to the f_2 axis. The software of modern spectrometers allows this procedure.

* Demonstration that this is the case is accomplished by spin-density matrix calculations, which will not be repeated here. Treatment of the subject will be found in reference *136*.

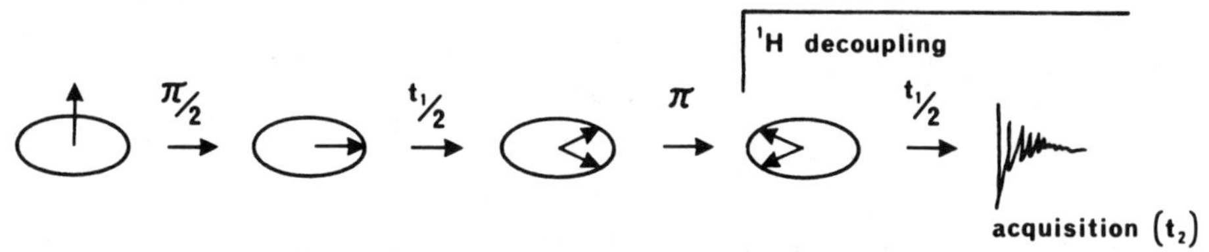

FIG. 9. Pulse sequence for two-dimensional J_{CH} spectra.

3. COSY

Determination of the accurate values of the coupling constants of the protons may uniquely indicate the protons with common couplings, and therewith, which protons are coupled to which. However, a two-dimensional experiment, which allows the direct demonstration of this relationship, is available in correlated spectroscopy (COSY) (*139*).

The pulse sequence that provides these spectra is shown in Fig. 11. After the 90° preparation pulse, the nuclear spins are allowed to develop for the evolution period t_1; a second 90° pulse now puts the y-components of the evolving nuclear spins along the z-axis. The x-components remain in the xy-plane and continue to evolve, developing a signal in the detector, which is now turned on. Evidently, at

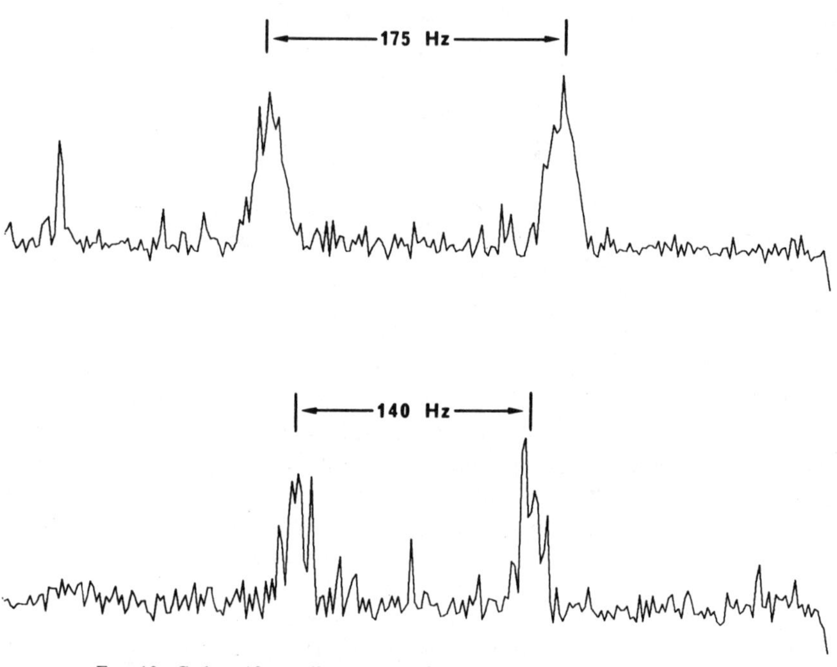

FIG. 10. Carbon-13 coupling patterns from the 2-D spectrum of undulatine.

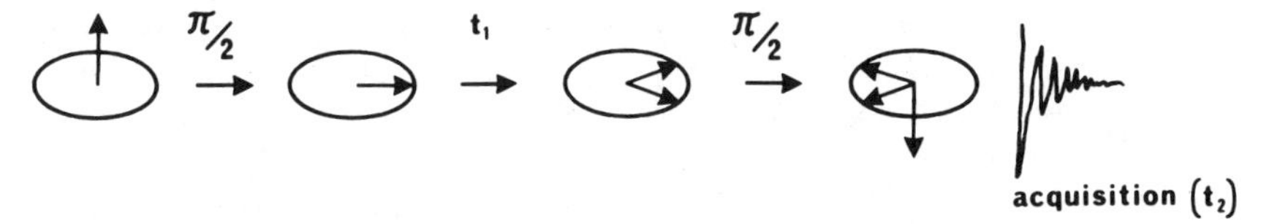

FIG. 11. Pulse sequence for COSY spectra.

a specific t_1, nuclear spins whose offset from the transmitter frequency results in the vector aligned with the x-axis before the second pulse will give rise to the maximum signal. Thus the sequence traces the spectrum along the diagonal of the f_1,f_2 array. Because coupled nuclei possess common energy levels, their

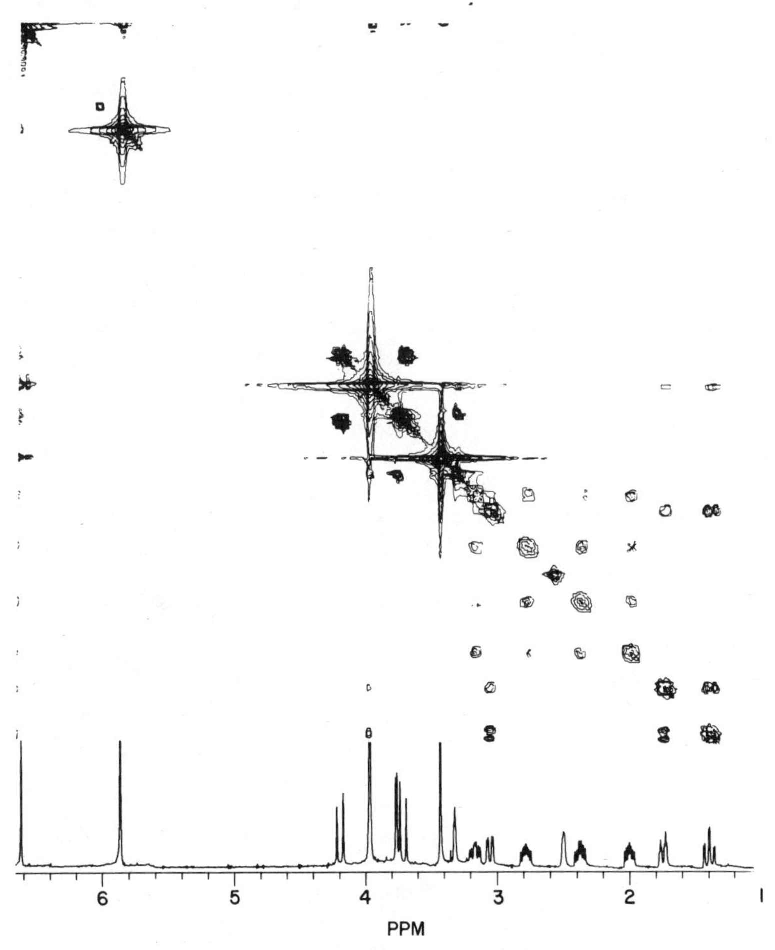

FIG. 12. COSY spectrum of undulatine.

maxima will reflect the chemical shift of both and give rise to off-diagonal peaks corresponding to both chemical shifts. The resulting spectrum is most conveniently presented as a contour map of peak intensities.

The COSY spectrum of undulatine is shown in Fig. 12. The off-diagonal peaks show the couplings around ring A, including that to H-3, itself obscured in the one-dimensional spectrum beneath the methoxyl peak at 3.9 ppm. The plot establishes the connectivity of the spin system as would a series of one-dimensional decoupling experiments. The assignments are achieved in a single experiment, and with superior resolution.

4. Chemical Shift Correlation Mapping

Chemical shift correlation mapping (CSCM) produces a plot with the chemical shift of the carbon atoms along one axis and those of the proton on the other. Examples of the use of the experiment on several ormosia alkaloids are discussed below.

Following a 90° pulse at the proton frequency, the proton nuclei are allowed to develop for a period $t_1/2$, at which point a 180° pulse at carbon frequencies inverts the spin states of the carbon-13 nuclei; the protons coupled to them will continue to evolve according to their chemical-shift differences, but the contributions from the carbon-proton couplings have been reversed; at the end of another period of $t_1/2$, the coupled isochronats have refocused at whatever point their chemical shifts have determined. Evolution is now allowed to continue for a period $1/2J_{CH}$, during which the vectors of the coupled protons evolve through π radians. Simultaneous 90° pulses at proton and carbon frequencies now place the proton spins aligned with the z-axis, with populations that reflect their chemical shifts. Carbon spins are placed in the xy-plane with intensities modulated by the disturbed populations of the coupled protons with which they share a common energy level. Their evolution is allowed to continue a further period of $1/2J_{CH}$ to allow refocusing before proton decoupling and detection are initiated. Peaks appear at coordinates δ_C,δ_H when the corresponding carbon and proton nuclei are coupled.

The experiment thus provides the same information as a series of single-proton, spin-decoupling experiments, again with resolution superior to that obtained by the decoupling experiment.

5. Carbon–Carbon Connectivities

A two-dimensional experiment, INADEQUATE,* which is potentially very useful, allows the demonstration of carbon connectivities by the examination of

* Incredible natural abundance double quantum transfer experiment.

carbon-13 satellites of carbon-13 resonances (*140*). However, because this requires the observation of signals with a strength 0.5% or less than normal carbon-13 signals, the experiment is limited to samples of perhaps three millimoles, soluble to the extent of 0.5 *M*. It has been used in the study of panamine in connection with the investigation of ormosinine described below (*141*) and shown to be a great aid in assigning the carbon-13 chemical shifts. An optimistic extrapolation of current developments may anticipate that new experiments will be devised that achieve this end with acceptable sensitivity. A multiple quantum experiment that promises to achieve this depends on the coupling of vicinal protons (*142*). However, the connectivities are blocked by a quaternary carbon, and use of the procedure on materials of any complexity has not been reported.

6. Instrumental Requirements

Two-dimensional Fourier transform experiments obviously require a spectrometer equipped to provide pulses at both proton and carbon frequencies under computer control. Such instruments have been commercially available for several years. Because the CSCM experiments (and the 2DJCH) provide the dispersion of the carbon chemical shift scale, low-field instruments are perfectly suitable. COSY and HOMOJ experiments will benefit from the greater dispersion of proton signals at high field. In point of fact, inasmuch as most instruments equipped to perform these experiments are built around a superconducting magnet, the point becomes academic.

The computer programs that effect the collection of the array of free induction decays and the two-dimensional Fourier transforms differ widely in the facility with which these tasks are performed, but they are rapidly being improved. Spectroscopists can anticipate that these experiments will soon be very easily performed. A major limitation at present lies in the limited memory available to collect the arrays. At present, experiments are limited, in practice, to about a million words of memory. If the COSY experiment is set up as a square array, each of the 1000 FIDs is limited to 1000 words. Thus, after Fourier transform, the range of proton shifts is represented by 500 data points. Ideally, the range used would be 10 ppm, or 200–400 Hz over 50 words. Clearly, the digital resolution is much less than that of the spectrometer. However, as instruments utilizing much more capacious disks become available, this problem will be much eased.

7. Applications

These methods are new enough that they have not yet been often applied. An impressive instance, however, has appeared in the study of the dimeric *Ormosia* alkaloid ormosinine (*141*). The mass spectrum of ormosinine provides a mo-

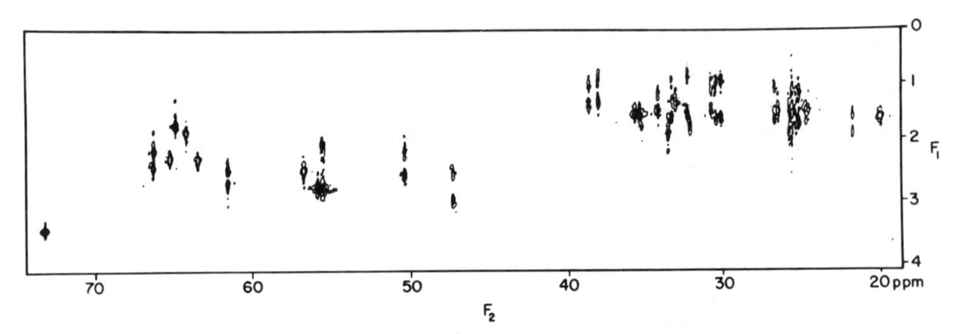

FIG. 13. CSCM spectrum of ormosinine.

lecular ion corresponding to $C_{40}H_{66}N_6$; sublimation produced panamine (**104**). The highly coupled proton-NMR spectra at 200 MHz provided very little useful information; one-dimensional carbon-13 spectra allowed the counting and classification of the carbon atoms but provided little structural detail. However, a CSCM spectrum of ormosinine (Fig. 13) allowed the preparation of a plot of carbon-13 chemical shifts versus those of the attached protons Fig. 14. Comparison of this plot with a similar plot from panamine verified the existence of the panamine residue in ormosinine. The remaining points of the plot could be compared with a plot from ormosanine (**105**) to demonstrate the presence of this residue and therewith structure **106** for ormosinine.

104 Panamine

105 Ormosanine

106 Ormosinine

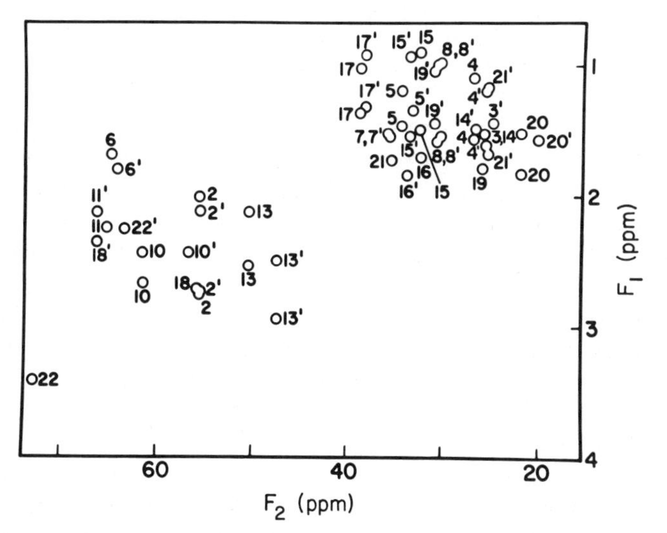

FIG. 14. Plot of δ_C versus δ_H for ormosinine. We are indebted to Dr. G. A. Morris for Figs. 13 and 14.

It seems clear that the two chemical shift correlation experiments discussed above, COSY and CSCM, will replace proton decoupling experiments for routine work, as suitable spectrometers become routinely available. In each case, a single experiment provides the information of a series of decoupling experiments, with superior resolution. The newer methods are particularly advantageous as the number of carbon atoms increases much beyond 20. The HOMOJ experiment will be called upon when the resolution of a conventional one-dimensional experiment proves inadequate. As was seen in the discussion above, the possibility of determining carbon–proton coupling constants readily provides a new tool, which will prove very useful in specific instances.

G. The Nuclear Overhauser Effect

Observation of the nuclear Overhauser effect (NOE) provides a means of estimating the distance between nuclei in a molecule (*143*). It has proved a powerful and popular tool for the investigation of alkaloid structures and conformations.

The NOE has its origins in relaxation phenomena. It can be shown that if a nucleus A is relaxed solely by a dipole–dipole interaction with nucleus B, saturation of B by irradiation at its frequency will cause an increase in the integrated intensity of the signal of A by an amount

$$\epsilon = \gamma_A / 2\gamma_B \tag{6}$$

in which the γs are the respective gyromagnetic ratios. This expression does not involve the internuclear distance, but represents the maximum NOE that could be observed. In alkaloid structural investigations, both nuclei are protons, and the maximum observable effect is 50%. To observe the effects of relaxation of specific nuclei, relaxation by paramagnetic contaminants must be eliminated by filtering the sample and degassing to remove oxygen by several freeze–thaw cycles. Integrals of the observed signals are compared with the irradiating frequency at that of the relaxing nucleus and with the irradiation substantially displaced to eliminate electronic artifacts. A recent study has shown the signal-to-noise ratio required for the accuracy desired (*144*).

Protons in molecules as complex as alkaloids are subject to many other competing relaxation mechanisms, with the result that maximum NOEs are never observed. Indeed, against this ''background relaxation,'' substantial NOEs are observed only when the two protons are quite close, because the contribution that proton B makes to the relaxation of A is inversely proportional to the sixth power of the distance between the nuclei. Bell and Saunders exploited this property in a study of the NOE of a variety of complex molecules of known structure in which one proton could be anticipated to be relaxed by another single proton or a methyl group (*145*). Plots of the degree of enhancement against the inverse of the sixth power of the internuclear distance showed the linear relationship, but the proposal that internuclear distances in an unknown molecule can be measured by the observed NOE has met a certain amount of skepticism (*146*). However, the technique has been widely used. In most investigations of alkaloid structures it is not necessary to estimate internuclear distances; in the context of other structural knowledge, the purpose of the observation is simply to determine which of several nuclei is close to that being irradiated.

An early use of this technique is found in the investigation of a *Fumaria* alkaloid, fumaricine, by a Canadian group (*147*). Mass spectra of the alkaloid suggested an ochotensimine skeleton, and proton NMR spectra at 100 MHz showed the aromatic rings substituted as shown in structure **107.** Confirmation and differentiation of the various possible substitution patterns rested on the NOE studies (*148*). Irradiation at frequencies corresponding to H-5 and the C-3 methoxyl both produced an increase of the intensity of H-4, showing that the methox-

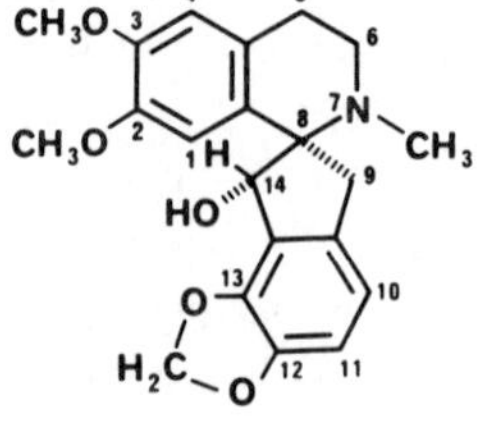

107 Fumaricine

yl groups are on ring A. Irradiation of the C-2 methoxyl similarly increased the intensity of H-1. Irradiation of the methylene protons of C-9 increased the intensity of both H-1 and H-10, establishing the location of the methylenedioxy group and of the spiro center. The study exemplifies the use of NOE observations to link spin systems not linked by couplings.

A major difficulty in the use of the NOE is that it requires the use of integrated signals; the integral has always been the least satisfactory datum yielded by an NMR spectrometer. It is unfortunate that NOEs are invariably reported without an indication of a mean value or error. Nonetheless, instances of erroneous NOE observations leading to erroneous structures are rare, no doubt because of its use within the context of other structural knowledge.

The advent of pulse spectrometers controlled by computers has made possible the study of NOE by difference spectra (*149*). The computer is programed to take groups of free induction decays with the irradiating frequency on the absorption to be studied and alternating groups with the irradiation much displaced. The Fourier transform of the difference of these two groups shows the NOE. The irradiated proton appears as a negative peak, which is taken as unity. This last assumption precludes determining absolute values, but using the difference eliminates many instrument artifacts, such as small changes of phase and intensity variations arising from changes in temperature. Observations of less than a percent are reported with confidence.

A recent study of repanduline has demonstrated the utility of this technique (*150*). The modified bisbenzylisoquinoline structural formula **108** had been deduced from degradative and spectral studies. The NMR spectra were now obtained at 250 MHz, and the resonances corresponding to each proton were assigned by the use of difference NOE, with the aid of a Dreiding model. In a typical observation, irradiation of $H_{8'}$ (5.11 ppm) produced a strong enhancement of $H_{2''}$; the models showed that only these two protons on rings C and E could be close enough to account for this observation. In earlier studies, the structure of ring G could be inferred only from biogenetic considerations. The

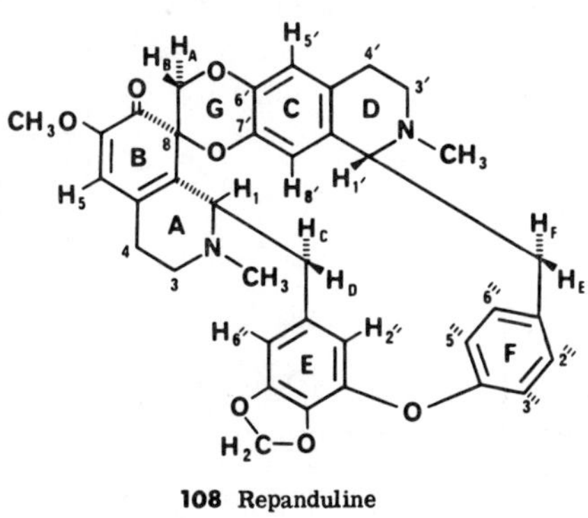

108 Repanduline

alternative structure, in which the spiro atom is that attached to the oxygen of 6′, could now be eliminated by the experiment shown in Fig. 15. Irradiation of H_A (4.15 ppm) produced a clear enhancement (0.18%) of H-5′ but negligible effect on H-8′. The configuration at C-1 of ring A was established by the experiments shown in Fig. 16. In the study of a solution containing 20% benzene-d_6 to effect separation of the chemical shifts of H-1, H_C, and H_D, irradiation at H_B, $H_{2''}$, and $H_{6''}$ produced the multiplets shown, readily assigned to H_1, H_C, and H_D. The observation that irradiation of H_A produced enhancement of $H_{5'}$ but not $H_{8'}$ implies that ring G is quite rigid. Similar observations throughout the ring system implied the conformation shown in Fig. 17. Throughout the study, enhancements approaching 0.1% were readily reproducible, sometimes requiring collections throughout a weekend.

H. Nitrogen-15 NMR (*151*)

Nitrogen-15 has a spin of ½ and is suitable for high-resolution NMR spectroscopy. However, its gyromagnetic moment is low, giving it a resonance of 10 MHz in a field of 2.34 Tesla (proton at 100 MHz), and as a consequence, a

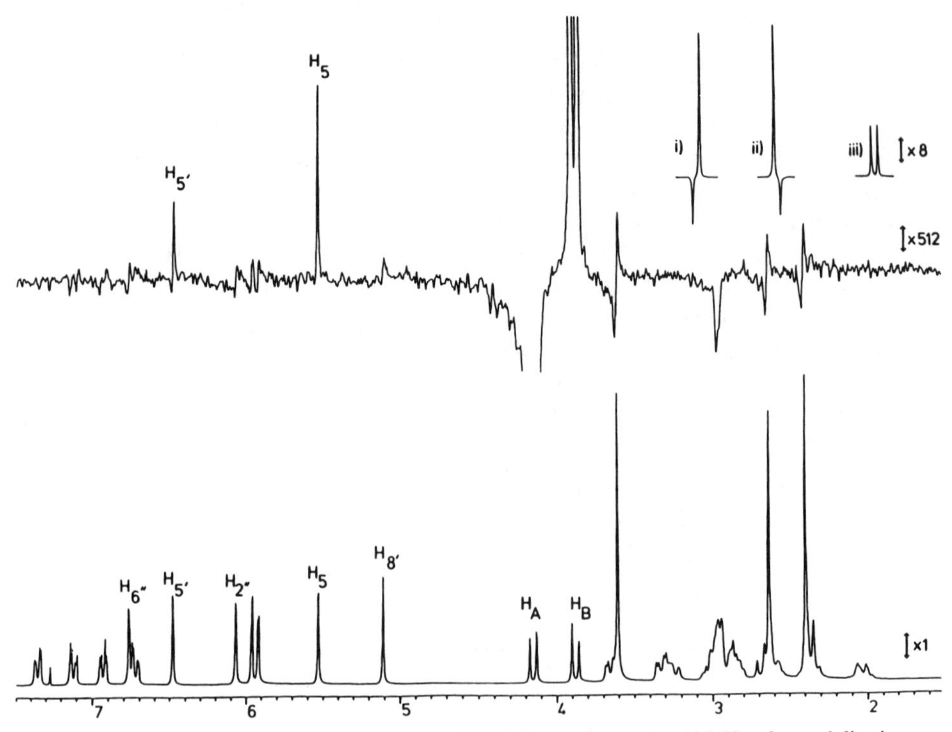

Fig. 15. The normal (lower) and long-range NOE difference spectra in which H_A of repanduline is irradiated.

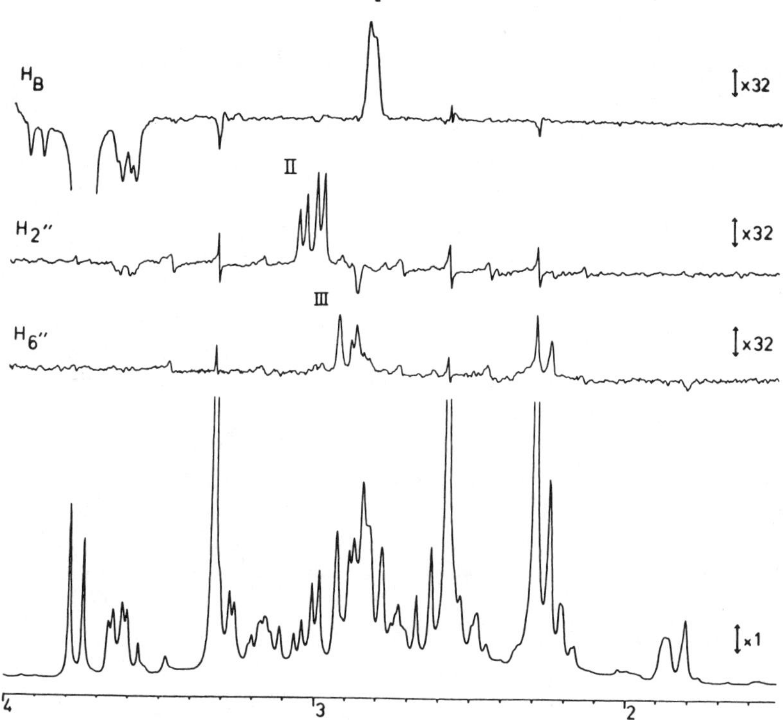

Fig. 16. Normal (lower) and NOE difference (upper) spectra of repanduline with irradiation at the sites noted.

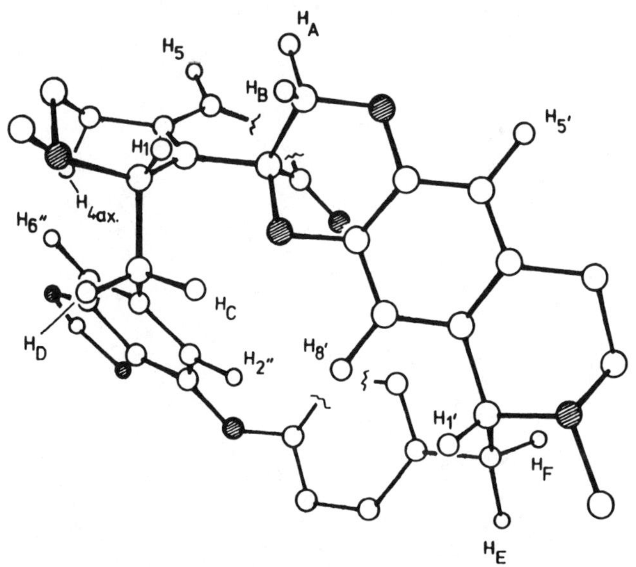

Fig. 17. The conformation of repanduline implied by the NOE difference spectra. We are indebted to Dr. D. Neuhause for Figs. 15, 16, and 17.

sensitivity per nucleus 0.1% that of proton. Its natural abundance is 0.37%; the two factors combine to produce a problem in sensitivity 40 times worse than that of carbon-13. As a result, study of nitrogen-15 NMR spectra has been limited to material available in approximately 1 mmol quantities, and capable of forming ~0.5*M* solutions.

The ratio of $\gamma_H/\gamma_N = -10$ produces a nuclear Overhauser effect of -5, or a fourfold maximum overall signal enhancement. However, its spin–lattice relaxation time in small molecules is often long, requiring long delays between pulses. Furthermore, if the full NOE is not obtained, the signal enhancement may be much less, or the signal may even be nulled. As a consequence, samples are often doped with chromium acetylacetonate (CrAcAc) to eliminate the NOE and to allow rapid pulsing.

The study of the relationship of nitrogen-15 NMR parameters to structure on materials available in large quantities yielded results assuring the value of the spectra to alkaloid chemists, once the problem of sensitivity is overcome. The results have been summarized in a lucid and comprehensive reference work (*151*). The relationship of chemical shift to substitution and environment resembles that of carbon-13, with the additional factor of further effects from the nitrogen lone electron pair. Thus aniline hydrochloride shows a chemical shift of 46.7 ppm (downfield from ammonia), while for the free base the value is 55.4 ppm. For pyridine, the corresponding values are 205 (in CF_3COOH–$CDCl_3$) and 319 ppm (in C_6D_6).

One of the few studies of alkaloids has shown that the stereochemistry of fused ring systems with nitrogen atoms is reflected in the chemical shift of the nitro-

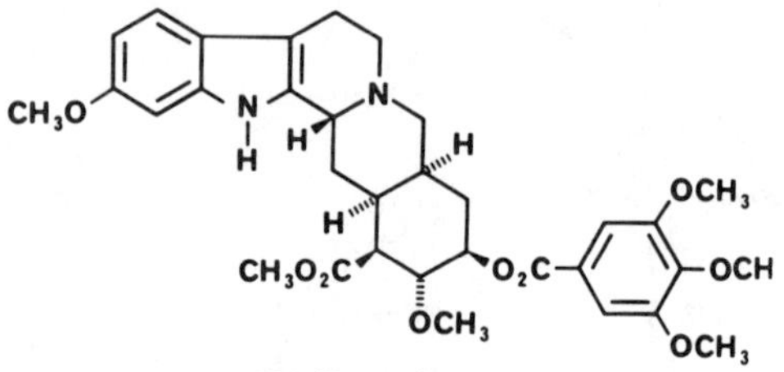

109 Reserpine

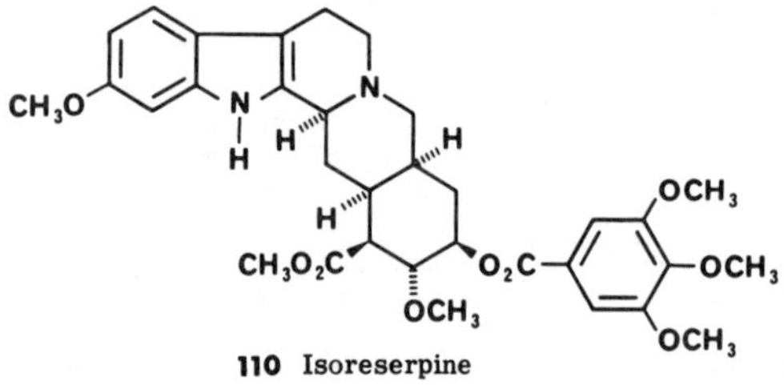

110 Isoreserpine

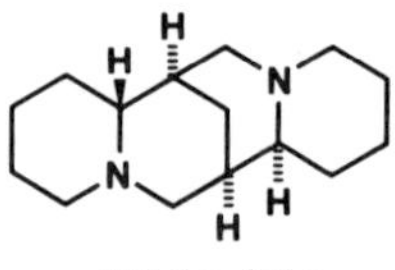

111 Sparteine

gen-15 (*152*). Comparison of isomeric pairs of quinolizidines, such as reserpine (**109**) and isoreserpine (**110**) showed that the N-15 nucleus in the cis–fused system was substantially more shielded than that in the trans system—here 31.9 and 47.0 ppm, respectively. The effect is attributed to the greater C═N character in the trans–fused system. The close correspondence of the N-15 signals of sparteine (**111**) to that of *trans*-quinolizidine supports the assignment of a trans fusion to the ring system (*153*) rather than the cis fusion that has been suggested (*154*).

A recently developed pulse sequence allowing observation of the double quantum transition of nitrogen-15 nuclei directly bound to hydrogen improves the sensitivity spectacularly (*155*). By this method, a 12-mg sample of the hexapeptide gramicidin-S (16 m*M*; because of the symmetry, 32 m*N*) provides a satisfactory spectrum from a 3-hr collection (*156*). Modification of the method to allow the study of tertiary nitrogens should be possible and should result in routine and valuable alkaloid nitrogen-15 spectra.

I. Conclusion

The compendium of NMR methods described above and those still evolving provide the means of establishing almost any alkaloid structure without the context of established structures, provided enough material (~1 mmol) and instrument time are available. An impressive illustration of this is to be found in the investigation of amphimedine (*157*). High-resolution mass spectra established the molecular formula as $C_{19}H_{11}N_3O_2$. Proton NMR readily established the presence of an ortho-disubstituted aromatic system, a pair of ortho-situated aromatic protons, and two aromatic singlets. Single-proton decoupling experiments established the pairing of directly bonded carbons and protons and three-bond couplings. With the aid of these last couplings, it could be shown that the carbon of δ146.2 ppm was coupled to the downfield proton of the four-proton group (H-4 of **112**) and to the upfield proton of the ortho pair (H-6). The observation established the carbon connectivities of the carbons 1–6. Three-bond couplings observed between H-6 and C-7a and between H-5 and C-12c completed the right hand side of **112** and established carbons three bonds from H-9 and H-12.

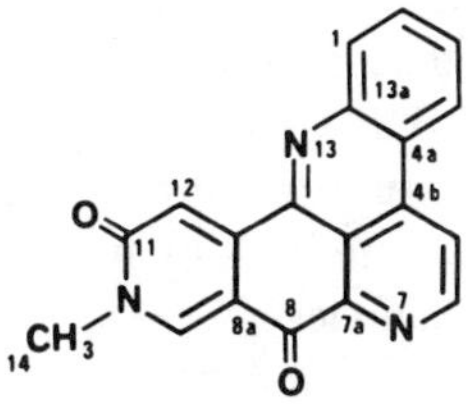

112 Amphimedine

The structural fragments thus established were joined by the demonstration of carbon–carbon connectivities by means of INADEQUATE. There resulted a set of carbon connectivities complete except for the second attachment of the two carbonyls. The sites of nitrogen attachment were inferred from the large one-bond couplings (186 and 196 Hz) and downfield shifts of the attached carbons. Finally, bonds were postulated for the attachments of the carbonyls to provide six-membered rings consistent with the observed carbonyl stretching frequencies (1690 and 1640 cm^{-1}).

IV. Infrared Spectroscopy

Because of their great powers, NMR and mass spectra have largely supplanted other spectral methods. However, infrared spectra provide certain structural details less readily available from these, or not at all.

The infrared spectra of solid samples give rise to a complex group of peaks between 1200 and 600 cm^{-1}, which provides a unique fingerprint of the compound (*158*). Comparison of the infrared spectrum of a solid-phase sample with that of a known sample remains one of the most secure means of establishing identity. Investigation of the alkaloids of *Cassia* spp. encountered the problem of distinguishing the *d*-cassine from the *dl*-congener (*159*). Crystallization of the hydrochlorides to samples providing a constant infrared spectrum of the mulls allowed the identification of the materials.

Certain functional groups are more readily detected by infrared spectra than by other means. Nitro groups give rise to strong peaks at 1530 and 1350 cm^{-1} but give no characteristic peaks in proton or carbon-13 NMR spectra; they can, however, be detected by mass spectrometry from the loss of oxygen and NO_2. Detection by nitrogen-15 NMR is commonly precluded by sensitivity. Whether this is significant in alkaloid investigations evidently depends on whether aristolochic acid derivatives are classified as alkaloids (*80*) or not (*160*). Nitriles give rise to no signal in proton NMR, and are easily missed in the carbon-13 NMR of weak samples because of their long relaxation time. Their absorption in the infrared spectrum near 2200 cm^{-1} is a useful means of identification.

The carbons of carbonyl groups are commonly readily detected in carbon-13 NMR spectra with shifts characteristic of their functionality and molecular situation. Carbonyl groups situated in five-membered rings absorb at substantially lower fields, but multiple substitution near the carbonyl in a six-membered ring can move it low enough to confuse the assignment of ring size. Their infrared spectra provide a particularly secure indication of their being in a five-membered or smaller ring. Thus cyclopentanones absorb at 1745 cm^{-1} and cyclobutanones at 1780 cm^{-1}, in contrast with cyclohexanones and aliphatic ketones at 1715 cm^{-1}. Lactones show similar shifts: β 1840, γ 1770, and δ 1735 cm^{-1}. Conjugated unsaturation commonly decreases these frequencies about 20 cm^{-1}, providing a datum to complement NMR observations.

Protons α to a nitrogen atom in a rigid ring give rise to characteristic absorptions in the infrared near 2700 cm^{-1} when they are situated antiperiplanar to the nitrogen lone electron pairs. First observed by Wenkert (*161*) they are commonly called "Bohlmann bands" (*162*). Their use aided the characterization of the lupin alkaloids, such as sparteine (**111**) (*163*), and matrin (**113**) (*164*).

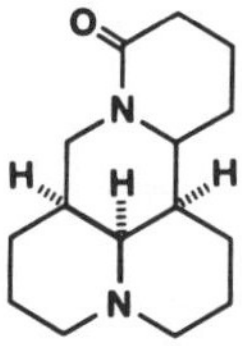

113 Matrin

The same molecular configuration that gives rise to these bands shields the axial protons α to the nitrogen to produce characteristic upfield positions (*165*). In alicyclic systems such as those of the lupin alkaloids, these protons are commonly invisible within the general envelope of proton absorption between 1 and 2 ppm, even at high field. However, they can generally be located through the carbon-13 spectrum, observing the α-carbon atoms near 50 ppm by single-proton spin decoupling or two-dimensional correlation maps. Such a study has been carried out on pumiliotoxin A (**101**) (*31*). Peaks in the carbon-13 spectrum at 54.6 and 53.3 ppm, chemical shifts characteristic of methylene groups attached to nitrogen, appeared as triplets in a coupled spectrum. That at 54.6 was converted to a doublet by irradiation (at 100 MHz) at 3.05 or 2.20 ppm; that at 53.3 responded similarly to irradiation 2.32 and 3.76 ppm. The substantial difference in chemical shifts of the two protons on each methylene group reflects the effect of the nitrogen lone pair in the rigid indolizidine system.

The detection of relatively weak intramolecular hydrogen bonds is best done by the infrared spectra of dilute solutions in nonpolar solvents. Strong hydrogen bonds, such as that of a salicylic acid derivative, are readily characterized by proton NMR, displaying a peak at low fields, 11–16 ppm. However, the down-field shift produced by a weaker bond is not readily detected. The piperidine alkaloid cassine (**114**) showed infrared absorption at 3530 cm^{-1} (*159, 166*). The

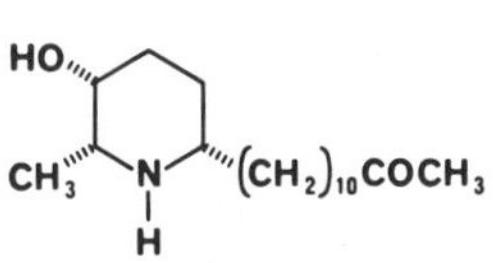

114 Cassine

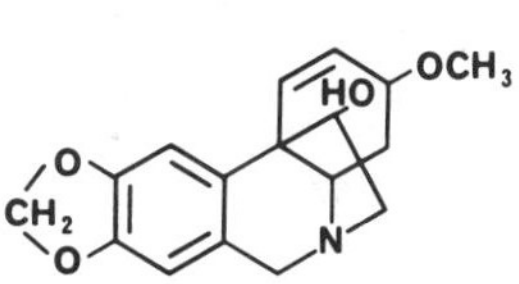

115 Haemanthamine

observation requires the stereochemistry shown in **114,** in which the 2,6-dialkyl groups are equatorial, which holds the hydroxyl group in the axial position most favoring hydrogen bonding (*167*).

The method is capable of subtle distinctions (*168*). The amaryllis alkaloid haemanthamine (**115**) showed a bathochromic shift of the hydroxyl stretching band of 27 cm^{-1} from the frequency of 3625 cm^{-1} anticipated of an allylic alcohol (*169*). This shift arises from hydrogen bonding to the Δ-1 double bond, since the dihydro derivative shows absorption at 3625 cm^{-1}. The C-11 epimer epihaemanthamine and its dihydro derivative both show absorption at 3560 cm^{-1}, the bathochromic shift arising from hydrogen bonding to the aromatic ring. The observations establish the stereochemistry of the C-11 hydroxyls.

Acknowledgment

We are indebted to Janet B. Wheeler for figures and structures.

References

1. I. L. Karle, *in* "The Alkaloids" (A. R. Brossi, ed.), Vol. 22, Chapter 2. Academic Press, New York, 1983.
2. K. Biemann, "Mass Spectrometry. Organic Chemical Applications." Mc-Graw Hill, New York, 1962.
3. F. H. Field, *Acc. Chem. Res.* **1,** 42 (1968).
4. J. W. Daly, *Fortschr. Chem. Org. Naturst.* **41,** 216 (1982).
5. J. W. Daly, private communication.
6. J. W. McCoy, M. R. Roby, and F. R. Sternmitz, *J. Nat. Prod.* **46,** 894 (1983).
7. M. S. B. Munson, *Anal. Chem.* **49,** 772A (1977).
8. H. R. Schulten, *Adv. Mass Spectrom.* **7A,** 84 (1978).
9. M. A. Posthumus, P. G. Kistemaker, H. L. C. Meuzelaar, and M. C. Ten Noever de Brauw, *Anal. Chem.* **50,** 985 (1978).
10. R. D. Macfarlane and D. F. Torgerson, *Science* **191,** 920 (1976).
11. A. Benninhoven and W. K. Sichtermann, *Anal. Chem.* **50,** 1180 (1978).
12. M. Dole, H. L. Cox, Jr., and J. Gieniec, *Adv. Chem. Ser.* **125,** 73 (1971).
13. E. R. Lory and F. W. McLafferty, *Adv. Mass Spectrom.* **8,** 954 (1980).
14. M. E. Tate, J. G. Ellis, A. Kerr, J. Tempe, K. E. Murray, and K. J. Shaw, *Carbohydr. Res.* **104,** 105 (1982).
15. D. T. Coxon, A. M. C. Davies, G. R. Fenwick, R. Self, J. L. Firmin, D. Lipkin, and N. F. James, *Tetrahedron Lett.* **21,** 495 (1980).
16. B. N. Meyer, J. S. Helfrich, D. E. Nichols, J. L. McLaughlin, B. V. Davis, and R. G. Cooks, *J. Nat. Prod.* **46,** 688 (1983).
17. G. R. Pettit, C. W. Holzapfel, G. M. Cragg, C. L. Herald, and P. Williams, *J. Nat. Prod.* **46,** 917 (1983).
18. M. Barber, R. S. Bordoli, R. D. Sedgwick, and A. N. Tyler, *J. Chem. Soc., Chem. Commun.* 325 (1981).
19. S. E. Unger, R. G. Cooks, R. Mata and J. L. McLaughlin, *J. Nat. Prod.* **43,** 288 (1980).

20. E. Stenhagen, S. Abrahamsson, and F. W. McLafferty, "Registry of Mass Spectral Data," Vol. 1, p. 367. Wiley, New York, 1974.
21. S. Pummanguru, J. L. McLaughlin, D. V. Davis, and R. G. Cooks, *J. Nat. Prod.* **45,** 277 (1982).
22. R. G. Powerll, C. R. Smith, R. D. Plattner, and B. E. Jones, *J. Nat. Prod.* **46,** 660 (1983).
23. I. L. Karle and J. Karle, *Acta Crystallogr., Sect. B* **B25,** 428 (1969).
24. R. D. Gilardi, *Acta Crystallogr., Sect. B* **B26,** 440 (1970).
25. T. Tokuyama and J. W. Daly, *Tetrahedron* **39,** 41 (1983).
26. J. W. Daly and C. W. Myers, *Science* **156,** 970 (1967).
27. J. W. Daly, T. Tokuyama, G. Habermehl, I. L. Karle, and B. Witkop, *Justus Liebigs Ann. Chem.* **729,** 198 (1969).
28. J. W. Daly, G. B. Brown, M. Mensah-Dwumah, and C. W. Myers, *Toxicon* **16,** 163 (1978).
29. T. Tokuyama, K. Uenoyama, G. Brown, J. W. Daly, and B. Witkop, *Helv. Chim. Acta* **57,** 2597 (1974).
30. J. W. Daly, T. Tokuyama, T. Fujiwara, R. J. Highet, and I. L. Karle, *J. Am. Chem. Soc.* **102,** 830 (1980).
31. T. Tokuyama, J. W. Daly, and R. J. Highet, *Tetrahedron* **40,** 1183 (1984).
32. A. M. Duffield, H. Budzikiewiez, and C. Djerassi, *J. Am. Chem. Soc.* **87,** 2926 (1965).
33. J. G. Leihr, P. Schulze, and W. J. Richter, *Org. Mass Spectrom.* **7,** 45 (1973).
34. O. Ekpa, J. W. Wheeler, J. C. Cokendolpher, and R. M. Duffield, *Tetrahedron Lett.* **25,** 1315 (1984).
35. J. W. Wheeler, O. Olubajo, C. B. Storm, and R. M. Duffield, *Science* **211,** 1051 (1981).
36. H. M. Fales, M. S. Blum, Z. Bian, T. H. Jones, and A. W. Don, *J. Chem. Ecol.* **10,** 651 (1984).
37. J. H. Tumlinson, R. M. Silverstein, J. C. Moser, R. G. Brownlee, and J. M. Ruth, *Nature (London)* **234,** 348 (1971).
38. J. H. Tumlinson, J. C. Moser, R. M. Silverstein, R. G. Brownlee, and J. M. Ruth, *J. Insect Physiol.* **18,** 809 (1972).
39. R. G. Riley, R. M. Silverstein, B. Carroll, and R. Carroll, *J. Insect Physiol.* **20,** 651 (1974).
40. J. H. Cross, R. C. Byler, U. Ravid, R. M. Silverstein, S. W. Robinson, P. M. Baker, J. S. de Oliveira, A. R. Jutsum, and J. M. Cherrett, *J. Chem. Ecol.* **5,** 187 (1979).
41. J. H. Cross, J. R. West, R. M. Silverstein, A. R. Jutsum, and J. M. Cherrett, *J. Chem. Ecol.* **8,** 119 (1982).
42. D. J. Pedder, H. M. Fales, T. Jaouni, M. Blum, J. MacConnell, and R. M. Crewe, *Tetrahedron* **32,** 2275 (1976).
43. M. S. Blum, T. H. Jones, B. Holldobler, H. M. Fales, and T. Jaouni, *Naturwissenschaften* **67,** 144 (1980).
44. T. H. Jones, M. S. Blum, R. W. Howard, C. A. McDaniel, H. M. Fales, M. B. DuBois, and J. Torres, *J. Chem. Ecol.* **8,** 285 (1982).
45. F. J. Ritter, I. E. M. Rotgans, P. E. J. Verweil, and C. J. Persoons, *in* "Pheromones and Defensive Secretions in Social Insects" (C. Noirot, P. E. Howse, and G. Le Manse, eds.), p. 103. Univ. of Dijon Press, Dijon, France, 1975.
46. J. W. Wheeler, T. Olagbemiro, A. Nash, and M. S. Blum, *J. Chem. Ecol.* **3,** 241 (1977).
47. G. R. Waller, "Biochemical Applications of Mass Spectrometry." Wiley (Interscience), New York, 1982.
48. H. M. Fales and G. R. Waller, private communication.
49. T. H. Jones and M. S. Blum, *in* "Alkaloids: Chemical and Biological Perspectives" (S. W. Pelletier, ed.), Vol. 1, Chapter 2, p. 46. Wiley, New York, 1983.
50. T. H. Jones, M. S. Blum, and H. M. Fales, *Tetrahedron* **38,** 1949 (1982).
51. J. G. MacConnell, M. S. Blum, and H. M. Fales, *Tetrahedron* **27,** 1129 (1971).

52. J. G. MacConnell, M. S. Blum, W. F. Buren, R. N. Williams, and H. M. Fales, *Toxicon* **14,** 69 (1976).
53. H. Schildknecht, D. Berger, D. Knauss, J. Connert, J. Gehlhaus, and H. Essenbreis, *J. Chem. Ecol.* **2,** 1 (1976).
54. T. H. Jones, M. S. Blum, H. M. Fales, and C. R. Thompson, *J. Org. Chem.* **45,** 4778 (1980).
55. J. A. Edgar and C. C. J. Culvenor, *Nature (London)* **248,** 614 (1974).
56. J. Meinwald, Y. C. Meinwald, J. W. Wheeler, T. Eisner, and L. P. Brower, *Science* **151,** 583 (1966).
57. J. Meinwald and Y. C. Meinwald, *J. Am. Chem. Soc.* **88,** 1305 (1966).
58. J. Meinwald, W. R. Thompson, T. Eisner, and D. F. Owen, *Tetrahedron Lett.* 3485 (1971).
59. J. Meinwald, Y. C. Meinwald, and P. H. Mazzochi, *Science* **164,** 1174 (1969).
60. J. Meinwald, C. J. Bonak, D. Schneider, M. Boppre, W. F. Wood, and T. Eisner, *Experientia* **30,** 721 (1974).
61. J. A. Edgar, C. C. J. Culvenor, and G. S. Robinson, *J. Aust. Entomol. Soc.* **12,** 144 (1973).
62. J. Meinwald, J. Smolanoff, A. T. McPhail, R. W. Miller, T. Eisner, and K. Hicks. *Tetrahedron Lett.* 2367 (1975).
63. J. Smolanoff, A. F. Kluge, J. Meinwald, A. McPhail, R. W. Miller, K. Hicks, and T. Eisner, *Science* **188,** 734 (1975).
64. F. J. Ritter, I. E. M. Rotgans, E. Talman, P. E. J. Verweil, and F. Stein, *Experientia* **29,** 530 (1973).
65. F. J. Ritter, I. E. M. Bruggeman-Rotgans, P. E. J. Verweil, and C. J. Persoons, *in* "Pheromones and Defensive Secretions in Social Insects" (C. Noiret, P. E. Howse, and G. Le Masne, eds.), pp. 99–103. Univ. of Dijon Press, Dijon, France, 1975.
66. T. H. Jones, R. J. Highet, M. S. Blum, and H. M. Fales, *J. Chem. Ecol.* **10,** 1233 (1984).
67. P. A. Hedin, R. C. Gueldner, R. D. Henson, and A. C. Thompson, *J. Insect Physiol.* **20,** 2135 (1974).
68. B. Tursch, J. C. Braekman, D. Daloze, C. Hootele, D. Losman, R. Karlson, and J. M. Pasteels, *Tetrahedron Lett.* 201 (1973).
69. B. Tursch, D. Daloze, J. C. Braekman, C. Hootele, and J. M. Pasteels, *Tetrahedron* **31,** 1541 (1975).
70. B. P. Moore and W. V. Brown, *Insect Biochem.* **8,** 393 (1978).
71. J. W. Wheeler, J. Avery, O. Olubajo, M. T. Shamim, C. B. Storm, and R. M. Duffield, *Tetrahedron* **38,** 1939 (1982).
72. J. W. Wheeler and M. S. Blum, *Science* **182,** 501 (1973).
73. J. J. Brophy, G. W. K. Cavill, and W. D. Plant, *Insect Biochem.* **11,** 307 (1981).
74. G. W. K. Cavill and E. Houghton, *Aust. J. Chem.* **27,** 879 (1974).
75. A. Hefetz and S. W. T. Batra, *Comp. Biochem. Physiol. B* **65B,** 455 (1980).
76. B. Maurer and G. Ohloff, *Helv. Chim. Acta* **59,** 1169 (1976).
77. J. J. Brophy and G. W. K. Cavill, *Heterocycles* **14,** 477 (1980).
78. Z. Valenta and A. Khaleque, *Tetrahedron Lett.* No. 12, 1 (1959).
79. E. Lederer, *J. Chem. Soc.* 2115 (1949); *Perfum. Essent. Oil Rec.* **40,** 353 (1949).
80. G. A. Cordell, "Intoduction to Alkaloids. A Biogenetic Approach." Wiley, New York, 1981.
81. M. Karplus, *J. Chem. Phys.* **30,** 11 (1959).
82. R. U. Lemiux, R. K. Kulling, H. G. Bernstein, and W. G. Schneider, *J. Am. Chem. Soc.* **80,** 6098 (1958).
83. J. C. N. Ma and E. W. Warnhoff, *Can. J. Chem.* **43,** 1859 (1965).
84. H. M. Fales and K. S. Warren, *J. Org. Chem.* **32,** 501 (1967).
85. R. J. Highet and P. F. Highet, *J. Org. Chem.* **30,** 902 (1965).
86. H. A. Lloyd, E. A. Kielar, R. J. Highet, S. Uyeo, H. M. Fales, and W. C. Wildman, *J. Org. Chem.* **27,** 373 (1962).
87. K. G. Pachler, R. R. Arndt, and W. H. Baarschers, *Tetrahedron* **21,** 2159 (1965).

88. L. M. Jackman and S. Sternhell, "Applications of Nuclear Magnetic Resonance Spectroscopy in Organic Chemistry," 2nd ed. Pergamon, Oxford, 1969.
89. E. D. Becker, "High Resolution NMR," 2nd ed. Academic Press, New York, 1980.
90. E. Pretsch, J. Seibl, W. Simon, and T. Clerc, "Spectral Data for Structure Determination of Organic Compounds." Springer-Verlag, Berlin and New York, 1983.
91. C. K. Yu, J. K. Saunders, D. B. MacLean, and R. H. F. Manske, *Can. J. Chem.* **49,** 3020 (1971).
92. G. Blasko, N. Murugesan, S. F. Hussain, R. D. Minard, and M. Shamma, *Tetrahedron Lett.* **22,** 3135 (1981).
93. E. W. Warnhoff and W. C. Wildman, *J. Am. Chem. Soc.* **82,** 1472 (1960).
94. J. A. Pople and A. A. Bothner-By, *J. Chem. Phys.* **42,** 1339 (1965).
95. R. C. Cookson, T. A. Crabb, J. J. Frankel, and J. Hudec, *Tetrahedron* **22,** Suppl. 7, 355 (1966).
96. T. A. Crabb, R. F. Newton, and J. Rouse, *Org. Magn. Reson.* **20,** 113 (1982).
97. M. Tits, D. Tavernier, and L. Angenot, *Phytochemistry* **19,** 1531 (1980).
98. T. A. Crabb and J. Rouse, *Org. Magn. Reson.* **21,** 683 (1983).
99. P. Ruedi and C. H. Eugster, *Helv. Chim. Acta* **55,** 1994 (1972).
100. A. A. Bothner-By, *Adv. Magn. Reson.* **1,** 195 (1965).
101. E. W. Garbisch, Jr., *J. Am. Chem. Soc.* **86,** 5561 (1964).
102. R. C. Clark, F. L. Warren, and K. G. R. Pachler, *Tetrahedron* **31,** 1855 (1975).
103. J. B. Stothers, "Carbon-13 NMR Spectroscopy." Academic Press, New York, 1972.
104. P. E. Hansen, *Prog. Nucl. Magn. Reson. Spectrosc.* **14,** 175 (1981).
105. N. Muller and D. E. Pritchard, *J. Chem. Phys.* **31,** 768 (1959).
106. J. N. Shoolery, *J. Chem. Phys.* **31,** 1427 (1959).
107. A. J. Jones, C. C. J. Culvenor, and L. W. Smith, *Aust. J. Chem.* **35,** 1173 (1982).
108. R. Aydin, J. P. Loux, and H. Gunther, *Angew. Chem., Int. Ed. Engl.* **21,** 449 (1982).
109. U. Voegeli and W. von Philipsborn, *Org. Magn. Reson.* **7,** 617 (1975).
110. R. J. Highet, *Org. Magn. Reson.* **22,** 136 (1984).
111. R. S. Norton and R. J. Wells, *J. Am. Chem. Soc.* **104,** 3628 (1982).
112. E. Breitmaier and W. Voelter, "^{13}C NMR Spectroscopy." Verlag Chemie, New York, 1978.
113. T. A. Crabb, *Annu. Rep. NMR Spectrosc.* **6a,** 249 (1975).
114. T. A. Crabb, *Annu. Rep. NMR Spectrosc.* **8,** 2 (1978).
115. T. A. Crabb, *Annu. Rep. NMR Spectrosc.* **13,** 59 (1982).
116. D. A. Okorie, *Tetrahedron* **36,** 2005 (1980).
117. M. LeBoeuf, M. Hamonniere, A. Cavé, H. E. Gottlieb, N. Kunesch, and F. Wenkert, *Tetrahedron Lett.* 3559 (1976).
118. C. P. Falshaw, T. J. King, and D. A. Okorie, *Tetrahedron* **38,** 2311 (1982).
119. N. K. Wilson and J. B. Stothers, *J. Magn. Reson.* **15,** 31 (1974).
120. E. L. Eliel and K. M. Petrusiewicz, *Top. Carbon-13 NMR Spectrosc.* **3,** 172 (1979).
121. N. A. B. Gray, *Prog. NMR Spectrosc.* **15,** 201 (1984).
122. M. Shamma and D. M. Hindenlang, "Carbon-13 NMR Shift Assignments of Amines and Alkaloids." Plenum, New York, 1979.
123. T. A. Broadbent and E. G. Paul, *Heterocycles* **20,** 863 (1983).
124. A. J. Jones and M. H. Benn, *Can. J. Chem.* **51,** 486 (1983).
125. S. W. Pelletier, N. V. Mody, A. J. Jones, and M. H. Benn, *Tetrahedron Lett.* 3025 (1976).
126. J. Finer-Moore, N. V. Mody, S. W. Pelletier, N. A. B. Gray, C. W. Crandell, and D. H. Smith, *J. Org. Chem.* **46,** 3399 (1981).
127. R. Benn and H. Gunther, *Angew. Chem., Int. Ed. Engl.* **22,** 350 (1983).
128. D. J. Cookson and B. E. Smith, *Org. Magn. Reson.* **16,** 111 (1982).
129. S. I. Patt and J. N. Shoolery, *J. Magn. Reson.* **46,** 535 (1982).
130. C. LeCocq and J.-Y. Lallemand, *J. Chem. Soc., Chem. Commun.* 150 (1981).

131. T. T. Nakashima and D. L. Rabenstein, *J. Magn. Reson.* **47,** 339 (1982).
132. P. Feng-kui and R. Freeman, *J. Magn. Reson.* **48,** 318 (1982).
133. H. J. Jakobsen, O. W. Sorensen, W. S. Brey, and P. Kanyha, *J. Magn. Reson.* **48,** 328 (1982).
134. G. A. Morris and R. Freeman, *J. Am. Chem. Soc.* **101,** 760 (1979).
135. J. W. Blunt and P. J. Steel, *Aust. J. Chem.* **35,** 2561 (1982).
136. A. Bax, "Two-Dimensional Nuclear Magnetic Resonance in Liquids," Reidel Publ., Boston, Massachusetts, 1982.
137. R. Freeman, S. P. Kempsell, and M. H. Levitt, *J. Magn. Reson.* **34,** 663 (1979).
138. W. P. Aue, J. Karhan, and R. R. Ernst, *J. Chem. Phys.* **64,** 4226 (1976).
139. A. Bax and R. Freeman, *J. Magn. Reson.* **44,** 542 (1981).
140. A. Bax, R. Freeman, T. A. Frenkiel, and M. H. Levitt, *J. Magn. Reson.* **43,** 478 (1981).
141. N. S. Bhacca, W. F. Balandrin, A. D. Kinghorn, T. A. Frenkiel, R. Freeman, and G. A. Morris, *J. Am. Chem. Soc.* **105,** 2538 (1983).
142. P. H. Bolton, *J. Magn. Reson.* **54,** 333 (1983).
143. J. H. Noggle and R. E. Schirmer, "The Nuclear Overhauser Effect." Academic Press, New York, 1971.
144. J. A. Ferretti and G. N. Weiss, *J. Magn. Reson.* **55,** 397 (1983).
145. R. A. Bell and J. K. Saunders, *Can. J. Chem.* **48,** 1114 (1970).
146. F. A. L. Anet and R. Anet, *in* "Determination of Organic Structures by Physical Methods" (F. C. Nachod and J. Zuckerman, eds.), p. 406. Academic Press, New York, 1971.
147. J. K. Saunders, R. A. Bell, C.-Y. Chen, and D. B. MacLean, *Can. J. Chem.* **46,** 2873 (1968).
148. J. K. Saunders, R. A. Bell, C.-Y. Chen, D. B. MacLean, and R. H. F. Manske, *Can. J. Chem.* **46,** 2876 (1968).
149. G. E. Chapman, B. D. Abercrombie, P. D. Cary, and E. M. Bradley, *J. Magn. Reson.* **31,** 459 (1978).
150. D. Neuhaus, R. N. Sheppard, and I. R. C. Bick, *J. Am. Chem. Soc.* **105,** 5996 (1983).
151. G. C. Levy and R. L. Lichter, "Nitrogen-15 Nuclear Magnetic Resonance Spectroscopy." Wiley, New York, 1979.
152. S. N. Y. Fanso-Free, G. T. Furst, P. R. Srinivasan, R. L. Lichter, R. B. Nelson, J. A. Panetta, and G. W. Gribble, *J. Am. Chem. Soc.* **101,** 1549 (1979).
153. P. W. Jeffs, *Experientia* **21,** 690 (1965).
154. J. Skolik, P. J. Krueger, and M. Wiewiorowski, *Tetrahedron* **24,** 5439 (1968).
155. A. Bax, R. H. Griffey, and B. L. Hawkins, *J. Magn. Reson.* **55,** 301 (1983).
156. A. Bax, R. H. Griffey, and B. L. Hawkins, *J. Am. Chem. Soc.* **105,** 7188 (1983).
157. F. J. Schmitz, S. K. Agarwal, S. P. Gunasekera, P. G. Schmidt, and J. N. Shoolery, *J. Am. Chem. Soc.* **105,** 4835 (1983).
158. K. Nakanishi and P. H. Solomon, "Infrared Absorption Spectroscopy." Holden-Day, San Francisco, California, 1977.
159. R. J. Highet, *J. Org. Chem.* **29,** 471 (1964).
160. S. W. Pelletier, "Alkaloids. Chemical and Biological Perspectives," Vol. 1. Wiley, New York, 1981.
161. E. Wenkert and D. Roychaudhuri, *J. Am. Chem. Soc.* **78,** 6417 (1956).
162. F. Bohlmann, *Angew. Chem.* **69,** 541 (1957).
163. M. Wiewiorowski, O. E. Edwards, and M. D. Bratek-Wiewiorowska, *Can. J. Chem.* **45,** 1447 (1967).
164. F. Bohlmann, W. Weise, D. Rahtz, and C. Arndt, *Chem. Ber.* **91,** 2176 (1958).
165. T. A. Crabb, R. F. Newton, and D. Jackson, *Chem. Rev.* **71,** 109 (1971).
166. R. J. Highet and P. F. Highet, *J. Org. Chem.* **31,** 1275 (1966).
167. M. Tichy and J. Sicher, *Collect. Czech. Chem. Commun.* **23,** 2081 (1958).
168. H. M. Fales and W. C. Wildman, *J. Am. Chem. Soc.* **85,** 784 (1963).
169. H. M. Fales and W. C. Wildman, *J. Am. Chem. Soc.* **82,** 197 (1960).

INDEX

B

C

D

E

F

G

N

Q

R